I0868753

Agrifood Economics and Sustainable Development in Contemporary Society

Gabriel Popescu
Bucharest Academy of Economic Studies, Romania

A volume in the Advances in Environmental Engineering and Green Technologies (AEEGT) Book Series

IGI Global
DISSEMINATOR OF KNOWLEDGE

Published in the United States of America by
IGI Global
Engineering Science Reference (an imprint of IGI Global)
701 E. Chocolate Avenue
Hershey PA, USA 17033
Tel: 717-533-8845
Fax: 717-533-8661
E-mail: cust@igi-global.com
Web site: http://www.igi-global.com

Copyright © 2019 by IGI Global. All rights reserved. No part of this publication may be reproduced, stored or distributed in any form or by any means, electronic or mechanical, including photocopying, without written permission from the publisher. Product or company names used in this set are for identification purposes only. Inclusion of the names of the products or companies does not indicate a claim of ownership by IGI Global of the trademark or registered trademark.

Library of Congress Cataloging-in-Publication Data

Names: Popescu, Gabriel, 1954- editor.
Title: Agrifood economics and sustainable development in contemporary society / Gabriel Popescu, editor.
Description: Hershey, PA : Engineering Science Reference, [2019]
Identifiers: LCCN 2017054765| ISBN 9781522557395 (hardcover) | ISBN 9781522557401 (ebook)
Subjects: LCSH: Agriculture--Economic aspects--Europe. | Sustainable agriculture--Europe. | Sustainable development--Europe.
Classification: LCC HD1917 .A395 2019 | DDC 338.1094--dc23 LC record available at https://lccn.loc.gov/2017054765

This book is published in the IGI Global book series Advances in Environmental Engineering and Green Technologies (AEEGT) (ISSN: 2326-9162; eISSN: 2326-9170)

British Cataloguing in Publication Data
A Cataloguing in Publication record for this book is available from the British Library.

All work contributed to this book is new, previously-unpublished material. The views expressed in this book are those of the authors, but not necessarily of the publisher.

For electronic access to this publication, please contact: eresources@igi-global.com.

Advances in Environmental Engineering and Green Technologies (AEEGT) Book Series

Sang-Bing Tsai
University of Electronic Science and Technology of China Zhongshan Institute, China
Ming-Lang Tseng
Lunghwa University of Science and Technology, Taiwan
Yuchi Wang
University of Electronic Science and Technology of China Zhongshan Institute, China

ISSN:2326-9162
EISSN:2326-9170

Mission

Growing awareness and an increased focus on environmental issues such as climate change, energy use, and loss of non-renewable resources have brought about a greater need for research that provides potential solutions to these problems. Research in environmental science and engineering continues to play a vital role in uncovering new opportunities for a "green" future.

The **Advances in Environmental Engineering and Green Technologies (AEEGT)** book series is a mouthpiece for research in all aspects of environmental science, earth science, and green initiatives. This series supports the ongoing research in this field through publishing books that discuss topics within environmental engineering or that deal with the interdisciplinary field of green technologies.

Coverage

- Green Technology
- Contaminated Site Remediation
- Waste Management
- Green Transportation
- Cleantech
- Sustainable Communities
- Electric Vehicles
- Renewable energy
- Policies Involving Green Technologies and Environmental Engineering
- Air Quality

IGI Global is currently accepting manuscripts for publication within this series. To submit a proposal for a volume in this series, please contact our Acquisition Editors at Acquisitions@igi-global.com or visit: http://www.igi-global.com/publish/.

The Advances in Environmental Engineering and Green Technologies (AEEGT) Book Series (ISSN 2326-9162) is published by IGI Global, 701 E. Chocolate Avenue, Hershey, PA 17033-1240, USA, www.igi-global.com. This series is composed of titles available for purchase individually; each title is edited to be contextually exclusive from any other title within the series. For pricing and ordering information please visit http://www.igi-global.com/book-series/advances-environmental-engineering-green-technologies/73679. Postmaster: Send all address changes to above address. Copyright © 2019 IGI Global. All rights, including translation in other languages reserved by the publisher. No part of this series may be reproduced or used in any form or by any means – graphics, electronic, or mechanical, including photocopying, recording, taping, or information and retrieval systems – without written permission from the publisher, except for non commercial, educational use, including classroom teaching purposes. The views expressed in this series are those of the authors, but not necessarily of IGI Global.

Titles in this Series

For a list of additional titles in this series, please visit: www.igi-global.com/book-series

Innovations and Trends in Environmental and Agricultural Informatics
Petraq Papajorgji (University of New York Tirana, Albania) and Francois Pinet (Irstea Centre de Clermont-Ferrand, France)
Engineering Science Reference • copyright 2018 • 340pp • H/C (ISBN: 9781522559788) • US $205.00 (our price)

Effective Solutions to Pollution Mitigation for Public Welfare
Ahmet Ozan Gezerman (Yildiz Technical University, Turkey) Burcu Didem Corbacioglu (Yildiz Technical University, Turkey) and Bhola R. Gurjar (Indian Institute of Technology Roorkee, India)
Engineering Science Reference • copyright 2018 • 274pp • H/C (ISBN: 9781522533795) • US $180.00 (our price)

Handbook of Research on Microbial Tools for Environmental Waste Management
Vinay Mohan Pathak (Gurukul Kangri University, India) and Navneet (Gurukul Kangri University, India)
Engineering Science Reference • copyright 2018 • 496pp • H/C (ISBN: 9781522535409) • US $215.00 (our price)

Food Systems Sustainability and Environmental Policies in Modern Economies
Abiodun Elijah Obayelu (Federal University of Agriculture – Abeokuta (FUNAAB), Nigeria)
Engineering Science Reference • copyright 2018 • 371pp • H/C (ISBN: 9781522536314) • US $195.00 (our price)

Microbial Biotechnology in Environmental Monitoring and Cleanup
Pankaj (G. B. Pant University of Agriculture and Technology, India) and Anita Sharma (G. B. Pant University of Agriculture and Technology, India)
Engineering Science Reference • copyright 2018 • 427pp • H/C (ISBN: 9781522531265) • US $235.00 (our price)

Handbook of Research on Renewable Energy and Electric Resources for Sustainable Rural Development
Valeriy Kharchenko (Federal Scientific Agroengineering Center VIM, Russia) and Pandian Vasant (Universiti Teknologi PETRONAS, Malaysia)
Engineering Science Reference • copyright 2018 • 672pp • H/C (ISBN: 9781522538677) • US $345.00 (our price)

Green Production Strategies for Sustainability
Sang-Binge Tsai (University of Electronic Science and Technology of China (Zhongshan Institute), China & Civil Aviation University of China, China) Bin Liu (Shanghai Maritime University, China) and Yongian Li (Nankai University, China)
Engineering Science Reference • copyright 2018 • 325pp • H/C (ISBN: 9781522535379) • US $235.00 (our price)

IGI Global
DISSEMINATOR OF KNOWLEDGE

701 East Chocolate Avenue, Hershey, PA 17033, USA
Tel: 717-533-8845 x100 • Fax: 717-533-8661
E-Mail: cust@igi-global.com • www.igi-global.com

Table of Contents

Detailed Table of Contents

two phases, the period before 1991 and the period after 1991. In the first phase (before 1991), the village was marked by severe livelihood crisis owing to acute water scarcity. The second phase (after 1991) is known for livelihood improvement realized through community engagement and planned investment initiatives for watershed development and organic farming.

Lucia Ovidia Vreja, University of Economic Studies, Romania
Sergiu Bălan, University of Economic Studies, Romania

This chapter presents the role of nature and nurture in shaping the behavior of human beings toward sustainability identifying instances of both dramatic extinctions of species and collapse of entire societies, as well as successful, peaceful, and healthy adaptation of human communities to their environment, in an attempt to presents the imperative conditions necessary for attaining sustainable development. A very long and intriguing history reveals that from the nature's point of view humans are rather destructive, interested in their own short-term survival. Nevertheless, the same long history of human species bears valuable lessons and examples of adaptive behaviors grounded by nurture, and based on these examples, the chapter aims at advancing a new perspective of thinking sustainable development that could lay the foundation of a new education curriculum.

Carmen Lenuta Trica, Bucharest University of Economic Studies, Romania
Luminita Ghita, Ministry of Environment, Romania

At present, legal provisions and environmental policy regulate the possibilities of using environmental considerations in the development of award criteria, as well as in the performance clauses of procurement contracts. The first part of the chapter analyzes the concept of green procurement and product categories for which green procurement can be used. The second part of the chapter presents the benefits of using green procurement. The third part of the chapter will include assessing the progress and impact of using green procurement. The fourth part of the chapter analyzes the legal framework for public procurement in Romania, as well as the capacity of the market to offer and develop products and services that include minimum environmental requirements and criteria. In the fifth part of the chapter, the authors analyze the possibility of implementing a mechanism for the operation and implementation of the legal provisions in Romania in order to improve the quality of the services and optimize the costs of the public procurement.

Ancuta Marin, Research Institute for Agricultural Economics and Rural Development, Romania

Romania's EU integration has determined major economic changes, including the transition to a functional market economy which meant the existence of a competitive environment, free commercial exchanges, and free movement of people, capital, and services. To solve the existent delays, Romania has benefit from the European Community's funds for investment in agriculture and rural areas through the financial mechanism which assures non-reimbursable funds for farmers, private societies, and local authorities. Without claiming that we are exhausting the issue of financing in Romanian agriculture, the topic addressed brings some issues regarding the financing under the National Rural Development Program (NRDP) and the possible explanations of the low absorption rate of these funds. NRDP is a program which allow people to access non-reimbursable funds from EU and Romanian Government for economic and social development of the rural area.

The level of socio-economic development, the types of relations, and the existing legislation at some point in society represent basic elements in the operation of public-private partnership. In the post-revolution period, a series of difficulties and discontinuities in agricultural policies are manifested in the Romanian society, even after 27 years. It has affected the configuration and development of associative structures, including public-private partnerships from rural areas. Currently, the evolution of the representative associative structures in public-private partnership and the diversification of operation record more than the progressive process, a correlation with the local development plans. Also, concerns on behalf of the central public administration are recorded regarding the update of the legislative and institutional set-up of associative type, because it is one of the main tools for operating the common agricultural policy in accordance with the rules of the European Union.

The chapter describes the situation from several EU countries on cooperation among farmers, with a focus on the approach of the cooperative concept related to legislation and function. The study comprises 10 member states from different parts of Europe: east, south, and north. The objective was to identify the differences and the existence of an integrative model for cooperative or producer organizations in agriculture. The member states included in the study were France, Italy, Spain, Denmark, Finland, Germany, Poland, Ireland, The Netherlands, and Romania. In order to find common features, the history and previous developing of cooperation were analyzed. Focused on the bibliographic research and comprising an analysis of the history and legislation, the author tried to underline some aspects that could facilitate the setting-up of new agricultural cooperatives and at the same time, a proper operation of the existing ones. The information gathered was presented and interpreted, in order to capture the situation of agricultural co-operative structures, legal type, and economic operation.

Traditional products market in Romania is starting to become competitive as a result of the adoption of new legislative framework (Order 724/2013), which requires new conditions regarding traditional foods definition and limitation of produced quantity. At the end of 2016, there were 556 registered traditional products nationwide as follows: 456 limited liability company, 64 individual, 14 family business, 17 individual, and 15 individual enterprise. Given that there is a limitation in terms of traditional foods production, we must consider the effects of land taxation over the entrepreneur's income in this field. The aim of this study is to identify an effective model of public-private partnership that could support the traditional producers, while taking into consideration the role of financial instruments in order to remove the barriers between the farmer and the government. One of these forms is the association or creating groups of manufacturers.

Section 3
Solutions for the Future

Chapter 12
European Agriculture in 2020: Solutions for a Sustainable Development 247
Silvia Elena Cristache, The Bucharest University of Economic Studies, Romania
Florica Georgeta Rotaru, The Bucharest University of Economic Studies, Romania
Elena Cristina Rotaru, Ovidius University, Romania

The chapter is structured in two parts: the first part presents the agricultural policy of the European Union in accordance with the Treaty of Lisbon and the Europe 2020 Strategy. At the same time, the National Rural Development Program of Romania is presented. In the second part of the chapter an econometric analysis of the agricultural sector is made. The linear multifactorial regression model presented in the second part of the chapter was based on data provided by Eurostat: surface cultivated with pesticides, GDP/capita, net investment in agriculture. The model proposed in this chapter explains and highlights the existence of a positive, linear, and significant relationship between the variables included in the model. The multiple linear regression model indicates that, with the increase of one million euros per GDP, the value of the surface cultivated with pesticides in Romania will grow by an average of 132.08 hectares.

Chapter 13
Organic Agriculture: Opportunities and Trends 275
Mirela Stoian, Bucharest University of Economic Studies, Romania
Diana Caprita, Bucharest University of Economic Studies, Romania

Promoting sustainability, including the production and consumption of food, is badly needed nowadays, given the fact that consumers are increasingly concerned about protecting their health, through a thorough verification of food quality. From this perspective, organic food may represent a viable solution for a healthier future. Currently, we are witnessing a substantial increase in the number of countries, organizations, and companies encouraging organic farming, an economic activity that involves environmentally friendly agricultural practices. The main objective of this chapter is to reveal the growing importance of organic farming to the food markets. This research will also focus on presenting a very detailed analysis of the defining elements of organic agriculture, such as the evolution of certified organic surfaces, both contributory and disfavoring factors of the developing organic agriculture, and last but not least, overall outlook for global consumption of certified organic products.

Foreword

The modern society and economy has and is experiencing a reconfiguration of the basically systems and components where the resource economics and management started to become central pillars in stimulating and encouraging new value-added creation. Understanding the intimate mechanisms and growth springs represent a major challenge for researchers, which still requires concerted efforts to highlight the growth potentials. In this context of diverse challenges, the agro-food economy has become, during the recent years, a determinant factor in promoting and assuring a sustainable economic development, in the larger context of current economic paradigm evolution.

The research topic approached within this reference book is topical, with a high visibility and uniqueness of the scientific approach, being in line with the current research trends in the field of economics and resource management, with predilection on agro food and rural issues economics.

Tis book constitutes an authentic scientific approach, which brings together an important collection of chapters, containing a series of specialty studies in the field, which are constituted in distinct and correlated topics, presenting the character of an original scientific work of high scientific importance by choosing the contributors of chapters, which through the methodology used in collecting, analyzing, reviewing and promoting chapters contributes to the understanding and deepening of the specificities of this research.

The complexity of the topic—*Agrifood Economics and Sustainable Development in Contemporary Society*—required the use and capitalization of a wide range of methodologies and research methods specific to the main topic, which form a unique and unitary research.

Through the scale of the scientific approach undertaken in this reference book, the book editor manages to mobilize a large number of contributors and, at the same time, to capitalize scientifically on the complexity of the subject, which gives this book a profound interdisciplinary character. By its subject, agro-food economics and sustainable development in contemporary society represent a milestone in researching the complex influence of agro-food economics in valuing the rural and agricultural potential.

The opinions expressed in the chapters by the contributors, the own interpretations of the available data, the highlighting of the multiple arguments and the theme specific to each chapter, make this book an authentic scientific contribution in the field, completing the space of research conducted in the field.

This book provides valuable insights on agrifood economics and sustainable development topics in contemporary society transformations and developments and it covers the major and necessary components in providing an interesting perspective on the topics analyzed. The challenges in agrifood and resource economics are not only just difficult and interesting, but also require a deep attention.

In this new global and interconnected economy, understanding the agrifood economics and resource management, doubled by the interconnectivity and interdependence between researches problematic, imposes new thematic approaches and new methods in research. The *Agrifood Economics and Sustainable Development in Contemporary Society* could be considered as a new and solid step in this direction, which I warmly recommend.

Jean-Vasile Andrei
Petroleum-Gas University of Ploiesti, Romania

Jean-Vasile Andrei, *PhD, is Associate professor at Petroleum-Gas University of Ploiesti, Department of Business Administration and Ph.D. mentor in economics at The Bucharest University of Economic Studies, Romania. He holds a Ph.D. in Economics from the National Institute of Economics Research – Romanian Academy of Sciences. He has earned a B.A degree in Administrative Sciences (2005) and in Banks and Finances (2007) from the Petroleum-Gas University of Ploiesti. He has an M.A degree in Economics, Administrative and Business Management (2007) earned at the same university. Jean Andrei is Editor-in-Chief of International Journal of Sustainable Economies Management (USA), Associate Editor of Economics of Agriculture (Ekonomika poljoprivrede) published by Institute of Agricultural Economics (Serbia), scientific reviewer for International Business Information Management Association Conferences – IBIMA and numerous scientific journals in the field. Issues like, agricultural and resources economics, business administration, and valuing economic and human potential are among his research and scientific interests, where he has published articles over fifty, nine scientific books and numerous conferences presentations.*

Preface

Natural resources depletion and adverse impact of environmental degradation, including desertification, drought, degradation of land, freshwater deficit and loss of biodiversity and ecosystem services, add to and exacerbate the list of challenges which humanity faces (United Nations [UN], 2015). Potocnik (2017) considers that:

These challenges we are facing in the 21st Century are increasing our individual and collective responsibility. The international agreements, in particular the adoption of the Sustainable Development Goals (SDGs) and operationalisation of one of the goals through the Paris climate agreement, are important steps in the movement to the right direction. They are the recognition of the increased awareness of humans that the transition to a more sustainable trajectory is necessary and unavoidable and that we should work together to be able to achieve it. (p. 4)

According to the Food and Agriculture Organization of the United Nations (2017):

In adopting the 2030 Agenda for Sustainable Development, in 2015, the whole international community committed itself to eradicate hunger and poverty and to achieve sustainable agriculture, healthy and a decent work for all, reducing inequality, and making economic growth inclusive. Before the 2030 deadline, all the states must to deeply think to the Sustainable Development Goals which are to be reached.

The successful changes in all the countries will be driven not only by productivity growth in production and services, because it will be necessary to do more, to take into account other determinants of agricultural production - diversification of rural economies, existing partnerships between State and various social solidarity activities (including public-private partnerships specifically adapted to social and solidarity economy), public policies on finance, favorable tax measures, specific criteria for inclusive public procurement, appropriate legal frameworks, how to avoid or alleviate the present constraints, the access to education and progress, etc. In this context, agrifood economy development is increasingly relying on new generation of economic growth which still protecting the environment and rural communities. Rural economy, sustainable development and the new venture creation represent great research topics in the actual context of development. Developing a well functional rural economy, which is often based on rural communities and agriculture, requires a massive effort in the field.

According to a report of the International Resource Panel (United Nations Environment Programme [UNEP], 2016):

Current food systems have a significant contribution to the deterioration or depletion of natural resources. The actual food systems vary worldwide from 'modern' food systems in industrialized and emerging regions to more 'traditional' food systems in rural areas from developing countries. This diversity of food systems, in combination with the social and natural systems in which they operate, has important implications on the possible pathways towards sustainable food systems and on the logic of intervention.

In the last decades, significant progress has been made in understanding the relationship between agricultural systems and other socio-ecological systems. The environment and resource management challenges for agriculture have been acknowledged and well understood in Europe as well. Further adaptation of the Common Agriculture Policy (CAP) is necessary to help EU farming become a well-structured industry which is economically viable and environmentally sustainable, as it is mentioned in the report "CAP - Thinking Out of the Box" (Buckwell, Matthews, Baldock, & Mathijs, 2017). The authors consider also that:

The next 15 years will see a generational turnover amongst farmers. The new generation faces the exciting challenge of comprising the wealth of new technology based on precision, digitisation, big data sets and even robotics which can be applied in the field of plant and animal genetics and in nutrition process, can lead to productivity increase, and thus incomes. The new generation of farmers has to be competitive globally, to master new technologies and to meet new societal demands beyond providing their primary output, food.

The 2030 SDGs shape the framework for EU's long-term planning, including for different sectoral policies like the Common Agriculture Policy, for the next Multiannual Financial Framework period. Aligning the EU spending in a coherent manner alongside the SDGs will lead to more effective and efficient – non-contradictory – spending and overall economic benefits in the mid and long term by avoiding expensive follow-up costs of unsustainable development (European Commission [EC], 2018).

About the reform of the EU's rural development policy post-2020 in this moment on can mentions four long-term strategic objectives: improving the competitiveness of agriculture; safeguarding the sustainable management of natural resources under climate action; ensuring a balanced territorial development of rural areas; assuring the knowledge transfer and innovation.

Taking into account other structural and investment funds, EU Member States and their regions will design Rural Development Programmes (RDPs), which will follow a set of common priorities including "promoting social inclusion, poverty reduction and economic development in rural areas", more attention paying to environment and education. The future RDPs will be designed so: to reinforce the content of rural development measures; to simplify rules and/or reduce related administrative formalities; and link rural development policy with other financial resources (Eurostat, 2017). All these actions will be financed through the European Agricultural Fund for Rural Development (EAFRD) and several other EU funds that will provide financial support for development of rural areas, such as: the European Regional Development Fund (ERDF), the Cohesion Fund, the European Social Fund (ESF) and the European Maritime and Fisheries Fund (EMFF).

Romania is one of the European countries with the highest potential in agriculture, with the sixth biggest used agricultural area from EU, but has a low productivity due to several factors, according to an analysis by professional services firm Price Waterhouse Coopers Romania (2017). The share of agriculture in Romania's GDP dropped constantly in the last 20 years, from 22.6 in 1993 to under 5 percent of the

GDP in 2015, due to the structural transformation of Romanian economy, from an industrial-agricultural economy to one based on services. Although the agriculture contribution to GDP is decreasing, Romania still has the highest share of agriculture in GDP of all countries in the European Union, three and a half times higher than the European average (Bendre, 2017). This book looks at the experiences achieved in the field of agriculture and rural development by a country which has the 7-th place as surface in EU.

In contemporary economies, agrifood economy, environment and sustainable development represent challenging multi- and transdisciplinary research topic. Such subject is driven by rural and agricultural factors who are utilising green factors as their competitive advantage in competitive markets. They not only consider the agricultural economic resources, or the optimal combination in production process and sustainable consumption, but also provide innovative solutions in achieving a sustainable development. Fulfilling the 2030 Agenda depends crucially on progress in rural areas and investments for knowledge base improvement and human capital development.

The current book provides some holistic approaches regarding the agrifood and rural economy models and practices, as a specific organizational model in highly competitive economy in its path of adapting to the new challenges. The research subjects address to a very actual and debate full aspect as the rural economics, agrifood and environment and its paradigms as it is perceived in modern economies.

The overall objective of the book is to provide relevant theoretical frameworks and the latest empirical research findings in area of agrifood economics and environment and its impacts on competitive markets. The specific objectives of the book are: 1) to create a reference for practitioners, students, professors in field of agrifood economics and environment; 2) to present the newest findings in the field of agrifood economics and environment; 3) to provide a platform for agrifood and environment debate; 4) to create a working paradigm regarding the agrifood economics and environment, and; 5) to provide a theoretical framework in the field.

The book, *Agrifood Economics and Sustainable Development in Contemporary Society*, provides several experiences from the Romanian Rural Space and not only, strongly anchored in the United Nations and European Union strategic orientations of the global effort to build-up a sustainable future, which has co-shaped by the international institutions together with all its member states partners.

A full overview of this book shows how Romanian policies and actions contributed to the implementation of the National Rural Development Programmes and those outputs during the last ten years after country accession to the EU. Generally speaking, this book was developed on two working lines. The first direction concerned developments in agrifood, including sustainable development issues and institutional building. The second direction developed by the authors of this book was to present their visions of future development for the Romanian rural area.

The target audience of this book is composed of professionals and researchers working in the field of agrifood economics and environment. Since the analyzes have targeted interesting cases mainly from an economy in transition, their outcomes will be shared with international networks of stakeholders, including research institutions, universities, and individual researchers in such frames as: rural economics, green development strategies, economic models and patterns, market economy. The presented experiences gather the expertise of some prestigious professors, researchers and experts from different domains and offer to readers real inspiring models. Moreover, the book will provide insights and support executives for policy makers and investors.

The book is organized in three sections: "The Actual State," "Institutional Structures," and "Solutions for Future." A brief description of each of the chapters follows:

THE ACTUAL STATE

Chapter 1, "Determinants of Agricultural Production in Romania: A Panel Data Approach," examines a panel data approach to determine the contribution of several factors on the agricultural output in terms of value and yield. Different regression models were developed for analysis of agricultural production determinants at national scale territorial level in Romania. The negative influence of the excessive droughts years on the cereals yield has been found while it is not a statistical relevance for the influence of the extremely rainy years. Other studies related to the influence of the environmental and social factors on the agricultural economic output will complete an integrated approach in the next future.

Chapter 2, "The Competitiveness Constraints of Romanian Wine Sector and the EU-28 Agricultural Model," aims to examine the competitiveness constraints of Romanian wine sector. The chapter reveals that the wine production has suffered a dramatically improvements both on quantity and quality. The paradigm change imposed by replacement of the hybrid vines with noble vines one has triggered both a quality and competitiveness increase. Thus, the domestic wine producers have to adapt their production to the new market demands in order to fulfill the competitiveness constraints.

Chapter 3, "Sustainable Development through Field Vegetal Exploitation in Romania," highlights the technological, economic and environmental support, prerequisite and accessible to entrepreneurs in the crops agriculture. Based on rigorous analytical documentation, the chapter discovers and recommends appropriate technologies for sustainable farming.

Chapter 4, "The Social, Economic, and Environmental Impact of Ecological Beekeeping in Romania," analyses the beekeeping activity in Romania, which benefits from an exceptional melliferous base and favorable climatic conditions, Local beekeepers can supplement their income, or keep this activity as the main or sole source of income. After 1989, during the transition period to the market economy, beekeeping experienced a severe decline, but due to a good managerial experience within the Association of Beekeepers in Romania, the continuous development was possible. The analysis of the quantitative and qualitative indicators highlights the main economic, social and environmental effects of beekeeping practices in Romania.

Chapter 5, "Sustainable Rural Livelihood: A Case of Hiware Bazaar, Maharashtra, India," focuses on the rural development and its sustainable livelihood. It is the case of Hiware Bazar village located in the Ahmednagar district of Maharashtra, India. The village is known for its livelihood status, which can be distinctly categorized into two phases: before 1991 and the period after 1991. In the first phase, the village was marked by severe livelihood crisis owing to acute water scarcity. The second phase is known for livelihood improvement through community engagement and planned investment initiatives for watershed development and organic farming. The main drivers have been analyzed as well as appropriate solutions.

Chapter 6, "Sustainable Development From a "Nature vs. Nurture" Perspective," describes the role of nature and nurture in shaping the behavior of human beings toward sustainability, identifying instances of both dramatic extinctions of species and collapse of entire societies, as well as successful, peaceful and healthy adaptation of human communities to their environment. The analysis of a very long and intriguing history reveals that from the nature's point of view humans are rather destructive, interested in their own short-term survival. Nevertheless, the same long history of human species bears valuable lessons and examples of adaptive behaviors grounded by nurture, and based on these examples the chapter aims at advancing a new perspective of thinking sustainable development, that could lay the foundation of a new education curricula.

Chapter 7, "Green Public Procurement: Instrument of Sustainable Development," analyzes the concept of green procurement and product categories, their benefits and assesses the progress and impact of using green procurement. An analysis of the legal and regulatory framework for public procurement in Romania has been made, as well as an analysis of the market capacity to supply and develop products and services that include minimum environmental requirements and criteria. Recommendations and solutions for operationalization and costs optimization of the public procurement have been proposed.

Chapter 8, "Romanian Agriculture Funding: Approaches Regarding the Funding in Romanian Agriculture After EU Integration," analyses the financial support provided in agriculture and rural areas through mechanisms which assures non-reimbursable funds for farmers, private societies and local authorities. The analysis focuses on the National Rural Development Program, which manages The European Commission and Romanian Government financial support for the economic and social development of the rural area. The author identified also other drivers for an efficient and competitive agriculture and suggests changes that could improve this sector, like changes of people's perception related to cooperation and association, regulatory framework improvement to attract youth and to diminish unemployment and emigration rate.

INSTITUTIONAL STRUCTURES

Chapter 9, "Prerequisites for Relaunching Economic Growth in Romanian Agriculture by Promoting Association," analyses the configuration and development process of associative structures, including public-private partnerships from rural areas. The author asses the evolution of the representative associative structures in public-private partnership, in correlation with the Local Development Plans, and recommends legislative and regulatory framework updates in order to adjust the implementation of the Common Agricultural Policy at the local socio-economic conditions of Romania. The specific features of the rural areas allow and support the amplification of positive results and experiences through the diversification of opportunities, contributing to the well-being increase.

Chapter 10, "The Role of Agricultural Cooperatives Models Among Europe," describes and analyses the cooperation among farmers in several European countries, focusing on legislation and functionality of the cooperative concept. The author identifies the differences and integrative models for cooperatives or producers organizations in agriculture. The EU member states included in the study are France, Italy, Spain, Denmark, Finland, Germany, Poland, Ireland, The Netherlands and Romania. Focusing on the analysis of the history and legislation, the author tried to underline some aspects that could facilitate the setting-up of new agricultural cooperatives and a proper operation of the existing ones.

Chapter 11, "Public-Private Partnership Model for Supporting Traditional Producers," analyses the traditional products market in Romania. The authors propose to identify effective models of public-private partnership that could support the traditional producers, taking into consideration the role of financial instruments in removing the barriers between the farmers and government. The authors indicate the association of manufacturers as an appropriate tool for improving the traditional products market.

SOLUTIONS FOR FUTURE

Chapter 12, "European Agriculture in 2020: Solutions for a Sustainable Development," shows a detailed econometric analysis of the agricultural sector in the context of European and national policies requirements. The authors analyze the Eurostat database and develop linear multifactorial regression model considering relevant factors which influence the agriculture sector, like GDP/capita, net investment in agriculture pesticides used. The model proposed in the chapter explains and highlights the positive, linear and significant relationship between variables included in the model.

Chapter 13, "Organic Agriculture: Opportunities and Trends," analyses the growing importance of organic farming to the food markets. This is determined by the consumers concern for protecting their health, through a careful verification of food quality. The chapter focuses on detailed analysis of the relevant elements of organic agriculture and provide an outlook for global consumption of certified organic products. The authors highlight the huge interests of countries, organizations and companies for organic farming - an economic activity which involves environmentally friendly agricultural practices.

Chapter 14, "The Role of Higher Education in Supporting Sustainable Rural Entrepreneurship," provides an interesting analysis of the contribution of higher education to the development of entrepreneurship in rural areas. The authors analyze through an empirical research how the knowledge and skills acquired during undergraduate studies have been perceived by the graduates. The study outcomes could be important elements to be taken into consideration for improving the education system, its relationship with the socio-economic context and for supporting decision making in higher education.

Chapter 15, "The Effects of the Transition Period on the Knowledge Transfer Market in Romanian Agriculture," analyses the challenges of the future society leaded by globalisation and an increased demand for agri-food products, which need to incorporate as much technical progress as possible. In the process of producing agri-food products a key role is played by the knowledge transfer market. The authors analyze the size, quality and structure of its main components, supply of information and knowledge, demand of knowledge and linkage vectors between the two components. The chapter captures the evolution of the actors involved in the knowledge transfer market of the Romanian agriculture.

Chapter 16, "Technology Scouting and Inventions Patenting With Impact on the Agrifood Future: INACO – Institutional Innovation for Competitiveness in Romania," describes possible pattern of the future world in which any kind of business could be developed in a completely new human, technological, agricultural and commercial context. The author brings into attention the mega-technology trends that will reshape completely the future business and jobs, focusing on the agrifood industry. Also, the author highlights challenges to patent inventions in terms of costs and timing in Romania, relative to other countries, and suggests pragmatic recommendations.

The book theme will definitely impact local stakeholders, farmers, public and governmental authorities, agri-businesses, researchers and PhD students, and also European and global level stakeholders. The shared experiences and lessons learned will significantly contribute to new understanding and approaches for effective achievement of those SGDs related to resources management and for decoupling food system activities from resources use and environmental impact and move towards more resource-smart food systems.

The theme also impacts the editor's field of research, since its major part is related to sustainable agrifood economics and development of agricultural production through a range of activities, including green entrepreneurship, agrifood economics and public-private partnership. The publication indicates a continuation of editor's research on the role of agriculture in modern business and society.

Gabriel Popescu
The Bucharest University of Economic Studies, Romania

REFERENCES

Bendre, G. (2017). *Romania has significant potential in agriculture, but low productivity, says PwC Romania study*. Retrieved from http://www.business-review.eu/news/romania-has-significant-potential-in-agriculture-but-low-productivity-says-pwc-romania-study-132113

Buckwell, A., Matthews, A., Baldock, D., & Mathijs, E. (2017). *CAP - Thinking Out of the Box: Further modernisation of the CAP – why, what and how?* Brussels: RISE Foundation.

European Commission (EC). (2018). *Implementing the Sustainable Development Goals through the next Multi-Annual Financial Framework of the European Union.* Advisory report to the European Commission by the Multi-Stakeholder Platform on the Implementation of the Sustainable Development Goals in the EU, March 2018. Retrieved from https://ec.europa.eu/info/sites/info/files/adopted-position-paper-on-the-mff_en.pdf

Eurostat. (2017). *Statistics_on_rural_areas in_the_EU*. Retrieved from http://ec.europa.eu/eurostat/statistics explained/index.php/Statistics_on_rural_areas in_the_EU

Food and Agriculture Organization of the United Nations (FAO). (2017). *2017 The State of Food and Agriculture- Leveraging Food Systems for Inclusive Rural Transformation*. Rome: Food and Agriculture Organization of the United Nations.

Organisation for Economic Co-Operation and Development (OECD). (2006, October). Reinventing Rural Policy. *Policy Brief.*

Potocnik, J. (2017). Preface. In CAP - Thinking Out of the Box: Further modernisation of the CAP – why,what and how? (pp. 4-5). RISE Foundation.

PricewaterhouseCoopers. (n.d.). *Potential of agricultural sector development in Romania*. PricewaterhouseCoopers Management Consultants SRL. Retrieved from https://www.juridice.ro/wp-content/uploads/2017/03/Raport_PwC-agricultura.pdf

United Nations Environment Programme (UNEP). (2016). *Food Systems and Natural Resources. A Report of the Working Group on Food Systems of the International Resource Panel* . Author.

United Nations (UN). (2015). *Transforming our world: the 2030 Agenda for Sustainable Development*. Resolution adopted by the General Assembly on 25 September 2015. Author.

Section 1
The Actual State

Chapter 1
Determinants of Agricultural Production in Romania: A Panel Data Approach

Alina Zaharia
The Bucharest University of Economic Studies, Romania

Simona Roxana Pătărlăgeanu
The Bucharest University of Economic Studies, Romania

ABSTRACT

Agriculture plays an important part in the worldwide challenges, such as sustainable development, climate change, high level of greenhouse gas emissions, food security and safety, overpopulation, social welfare, and natural resource depletion. This chapter examines a panel data approach to determine the contribution of several factors on the agricultural output in terms of value and of yield. Different regression models were established for the analysis at territorial level in Romania. Some findings suggest a negative influence of the excessive drought years on the cereals yield while a statistical relevance could not be found for the influence of the excessively rainy years. Still, further studies should be conducted on analyzing the influence of the environmental and social factors on the agricultural economic output.

INTRODUCTION

The factors influencing agriculture are highly relevant for decision making in terms of their wide implications for food security and safety, sustainable management of natural resources, and climate change. While the global population is continually increasing, policy makers try to identify ways of ensuring food security and safety by investigating the relationship between agricultural inputs and outputs (Burja, 2012; Teryomenko, 2008). The liberalization of trade has increased competition on local markets and has required more attention given to production costs, resource management, farm size, agricultural policies. Despite increased competition, the trade liberalization could contribute to diminishing the degradation of natural capital by internalizing its effects on production (Lopez, 1994).

DOI: 10.4018/978-1-5225-5739-5.ch001

Copyright © 2019, IGI Global. Copying or distributing in print or electronic forms without written permission of IGI Global is prohibited.

Moreover, climate change impacts the agriculture and the land use and, consequently, policies regulate new sustainable agricultural practices. Several studies have indicated the positive effects of climate change on agriculture, such as extension of the arable area to the North, introduction of new crop species; negative effects have also been documented such as increase in temperature and extreme weather events (Vijayasarathy & Ashok, 2015; Olesen, 2006). Hence, analyses of agricultural determinants are explored by a vast literature which will be thoroughly discussed in the literature review section of this paper.

It becomes increasingly necessary to continually analyze the factors influencing agricultural production in order to better understand their impact on agricultural outputs. The impact of agricultural activities on environment and social dimensions should be considered. As a result, new sustainable strategies and practices might emerge.

The objective of the paper is to explore the relationships between the agricultural determinants and outputs in general, as well as, the cereals' ones in particular. The authors argue that economic and social indicators, technical and material capital, financial support, human capital, natural resources, climate variability represent the main determinants of agricultural output.

The main questions that underpin this research are: Which are the main drivers of the Romanian agricultural output? Which are the main influences of the agricultural output's drivers at territorial level in Romania?

To the best of authors knowledge, few studies focus on discussing the influencing factors of the Romanian agricultural output from an econometric point of view. Unlike previous studies, which focused on more specific relationships with fewer variables, this paper undertakes an overview of the Romanian agriculture sector over 1997-2014 periods. Hence, improved understanding of this subject will hopefully lead to identifying the directions for increasing the agricultural output.

BACKGROUND

The gross value added of agriculture, forestry and fishing represented 1.7% of the EU total gross value added while in Romania this indicator reached the EU peak of 6.4% of total gross value added in 2013 (Eurostat, 2016). The employment level is 2.75 million Romanian persons occupied in this sector in 2013 and 2.5 million persons self-employed – the highest in the EU – reaching to a share of 24.16% in 2013 (Eurostat, 2016).

In addition, food security and safety is a more and more discussed issue all over the world. In simplistic definitions, food security refers at assuring the necessary quantity of food for the population while food safety refers at assuring the so needed quality of food. It is interesting to emphasize the existence, in general, of enough food available for the population, still the lack of its quality and the access to it raise many problems. By access, the specialists refer mainly to the affordability (Hazell & Wood, 2008), to *the access to sufficient, affordable and nutritious food* (European Commission, 2016a). Still, by solving the problem of assuring food security by providing for the population enough money to afford the healthy and the necessary food, it might arise:

Income increases => land conversion for cultivation => overconsumption => overpopulation => environmental issues (natural resource depletion, loss of biodiversity, agricultural and food waste, pollution) => climate change (social, economic and environmental problems).

The climate change has intensified since the industrial revolution when fossil fuels started to be intensively used in all sectors of economy and pollution started to increase drastically. It is a worldwide known problem for which policy makers, experts, farmers, etc. identified the main causes and impacts and now they try to find solutions and apply policies for mitigating and adapting to climate change (Zaharia & Antonescu, 2014). Climate change affects the agricultural activities considerably, as the status of environmental inputs represents a vital condition for increasing the agricultural yields and agricultural output values, as well as for assuring social, environmental and economic stability.

Climate change produces changes in temperature, precipitation, availability of resources and cropping systems, as well as increased climatic variability (Olesen & Bindi, 2004). These global changes have positive and negative effects. Olesen (2006) states that, among the positive ones, we could find the emergence of new crop species and varieties, expansion of agricultural area, increased crop production, cultivation of some crops in winter. Among the negative ones, the following could be identified: water shortages, diminution of soil organic matter, emergence of extreme weather events, and variability in crop yields. In light of these positive and negative effects, the Southern areas will still emerge as the most affected regions (Olesen, 2006).

Equally, agricultural activities contribute to climate change through the emissions they generate (Olesen, 2006). The greenhouse gas emissions have drastically increased since the end of the 19th century, mainly due to massive industrialization and use of fossil fuels (Zaharia & Antonescu, 2014). According to the European Commission (2016b), currently, the EU share in the total global emissions is approximately 11%. Zaharia and Antonescu (2014) suggest that the EU policies on the reduction of pollution are contributing to the diminishing emissions on a downward trend, including the ones from agriculture. Thus, the EU GHG emissions registered decreases of 18.33% overall and of 23.12% in agriculture over 1990-2011 (Zaharia & Antonescu, 2014). In 2013, the European Union's agriculture contributed 9.55% to the overall EU GHG emissions while Romania's agriculture contributed 0.38% to the overall EU GHG emissions. In addition, Romania's agriculture contributed 4.004% to the GHG emissions produced by the EU agriculture and 6.31% to the overall Romanian GHG emissions (European Commission, 2016c).

As a result, the agricultural sector is very important for the three dimensions of sustainable development: social, economic and environmental ones. In these circumstances, European Union adopted and is implementing a series of sustainable policies, regulation and measures for increasing the overall performance of agriculture, the farmers' income and the agricultural yields and values, as well as for mitigating and adapting to climate change and assuring food security and safety.

One of these EU directions is the "Europe 2020 Strategy" (European Commission, 2010), by which the European Union establishes the targets on the reduction of GHG emissions by each Member State for contributing to climate change mitigation. In addition, this strategy establishes the guidelines for increasing the resource efficiency, for protecting the biodiversity and the natural capital and for increasing the practices of organic farming. European Commission presents indicators for all these issues. This important EU growth strategy is also applied in Romanian agriculture through the Rural Development Strategy 2014-2020 and financial supported by the National Rural Development Programme 2014–2020. The Romanian sustainable development measures introduced through these strategic documents are based on a series of 26 agricultural needs, such as: education and counseling of farmers and of agricultural occupied population; linking the agricultural research, production and market; modernization of farms, processing units, infrastructure; protection and improvement of natural resources; diminishing the pollution and adapting to climate change (The Romanian Ministry of Agriculture and Rural Development, 2016). Based on these needs this Programme gives financial supports for rural projects.

Another important policy for EU agriculture is the Common Agricultural Policy. In 1962 the European Union created the CAP which, over time, has been reformed repeatedly in order to face the challenges in the agricultural sectors of the member states (European Commission, 2014a). Currently, this policy focuses on improving and conserving the natural capital while developing a sustainable market-oriented agriculture. CAP seeks to ensure food security and safety, as well as farmers' income among its objectives (European Commission, 2016d; Popescu, 2014). Romania shall beneficiate of 20 billion euro for improving the agri-food sector and the rural areas during 2014-2020. CAP finances the agricultural activities through direct payments, market measures and rural investments (European Commission, 2014b).

In this context, many studies pay close attention to the influence of climatic factors, the development level of technical and material inputs, the educational level of the persons occupied in agriculture, the investments and financing of the agricultural sector, and the overall factors which could induce high values of agricultural outputs.

Agriculture has several agricultural drivers that could be divided according to the global, national and local scale. Hazell and Wood (2008) present as global drivers: trade liberalization, climate change, international agreements on these issues in all economic sectors, *rapid globalization of science and knowledge access*, as national drivers: legislative framework, population income, infrastructure and market access, and, as local drivers, the climate variability and the agricultural production system. In addition, the above-mentioned authors state that the international trade and the globalization of markets increase competition and increase the difficulties of small farms to enter and remain on the market. So, due to this matter, the costs of the small agricultural producers increase.

Zaharia and Antonescu (2014) consider the relationship of interdependency between agriculture and climate change because the agriculture produces GHG emissions and other negative effects which contribute to climate change and the latter influences negatively and positively the agricultural activities by weather changes, agricultural system changes, natural resources changes. Their study also concluded that even though "there is no dependency between the Romanian arable land and the Romanian GHG emissions, there are other factors which could influence the pollution level, such as: synthetic fertilizers, bad management of manure, type of cultivated crops, burning crop residues" (Zaharia & Antonescu, 2014). Figure 1 presents this relationship illustrated by the above mentioned authors.

Two main types of GHG emitted by agricultural activities have the highest values of all GHG emissions from agriculture: nitrogen protoxide and methane. These are very dangerous and powerful in creating the greenhouse effect which contributes to climate change. Moreover, the nitrogen protoxide is 300 times more harmful than the CO2 (Romanian National Institute of Statistics, 2016a). Howbeit, the agricultural GHG emissions are released by practicing a varied group of activities, such as applying pesticides and fertilizers, manure management, burning crop residues, cereals cultivation (mostly the rice) (Zaharia & Antonescu, 2014).

Figure 1. The interdependency between agriculture and climate change
Source: Zaharia and Antonescu, 2014.

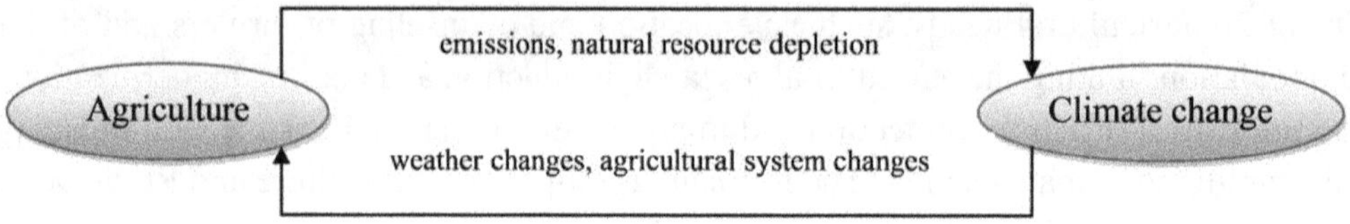

So far, the main looked upon determinants of agriculture were mainly the global ones, therefore, in the following, the authors discuss the main focus of this chapter: the national and local determinants of agriculture, which will be thoroughly investigated in this paper.

MAIN FOCUS OF THE CHAPTER

Analyzed Issues

Many studies show the importance of infrastructure for agriculture development as the productivity growth of crops is explained by increases in investments for public infrastructure (Mamatzakis, 2003). Expanding to the digital world, the development of geospatial applications is becoming more helpful for the farmers in planning the location of land operations (Wiebensohn & Jackenkroll, 2013). Moreover, water and sewage infrastructure contribute to the increases in the welfare of the rural inhabitants (Piasecki & Jurasz, 2015).

Nevertheless, the natural capital represents the main influencing factor of crop productivity and value. Nowadays, European Union follows the evolution of indicators on the state of natural resources in all member states. Soil and water quality, climate variability and land changes are continually monitored in the context in which the use of these resources should be planned for a better sustainability of agricultural inputs and for increasing their efficiency. The resource efficiency is an EU objective because it might,

reduce the risk of vulnerability to future resource volatility or scarcity, reduce the costs through productivity savings and mitigate the environmental and social problems linked with unsustainable use of resources, such as pollution. (European Commission, 2011)

The Romanian legislation regulates the investigation, evaluation, monitoring restoration, and conservation of natural resources. The soil has many vital functions, such as: food production, water storage, source of raw materials, biodiversity, geological and archaeological heritage, represents the physical environment for human activities (Romanian National Agency for Environmental Protection, 2016). In this context, EU tries to create databases with the evolution of soil quality indicators: texture, organic matter content, pH, electric conductivity, soil water retention, groundwater level. Nevertheless, the soil state encounters many problems, such as: erosion, organic matter degradation, soil biodiversity loss, compaction, salinization, contamination, landslides and floods, sealing (Romanian National Agency for Environmental Protection, 2016). So, soil productive potential should be emphasized and improved. Crop rotation helps maintaining soil fertility and control of pests and diseases, and it has a role in terms of crop diversification by increasing their efficiency and the conservation of plant genetic resources (Voicu & Dobre, 2003).

The water is also a vital natural resource because without it the agriculture faces real challenges in conditions of actual pollution and increasing desertification. The water depends on the environment conditions as one of the major aspects in providing the necessary water for healthy increasing proportions of agriculture products (Giovannucci Scherr, Nierenberg, Hebebrand, Shapiro, Milder, & Wheeler, 2012).

Climate parameters represent also an influencing factor of agricultural output in time and space. The states of these parameters determine the suitability of one culture or another, of the applied production system and the supplementary investments and expenses which have to be made for improving productivity. All of these would eventually impact the final agricultural outcome (Voicu & Dobre, 2003).

The agricultural producer can only act within the limits imposed by the natural conditions and, where possible, further improvements could be made by using the technical capital. This may conduct to increased costs, but so needed for the achievement of desired level of production. Studies indicate negative effects of technical capital on food safety and environment. The decrease of emissions from agricultural soils is due to diminishing the use of mineral and organic fertilizers and pesticides. The soil emissions are mainly related to fertilizers, waste decomposition, sewage sludge application, cultivated crop (Anderl, 2009).

All these presented factors should be combined according to the type of crop culture used and to the type of production system applied. The allocation decisions and the combination of inputs are made exclusively by the farmer, considering the past prices on the market (Voicu & Dobre, 2003).

The national and local drivers influence also the outcome of agriculture in what concerns the productivity and its value. Popescu and Zaharia (2015) analyzed the relationship between energy consumption, technical progress and value added in the European Union's agriculture and found that the value added of agriculture, forestry and fishing is correlated with the energy consumption, with tractors, and with fertilizers and pesticides consumption, all reported to 1000 ha agricultural area. The authors concluded that there is no statistical correlation to day between the Romanian area equipped for irrigation and the value added in agriculture, forestry and fishing.

Another study (Carvalho, 2006) emphasizes the importance of soil and water management, intensive irrigation, fertilizers, pesticides and agricultural land increases in assuring the future food security in a context in which the negative effects of the previous presented factors and the status of the environmental factors, such as the soil and water quality, would greatly impact the agricultural output. The application of pesticides and chemical fertilizers negatively influence the cost of production, the human and biodiversity health, as well as the natural capital in its entire. As the weeds, insects and other types of pests increase their resistance to control chemicals; the latter is increasing to be applied while raising the costs and damaging the agricultural product nutrient structure. Also, the pesticides and the fertilizers damage the quality of soils, water and atmosphere. Nevertheless, a better control and use of these substances could diminish their negative impact on human health and environment.

Several researchers have analyzed the determinants of agricultural production at national level. Abugamea (2008) investigated the agricultural production outcome in relation with the input cost, the labor force and the cultivated land by using time series econometrics procedures. The authors find a significant effect on agricultural production both from capital (negative) and labor (positive).

Moreover, Teryomenko (2008) explored several parametric and non-parametric methods for establishing and analyzing the determinants of agricultural productivity, namely the farm size, the farm specialization, the expenses on inputs such as electricity, repairs, labor. The results of Teryomenko's study indicate that farm size influence the agricultural productivity up to a point where more rented land could generate decreases of productivity and that the policies pursued by authorsities which might influence positively or negatively the analyzed variables could generate changes on the productivity level.

Another study (Ekbom, 1998) focuses its analysis on the material, human and capital inputs to explore the influence on agricultural productivity by using a Cobb-Douglas production function and the research concludes that in Kenya is statistically significant correlation between agricultural productivity and the following factors: farm size, distance, labor availability, costs of fertilizers, on-farm non-agricultural incomes, access to credits, and soil conservation quality. Likewise, it was found that the productivity of energy, substitution of labor by capital and crop output per hectare influence the mixed farming in EU. As previously seen, the evaluation of production efficiency is as a result of climatic conditions and variability of farms (Špička, 2014).

Burja (2012) concludes that the agricultural holdings from the West region are the least productive while the ones from the South Muntenia region are similar in efficiency as those from the EU. *Above average performance is recorded by the South-West Oltenia and South-East regions.* Other study (Nowak, Kijek, & Domańska, 2015) suggests that Romania has a low agricultural technical efficiency and that its determining factors are the age of the head of the household, the soil quality and the surcharges for investments while the farm size seems to be irrelevant. One negative influencing factor of the Romanian agriculture is the labor force which is characterized by many families who work for their subsistence farms, by the aging population with a low level of education (Tocco, Davidova, & Bailey, 2014).

Methodological Background

This study focuses on analyzing the determinants of agricultural production at national and territorial level in Romania by applying panel data and secondary data analyses in EViews 7 software for explaining the influencing factors of agricultural crop values on eight Romanian development regions during 1997-2014. The data was provided by the Romanian National Institute of Statistics, the European Commission (Eurostat and Farm accounting data network) and the ESPON database portal.

The indicators considered in this paper are: agricultural values, agricultural crop values, cereal yield – as endogenous variables -, irrigated area, the number of villages, final energy consumption in agriculture, physical agricultural tractors per agricultural ha, the labor cost per employee, civil economically active population, chemical and natural fertilizers, pesticides, greenhouse gas emissions from agriculture, gross domestic product per capita, population density, the actual heating degree-days per year, and the type of season to illustrate the climate variability (rainy, dry – dummy variables) – as explanatory variables.

The cereals crop was chosen as it represents one of the most important agricultural cultures worldwide, because of their importance in assuring food security. Thus, cereals are easy to cultivate, transport, store, and give more than 60% of the nutritional value required by the human body (Cereal Science and Technology, 2016). Cereals have economic, social and environmental significance. Also, corn and wheat are the two cereal cultures prevailing in Romania that are belonging to the cereals group.

The corn is the predominant culture in Romania due to the fact that it can be used both as food and as fodder. It is an expensive culture to set up; still, it needs no high degree of mechanization which makes it accessible to the numerous Romanian individual households. Unlike corn, the wheat requires a higher degree of mechanization, which leads to higher volume of payments made to service providers. (Gavrilescu & Giurcă, 2000: p. 111-113)

THE ROMANIAN LAND USE AND THE LAND IMPROVEMENTS

Since 1997 the land fund use has changed due to various factors. The shift from the socialist economy (centrally planned economy) to a market economy has influenced this change as the restructuring of the Romanian economy has negatively impacted the agricultural activities. The cropland abandonment has been affected due to several determinants, such as topography, migration, market access, tractors density, public policies (Müller, Leitão, & Sikor, 2013). According to the Romanian National Institute of Statistics (2016b), the total land of Romania is 23839071 hectares, of which 14630072 hectares agricultural land and 9208999 hectares non-agricultural land. The agricultural land has diminished by 1.11% while the non-agricultural land has increased by 1.81% in 2014 compared to 1997. The decreases of the Romanian agricultural land could be explained by the shift in the type of usage. So, although the arable land and the meadows land have registered small increases by 0.58% and by 4.39% during 1997-2014, the land with orchards, the vineyards and the pastures has significantly decreased by 25.87%, 26.86% and 4.04% during 1997-2014. Figure 2 illustrates the evolution of the Romanian agricultural land structure evolution.

The lowest shares of arable land of each regional agricultural land are found in Center region (40%) and in the North-West region (50%) in the analyzed time frame. The arable land of the Center region is the 2nd smallest region as arable surface of all the eight regions while the North-West region is placed on 6th. Even though the Bucharest-Ilfov region has the smallest area of arable land, this region has the highest share of arable land in the Bucharest-Ilfov agricultural land.

Equally, the highest regional shares of agricultural area are found in the South Muntenia region (16%) and in the South-East (15%) while, as expected, in the Bucharest-Ilfov region has the lowest share of 1%. The situation is similar when analyzing the arable land structure.

In addition, the land improvements and quality represent important factors which contribute to the cereal yield and, further, to the agricultural value. As all the EU member states, Romania registered loses in the agricultural land because of various factors. The Romanian land improvements have increased since 1997. The erosion and soil improvements have slightly increased as area, by 1.86%. Yet, on regional level the changes are a bit bigger as in the case of the irrigated area and the drainage works in hectares. The Romanian actual irrigated area has improved by 15.11% while the area with drainage works has increased by 8.17% during 1997-2014. The land use and the status of the natural resources' quality contribute to the agricultural crop outcome. Therefore, their analysis, in accordance with other determinants, is important for developing the performance of the Romanian agriculture.

Figure 2. The evolution of the Romanian agricultural land use during 1997-2014
Data source: Romanian National Institute of Statistics, 2016.

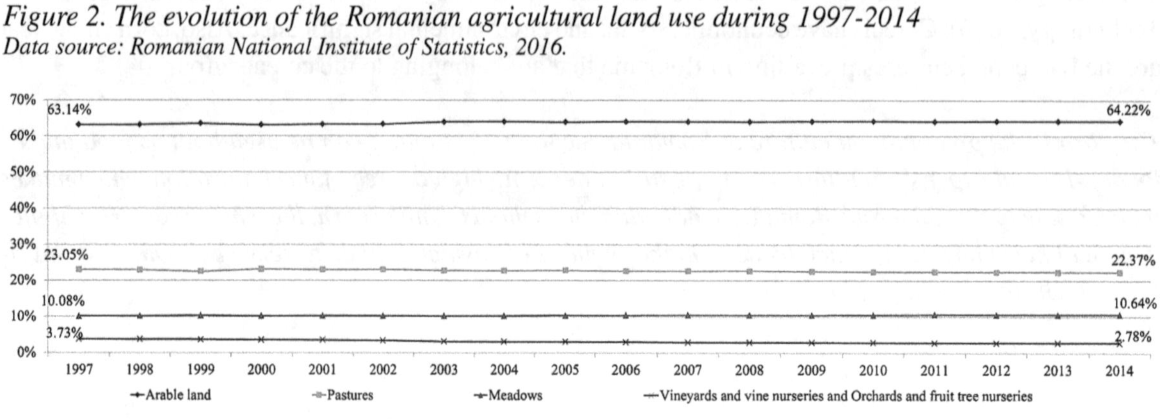

THE CAUSAL RELATIONSHIP BETWEEN THE AGRICULTURAL DETERMINANTS

The support for Romanian rural development has slightly increased in terms of financial support in 2013 since 2010, but this has slightly decreased in terms of supported farms. The number of Romanian farms, presented in Table 1, has decreased between 2005 and 2013, maybe due to the changes in the legislative framework as well as the economic crises. It is interesting to highlight the slight increase of the number of farms with more than 50% of production self-consumed by the holder until 2010, although the European Union supports agriculture directed to market. Nevertheless, the number of farms with more than 50% of production self-consumed by the holder has decreased less than the total number of farms since 2005.

Romania has more than half of its farms producing for at least 50% to self-consumption. In 2013, the lowest share of farms with more than 50% of production self-consumed by the holder in total number of holdings was registered in Bucharest-Ilfov region (72.99%) while the highest share was listed in the South-West Oltenia region; The Romanian average share was 87.57%. This indicator could show the development degree of holdings because when the self-consumption increases it could mean the economic downfall of the holder of the farm. In addition, it is well-known the Romanian farm structure: that there are the subsistence farms above EU average.

The total agricultural value (the agricultural output) and, in particular, the crop output (the agricultural crop value) followed a varying and increasing trend since 1997 that it could be explained by the modification of moneys' value as well as by the agricultural market in a specific year. Both indicators have augmented by 9 times in 2014 compared to 1997.

The agricultural greenhouse gas emissions have decreased by 15% while the final energy consumption by 53.29% during 1997-2014, as illustrated in Figure 3.

The population density recorded a slight decrease since 1997 while the GDP per capita has almost doubled during 1997-2013. The material and economic capital, which might influence the agricultural productivity, has also varied overtime. The tractors density has increased since 1997 as well as the technological development of the field. The fertilizers and pesticides have started to be carefully monitored

Table 1. Total number of Romanian farms by region

GEO/TIME	2005	2007	2010	2013	2005	2007	2010	2013
	Total No. of Farms				More Than 50% of Production Self-Consumed by the Holder (no.)			
Romania	4256150	3931350	3859040	3629660	3444760	3172280	3589530	3178490
North-West	591510	533770	528460	499860	465490	404800	489710	421780
Centre	440710	398540	394650	358470	348760	321640	348480	292020
North-East	854870	807460	790790	754530	701530	662920	749070	684120
South-East	532150	501420	460330	433040	414010	374010	420470	363620
South - Muntenia	847560	762890	800830	753590	697700	621860	738580	671780
Bucharest – Ilfov	63860	62410	33490	25320	48670	53880	29270	18480
South-West Oltenia	608160	580610	576600	557850	527040	518410	561030	525410
West	317330	284260	273890	247000	241560	214760	252930	201280

Source: (European Commission, Eurostat database, 2016).

Figure 3. Some key Romanian indicators relevant for agriculture
Source: Romanian National Institute of Statistics, 2016.
Note: AV/1000ha – agricultural value per 1000 ha, ACV/1000ha – agricultural crop value per 1000 ha, GHGa – agricultural greenhouse gas emissions, YC – cereal yield, LCE – labor cost per employee, FECa – final energy consumption in agriculture

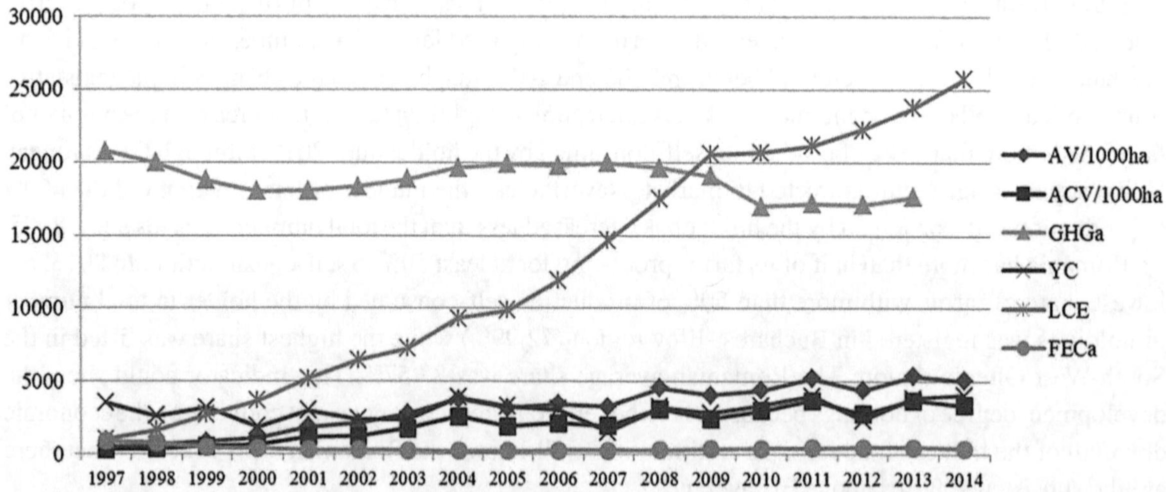

for a better protection of the environment and the human health. Both the chemical and the natural fertilizers have decreased 29.26% and 15.49% during 1997-2014. Likewise, the insecticides, the herbicides and the fungicides have significantly reduced by 76.84%, 45.74% and 67.02% since 1997. The labor cost per employee has gotten up during the analyzed period while the number of employees working in agriculture has significantly decreased, by 69.4%, during 1997-2014.

Again, the climate variability influence considerably the agricultural crop value and the cereal yield; this is why, in the analyses, the authors took into consideration also the type of the season (excessively dry, excessively rainy) and, as much as possible, the actual heating degree-days.

Further, in order to determine the correlation matrix between the variables influencing the crop output, it must be tested the stationarity of each variable for model validity. The authors chose the augmented Dickey-Fuller test for unit roots in order to eliminate the autocorrelation and to test the stationarity of data (Asteriou & Hall, 2007).

Testing for Units Roots by Using the Augmented Dickey-Fuller Test

Table 2 presents the results of the unit root tests briefly.

If the null test hypothesis (that the analyzed variable has a unit root, meaning that the data is non-stationary) is rejected, than the data is stationary and the further steps of the analysis could be followed. Otherwise, the data is non-stationary and, before continuing the analysis, the data must be processed by applying the required difference.

The analysis was initially performed in level with the option of trend and intercept at maximum lags 3. The pesticides data is stationary without trend and intercept at 5% significance level.

Table 2. The results of the unit root test for the analysis of agricultural output determinants

EViews Code	Variable Name	Null Test Hypothesis in Level	Conclusion
av	Agricultural output	accepted	non-stationary data
acv	Agricultural crop value (crop output)	Rejected (trend and intercept)	stationary data***
yc	Cereal yield	Rejected (trend and intercept)	stationary data***
agrl	Agricultural land	Rejected (trend and intercept)	stationary data***
sirr	Irrigated area	accepted	non-stationary data
feca	Final energy consumption agriculture	accepted	non-stationary data
ghga	GHG agriculture	accepted	non-stationary data
gdp	GDP pps per inhabitant	accepted	non-stationary data
lfc	Labour force total costs	accepted	non-stationary data
ceapop	Civil economically active population	accepted	non-stationary data
ir_fp	Inflation rate food products	Rejected (trend and intercept)	stationary data***
ur	Unemployment rate	accepted	non-stationary data
pd	Population density	accepted	non-stationary data
pdagrl	Population density on agricultural land	Rejected (trend and intercept)	stationary data***
pat	Physical agricultural tractors per agr.ha	accepted	non-stationary data
cf	Chemical fertilizers	accepted	non-stationary data
nf	Natural fertilizers	accepted	non-stationary data
p	Pesticides	Rejected (trend and intercept)	stationary data**
subv	Subsidies	accepted	non-stationary data
hdd	Actual heating degree-days per year	accepted	non-stationary data

*10% significance; ** 5% significance; ***1% significance
Note: See Appendix 1 for the legend

The Correlation Matrix for the Agricultural Output in Romania

The correlation for the agricultural output (expressed in thousands lei) takes into consideration the previous possible analyzed determinants, when it was conducted the unit root test for stationing the data.

The correlation matrix shows the influence of one variable to another. As it can be observed, the crop value is correlated with the yield of cereals, the agricultural area, the inflation rate of food products, the population density per agricultural land, the difference of the tractors' number and the pesticides. The cereal yield is correlated with the agricultural area, the irrigated area, the number of tractors and the 2nd difference of the population density.

Further, the Granger causality test is conducted to identify in which way these variables influence each other. The null hypothesis of this test is that the independent variable does not affect the dependent variable. Patra and Poshakwale (2006) recommend using 6 lags in order to determine the optimal results. However, in this case, as the dimension of the sample is quite small, the analysis will be made with 2 lags.

The difference of the agricultural value (dav) is influenced by the yield of the cereals (yc) and the difference of subsidies (dsubv) with 99% significance, as well as by the difference in the irrigated area (dsirr) by more than 95%. Also, the crop output (acv) is influenced by the yc, the dsubv and by the pes-

Table 3. The transformation of the non-stationary data into the stationary ones

EViews Code	Applied Difference	Significance	Trend	Intercept	None
dav	1st	1%	x	x	
dsirr	1st	1%			x
ddfeca	2nd	1%			x
ddghga	2nd	1%	x	x	
ddgdp	2nd	1%			x
dlfc	1st	1%	x	x	
dceapop	1st	1%			x
dur	1st	5%	x	x	
ddpd	2nd	5%			x
dpat	1st	5%	x	x	
dcf	1st	5%	x	x	
dnf	1st	1%			x
dsubv	1st	5%			x
dhdd	1st	5%	x	x	

Note: See Appendix 1 for the legend

ticides (p) with more than 90% significance. The yield of cereals (yc) is influenced by the difference of subsidies (dsubv) with more than 90% significance.

The 2nd level difference of the greenhouse gas emissions from agriculture (ddghga) are influenced by the difference of the unemployment rate (dur), by the 2nd level difference of the GDP per capita (ddgdp), by the difference of the labor force costs in agriculture (dlfc), by the difference of the chemical fertilizers (dcf) and by the difference of the agricultural value (dav) with a significance of more than 90%.

The 2nd level difference of the energy consumed in agriculture (ddfeca) is influenced by the agricultural crop value (acv), by the agricultural land and by the inflation rate at food products (ir_fp) with a significance of more than 89%.

These relationships could represent a focus for the decision makers at the level of political, legislative and agricultural activities.

THE RELATIONSHIP BETWEEN SOCIAL, ECONOMIC, AND ENVIRONMENTAL DIMENSIONS ON TERRITORIAL LEVEL

A series of Romanian indicators are explored at territorial level by using the panel data approach. This section presents the analysis conducted in EViews. Economic, social and environmental indicators are explored with the hope of shedding new light on the Romanian agriculture's performance at territorial level and its impact on environment.

For the purpose of this small scale research, it is important to understand what the development regions are. These are not administrative units of Romanian territory, but regions created for helping the planning and the distribution of the EU funds in Romania. This country has 8 development regions (Ministerul Dezvoltării Regionale şi Administraţiei Publice, 2016):

- **North - East Region – Counties:** Bacău, Botoşani, Iaşi, Neamţ, Suceava,Vaslui;
- **South - East – Counties:** Brăila, Buzău, Constanţa, Galaţi, Vrancea, Tulcea;
- **South - Muntenia – Counties:** Argeş, Călăraşi, Dâmboviţa, Giurgiu, Ialomiţa, Prahova, Teleorman;
- **South - West Oltenia – Counties:** Dolj, Gorj, Mehedinţi, Olt, Vâlcea;
- **West – Counties:** Arad, Caraş–Severin, Hunedoara, Timiş;
- **North - West – Counties:** Bihor, Bistriţa-Năsăud, Cluj, Sălaj, Satu–Mare, Maramureş;
- **Center – Counties:** Alba, Braşov, Covasna, Harghita, Mureş, Sibiu;
- **Bucureşti – Ilfov:** Bucharest municipality and the Ilfov county.

Except the Bucureşti-Ilfov region, which has around 0.8% of Romanian territory, the rest of the regions have around 12-15% of Romanian territory.

The Panel Unit Root Test and the Correlation Matrix

The panel unit root test provides the necessary results for establishing the existence of non-stationary data. These types of variables should be transformed to stationary data in order to have valid further analysis. It is used the summary method to apply this test at automatic leg length. Table 4 presents the results of the panel unit root test and it explains the data transformation in what concerns the stationarity.

Table 4. The panel unit root test for the data stationarity

EViews Code	Type of Data	Conclusion - Action
av, cf,	stationary data*	Null hypothesis rejected (trend and intercept) – summary
acv, yc, qc, sirr	stationary data***	Null hypothesis rejected (trend and intercept) – summary
yc	stationary data***	Null hypothesis rejected (trend and intercept) – summary
sc	stationary data**	Null hypothesis rejected (intercept) – summary
nov	stationary data*	Null hypothesis rejected (none) – summary or 1st difference, trend and intercept*** => *dnov*
p	stationary data**	Null hypothesis rejected (intercept) – summary
ceapop, pd, pdagrl	stationary data***	Null hypothesis rejected (none) – summary
pat	non-stationary data	1st difference, trend and intercept*** => *dpat*
nf	non-stationary data	1st difference, intercept*** => *dnf*
lce	non-stationary data	1st difference, intercept*** or trend and intercept* => *dlce*
lfc	non-stationary data	1st difference, trend and intercept*** => *dlfc*
rgdp	non-stationary data	1st difference, trend and intercept*** => *drgdp*
hdd	stationary data***	Null hypothesis rejected (intercept) – summary

*10% significance; ** 5% significance; ***1% significance
Legend: See Appendix 1

The data is non-stationary if the analyzed variable has a unit root (the null test hypothesis is accepted) by using the summary method. If the null hypothesis is rejected, then the data is stationary. Those variables which present non-stationary data have been transformed by difference at 1st, 2nd or 3rd level for model validity. The analysis was initially performed in level with the option of trend and intercept by using the summary test type in EViews.

In Appendix 2, the descriptive statistics of the panel data are presented and it could be noticed that the mean cereal yield is 3039.1 kg/ha during 1997-2014. The maximum of 4381 kg cereals/ha was registered in the Bucharest-Ilfov region in 2004 and the minimum of 834 kg cereals/ha was registered in the South-West region in 2007. The 2007 excessive draught year negatively influenced the cereal yield. Although, the majority of development regions registered favorable cereal yields in 2004, the Bucharest-Ilfov region registered a low agricultural crop value in 2004. This year did not have excessive draughts or excessive rains and the inputs used were relatively high in values.

In addition, big differences between the maximum values and minimum ones could be observed for several determinants, such as the difference in the labor costs.

After transforming the data to be able to use it, the correlation matrix is listed in Appendix 3. The correlation's coefficients are stained according to the maximum level of 40% and the probabilities are marked with *, ** or *** according to the 10%, 5% or 1% significance.

Moreover, the Granger causality test was performed to identify the possible causal relationships between the considered variables and their direction of action. The results show that the agricultural value (*av*) is influenced by the cereal yield (*yc*), the cereals production (*qc*), the civil economically active population (*ceapop*), the difference between the labor force costs (*dlfc*), the agricultural crop value (*acv*), the irrigated area of land (*sirr*), the population density on the agricultural land (*pdagrl*), the number of villages (*nov*) and the population density (*pd*) at a significance level of more than 99%, as well as by the difference of the labor cost per employee (*dlce*) at a significance level of more than 95% and by the pesticides (*p*) at a significance level of more than 90%. Agricultural value influences the *dlfc, acv*, the difference of the regional GDP (*drgdp*), *qc*, cultivated area with cereals (*sc*) and *yc*.

Equally, agricultural crop value (*acv*) is causally influenced by *yc, av, qc, dlfc, dlce, sirr* and *sc* at a significance level of more than 99%, as well as by *ceapop, nov, pdagrl* and *pd* at a significance level of more than 95%. Vice versa, the agricultural crop value influences the *dlfc, drgdp, av, qc, sc* and *yc*.

The cereal yield (*yc*) is influenced by the *dlce* and *p*, at a significance level of more than 99%, as well as by *av* and *acv* at a significance level of more than 95% and by *dlfc, ceapop* and *hdd* at a significance level of more than 90%. The cereal yield (*yc*) influences *av, acv, qc, drgdp, sc, pd, ceapop* and *dlce*.

The regional GDP (*rgdp*) is influenced by the agricultural value, including also the crop value, by the civil economically active population, by the population density (*pd and pdagrl*), by the number of villages, by the cereal yield.

The Panel Data Analysis

The panel data analysis includes 6 models considering the correlation matrix and the Granger causality test. Table 5 and table 6 present the determinants of cereal yield and of agricultural output value considered for the regression models and the related results.

The models analyzed in table 5 (M1, M2, M3 and M4) have been established without the dummy variables regarding the type of the season. After testing all the possible panel models, the suitable one was the fixed cross-sectional model with no period for the M2, M3 and M4 models. The fixed cross-

Table 5. Determinants of the cereal yield and of the crop output for the Romanian development regions – Linear regression model without dummy variables

Variable	M1 1998-2014	M2 1998-2014	M3 1998-2014	M4 1998-2009
Agricultural output	endogenous			
Crop output		endogenous		
Cereal yield	1645.439***	782.6529***	endogenous	endogenous
Irrigated area			-0.0063***	-0.0061***
The modification of the physical agricultural tractors per agricultural ha			434904.2***	416804.4000***
Chemical fertilizers			1018.661**	1069.3680**
The modification of the natural fertilizers			14.90256**	26.6795*
Pesticides			-100.6186***	-56.7003
The modification of the labor cost per employee	671.0702**	-124.0245		
Civil economically active population	-9434.669***	-30689.53***		
The number of villages	5181.596***	24222.01***		
Actual heating degree-days per year				0.2972
Intercept	-5038207.***	-27016382	3306.585***	2174.7330*
R-squared	0.521576	0.876810	0.406106	0.378747
Hausman test	-	Valid, but not statistically relevant	Valid, but not statistically relevant	Valid, but not statistically relevant
Redundant fixed effects (F)	-	80.2338***	2.8849***	2.0117*
Durbin-Watson	0.326846	1.2147	2.3231	2.6353
Chosen model	Cross-sectional: none; Period: none	Cross-sectional: fixed; Period: none	Cross-sectional: fixed; Period: none.	Cross-sectional: fixed; Period: none.

***1% significance **5% significance *10% significance
Note 1: See Appendix 1 for the legend.

sectional model indicates the consistency of the regression model, meaning that, as the sample size increases, the model is more accurate.

The R-squared tendency towards 1 shows the validity of each model and the tendency of Durbin-Watson test towards 2 indicates the existence of relatively small correlation among the model variables. So, the chosen panel model for M2, M3 and M4 are consistent and it could be relevant in practice.

As can be seen in M2, the crop output is positively correlated by the cereal yield and by the number of villages. The difference in the labor cost per employee and the number of civil economically active population influence negatively the agricultural crop value. These findings are sustained by the reality practices because, sometimes, as the employees and their related costs are increasing, the agricultural crop value might be negatively influenced due to price rises.

The positive influence of the labor cost per employee on total agricultural output from model 1 could be explained by the increase in its productivity. Maybe, the increase of this cost is generated by the improvements in education and/or experience which, further, it leads to productivity rises, and, therefore, fewer number of employees. The other 3 variables are similar to the ones from M1.

As the number of persons working in agriculture or the ones depending on a lower surface of agricultural land increases, the value of agricultural production is smaller. The population density per agricultural area was considered in the analysis as an indicator of food security. Still, due to its high correlation with many considered variables, there were no concluding results.

The 3rd and 4th regression models display the positive determinants of cereal yield (the difference of the physical agricultural tractors used in agriculture per ha, the chemical fertilizers, the difference of the natural fertilizers and the actual heating degree-days) and the negative ones (the irrigated area and the pesticides). Yet, the 4th model was analyzed on short timeline. Table 6 illustrates the negative effects which appeared in the North-West and South-West Oltenia regions in what concerns the cereal yield.

As the cereal yield depends on the climate variability, 2 models of regressions are analyzed in Table 7 by considering 2 scenarios: model 5 (M5) indicates the regression results by taking into account the years with excessive draughts; model 6 (M6) indicates the regression results by taking into account the excessively rainy years.

The negative and positive determinants of cereal yield explained by the M5 and M6 regressions are as the ones from M3 and M4. The dummy variable of M5 (the excessive draughts years) indicates a negative influence on the cereal yield. This result could be correlated with the reality. With an opposite effect, the dummy variable of M6 (the excessively rainy years) did not registered a statistical relevance as the p-value is 43.7%, which is a lot more than the 5% accepted significance level.

As previous, all possible regression models were tested in order to choose the right one, that of the fixed cross-sectional model without random period for the 2 explored models. The R-squared tendency towards 1 shows the validity of each model and the Durbin-Watson test indicates the existence of relatively small correlation among the model variables. So, the chosen panel model for M5 and M6 is consistent and it could be relevant in practice. The redundant fixed effects test (F) indicates the effects on each region. These are presented in the Table 8 which summarize the results regarding the cross-section fixed effects (at regional level).

Table 6. Cross-section fixed effects of model 3

Romanian Development Region	M3 Effect
North - West	-153.0868
Center	52.73498
North - East	32.58526
South - East	741.7025
South - Muntenia	332.7986
Bucharest - Ilfov	174.8243
South - West Oltenia	-1528.382
West	321.1813

Table 7. Determinants of cereal yield considering the type of the yearly season - Linear regression model

Variable	M5 1998-2014	M6 1998-2014
Yield cereal	endogenous	endogenous
Irrigated area	-0.005743***	-0.005996***
The modification of the physical agricultural tractors per agricultural ha	397786.1***	424095***
Chemical fertilizers	874.9874**	1029.805**
The modification of the natural fertilizers	15.17281***	15.35443**
Pesticides	-85.48199**	-94.75368**
Ddry	-428.392***	
Drainy		122.3217 (p-0.437)
Intercept	3609.936***	3238.207***
R-squared	0.443735	0.409550
Adjusted *R*-squared	0.374202	0.335744
Durbin-Watson test	2.5136	2.3070
Hausman test	Not statistically valid	Not statistically valid
Redundant fixed effects(F)	2.4814**	2.7174**
Chosen model	Cross-sectional: fixed; Period: none	Cross-sectional: fixed; Period: none

***1% significance **5% significance *10% significance

Note 1: The analyzed determinants are estimated according the type of season, which is represented by dummy variables for the years with excessive draughts and excessive rain (Ddry and Drainy).

Table 8. Cross-section fixed effects for M5 and M6

No.	Romanian Development Region	M5 Effect	M6 Effect
1	North - West	-155.5458	-109.5532
2	Center	43.11479	64.89199
3	North - East	15.08060	37.98230
4	South - East	647.9279	701.3351
5	South - Muntenia	297.3141	322.6093
6	Bucharest - Ilfov	138.9700	180.3643
7	South - West Oltenia	-1356.052	-1531.176
8	West	331.4499	339.4943

The fixed cross-sectional model indicates the consistency of the regression model, meaning that, as the sample size increases, the model is more accurate. In addition, the error term from the equation is not constant. The models register different effects on regional level.

CONCLUSION AND FUTURE RESEARCH DIRECTIONS

The agriculture is the new old key for continuous economic development in Europe as it can be a vital tool for reducing unemployment while achieving food security and safety. Agriculture is a strong option to stimulate growth, overcome poverty, and enhance food security which helps in accelerating growth within an economy. Moreover, agriculture plays an important part in worldwide challenges, such as climate change, high level of greenhouse gas emissions, food security and safety, overpopulation and natural resource depletion.

This chapter presented an overview of the Romanian agriculture sector over 1997-2014 periods and explores the influences of the agricultural output determinants at the Romanian territorial level. The results indicate a mean cereal yield of 3039.1 kg/ha during 1997-2014 while the maximum yield registered 4381 kg cereals/ha in the Bucharest-Ilfov region in 2004 and the minimum yield registered 834 kg cereals/ha in the South-West region in 2007. The majority of development regions registered favorably yields of cereals in 2004 as many agricultural inputs were used and no excessive draughts and rains were recorded. However, as expected, 2007 was a year with excessive draught, which negatively influenced the yield from cereals.

The Granger causality test suggests that the value of the agricultural output influences the difference of the regional GDP, the difference between the labor force costs, the cereals quantity and surface, as well as the cereal yields while it is causally influenced by the cereal yield, the civil economically active population, the difference between the labor force costs, the irrigated area of land, the population density and the number of villages at a significance level of 99%. The cereal yields influence the agricultural values, the difference of the regional GDP and the difference of the labor cost per employee while the difference in the labor cost per employee and the pesticides determine the yields of cereals by considering the significance error of 1%.

Moreover, the panel data approach consisted in analyzing 6 models based on the least correlated variables explained by the correlation matrix. 5 out of 6 problems were finally analyzed with the fixed cross-sectional model which indicates consistency of the regression models. Also, the findings suggest a negatively influence of the excessive draughts years on the cereal yield while it could not be found a statistical relevance for the influence of the excessively rainy years.

However, the limit of this research is the lack of available data on many agricultural indicators during time and/or at territorial level. The regression models indicated a high level of correlation which raises a question mark regarding the validity of the models in practice. Anyhow, the determinants could represent a starting point in analyzing and forecasting the farm activities. Future research might explore further the determinants of the agricultural production on every category of crop and of livestock. Likewise, it should be considered into the analysis more socio-environmental indicators for sustainable management of agricultural resources.

REFERENCES

Abugamea, G. H. (2008). A Dynamic Analysis for Agricultural Production Determinants in Paletsine: 1980-2003. In *Proceedings of International Conference on Applied Economics, ICOAE* (pp. 3-10). Academic Press.

Anderl, M. (2009). Emisiile în aer produse de agricultură în Austria şi România. *Pro Environment, 2*, 84–92.

Asteriou, D., & Hall, S. G. (2007). *Applied econometrics: A modern approach using eviews and microfit* (revised edition). Palgrave Macmillan.

Burja, C. (2012). Determinants of the agricultural productivity growth among Romanian regions. *Annales Universitatis Apulensis Series Oeconomica, 1*(14).

Carvalho, F. P. (2006). Agriculture, pesticides, food security and food safety. *Environmental Science & Policy, 9*(7), 685–692. doi:10.1016/j.envsci.2006.08.002

Cereal Science and Technology. (2016). *The economic importance of cereal grains*. Retrieved from http://cereal-scientech.blogspot.ro/2011/12/economic-importance-of-cereal-grains.html

Ekbom, A. (1998). Some determinants to agricultural productivity: An application to the Kenyan highlands. In *World Conference of Environmental Economics* (pp. 25-27). Academic Press.

ESPON Database Portal. (2016). Retrieved from http://database.espon.eu/db2/home;jsessionid=b7c0e44c4a7260687a2df61e4e09

European Commission. (2010). Europe 2020 - A European strategy for smart, sustainable and inclusive growth. Brussels: EC.

European Commission. (2011). *Analysis associated with the Roadmap to a Resource Efficient Europe*. Part I, Commission staff working paper, Brussels, 20.9.2011, SEC(2011) 1067 final, Accompanying the document Communication from the Commission to the European Parliament, the Council, the European Economic and Social Committee and the Committee of Regions, Roadmap to a Resource Efficient Europe, COM(2011) 571 final, SEC(2011) 1068 final.

European Commission (2014a). *Agriculture and Rural Development, The CAP, Overview, The EU explained: Agriculture*. Retrieved from http://ec.europa.eu/agriculture/index_en.htm

European Commission (2014b). *Agriculture and Rural Development, Romania, Common Agricultural Policy*. Retrieved from http://ec.europa.eu/agriculture/cap-in-your-country/index_ro.htm

European Commission (2016a). *International Cooperation and Development, Food and agriculture, Food and nutrition security*. Retrieved from https://ec.europa.eu/europeaid/sectors/food-and-agriculture/food-and-nutrition-security_en

European Commission. (2016b). *Climate Action-Policies-Greenhouse gas emission*. Retrieved from http://ec.europa.eu/clima/policies/g-gas/index_en.htm

European Commission (2016c). *Sustainable Development indicators, Climate change, Greenhouse gas emissions by sector.* Retrieved from http://ec.europa.eu/eurostat/data/database

European Commission. (2016d). Overview of CAP Reform 2014-2020. Agricultural Policy Perspectives Briefs, no.5, December 2013, pp 1-10. Retrieved from http://ec.europa.eu/agriculture/policy-perspectives/policy-briefs/index_en.htm

Eurostat. (2016). Retrieved from http://appsso.eurostat.ec.europa.eu/nui/submitViewTableAction.do

Gavrilescu, D., & Giurcă, D. (Eds.). (2000). Economie agroalimentară. editura Expert şi Bioterra, 111-113.

Giovannucci, D., Scherr, S. J., Nierenberg, D., Hebebrand, C., Shapiro, J., Milder, J.,Wheeler, K. (2012). *Food and agriculture: The future of sustainability. The sustainable development in the 21st century (SD21).* Report for Rio, 20.

Hazell, P., & Wood, S. (2008). Drivers of change in global agriculture. *Philosophical Transactions of the Royal Society of London. Series B, Biological Sciences, 363*(1491), 495–515. doi:10.1098/rstb.2007.2166 PMID:17656343

Lopez, R. (1994). The environment as a factor of production: The effects of economic growth and trade liberalization. *Journal of Environmental Economics and Management, 27*(2), 163–184. doi:10.1006/jeem.1994.1032

Mamatzakis, E. C. (2003). Public infrastructure and productivity growth in Greek agriculture. *Agricultural Economics, 29*(2), 169–180. doi:10.1111/j.1574-0862.2003.tb00155.x

Ministerul Dezvoltării Regionale şi Administraţiei Publice. (2016). *Dezvoltare regionala, Programe europene 2007-2013, Programul Operational Regional 2007-2013, Regiuni de dezvoltare.* Retrieved from http://www.mdrap.ro/dezvoltare-regionala/-2257/programul-operational-regional-2007-2013/-2975

Müller, D., Leitão, P. J., & Sikor, T. (2013). Comparing the determinants of cropland abandonment in Albania and Romania using boosted regression trees. *Agricultural Systems, 117*, 66–77. doi:10.1016/j.agsy.2012.12.010

Nowak, A., Kijek, T., & Domańska, K. (2015). Technical efficiency and its determinants in the European Union agriculture. [Zemědělská Ekonomika]. *Agricultural Economics, 61*(6), 275–283.

Olesen, J. E. (2006). *Climate change as a driver for European agriculture. SCAR-Foresight in the field of agricultural research in Europe.* Expert paper.

Olesen, J. E., & Bindi, M. (2004). Agricultural impacts and adaptations to climate change in Europe. *Farm Policy Journal, 1*(3), 36–46.

Patra, T., & Poshakwale, S. (2006). Economic variables and stock market returns: Evidence from the Athens stock exchange. *Applied Financial Economics, 16*(13), 993–1005. doi:10.1080/09603100500426523

Piasecki, A., & Jurasz, J. (2015). Development of water and sewage infrastructure on rural areas in Poland. *Development, 15*(3).

Popescu, G. (2014). The common agricultural policy-in the traps of land property. *Calitatea, 15*(8), 8-15.

Popescu, G., & Zaharia, A. (2015). Analysis of technical progress, energy consumption and value added in European Union's agriculture. *2nd International Multidisciplinary Scientific Conference on Social Sciences and Arts*, 3, 49 - 56.

Romanian National Agency for Environmental Protection (2016). *Areas-Soil.* Retrieved from http://www.anpm.ro/sol-subsol

Romanian National Institute of Statistics. (2016a). Retrieved from http://statistici.insse.ro/shop/?page=tempo3&lang=ro&ind=PMI114B

Romanian National Institute of Statistics. (2016b). Retrieved from http://statistici.insse.ro/shop/

Špička, J. (2014). The regional efficiency of mixed crop and livestock type of farming and its determinants. *AGRIS On-Line Papers in Economics and Informatics, 6*(1), 99–109.

Teryomenko, H. (2008). *Farm Size and Determinants of Agricultural Productivity in Ukraine* (Doctoral dissertation). National University.

The Romanian Ministry of Agriculture and Rural Development. (2016). *National Rural Development Programme 2014 – 2020, The Romanian village has future!* Retrieved from http://www.madr.ro/pndr-2014-2020/implementare-pndr-2014-2020/documente-aprobate.html

Tocco, B., Davidova, S., & Bailey, A. (2014). labor adjustments in agriculture: Evidence from Romania. *Studies in Agricultural Economics (Budapest), 116*(2), 67–73. doi:10.7896/j.1406

Vijayasarathy, K., & Ashok, K. R. (2015). Climate Adaptation in Agriculture through Technological Option: Determinants and Impact on Efficiency of Production. *Agricultural Economics Research Review, 28*(1), 103. doi:10.5958/0974-0279.2015.00008.7

Voicu, R., & Dobre, I. (2003). *Organizarea şi strategia dezvoltării unităţilor agricole*. Bucureşti: Editura ASE.

Wiebensohn, J., & Jackenkroll, M. (2013). Evaluation and modelling of a standard based spatial data infrastructure for precision farming. In *Proceedings of the EFITA-WCCA-CIGR Conference* (*Vol. 2427*, p. C0107). Academic Press.

Zaharia, A., Antonescu, A. (2014). Agriculture, greenhouse gas emissions and climate change. *14th International Multidisciplinary Scientific GeoConference SGEM*, 3.

KEY TERMS AND DEFINITIONS

Agricultural Crop Value: The value generated by all the crop products in a year.

Agricultural Inputs: All type of entry resources for a process, activity, and business.

Agricultural Outputs: All type of results generated by a process, activity, and business.

Cereal Yield: The ratio between the cereal production and surface.

Cross-Section Fixed Effects: The effects determined by the differences between different spaces.

Panel Data: It represents an econometric technique used to estimate cross-sectional and time effects of several factors on a variable.

Romanian Development Regions: Regions created for helping the planning and the distribution of the EU funds in Romania.

APPENDIX 1: DATA LEGEND

Table 9.

EViews Code	Variable Name	Data Source	Website
av	Agricultural output	The Romanian National Institute of Statistics	www.insse.ro
acv	Agricultural crop value (crop output)	The Romanian National Institute of Statistics	www.insse.ro
yc	Cereal yield	The Romanian National Institute of Statistics	www.insse.ro
agrl	Agricultural land	The Romanian National Institute of Statistics	www.insse.ro
sirr	Irrigated area	The Romanian National Institute of Statistics	www.insse.ro
feca	Final energy consumption agriculture	The Romanian National Institute of Statistics	www.insse.ro
ghga	GHG agriculture	The Romanian National Institute of Statistics	www.insse.ro
gdp	GDP pps per inhabitant	European Commission	eurostat
lfc	Labour force total costs	The Romanian National Institute of Statistics	www.insse.ro
lce	The labor cost per employee	European Commission	eurostat
ceapop	Civil economically active population	The Romanian National Institute of Statistics	www.insse.ro
ir_fp	Inflation rate food products	The Romanian National Institute of Statistics	www.insse.ro
ur	Unemployment rate	The Romanian National Institute of Statistics	www.insse.ro
pd	Population density	The Romanian National Institute of Statistics	www.insse.ro
pdagricl	Population density on agricultural land	The Romanian National Institute of Statistics	www.insse.ro
pat	Physical agricultural tractors per agricultural ha	The Romanian National Institute of Statistics	www.insse.ro
cf	Chemical fertilizers	The Romanian National Institute of Statistics	www.insse.ro
nf	Natural fertilizers	The Romanian National Institute of Statistics	www.insse.ro
p	Pesticides	The Romanian National Institute of Statistics	www.insse.ro
subv	Subsidies	European Commission	eurostat
hdd	Actual heating degree-days per year	European Commission	eurostat
qc	Production of cereals (tonnes)	The Romanian National Institute of Statistics	www.insse.ro
sc	Cultivated area with cereals	The Romanian National Institute of Statistics	www.insse.ro
rgdp	Regional GDP	The Romanian National Institute of Statistics	www.insse.ro
nov	The number of villages	The Romanian National Institute of Statistics	www.insse.ro
ddry	Years with excessive draughts	Dummy variable – introduced by authors	
drainy	Years with excessive rain	Dummy variable – introduced by authors	

APPENDIX 2: DESCRIPTIVE STATISTICS OF PANEL VARIABLES

See Table 10.

APPENDIX 3: MATRIX CORRELATION

The matrix correlation was performed for 1997-2014 time frames; 144 observations included after adjustments, unbalanced sample, pairwise samples (pairwise missing deletion) (see Table 11).

Table 10.

Indicators	AV	ACV	YC	SC	SIRR	QC	DPAT	CF	DNF	P	DLCE	DLFC	CEAPOP	PD	PDAGRL	DRGDP	HDD	NOV
Mean	5779436.0	3738187.0	3039.1	698465.0	31623.1	2108581	0.0	0.2	1.5	4.3	1452.7	17749.8	349.9	2.4	3.9	783.0	2936.8	1625.7
Median	6049575.0	3816415.0	3127.5	700953.0	3582.5	2051773	0.0	0.1	0.0	3.9	1248.0	16659.4	355.7	0.8	1.4	650.0	2858.2	1800.0
Maximum	15613379.0	11539817.0	4381.0	1412613.0	305404.0	5588606	0.0	2.0	64.9	11.2	3048.0	107920.3	721.8	13.7	23.9	6600.0	3754.8	2445.0
Minimum	138343.4	75522.2	834.0	29782.0	12.0	47194	0.0	0.0	-18.7	1.2	96.0	-76707.6	35.6	0.6	1.1	-3500.0	2410.9	91.0
Std. Dev.	3896944.0	2682861.0	792.5	375581.2	56884.8	1302584	0.0	0.4	10.2	2.0	857.7	25417.5	164.3	4.2	6.8	1115.7	361.0	665.7
Skewness	0.25	0.51	-0.54	-0.12	2.47	0.52	-2.81	2.60	3.94	1.06	0.42	-0.42	-0.21	2.26	2.28	1.53	0.53	-1.26
Kurtosis	2.14	2.66	2.94	2.18	9.44	2.98	19.65	8.59	23.77	4.44	2.11	7.16	2.71	6.14	6.27	12.97	2.22	3.85
Jarque-Bera	5.98	6.91	6.89	4.44	346.56	6.55	1750.7	349.93	2795.2	39.5	8.51	101.95	1.61	182.18	189.3	398.51	7.58	42.27
Probability	0.05	0.03	0.03	0.11	0.00	0.04	0.00	0.00	0.00	0.00	0.01	0.00	0.45	0.00	0.00	0.00	0.02	0.00
Observations	144	144	144	144	126	144	136	144	136	144	136	136	144	144	144	88	104	144

Source: own computation with Eviews, 2016

Table 11.

Correlation Probability	AV	ACV	YC	SC	SIRR	QC	DPAT	CF	DNF	P	DLCE	DLFC	CEAPOP	PD	PDAGRL	DRGDP	HDD	NOV	DNOV
av	1																		
	-----																		
acv	0.9905	1																	
	***	-----																	
yc	0.3061	0.3552	1																
	***	***	-----																
sc	0.4139	0.4270	-0.0493	1															
	***	***	0.5574	-----															
sirr	0.1464	0.1536	-0.2346	0.5390	1														
	0.1018	*	***	***	-----														
qc	0.5166	0.5601	0.3945	0.8640	0.2986	1													
	***	***	***	***	***	-----													
dpat	0.2198	0.2069	0.3623	0.1260	-0.0408	0.1719	1												
	**	**	***	0.1438	0.6608	**	-----												
cf	-0.0164	0.0044	-0.0496	0.2506	-0.0075	0.2007	0.0723	1											
	0.8449	0.9583	0.5550	***	0.9332	**	0.4027	-----											
dnf	0.2087	0.2375	0.2040	0.1310	0.0673	0.2635	0.0027	0.1794	1										
	**	***	**	0.1284	0.4692	***	0.9754	**	-----										
p	-0.3914	-0.3942	-0.2090	0.2387	0.0374	0.1245	-0.1151	-0.1405	-0.0881	1									
	***	***	**	***	0.6778	0.1370	0.1822	*	0.3080	-----									
dlce	0.1795	0.1492	-0.0434	-0.0194	0.0857	-0.0342	-0.0679	-0.0673	0.0193	-0.2767	1								
	**	*	0.6158	0.8225	0.3559	0.6927	0.4324	0.4366	0.8232	***	-----								
dlfc	0.1645	0.1242	-0.0165	0.2300	0.1901	0.1796	-0.0101	-0.0816	0.0742	0.0237	0.4430	1							
	*	0.1496	0.8487	***	**	**	0.9067	0.3451	0.3908	0.7839	***	-----							
ceapop	0.2369	0.2069	-0.1738	0.7028	0.1767	0.5365	0.1679	0.1421	-0.0195	0.5410	-0.1189	0.1604	1						
	***	**	**	***	**	***	*	*	0.8221	***	0.1678	*	-----						

continued on following page

Table 11. Continued

Correlation Probability	AV	ACV	YC	SC	SIRR	QC	DPAT	CF	DNF	P	DLCE	DLFC	CEAPOP	PD	PDAGRL	DRGDP	HDD	NOV	DNOV
pd	-0.5012	-0.4748	-0.0014	-0.6446	-0.1836	-0.5628	-0.2920	-0.1975	-0.0394	-0.1487	-0.0001	-0.1642	-0.6872	1					
	***	***	0.9868	***	**	0.0000	***	**	0.6488	*	0.9994	*	***	-----					
pdagrl	-0.5009	-0.4749	0.0064	-0.6531	-0.1918	-0.5700	-0.2748	-0.1980	-0.0418	-0.1622	0.0007	-0.1691	-0.6889	0.9984	1				
	***	***	0.9391	***	**	***	***	**	0.6293	*	0.9938	**	***	***	-----				
drgdp	-0.3693	-0.3336	-0.0076	-0.3031	-0.0883	-0.2773	-0.0862	-0.1072	-0.1398	-0.0505	0.0883	-0.2294	-0.3729	0.4166	0.4123	1			
	***	***	0.9440	***	0.4545	***	0.4248	0.3204	0.1941	0.6403	0.4135	**	***	***	***	-----			
hdd	0.0365	0.0079	0.1359	-0.1845	-0.3202	-0.0938	0.1905	-0.1828	0.1327	0.3084	-0.3110	-0.0586	0.3072	-0.2869	-0.2804	-0.1935	1		
	0.7129	0.9363	0.1689	*	***	0.3438	*	*	0.1975	***	***	0.5705	***	***	***	0.1530	-----		
nov	0.5018	0.4580	-0.0668	0.6037	0.0045	0.4979	0.2650	0.3037	0.0387	0.2408	-0.0039	0.1538	0.8558	-0.8599	-0.8573	-0.4087	0.4097	1	
	***	***	0.4263	***	0.9604	***	***	***	0.6551	***	0.9639	*	***	***	***	***	***	-----	

Source: Own computation with Eviews, 2016.

***1% significance **5% significance *10% significance

Note: Accepted correlation for the regression models at more than 0.4 coefficient; Rejected correlation for the regression models

Chapter 2
The Competitiveness Constraints of Romanian Wine Sector and the EU-28 Agricultural Model

Ladaru Georgiana-Raluca
The Bucharest University of Economic Studies, Romania

Boboc Dan
The Bucharest University of Economic Studies, Romania

ABSTRACT

Romanian agrifood sector has experienced a profound paradigm change during the last period, especially after the accession to the European Union (EU). Starting from the need to be convergent to the EU agricultural model requirements and valuing the domestic agricultural potential, numerous developments and constraints have occurred. The objective of the study is to further examine the competitiveness constraints of Romanian wine sector. The study reveals that the wine production has suffered a dramatic improvement both on quantity and quality. The paradigm change imposed by replacement of the hybrid vines with noble vines has triggered both a quality and competitiveness increase. The domestic wine producers have adapted their production to the new market demands. Now they must fulfill both the national and European competitiveness constraints and they should stay as a new quality wine promoter.

INTRODUCTION

Romania has a massive wine production potential which should be valued as close as possible to the domestic consumption patterns and, in the same time, to be convergent to the EU pattern. In this context, the main objectives of the chapter are connected to the following research constraints:

- A short analysis of the Romanian wine potential in context of the last agricultural transformations.

DOI: 10.4018/978-1-5225-5739-5.ch002

Copyright © 2019, IGI Global. Copying or distributing in print or electronic forms without written permission of IGI Global is prohibited.

- A sectorial analysis in a tight connection to the domestic economic branches, both from the agricultural economic perspective and, also, from the integrative point of view.
- A comparative analysis between the traditional and greatest wine producers and their strengths in the market, technologies and knowledge transfer.
- Revealing the convergence between the domestic wine producers and foreigner ones both from the EU agricultural model requirements and the behavioral patterns.

Taking into account all the dimensions presented above, the chapter contain a short literature review in the field, the methodological aspects, sectorial analysis and conclusions. The main aim of the chapter consist of an analysis of the wine sector competitiveness in Romania, taking into account foreigner determinants, market constraints, production technologies and knowledge transfer.

Over the last decade, globalisation of the wine industry has boosted competition between Old and New World producers (Velikova & Dodd, 2016) and is currently facing profound changes both in terms of consumption and production (Hristov & Kuhar, 2015). Also, during the final two decades of the 20th century, the internationalisation of wine production and consumption continued apace (Fleming, Mounter, Grant, Griffith, & Villano, 2014).

In recent decades wine has become a widely consumed product, generating trade flows which affect not only traditional producing countries and consumers. In this context, the EU remains by far the world's largest market as regards wine production and consumption, as well as the chief exporter and importer (Lombardi, Dal Bianco, Freda, Caracciolo, & Cembalo, 2016). The analysis of competitive market structure is an important area of marketing research due to its significance in explaining the nature and extent of competition among companies and their products, including the identification of competitors, market segmentations, product positioning, and pricing (Elrod, Russell, Shocker, Andrews, Bacon, Bayus, ... & Mazanec, 2002). The intense competition affecting the wine industry in recent decades has forced wineries and retailers to reshape their marketing strategies on the basis of consumer preferences (Pomarici, Lerro, Chrysochou, Vecchio, & Krontalis, 2017). Vlachos (Vlachos, 2017) said that several countries in the EU have not fully recovered from the recent economic crisis and their economic policies aim to improve the government budget balance and the productivity and competitiveness of certain industries. It could be an example of wine Romanian economy. Also, global warming is typically thought of as a major driver of new investment in cool climate wine regions, including from producers in warmer areas seeking to supplement supplies that can help them maintain their current styles of wines as well as add new ones (Ashenfelter anf Storchmann, 2016). In this context, producers adopt various differentiation strategies such as regions of origin, varieties, wine styles and vintages. This makes wine one of the most differentiated products on the food market and consumers face many different cues on wine labels and a wide range of prices (Schäufele & Hamm, 2018). Also, predictable and transparent laws, policies, and support structures would improve the business climate (Rendleman, 2016).

ROMANIAN WINE MARKET ASPECTS

Romania is among the most important European wine producing countries due to its geographical location, relief and favorable climatic conditions. With a historical past and many traditions, the vineyard culture has developed continuously, being recognized as one of the main agricultural branches of our country. Today, Romania faces various changes trying to build a future that is in line with the requirements of the European Union and implicitly to enjoy an important status as a producer of high quality wines.

The Supply

Romania is a wine-growing country, a member of the International Organisation of Vine and Wine since 1927. Romania's wine-growing area is divided into eight regions corresponding to the European wine-growing areas: Transylvania (wine-growing zone B); Moldavian Hills, Oltenia, Banat, Crişana and Maramureş Hills (wine zone CI); Muntenia Hills, Dobrogea Hills and Danube Terraces (CII wine area). The surface of the vineyards in Romania has progressively increased until 1972, when it exceeded 300,000 hectares, after which it declined steadily. In 2016, the area occupied with vineyards was 178,151 hectares and placed our country on the 11th place in the world and 5th place at European level. However, half of the area is cultivated by the country's major producers, making Romania the 13th largest producer of wine in the world. As can be seen in Figure 1, the area of vineyards in Romania has not exceeded 180,000 hectares over the last 6 years. In 2012, the area measured 178,654 hectares, with 536 hectares more than in 2015, the year 2016 being similarly and increased by only 33 hectares. Also, a decrease in the area of vines occurred in 2014, when Romania lost 1,703 hectares compared to the previous year.

On a more thorough analysis of the areas occupied by vines in Romania we can see that they are divided according to the type of grapevine and depending on the destination of the grapes, in four categories: grafted vines, hybrid vines, respectively vines that produce grapes for wine or for table. Of all these types of vineyards, vineyards grafted and those producing grapes for wine occupy the largest area in Romania, as shown in Table 1.

It can also be noticed that the areas occupied with wine grapes during the period 2011-2016 were not constant, increasing from 2011 to 2016 by 3069 ha. From this year, wine grape areas are beginning to grow, with 744 ha in 2011, another 1790 ha were occupied in 2012, and 374 ha in 2013. All this is due to the National Support Program for Viticulture from 2009-2013. The financial support allocated to this agricultural sector amounted 42.1 million euro/year and was absorbed in proportion of 100%. This program supports, among other things, the restructuring and conversion of vineyards, thus explaining that area decline in the first years of project implementation. Although the areas with vines have decreased,

Figure 1. The surface of vineyard in Romania (hectars)
Source: Author own processing based on data from http://www.insse.ro/

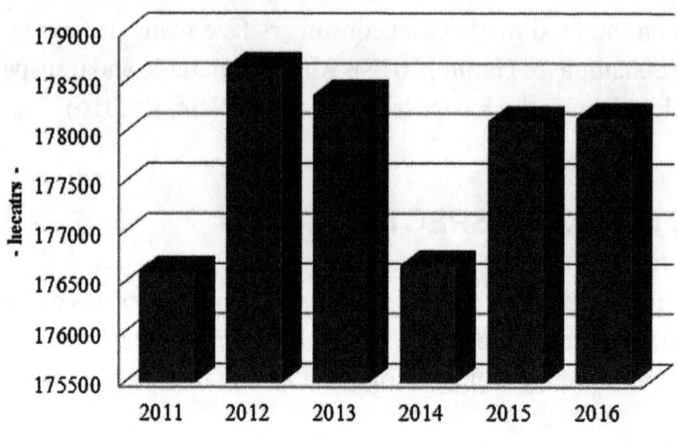

Table 1. The area occupied by vineyards in Romania (hectars)

Categories of Vines	2011	2012	2013	2014	2015	2016
Grafted vines	88047	89735	89735	90019	92495	92697
Hybrid vines	88569	88919	88643	86656	85623	85454
Vines producing table grapes	8452	8700	8050	7183	6786	6918
Vines producing grapes for wine	168164	169954	170328	169492	171332	171233
TOTAL	176616	178654	178378	176675	178118	178151

Source: Author own processing based on data from http://www.insse.ro/

their quality has improved. Romania enjoys a diversified vineyard made up of both traditional and local varieties and foreign varieties, which are highly appreciated from the point of view of oenological aptitudes. In Romania, varieties of vine for white wines are cultivated, including Romanian varieties such as Fetească Albă, Fetească Regală, Galbenă de Odobeşti, as well as varieties of foreign origin: Italian Riesling, Aligote, Sauvignon, Pinot Gris, Chardonnay, Traminer pink. Among the varieties of grapes for red wines, Băbească Neagră and Fetească Neagră are valuable native varieties in our country, and Merlot, Cabernet Sauvignon, Pinot Noir and Burgund are foreign varieties with special oenological features. At the same time, for the production of aromatic wines, our country can pride with varieties recognized as Tămâioasă Românească and Busuioacă de Bohotin, but also with Muscat Ottonel as an external variety.

The vineyard in our country is cultivated in the hilly areas, on the sand, but also on other lands that are favorable to its development, called wine-growing areas. From the territorial point of view, vineyards in our country are grouped in: wine-growing areas, wine-growing regions, wine-growing vineyards, wine-growing centers and vineyards. In Romania, there are eight wine-growing regions that bring together about 37 vineyards with a total of 170 wine centers where a wide range of wines is produced. The largest and most renowned wine region is that of Moldova's Hills covering about 70,000 hectares. As mentioned above, the Moldavian Hills Region is the most extensive with 12 vineyards: Cotnari, Iaşi, Huşi, Dealul Bujorului, Tutova Hill, Covurlui, Nicoreşti, Zeletin, Iveşti, Panciu, Odobeşti and Coteşti. The region's profile is characterized by the production of white wines in a very wide range, generally sweet and less of red wines. The wines produced here are designed for both current consumption but can also achieve remarkable qualities such as Cotnari wine. Odobeşti, Coteşti and Panciu vineyards are renowned for dry wines. The hills of Wallachia and Oltenia are a vineyard region with 8 vineyards and 32 wine centers. With the exception of the Sâmbureşti vineyard, which is profiled on the production of red wines, white wines are produced mostly. Five viticultural vineyards with 17 wine-growing centers belong to the Transylvanian Plateau, a wine-growing region with real possibilities for expanding the vineyard culture. A wide range of white wines are also produced here, from current consumption to superior, such as Jidvei sparkling wine, also recognized across the country.

The climatic conditions of Crişana and Maramureş Hills allow the production of a large range of red and white wines. Sparkling wines are characteristic of the Şimleul Silvaniei vineyard. The wine region of the Dobrogea Columns is about 18,000 hectares. The natural environment is one of the most favorable for the expansion and development of viticulture in this sector. Notorious wines are produced here, as is the case with Murfatlar. The Danube Terraces with its two vineyards, Ostrov and Greaca, are characterized by the production of table grapes. Concerning wine production, an important share is attributed to white wine consumption. Above we were able to observe the area occupied by the vine, and in the following pages we will analyze the production obtained on these surfaces.

Following the data in Table 2, we can see that the largest grapes production was recorded in 2013. From this total grapes production, the largest amount is represented by grapes for wine. Due to the decrease of the cultivated areas, because of the implementation of the National Support Program, the surface and the production were affected. In 2011 there was a rise in the production of wine grapes by 133,066 tons compared to the previous year. Despite the slight increase in surfaces cultivated with wine grapes in 2012, grapes production declined by 126,175 tons, continuing to increase by 2013 with 23,8292 tons. Overall, during the period 2011-2016 the production of grapes decreased from 879,487 tons to 736,892 tons while improving its quality due to the conversion of vineyards within the National Support Program. In the table below we can see that the eight regions have different productions. The South-East Region can boast the highest grapes production over the last 5 years. The most important production was obtained in 2013, when it exceeded 425,000 tons of grapes. Northeast Region is ranked second with about 150,000 tons. The lowest production is registered in the Bucharest-Ilfov Region, which surpassed very little 4,300 tons.

Worldwide, Romania ranks 13th in terms of wine production in 2015 with a production increase of up to 9% compared to the quantity obtained in the previous year. In 2014, due to the rains that continued throughout the spring, but also for a long time in the summer, Romania suffered a significant loss

Table 2. Grapes production in Romania (tons)

Categories of Grapes	2011	2012	2013	2014	2015	2016
Grapes from grafted vines	497380	421930	529859	466197	476385	451722
Grapes from hybrid vines	382107	324455	461700	317493	322380	285170
Table grapes	55959	49032	55377	35959	42088	38806
Grapes for wine	823528	697353	936182	747731	756677	698086
TOTAL	879487	746385	991559	783690	798765	736892

Source: author own processing based on data from http://www.insse.ro/

Table 3. Total grape production by region (tons)

Regions	2011	2012	2013	2014	2015	2016
North-West Region	29329	30160	35062	37104	41345	32581
Center Region	28343	19807	31276	35603	26702	29631
North-East Region	151373	111715	155116	140452	148768	133840
South-East Region	393218	334386	425900	335486	345508	327282
South-Muntenia Region	107413	96743	147308	98580	103000	84829
Bucharest-Ilfov Region	6018	5368	8185	4326	4387	4105
South-West Region	143227	125828	162052	103802	102234	95559
West Region	20566	22378	26660	28337	26821	29065
Total	879487	746385	991559	783690	798765	736892

Source: author own processing based on data from http://www.insse.ro/

of wine production by about 20%, so Bachus's liquor failed to overcome four million hectoliters. The highest production was registered in 2013, of 5.93 million hectoliters, which was Romania's 12th place in the world.

Based on the data in Table 4, it can be seen that wine production in Romania had a downward trend in the period 2011-2012. 2013 was the year with the highest wine production in the analyzed period. In 2014, the production decreased significantly compared to the previous year, with an amount of 1363.3 thousand hl. In Europe, statistical data show that we are clearly overtaken by countries with a tradition in viticulture such as Italy, which has again become the world's leading producer of 48.9 million hectoliters, followed by France, where wine production exceeds 44 million hectoliters and Spain, with 36.6 million hectoliters of wine, a production lower with 4% than in 2014.

The Demand

Our country registered an average consumption of 24.26 liters per capita in 2015, rising over the previous years. With this value, Romania is only half the consumption of wine than in other more developed countries, where it stands at 54 liters / inhabitant in the Vatican, Europe's largest consumer of wine. However, Romania is ranked 20th in the world after Greece and Germany and ahead of Hungary, Malta, Belgium and Great Britain.

In our country, the maximum of wine consumption are recorded in the North-East and South-East Regions. The highest consumption was recorded in 2011 by 1,325 liters/month of wine per person. This consumption is also due to the fact that here we meet one of the most famous and productive wine regions in Romania. A high consumption is also recorded in the South-West Region and the Western Region, about 1 liter/month/person. The minimum consumption is recorded in Bucharest-Ilfov and Muntenia Regions where it does not exceed 0.6 liters/month/person. In 2015, in these regions wine consumption did not exceed 0.575 liters of wine per person, the lowest value in the last five years.

According to the data presented in the table below, the highest consumption of wine is attributed to farmers. They consume around 1.5 liters of wine per month and the consumption trend is on the rise, the year 2014 being the time when the highest wine volume of 1,551 liters per person was consumed in a month. Retirees are another social category that uses the Bachus spirits, accounting for an average

Table 4. Wine production in Romania (thousand hectoliters)

Country	2011	2012	2013	2014	2015	2016
Romania	4058,1	3310,6	5113,3	3750,0	3627,0	3267,0

Source: author own processing based on data from http://www.madr.ro/

Table 5. Average annual wine consumption per capita (liters)

Main Food and Beverages	Years				
	2011	2012	2013	2014	2015
Wine and wine products	21,3	21,1	21,7	22,6	24,26

Source: Author own processing based on data from http://www.insse.ro/

Table 6. Average monthly consumption of wine per person per region (liters)

Regions	Years				
	2011	2012	2013	2014	2015
North-West and Center Regions	0,471	0,488	0,56	0,552	0,625
North-East and South-East Regions	1,325	1,271	1,281	1,277	1,307
South-Muntenia and Bucharest-Ilfov Regions	0,583	0,61	0,581	0,593	0,575
South-West and West Regions	1,028	1,098	1,05	1,161	1,035

Source: author own processing based on data from http://www.insse.ro/

monthly consumption of about 1 liter of wine per person. However, the trend is slightly declining in 2013 when the registered quantity was 1,051 liters and in 2015 only 0.953 liters of wine/person/month. Employees, self-employed in non-agricultural activities and unemployed people consume around 0.6-0.7 liters in a month, being considered the social categories with the lowest monthly average consumption of wine in Romania.

As regards the quantity of wine bought, one can see an increase over the five years of the monthly quantity of wine purchased by one person per month. By comparing the two environments, a person in the urban environment tends to buy over 0.26 liters of wine per month, while a rural person does not exceed the 0.20 liter level over the same period.

When it comes to the country of origin of the wine, the Romanians choose to remain faithful to the local brands. According to statistics, most Romanians prefer to consume the Romanian wine to the detriment of the foreigner. At present, the market share value does not exceed 10%. Romanian preferences for imported wine are moving to countries like Italy, France, Chile and the New World - California, Australia, South Africa, New Zealand.

Table 7. Average monthly consumption of wine per person per social category (liters)

The Main Social Categories	Years				
	2011	2012	2013	2014	2015
Employees	0,654	0,673	0,657	0,662	0,733
Self-employed in non-agricultural activities	0,793	0,729	0,728	0,648	0,682
Farmers	1,43	1,511	1,369	1,551	1,541
Unemployed	0,608	0,569	0,576	0,584	0,67
Retirees	0,972	0,98	1,051	1,033	0,953

Source: Author own processing based on data from http://www.insse.ro/

Table 8. The average monthly quantity of wine bought by a person by place of residence (liters)

Places of Residence	2011	2012	2013	2014	2015
Urban	0,21	0,213	0,211	0,231	0,269
Rural	0,153	0,156	0,146	0,151	0,19

Source: Author own processing based on data from http://www.insse.ro/

Foreign Trade

Wine is a very important agro-food product in our country; it has at its disposal the majority of the resources needed to reach the maximum potential in Romania, becoming one of the country's main export products. Simultaneously with the provision of these resources, this wine sector must be financed in order to increase its competitiveness and, implicitly, that of the product on the foreign market.

Import of Wine

As for the quantity of wine imported into Romania, there were oscillations during the period 2011-2016, in the first 4 years, the imported quantity decreased from year to year, but it stabilized in the last two years of the period at the volume of about 50 of thousands of tons.

From the figure presented above, it is noted that the largest quantity of wine imported by Romania was registered in 2011 when it reached 90769 tons. This increase was followed by a sharp decline in 2012 when imports were down 36240 tons less, to 54529 tons. Similarly, for the years 2013 and 2014, imports gradually decreased, so that the smallest quantity of imported wine was recorded in 2014, when it exceeded 34 thousand tons. One of the factors that influenced this decrease in imports was the grape production, which registered in 2013 the highest value of almost 1 million tons, while in 2014 there was a significant production of 783 of thousands of tons. Romania's wine imports began to stabilize in terms of quantity in the years 2015 and 2016, respectively to 50726 tons and 50233 tons. Overall, during the analyzed period, Romania's wine imports fell on average by 4.69%.

As regards the value of wine imports from Romania, its evolution keeps the trend of imported quantities, so imports of wine have even reached about 70 million dollars in 2011 due to the quantity of over 90000 tons imported. There was a decreasing trend of the value of imports, reaching in 2014 only 44602 thousands of dollars, the lowest value of imports from the analyzed period. In 2012, the value of wine imports from intra-Community and extra-Community countries raised nearly 54 million dollars, decreasing with 22% compared to import value of about 70 million dollars spent in 2011, but for a much larger quantity, respectively 90,769 tonnes of wine. In 2013, Romania spent a little over 50 million dollars on imported quantity of 36,747 tons of wine. Years 2015 and 2016 marked an increase in the value of wine imports to 47 million dollars and 52.4 million dollars, respectively. On average,

Figure 2. The quantity of wine imported into Romania (tons)
Source: Author own processing based on data from http://www.intracen.org

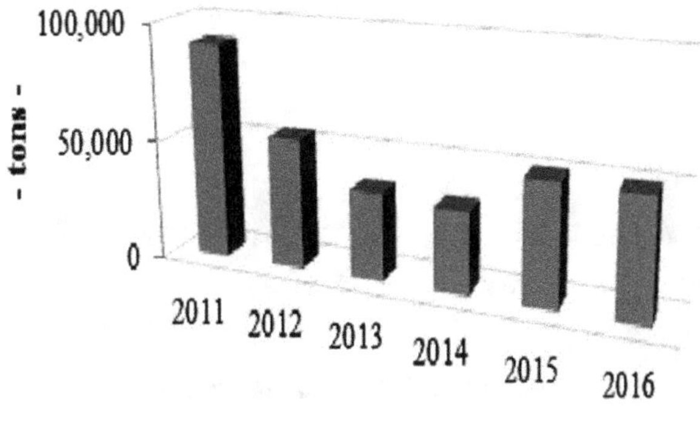

Figure 3. Value of wine imports in Romania (thousands of dollars)
Source: Author own processing based on data from http://www.intracen.org

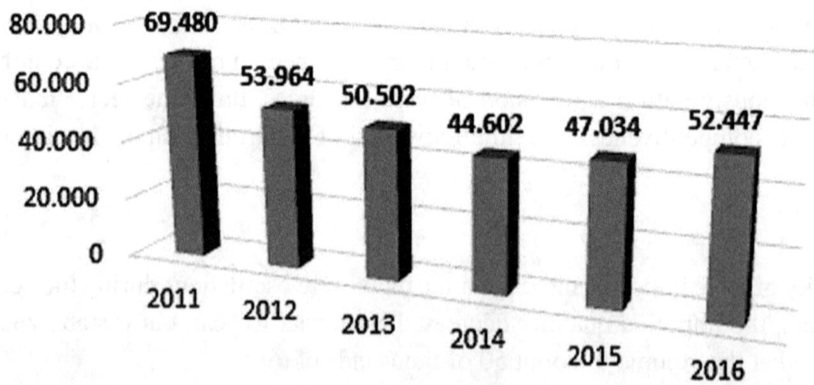

during the whole period analyzed, the evolution of the value of wine imports was decreasing, with an average growth rate of -6.4%.

In the above figure are presented the main countries that Romania uses when it comes to the import of wines. In 2015, wines came mainly from Spain, which, after the fall of 2013, started to be appreciated again on the Romanian market, exceeding a little over 25,000 tons. Imports of Italian wine have fallen over the five years under review and those from France have remained relatively constant.

Along with the liberalization of wine imports from the Republic of Moldova, they saw an increase in 2016 when they exceeded 10000 tons. Wine imports from Hungary gradually increased, accounting for around 3,000 tonnes of total imports in 2016. These five countries, according to the value of imports in Romania in 2016, account for a share of the total value of Romanian wine imports of over 86%, so in the last year Spain exported to Romania a share of the value of wine of 25.15%, Italy 19.88%, France 17.37%, Republic of Moldova 16.89%, and Hungary contributed 7.08% to the value of imported wine in Romania.

The average import price of wine in Romania increased in 2011-2013, from $ 0.77 per liter to $ 1.37 per liter, respectively 77.9%. It dropped to $ 0.93 per liter in 2015, and in the last year of the period it was $ 1.04 per liter. Thus, on average, the average import price was $ 1.07 per liter. As can be seen

Figure 4. The main exporting countries for Romania (thousands of dollars)
Source: Author own processing based on data from from http://www.intracen.org

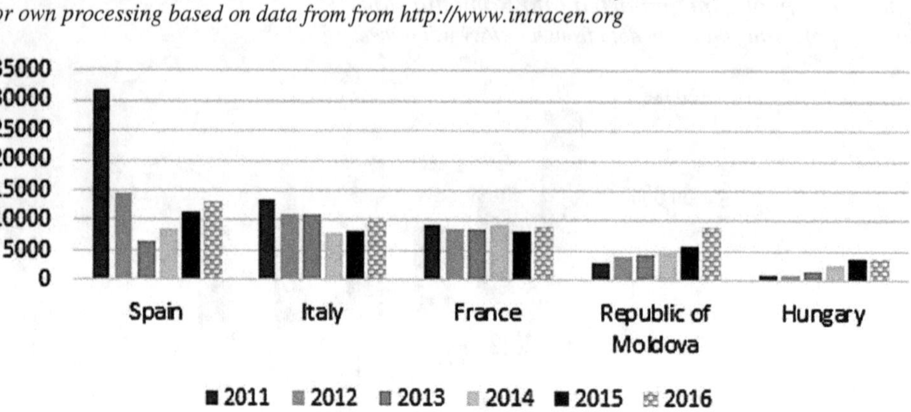

Figure 5. Average import price of wine in Romania (USD/liter)
Source: Author own processing based on data from http://www.intracen.org

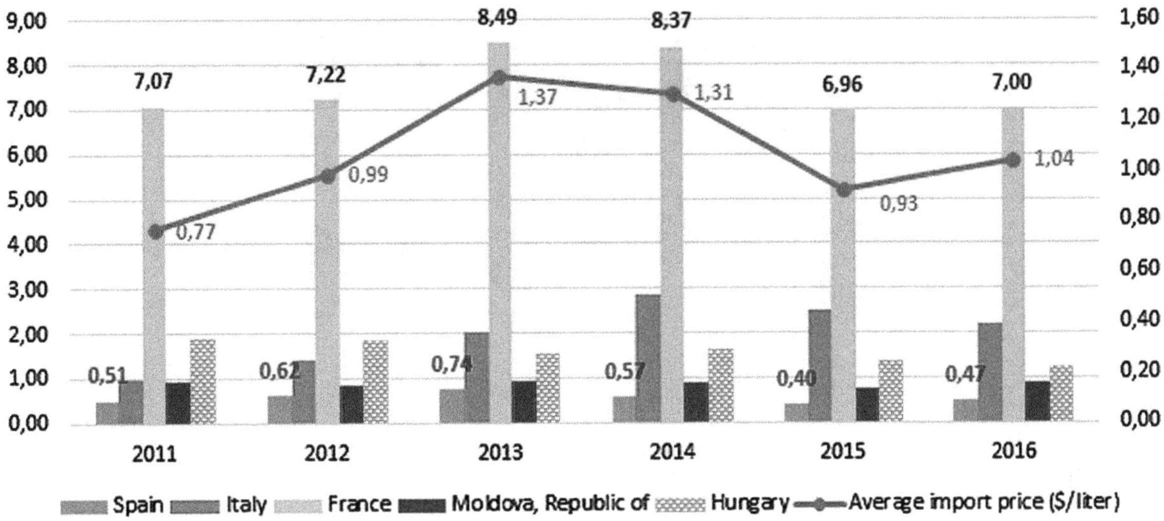

from the figure above, wine from France arrives in Romania at a very high price, about 7 times higher; registering a maximum of $ 8.49 per liter in 2013, and the lowest value for one liter of wine was $ 6.96.

Regarding the country exporting to Romania at the lowest average import price, it can be noticed that Spain occupies this position, and exported to Romania with $ 0.74/liter (the most expensive) and $ 0.4/liter (cheapest). Thus, the average import price of Spanish wine is 48% lower than average of total prices. It is worth mentioning that, although Romania imports a rather small quantity of wine from the Netherlands, compared to the other countries, namely 48 tonnes (0.1% of total imports), the value of this quantity represents 0.8% of the value of total imports, leading to the formation of an average import price over the whole analyzed period of USD 11.17 per liter of wine.

Export of Wine

Analyzing the quantity of wine exported by Romania, it can be estimated that it was increasing during the analyzed period, registering an average growth rate of 5.08%.

The wine exports from our country totaled in 2015 the highest quantity, namely 13,882 tons of wine. Compared to the beginning of the analyzed period, wine exports increased by 3407 tons of wine, an important increase given that in the other years the quantity of exported wine remained somewhat constant, exceeding by 10500 tons in 2014. The exception is the year 2012 when the exported quantity registered 11383 tons, about 900 tons more than the quantity exported in the years 2011 and 2013. In 2016 there was a decrease compared to the previous year, of 992 tons, respectively by 7.14%.

As can be seen in the figure presented, the value of Romanian wine exports has risen steadily from year to year until 2015, when the highest export value of 24.67 million dollars was registered; compared to the first year of the analyzed period, there was an increase of 23.52%. Over the last year, the value of exports has decreased almost directly in proportion to the drop in exported quantity, so in 2016 it was exported from Romania wine worth of 22.7 million dollars, 7.9% less than last year. On average, during

Figure 6. The quantity of wine exported by Romania (tons)
Source: Author own processing based on data from http://www.intracen.org

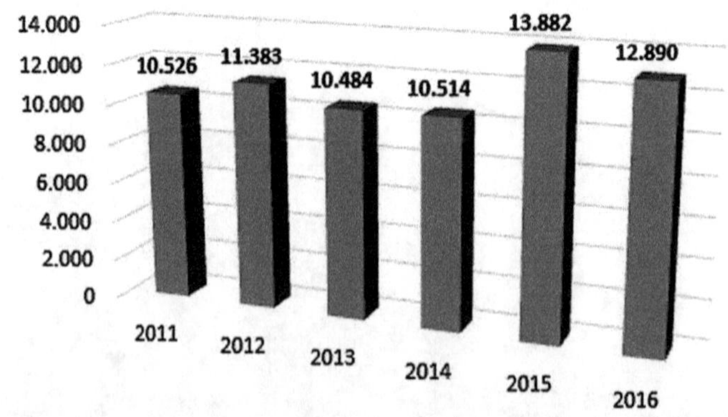

Figure 7. Value of wine exports from Romania (thousands of dollars)
Source: Author own processing based on data from http://www.intracen.org

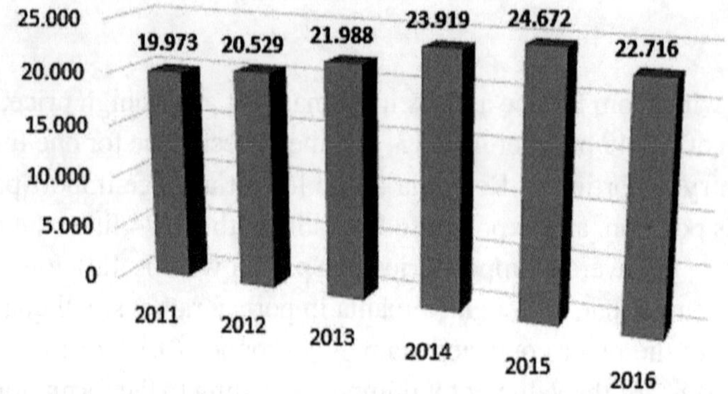

the reference period, the evolution of export value was increasing with an average annual growth rate of 2.77%.

The most important export markets for wines produced in Romania are Great Britain, China, Germany, the Netherlands, USA, Spain and Italy. Local wine growers are starting to orient and choose China as their main export country because they believe that this will be one of the most profitable markets in the coming years compared to traditional markets in Europe in the near future. Outside the European area, besides China, Romanian producers are increasingly focusing on the United States market, which has imported more than $ 1,500,000 worth of wine from Romania. In 2016, out of the total export value of $ 22.71 million, UK had the largest share of 20%, followed by China with 14.69%, Germany with 12.77%, the Netherlands with 8,16%, the United States with 6.4%, Spain with 6.33% and Italy with 6.13%, so these first seven countries account for 74.5% of Romania's total wine exports.

On average, the export price was somewhat constant, with increases in the years 2013 and 2014 when it reached the level of $ 2.27 per liter, at the same time being the highest price of the reference period, and it decreased in the following years, reaching the other extreme, in 2016 being only $ 1.76 per liter. Overall, the average export price was $ 1.94 per liter. Compared to the average, the lowest average price was recorded for Germany, which paid an average of $ 1.25 per liter of Romanian wine, representing a

Figure 8. Romania's main importing countries (thousands of dollars)
Source: Author own processing based on data from http://www.intracen.org

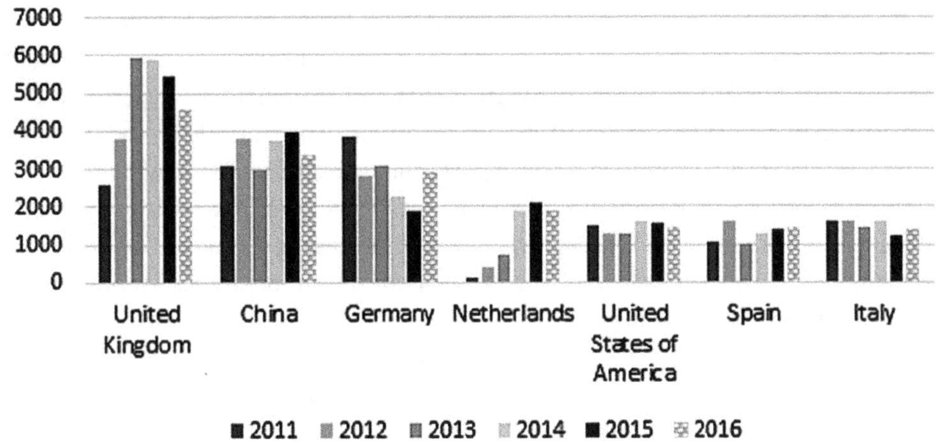

Figure 9. Average export price of wine in Romania (dollars/liter)
Source: Author own processing based on data from http://www.intracen.org

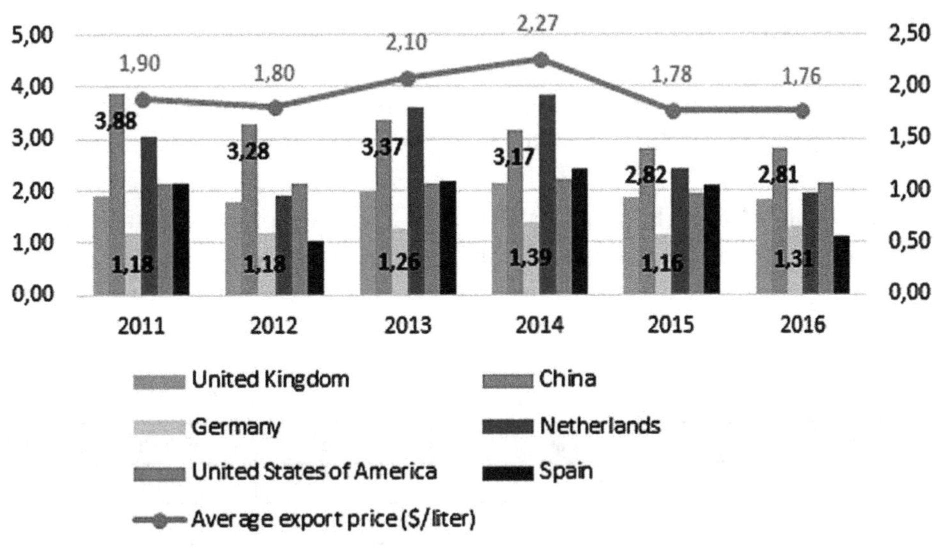

35% difference. On the opposite side, it can be seen that the highest price was paid by China, on average, of $ 3.22 per liter, 66% higher than its average.

Trade Balance

Although they have grown slightly in recent years, exports are failing to keep up with imports that have exploded after EU integration. The value of wine imports from intra-Community and extra-Community countries totaled $52.45 million in 2016, rising from $47 million in 2015 for a smaller quantity of 50.2 tonnes wine. As far as exports are concerned, they grew by 3.3 thousand tons in 2015 compared to the previous year and worth about $24.6 million. Thus, for the year 2015, the trade balance registered a deficit of 36.8 thousand tons of wine compared at about 22.3 million dollars.

In 2011, Romania recorded the largest deficit both in terms of quantity and value. The imported wine quantity exceeded by about 80.2 thousand tons exports and their value meant a loss of about 49.5 million dollars for our country. For 2012, imports are reduced but still remain above exports by 43.1 thousand tons, which means high expenses of no less than $33.4 million. The years 2013 and 2014 have meant a decrease both in the value and quantity level of imports and exports of wine in our country. In 2014, for an imported quantity of 34.1 thousand tons, there was spent $44.6 million, which led to a deficit value of 20.68 million euros compared to the same year's exports, representing the lowest deficit in the period taken in the analysis. The import for 2016 was higher than the export of 37.34 thousand tons, which meant a deficit of 29.7 million dollars.

From a graphical point of view, this shows the situation of imports, wine exports and the trade balance in Romania. Wine exports from our country are too low compared to imports, which leads to a trade deficit in the wine trade. Although, compared with the beginning of the period considered, the deficit has fallen to around 46% in terms of quantity and 6% in value, on average; Romania's wine imports are more than 4.55 times higher than its exports in terms of quantity and 2.37 times in terms of value.

THE COMPETITIVENESS OF ROMANIAN WINE MARKET

In order to determine the competitiveness of Romanian wine on the foreign market, the authors analyzed the competitiveness of foreign trade with eight indicators of external trade performance assessment, such as:

- **The Export Benefit Index (RXA):** Establishes the advantage or disadvantage owned by a country for the production of that good, so this indicator has registered sub-unit values in the period

Table 9. Intra and extra-Community import and export of wines

Years	U.M	Import	Export	Trade Balance
2011	Thousands tons	90.77	10.53	-80.24
	Million USD	69.48	19.97	-49.51
2012	Thousands tons	54.53	11.38	-43.15
	Million USD	53.96	20.53	-33.44
2013	Thousands tons	36.75	10.48	-26.26
	MillionUSD	50.5	21.99	-28.51
2014	Thousands tons	34.16	10.51	-23.64
	MillionUSD	44.6	23.92	-20.68
2015	Thousands tons	50.73	13.88	-36.84
	MillionUSD	47.03	24.67	-22.36
2016	Thousands tons	50.23	12.89	-37.34
	MillionUSD	52.45	22.72	-29.73

Source: Author based on data from http://www.intracen.org

Figure 10. Wine import and export in Romania
Source: Author own processing based on data from http://www.intracen.org

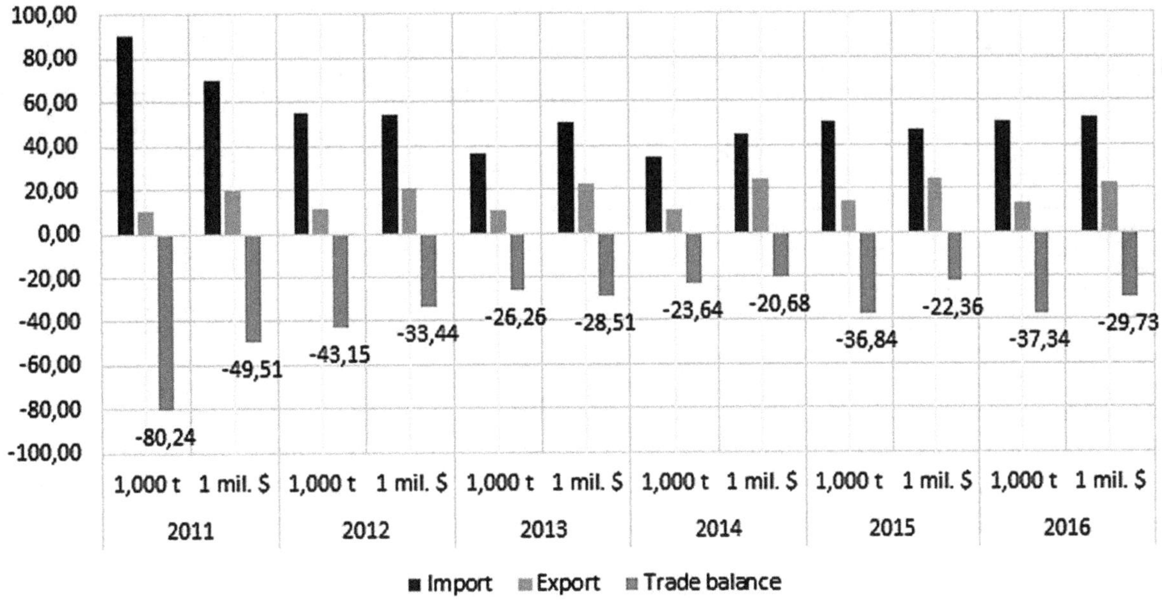

2011-2016 ranging from 0.1749 and 0.2112, therefore Romania has a comparative disadvantage in the production of wine.

- **The Relative Import Penetration Index (RMP):** Measurement result is sub-unit, throughout the analyzed period, it has reached values between 0.3040 and 0.4913, which means that Romania has a comparative advantage for wine.
- **The Relative Trade Benefit Index (RTA):** Measures which of the two components of the trade balance dominates, thus in the present case for each analyzed year it is negative taking values between -0.3164 and -0, 1173, which establishes that wine imports have a comparative advantage greater than export.
- **The Comparative Advantage Revealed (RCA):** Shows that this product (wine) does not present this comparative advantage, as the result of this indicator is negative for the whole reference period, ie between -1,0494 and -0,5031.
- **The Grübel-Lloyd Index (GL):** Determines the balance between imports and exports; in the case of Romania, this indicator is very close to the value of 1, ranging from 0.9993 to 0.9997, so the value of exported wine almost balances it the one imported; this value (of 1) of this indicator shows the character of foreign trade in wine, namely one that focuses on trade within the industry and not between two or more industries.
- **The Export Market Share (EMS):** Recorded each year from 2011 to 2015 increases for wine, reaching the share of 0.0775%, and in 2016 it decreased to the value of 0.0702%.
- **Import Market Share (IMS):** For the wine product is higher on average than that of exports 2.3 times, reaching in the last year 0.16% of the import value of wine world level.

Table 10. Indicators of the performance of wine trade

Indicator	2011	2012	2013	2014	2015	2016
RXA	0.1749	0.1970	0.1812	0.1868	0.2112	0.1751
RMP	0.4913	0.4249	0.3683	0.3040	0.3444	0.3444
RTA	-0.3164	-0.2279	-0.1871	-0.1173	-0.1332	-0.1694
RCA	-1.0494	-0.7731	-0.7227	-0.5146	-0.5031	-0.6768
GL	0.9993	0.9995	0.9996	0.9997	0.9997	0.9995
EMS (%)	0.0607	0.0618	0.0630	0.0688	0.0775	0.0702
IMS (%)	0.2048	0.1605	0.1426	0.1247	0.1445	0.1600
NEI	-0.5534	-0.4488	-0.3934	-0.3018	-0.3119	-0.3956

Source: Author own calculation based on data from http://www.intracen.org

- **The Net Export Index (NEI):** Determines to what extent this trade balance is inclined, thus, during the analyzed period, this indicator varied between -0.5534 and -0.3018; these negative values characterize foreign trade with wine, in the sense that Romania develops mainly import activities.

CONCLUSION

Romania has a significant agricultural potential and a comparative advantage of natural resources. Under an open economy, competitiveness issues acquire complex valences. The prosperity or failure of Romanian agriculture, after joining the EU, depends on the level of competitiveness. The performances of the agri-food sector in Romania also depend on the quantitative and qualitative changes of the agriculture and food industry, the level of exports and imports, the prices on the world market. The wine sector is one of the competitive agri-food sectors that meets the requirements of the European Union market. Considering the wine consumption trend in countries where this product is preferred to wine-based or beer-based products, stricto-senso, a comparative analysis should be made on the structure of grape varieties, the proportions of table and quality wines, and the relationship between red and white wines. These issues should be the starting point for addressing the long-term quality wine survey.

Being a culture that has been in operation for decades, choosing the vineyard assortment must be the result of a forward-looking study that takes into account the whole set of socio-economic factors, combined on a global scale, parallel to the overall strategy for wine promotion of quality, accepted relatively united by all the world's largest wine producers.

Generally, productivity levels in Romania for agricultural production are below those achieved in the EU. Moreover, while the level of productivity of farms in the EU has been increasing, in Romania, since the start of the transition, the productivity level has decreased, leading in time to an increase in the gap between Romania and the member countries of the EU. This situation, with major economic and social implications, requires special attention and urgent intervention by the responsible actors, to improve productivity that will support and increase the quality of production and agricultural products and will strengthen the competitive position of their processing units.

For Romania, wine reform can be said to be an opportunity through the measures it aims at, such as the simplification of the provisions on the quality of extending the role of professional organizations, alignment with international oenological practices, simplification and adaptation of the labeling policy. In this context, it can be said that Romania, through the producers, can promote on the European market several varieties of quality wine, such as Fetească Albă, Fetească Regală, Tămâioasă Românească, Fatca Cotnari, Fetească Neagră, Băbească Neagră, Cadarcă, Braşcuşă, Busuioaca de Bohotin.

Currently, for Romanian farmers, both opportunities and risks have increased. Rising fuel prices will further exacerbate the risks faced by farmers. The evolution of European agriculture support policy is a short- and medium-term need, increasing productivity and competitiveness of agriculture being a concern at European level. The use of innovation and agricultural research can prove to be an optimal solution, starting from the premise that, in a competitive context, innovation is an important competitive advantage, being the only factor of differentiation in a food market.

Integration into the EU means both opportunities and risks for Romanian wine producers. Therefore, capitalizing on Romania's wine-growing potential in order to obtain sustainable competitive advantages on foreign markets requires the creation and promotion of a Romanian wine identity that will make it visible on the world market.

REFERENCES

Ashenfelter, O., & Storchmann, K. (2016). Climate change and wine: A review of the economic implications. *Journal of Wine Economics*, *11*(1), 105–138. doi:10.1017/jwe.2016.5

Elrod, T., Russell, G. J., Shocker, A. D., Andrews, R. L., Bacon, L., Bayus, B. L., ... Mazanec, J. A. (2002). Inferring market structure from customer response to competing and complementary products. *Marketing Letters*, *13*(3), 221–232. doi:10.1023/A:1020222821774

Fleming, E., Mounter, S., Grant, B., Griffith, G., & Villano, R. (2014). The New World challenge: Performance trends in wine production in major wine-exporting countries in the 2000s and their implications for the Australian wine industry. *Wine Economics and Policy*, *3*(2), 115–126. doi:10.1016/j.wep.2014.12.002

Hristov, H., & Kuhar, A. (2015). Subjective knowledge as a determinant of young adult consumers wine behaviour. *British Food Journal*, *117*(12), 2930–2946. doi:10.1108/BFJ-04-2015-0163

Lombardi, P., Dal Bianco, A., Freda, R., Caracciolo, F., & Cembalo, L. (2016). Development and trade competitiveness of the European wine sector: A gravity analysis of intra-EU flows. *Wine Economics and Policy*, *5*(1), 50–59. doi:10.1016/j.wep.2015.12.002

Pomarici, E., Lerro, M., Chrysochou, P., Vecchio, R., & Krontalis, A. K. (2017). *One size does (obviously not) fit all: using product attributes for wine market segmentation*. Wine Economics and Policy.

Rendleman, C. M., Hoemmen, G. A., Altman, I., Taylor, B., Moon, W., & Smith, S. (2016). Wine Industry Competitiveness: A survey of the Shawnee Hills American Viticultural Area. *Wine Economics and Policy*, *5*(1), 4–13. doi:10.1016/j.wep.2016.03.002

Schäufele, I., & Hamm, U. (2018). Organic wine purchase behaviour in Germany: Exploring the attitude-behaviour-gap with data from a household panel. *Food Quality and Preference*, *63*, 1–11. doi:10.1016/j.foodqual.2017.07.010

Velikova, N., & Dodd, T. (2016). Sustainability of the wine market though emerging consumer segments: The case of US Hispanic consumers. *Agriculture and Agricultural Science Procedia*, *8*, 81–87. doi:10.1016/j.aaspro.2016.02.011

Vlachos, V. A. (2017). *A macroeconomic estimation of wine production in Greece*. Wine Economics and Policy.

ADITIONAL READING

Ambroziak, L. (2015). The competitive position of the New EU Member States In Trade In Food Industry Products. *Competitiveness of Agro-Food and Environmental Economy*, 55.

Čačić, J., Kljusurić, J. G., Karpati, L., & Čačić, D. (2010). Collective wine export promotion in transitional countries: The case of Croatia. *Journal of Food Products Marketing*, *16*(3), 262–277. doi:10.1080/10454446.2010.484739

Carmichael, B. A., & Senese, D. M. (2012). Competitiveness and sustainability in wine tourism regions: The application of a stage model of destination development to two Canadian wine regions. In *The Geography of Wine* (pp. 159–178). Springer Netherlands. doi:10.1007/978-94-007-0464-0_9

Chen, D., & Zhan, J. (2011). Study on the Change of Grape Wine Industrial Pattern in Asian Countries in Recent Ten Years & the Evolution of Grape Wine Industry Competitiveness in These Countries. *Liquor-Making Science & Technology*, *6*, 047.

Crescimanno, M., & Galati, A. (2014). Competitiveness of Italian wines in the international market. *Bulgarian Journal of Agricultural Science*, *20*(1), 12–22.

Elliot, S., & Barth, J. E. (2012). Wine label design and personality preferences of millennials. *Journal of Product and Brand Management*, *21*(3), 183–191. doi:10.1108/10610421211228801

Esterhuizen, D., & Van Rooyen, C. J. (2006). An inquiry into factors impacting on the competitiveness of the South African wine industry. *Agrekon*, *45*(4), 467–485. doi:10.1080/03031853.2006.9523758

Foster, M., & Spencer, D. (2002). *World wine market: Barriers to increasing trade*. Australian Bureau of Agricultural and Resource Economics.

Fulconis, F., & Viviani, J. L. (2006). Segmentation of the Romanian wine market: An exploratory study. In *3rd International Wine Business and Marketing Research Conference*.

Gavrilescu, C., & Voicilas, D. M. (2014). Competitiveness of the Romanian agri-food trade and the new agricultural policies. *Zagadnienia ekonomiki rolnej*, *4*, 106-116.

Geana, E. I., Marinescu, A., Iordache, A. M., Sandru, C., Ionete, R. E., & Bala, C. (2014). Differentiation of Romanian wines on geographical origin and wine variety by elemental composition and phenolic components. *Food Analytical Methods*, *7*(10), 2064–2074. doi:10.100712161-014-9846-2

Hollebeek, L. D., Jaeger, S. R., Brodie, R. J., & Balemi, A. (2007). The influence of involvement on purchase intention for new world wine. *Food Quality and Preference*, *18*(8), 1033–1049. doi:10.1016/j.foodqual.2007.04.007

Iordachescu, A., Moore, A., & Iordachescu, G. (2010). Consumer perceptions of organic wine. *The Annals of the University Dunarea de Jos of Galati. Fascicle VI, Food Technology*, *34*(1), 51.

Lădaru, G. R., & Beciu, S. (2015). Trends of wine market in Romania: Competiveness and development opportunities in the context of markets globalization. *Scientific Papers. Series Management, Economic, Engineering in Agriculture and Rural Development*, *15*(2), 189–192.

Lădaru, G. R., Beciu, S., Vlad, I. M., & Pădure, M. C. (2014). Aspects regarding wine production and wine sector competitiveness in Romania. *Scientific Papers. Series Management, Economic, Engineering in Agriculture and Rural Development*, *14*(3), 134.

Marks, D. (2011). Competitiveness and the market for Central and Eastern European wines: A cultural good in the global wine market. *Journal of Wine Research*, *22*(3), 245–263. doi:10.1080/09571264.2011.622517

Noev, N. (2007). Land, Wine, and Trade: The Transition of the Romanian Wine Sector. *Eastern European Economics*, *45*(3), 76–114. doi:10.2753/EEE0012-8775450304

Popa, D. (2010). Romanian wine growing sector in the context of European integration, competitive and sustainable development. *Bulletin of University of Agricultural Sciences and Veterinary Medicine Cluj-Napoca. Horticulture*, *66*(1), 341–344.

Steiner, B. (2002, July). The valuation of labelling attributes in a wine market. In *2002 AAEA-WAEA Annual Meeting*.

Veith, C., & Nămoloşanu, I. (2012). Studies regarding the wine consumption in Germany and the perspective of increasing the Romanian wine export on this market. *Scientific Papers-Series B, Horticulture*, (56), 489-494.

Voicilas, D. M. (2013). Gains and loses of Romanian agrifood products on EU intra-trade market. *Competitiveness of Agro-Food and Environmental Economy*, 401.

KEY TERMS AND DEFINITIONS

Competitiveness: A complex concept that has long been debated by economic experts around the world, competitiveness refers to the ability or ability of individuals, firms, national economies to stay in competition both internally and internationally and to gain economic benefits from this.

Demand: The aggregate quantity of a product or service estimated to be bought at a particular price.

Export: Goods or services produced in one country to another country.

Import: Goods and services purchased from abroad.

Price: The value that is received by an organization in exchange of its products or services.

Supply: The total amount of a product available for purchase at any specified price.

Trade Balance: Measure of international trade in which a country's imports exceeds its exports.

Chapter 3
Sustainable Development Through Field Vegetal Exploitation in Romania

Mariana Bran
The Bucharest University of Economic Studies, Romania

Iuliana Dobre
The Bucharest University of Economic Studies, Romania

Stefania Daniela Bran
The Bucharest University of Economic Studies, Romania

ABSTRACT

The anthropic factor, expanding its vital area, has disrupted the natural system of existence. Over time, the evolution has been managed to meet the current needs of the human being in order to reach economic and geopolitical control of interest groups. The study is based on research into the living conditions and productivity of natural and cultivated ecosystems, as a scientific basis for plant production. The local natural environment exploited by field crops has been analyzed, focusing on species with a weight in agriculture. Romania has higher agricultural potential compared to the world's agricultural potential, but there are some which sporadically influences economic development. The chapter, based on rigorous analytical documentation, highlights the technological, economic, and environmental support necessary and accessible to entrepreneurs in the vegetable sector. In order to develop a sustainable farming, the chapters try to find out what types of technology are appropriate to that.

INTRODUCTION

Branch of material and economical production, the farming manages the ecosystems. These are defined and known at all levels, so, on both local and worldwide level, a huge campaign to monitoring the life quality is under construction. The life, as existence form of matter at high organization level, is permanently in interdependence with abiotic environment. The vegetal and animal associations, as biosphere

DOI: 10.4018/978-1-5225-5739-5.ch003

Copyright © 2019, IGI Global. Copying or distributing in print or electronic forms without written permission of IGI Global is prohibited.

functioning units, cumulate solar energy and different substances of environment and transform them into vegetal, then animal biomass. Based on biogeokinetic reactions, step-by-step, the biomass is given back to the nature as available forms, the matter circuit is reloaded. Under these conditions, the ecosystem is fully presented, with dynamic balance and self-adjustment.

The anthropic factor, expanding its vital area, has disrupted this natural system of existence. Over time, the evolution has been managed to meet the current needs of the human being in order to reach economic and geopolitical control of interest groups. The proposed study is based on research into the living conditions and productivity of natural and cultivated communities in ecosystems, as a scientific basis for plant production. Activity as such takes place in farms. At the same time, the agriculture must take responsibility for protecting the soil and other environmental resources that it can degrade.

The concept of sustainability in farming implies choosing those technologies that will not compromise decisions during future. The major objectives of sustainable farming are to optimize productivity, preserve the basic natural resources, minimize and even eliminate the negative effects of intensifying agriculture. Thus, in order to ensure the quality of the environment and promote the sustainable economy, the balance between inputs and outputs, between investments and benefits must be maintained in farming systems. Increasing the level of soil productivity and preserving a healthy environment are compatible concepts if optimization of the factors of production is closely linked to the requirements of the species cultivated against them.

Currently, there is a particular interest in the conservative farming system, which, taking into account all the components of the agricultural technology system (soil tillage, plant vegetation management, crop rotation, fertilization, irrigation, crop protection, harvesting and transport) of vegetal remain, ensures long-term sustainable land use, preventing and minimizing soil degradation by restoring both its productive capacity and life support processes.

It is appreciated that only such farming can contribute to the improvement and conservation of various environmental resources, including biodiversity, leading to sustainable social development.

The vegetal biodiversity cultivated for economic purposes in Romania is diverse and includes annual species (with a few exceptions: perennial, medicinal, sugar beet and hops) that lend themselves to local conditions. Diversification, as a form of production structure, is one of the factors that influence the eco-economic performance of the farm.

The paper, based on rigorous analytical documentation, highlights the technological, economic and environmental support necessary and accessible to entrepreneurs in the vegetable sector.

BACKGROUND

The World Conservation Strategy, published in 1980 by the International Union for the Conservation of Nature and Natural Resources, included a first reference to sustainable development as a global priority (Sachs & Jeffrey, 2015).

The modern concept of sustainable development is based mainly on the Brundtland Report of 1987:

Our Common Future. Conservation and Sustainable Use - States shall maintain ecosystems and ecological processes essential for the functioning of the biosphere, shall preserve biological diversity, and shall observe the principle of optimum sustainable yield in the use of living natural resources and ecosystems. (United Nations World Commission on Environment and Development, WCED)

As concept, sustainable farming is farming ecologically by promoting methods and practices that are economically viable, environmentally sound and protect public health (https://www.conserve-energy-future.com/sustainable-farming-practices.php).

In this respect, is needed to promote sustainable technologies as Organization for Economic Co-operation and Development (OECD, 2001) propose. The ways in which technologies are developed and adopted differ between countries.

Research efforts, farmer education and training, advice and information are shifting towards balancing economic efficiency with environmental and social sustainability. Technologies are increasingly being developed in a global market, applied at the farm level but impacting on sustainability beyond the farm. Adopting technologies for sustainable farming systems is multi-disciplinary. Adopting technologies for sustainable farming systems involves uncertainty and tradeoffs.

Also, are different factors such as economic, structural, behavioral and policy which combines in a wide range of situation. These aspects make that farmer to adopt specific technologies.

Viatte Gérard (2001) said that solving environmental problems in agriculture requires developing and diffusing new technologies. The author explains the importance of technologies for create a sustainable farming. *Until recently, the choice of technologies available to farmers was largely determined by the need to increase production, profits and productivity.* Now, agriculture must to be competitive, to obtain agricultural products of high quality while meeting sustainability goals. *Farmers also need to change their production and management practices in response to agricultural policies that include environmental conditions and I am confident that farmers have the capacity to do so.* Viatte sustains sustainable technologies because their environmental health and food. Also, the relevance of future technologies consists into develops scientific farming which use efficient pests, nutrients, water. In the same time, resorting to that will take place a management based on good practices and dissemination of knowledge to farmers.

Agricultural technology innovation and diffusion, particularly to small holders, will be key to boosting yields, productivity sustainably (Ban Ki-moon, 2008).

SUSTAINABLE AGROSYSTEMS: FEATURES AND SUITABILITY

Nature is defined as everything that is not directly subject to human intervention. Fortunately, as recognized by current Romanian visitors, there are still natural communities (far from human, at altitude and on the Danube Delta), with ecosystem load favorable and stimulating evolutionary knowledge; statistics show that there are 661939 ha of biosphere reserves in 2016 (Figure 1: the Danube Delta is located in the South East Development Region).

European Union (EU) documents distinguishes the environmental elements (air and atmosphere, soil, land, landscape and natural areas, including wetlands, marine and coastal areas, biodiversity and its components, including genetically modified organisms, as well as interaction between them) environmental factors (substances, energy, noise, radiation, waste, including radioactive waste, emissions, discharges and other discharges that may affect environmental elements).

The area where we live in is directly conditioned by natural factors such as climate, relief and biodiversity.

Figure 1. Area from the development regions occupied by protected natural area, %
Source: Data processing National Institute of Statistics (NIS) Romania and Agenţia Naţională pentru Protecţia Mediului, 2012; http://www.insse.ro/; http://www.anpm.ro/

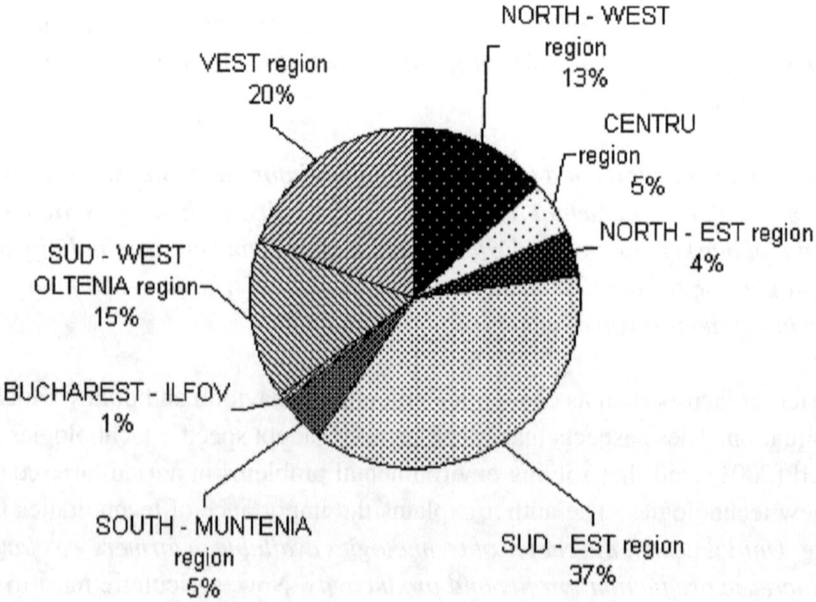

The continental temperate climate shows variations in temperature according to the four seasons (exceptions: 8-12 months a year with positive temperatures in southern and seaside areas compared to 4 months in high mountain areas). For example, in the area where National Agricultural Research and Development Institute (NARDI) Fundulea, as the central manager of the Romanian agricultural research, is located, the temperature and rainfall variations were registered, according to the graphs in Figure 2; winter season (October-March), and in growing season (April-September).

The rainfall, at country level, scores a multiannual average of 640 mm, with different repartition on both relief and time.

Climatic conditions are directly responsible for the achievement of photosynthetic yield (net yield of terrestrial vegetation). From the net yield of terrestrial vegetation, human energy is about 7% (Muntean, Roman, Borcean, & Axinte, 2003).

Relief presents unique features in Europe and rare all over the world: 28% of mountains (altitude over 1,000 meters), 42% of hills and plateaus (300 to 1000 meters altitude) and 30% of plains (altitude below 300 m; National Strategy for Sustainable Development of Romania, 2008). Mountains, the least modified anthropic part, have 12 of the 13 national parks and 9 of the 14 natural parks (National Strategy for Sustainable Development of Romania, 2008). The hills and plateaus are animated by urban and rural settlements and infrastructure elements, as well as by specific economic activities: agriculture (plantations of vines and fruit trees, technical plant and grain crops, livestock breeding), forest exploitation, hydrocarbon extraction, mining, industrial enterprises. The fields are the most populated and heavily exploited, which is why the natural areas are exceptions (minimum human intervention: the Danube Delta, the lagoon systems and the meadows of some rivers).

Figure 2. (a) Temperature and; (b) rainfalls registered in a very favorable area of NARDI Fundulea (1960 – 2015)
Source: Cociu and Alionte, 2017; http://www.incda-fundulea.ro/rar/nr34/rar34.8.pdf

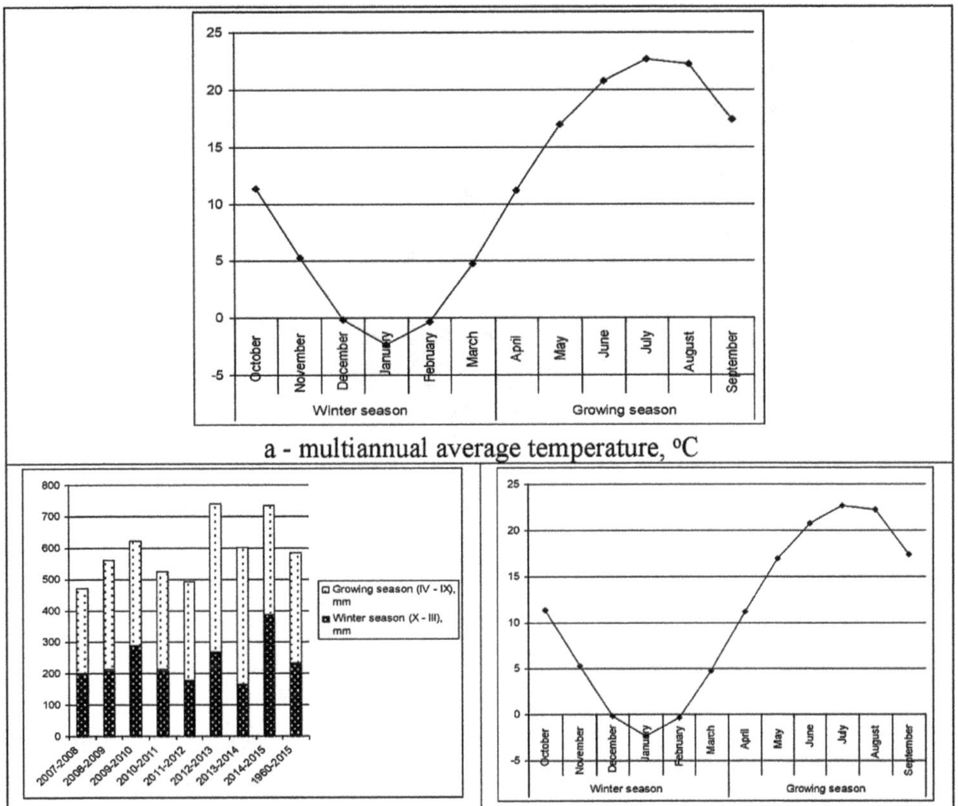

a - multiannual average temperature, °C

b' – annualy and seasonal reinfulls, mm

b" – miltiannal average of reinfulls, mm

Soil has the role of accumulating and supplying nutrients and energy to living organisms, but also to provide the other conditions favorable to the development of these organisms. Through this role, the soil is compared to a huge plant, which produces permanently on a planetary basis, using automorphe, phytomatous processes, which are the basis for the development of heterotrophic organisms, including humans. Without the phytosomal assurance of nutrition with carbohydrates, proteins and other compounds, as well as the necessary energy, life on the Earth would not exist and could not take place (Florea, 2003).

Soil is also a renewable resource, as long as its use by humans does not adversely affect its functionality. Soil regeneration is mostly done by vegetation, which, under the microorganisms and climatic factors action, contributes to the formation of clay, supporter of fertility (clay is the soil component with influence on the physical, chemical and biological properties of soils, with implications for the state production potential). The mode of supply of soil with vegetal residues is different depending on the type of natural or cultivated biocenosis. In meadows, forests and agricultural crops, the soil accumulates annually, on the surface and inside, different amounts of dead organic, vegetal residues as well as microorganisms (Figure 3).

Figure 3. Structure of the soil elements
Source: Interpretation after study bibliographic materials

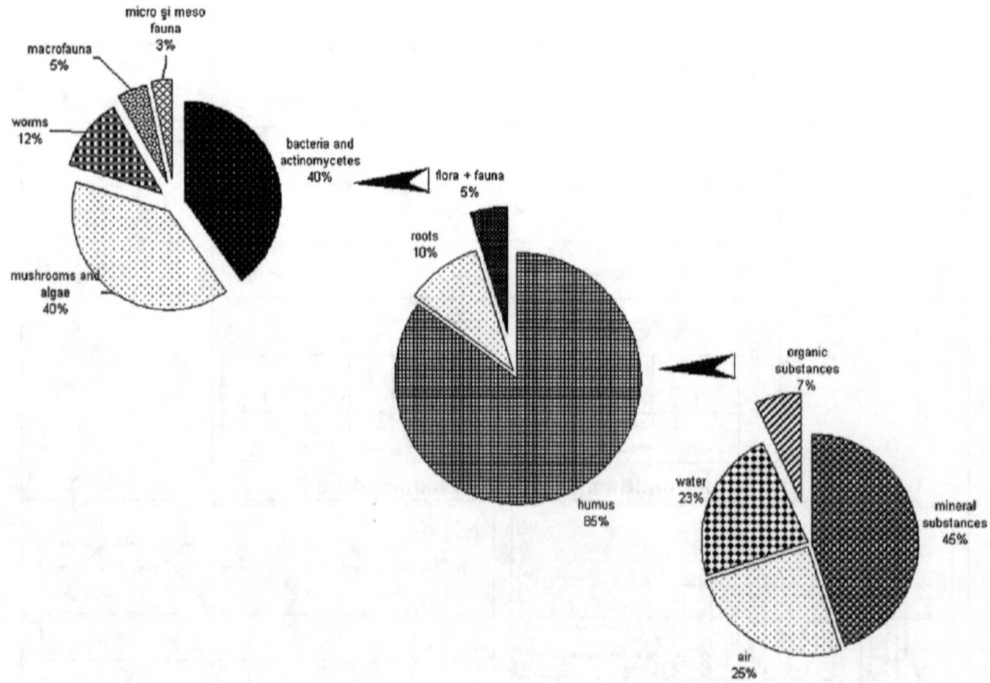

Analyzes on plant residues revealed that carbon, oxygen and hydrogen were predominant (90-95%) by weight of organic matter. In addition, ash and smoke produced during burning (5-10% of dry matter) provide many nutrients in small quantities but have a vital role in the nutrition of plants and animals.

The soil resource, as a support with trophic attributes for flora, has formed and evolved during time only under the influence of cumulative abiotic and biotic factors. As a result, the temperate continental climate has contributed to the formation of many types of soils in Romania, which are included in 12 classes, which are also identified in other places of Europe and which are suitable for the regeneration of the above-mentioned biodiversity. The presence of plant species indicates the presence of nutrients in the soil. For example, *Lamium maculatum* indicates soils with very high trophicity; the trophicity decreases when *Valeriana officinalis*, *Galium aparine*, *Rhytidiadelphus triquetrus* are present and, at low level, in case of the presence of the species *Vaccinium vitis-idaea*. A special situation is marked by the euritrophe soils, which have high ecological amplitude to the soil trophicity. The situation is highlighted by many species, the most well-known being *Dactylis glomerata*. The presence of ammonium cations (NH^{4+}) encodes the nettle (*Urtica dioica*) among the bioindicators, and *Leontopodium alpinum* has adapted to develop on soils with a large amount of calcium (Ca^{2+}) cations.

The suitability of the soil resource for supporting the vegetal, natural and anthropic vegetation cover is manifested by the physical-mechanical and chemical attributes, which are at the interference of all the factors of formation and, last but not least, by the management of the economic activities. During time, the compaction, soil structure degradation, depletion of nutrients, which contributed to the decrease of fertility of the agricultural area were found: 52% have low or very low fertility (2007), 20.7% moderate fertility and the rest, high fertility and very high (National Strategy for Sustainable Development of Romania, 2008).

Seen in its entirety, the nature regroups sites, landscapes and ecosystems.

In Romania, geographical positioning has made possible to bring together several biogeographical regions differentiated by the existence of unique biodiversity, both at the level of ecosystems and species, and at the genetic level. These represent for the world natural heritage a genuine *natural gene bank*.

Under the influence of the climate and its differentiation, the rocks in the substrate (the chemical composition of the soil and subsoil) have led to the emergence of a large number of habitats, both in the arid areas in the east or in the oceanic areas of the West, as well as in the defined plains and mountains by the altitude of the relief.

The habitats (continental, 53%, alpine, 23%, steppe, 17%, panonic, 6%, pontic, 1%; in The state of the environmental factors in Romania, 2010) are characterized by a certain composition of the flora (3630 plant species, of which 23 are declared monuments of the nature) and fauna, components of biocenosis and are influenced by various climatic or edaphic factors. Thus, if we refer only to the Danube Delta Biosphere Reserve (580000 ha: water formations, wetlands, forests, bushes and grass vegetation, landscapes, cities and villages), there were identified 30 types of ecosystems (of which 23 are natural and 7 anthropic) where the biodiversity consists of 7405 species (of which 2383 plants, 4029 animals, microorganisms, protozoa and fish).

Over the time, there have been significant changes in the structure and functioning of ecosystems, both qualitatively and quantitatively. It is possible to list the identified manifestations of erosion of biological diversity:

- Fragmentation of habitats (barrage of watercourses, dredging of floodplains and drafting of roadways with repercussions in blocking or restricting the migration routes of fish and animal species);
- Restriction or elimination of some types of habitats or ecosystems under transition areas (forest curtains, wetlands in large farm structures), minimizing or even canceling control functions on pollution, soil erosion, biological control of pests for crops Agricultural;
- Changing the configuration of river basins and watercourses;
- Simplifying the functionality of ecological formations by expanding intensive agriculture;
- Changing the landscape through the structure of cultural elements, biodiversity and geomorphologic structure.

The rhythm, forms and forces of socioeconomic development have put their mark on the situation discussed at The Earth Summit in Rio de Janeiro in 1992 when it came to finding ways to stop the destruction of natural resources and the pollution of the planet.

The threats of direct and indirect anthropogenic pressure on biodiversity and, implicitly, ecosystems have intensified through industrialization. The consequences, at a global level, bring policy makers together more and more to recognize the importance of natural capital for the sustainable development of human communities (providing renewable resources: water, air, food, clothing, medicines, air and water regeneration), landscape importance and recreation, protection and environmental balance to maintain a healthy environment. As a result, the necessary condition for the harmonious development of future generations is the biodiversity preservation.

As historical perspective of mankind - the crisis of nutritional and fuel means - biochemistry and bioeconomics must clarify the complexity and diversity of the connections between the systems and the surrounding bio-systems. And, as an example, the plant kingdom is characterized by the ability to synthesize carbon, lipids, antidotes, but also to filter and enrich the air with O_2. For these reasons, it is at the center of the ecosystem. *Services* of nature are economically free and beneficial.

Life, as a form of existence of matter at the upper level of organization, is in constant interdependence with the abiotic environment. The plant and animal associations, as functional units of the biosphere, store solar energy and various substances in the environment, transforming them into plant and animal biomass. Plant biodiversity sums up, on a global scale, hundreds of thousands of species (known and still unchecked but also some of them endangered) that have been able, during time, to generate the organic matter of human life (for human food, 25 species are very important).

The agro-productive ecosystem is part of a territorial one. It was defined by Christiane Rolland-May (2000) as *a system with a high degree of organization, increased complexity both in terms of internal structure, environmental relations, connection with sub-systems and integrative ones and ability to define its own objectives or means to reach them*. Under this context, the Figure 4 shows the differentiation of the anthropic systems from the natural ones.

The natural systems are physionomically more homogeneous (Ianoş & Heller, 2006) and have very clear interconnection relationships between the lithosphere, hydrosphere, biosphere and atmosphere, compensation relations, inter-conditioning and cooperation in the flow of matter, energy and information, ensuring the coherence of the system; on a territorial level, homogeneity results from the physiological and functional similarity of the components and not from their interconnection (Ianoş & Humeau, 2000).

The anthropogenic systems are heterogeneous, but they show spatial coherence and synergy. Their functional coherence is due to the territorial imbalance (Magurran, 1998).

Despite these differences, the territorial system results from the interrelationships between the two sub-systems, the physionomy and its functionality depending on the intensity and the forms of relations between them (Law no. 58 of July 13, 1994).

Figure 4. Regional system (territorial system conceptual adaptation)
Source: Data processing after Musila, Todt, Uster, and Dalitz, 2005; https://link.springer.com/chapter/10.1007/0-387-24320-8_40

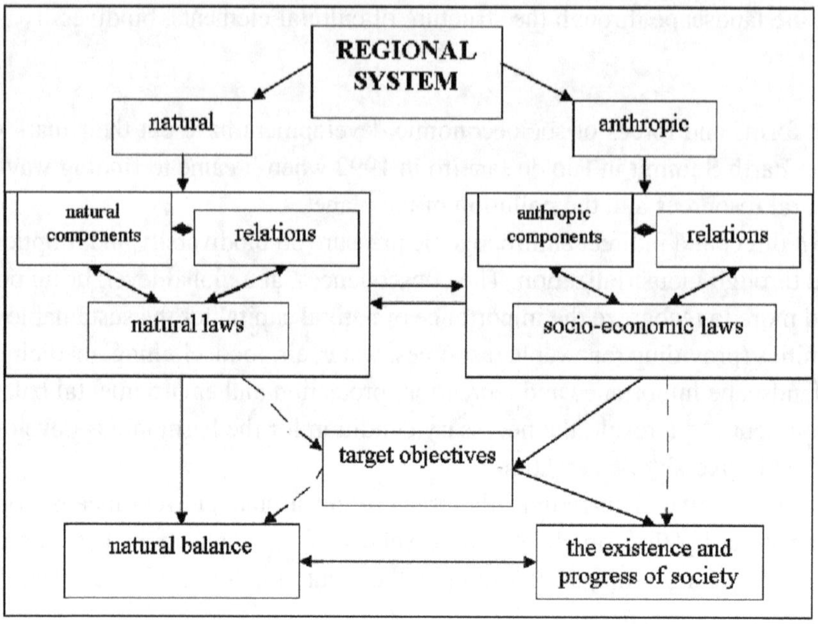

The coupling of the socio-economic systems to the natural ones is done under all functional aspects. Thus, the flow of matter is taken from the natural systems, in whose trophic chains the human is consumer. The supply of resources is made directly from these systems or from their transformation into anthropic systems, respectively agroecosystems.

The collection of energy (and resources) is done through technologies developed by the human species (Sârbu, 2006). From an energy point of view, the socio-economical systems discharge natural energy, which is introduced in different ways (fertilizers, pesticides, soil processing, green areas, food) in quantities far exceeding the input of primary producers (Petrişor, 2008). The increasing of the quantity absorbed by the anthropic systems is achieved by increasing of the complexity of absorption channels of the resources by the human society, highlighting the structural character of the human activities on the geographical space (Sârbu, 2006).

Changing biogeochemical circle and reducing biodiversity causes a decline in the stability of socio-economical systems, their self-regulation being dependent on human intervention.

The communities, regardless of their status, are a pool of organisms and factors of favoritism for initiation and maintenance of life. Under evolution, it followed specific cycles with qualitative transformations at the beginning. The quantity was marked with trade and desire for better (emerging economics).

For human communities, the surroundings are structured in units of the land fund, as a result of anthropic activity (Figure 5).

The practical interpretation of the research of the natural conditions for the needs of agricultural production distinguishes two aspects: the qualification and the technological characterization of the crops. Land use is a complex research and quantitative assessment of the main conditions that lead to the growth and yield of plants, to the degree of favorability of these conditions for each use and crop (a land may be unfavorable for certain crops and uses, but favorable for others soil resource; Dincu & Bran, 2002). The expression of favorability for agricultural use is made by means of a rating Table 1.

Besides the fact that the bonitation serves to establish the correct uses, both in terms of production levels and economic efficiency, it also helps to substantiate investments, labor input in agriculture, technologies (Dincu & Bran, 2002).

Figure 5. Structure of land fond, Romania
Source: author own processing based on data from http://www.insse.ro/

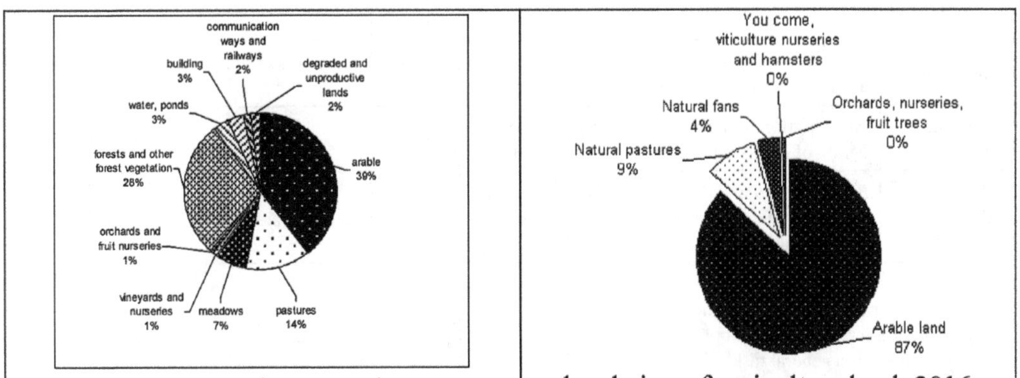

a - land structure in Romania, 2014

b - draing of agriculture land, 2016

Table 1. Classification of agricultural land in quality classes, according to the rating scale, by country: Romania, in 2010 (partly)

Use	The Total Map Area		Of Which on Quality Classes,% of Total Use				
	Ha	% of Total Agricultural	I	II	III	IV	V
Arable	9236031.59	63.40	6.70	28.69	38.19	19.10	7.32
Pastures + Hay	4811360.70	33.03	1.61	8.09	27.11	38.91	24.28
Vii	269488.41	1.85	3.13	23.51	31.65	32.26	9.45
Orchards	250523.74	1.72	0.52	9.72	31.86	43.92	13.98
Total agricultural	14585690.02	100.00					

Source: Author own processing based on data from ANMP (National Agency for Environmental Protection) citing ICPA (National Research - Development Institute for Soil Science and Agricultural Chemistry Environmental Protection) and OJSPA (County Offices for Pedological and Agrochemical Studies)

The agricultural land management has as main aim the supply food, ignoring ecosystem services and destroying traditional agricultural landscapes that, as a cultural heritage, would attract tourists for recreation.

In some regions, the agriculture intensifies and, in others, abandons the lands with repercussions on land conversion, leading to the biodiversity loss and soil degradation.

Because the biodiversity is the heart of the cycle for the raw matter transformations, any disruption would create problems, some of them being already found and globally assessed (Table 2).

Transformation of natural / semi-natural ecosystems into arable land was possible through the development and application of intensive production technologies. Thus, the floodplain floodplains of the main rivers and the Danube Meadow were embanked and transformed into intensive agricultural ecosystems such as steppe grassland pastures, excess moisture fields, forest curtains and forest bodies in the plain area after deforestation (National Strategy for Sustainable Development of Romania, 2008).

Inappropriate agricultural use of land severely affects the soil due to losses (such as erosion, erosion has other causes too, and it will be corrected by land reclamation – Figure 6) of its main constituents (firstly organic matter); the degree of loss of the arable layer is within wide limits (3.2 - 41.5 tons / ha / year; Negulescu, Vaicum, & Pătru, 1995), due primarily to diversity of relief, fragmentation in small

Table 2.

Social Situation	About 1.3 billions of humans live in absolute poverty, 40 billions/year starve to death, urbanization reunite already 50% of earth habitants
Biodiversity State of Play	Afforestation have an annual average of about 2%; 5-14% of diverse species are strongly threatened and so on
Water Resource	*humans will firstly died of thirst before starving*; during last decade, over 1 billion of humans had no access to potable water, while about 10 millions annually died of water lack
Observations: It will be necessary to bring solutions from biodiversity preservation as a whole! Biological diversity preservation studies not only species and habitats, but protection active measures. Biodiversity preservation is a general task! Biodiversity preservation brings added values to the next generations!	

Figure 6. Soil erosion: (a) The degree of loss of the arable layer; (b) erosion by categories of agricultural use
Source: Author own processing based on data from http://www.insse.ro/

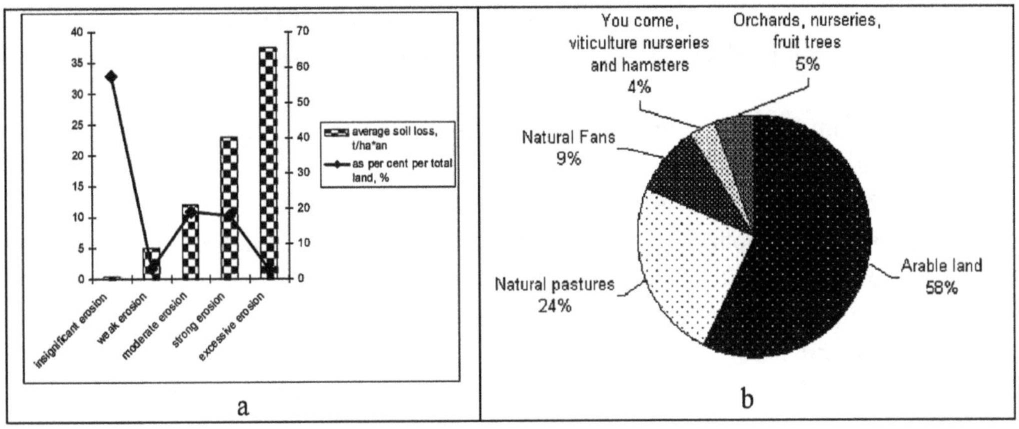

farms, short-term exploitation and applied technologies. Values are higher than world tolerable losses (5 t soil / ha / year; Ţuţuianu, 2011). Consequently, the necessity of land reclamation is imposed on 2/3 of the agricultural area of the country. For this, Măgdălina (2006) specifies soil tillages: exploitation, rehabilitation and modernization of existing systems; the expansion of landscaped areas in order to increase the performance of agriculture and to protect the environment (water, soil).

The farming itself (conventional or organic), by taking over natural deforested habitats, has a considerable impact on biodiversity. But this impact is not always negative. The cultivation of plant species produces new ecosystems that can enrich biodiversity (ranging as type and number of living organisms in a given place). The species which exploit the arable land are differently spread on the territory of Romania, according to their specific requirements towards the vegetation factors. Knowing the plant-environment relationship, measures can be established in order to harmonize this relationship with the goal of producing high quality and high yields. Optimizing the factors in this relationship, ie creating the most favorable conditions for the intensification of the photosynthesis process, is solved differently. The inorganic compounds, taken from the environment and processed by the plants in the photosynthesis process, are the basis for the synthesis of organic compounds. The amounts of organic substances contained in plants differ not only from one species to another, but also from the same plant, being influenced by the variety, place where the plants develop, used soil tillage, application of fertilizers and so on. Neamţu et al. (1983) identifies over 60 chemicals in plant organisms: 12 - macroelements, plastics or constituents - account for 99% of the weight of organisms, and trace elements which represent less than 1%. Determining the chemical composition of a plant, the amount of nutrients removed from the soil together with the crop should be compensated from fertilizers in the case of the plant. The quantitative and qualitative determination of minerals is done by chromatographic, spectrophotometric, methods, known as rapid and precision ones. At the same time, it recognizes the direct participation of water in all the reactions taking place in living organisms.

The organic matter, as result of photosynthetic effect on leaf (*plant kitchen*), is diverse and specific. Thus, for the cultivated species, it is exemplified:

- Cellulose accumulates mostly in cotton textile fiber (99.8%), but, as a supportive role in plants, it is found in the largest proportion in the world's vegetation (Neamţu, Popescu, Lazăr, Burnea, Brad, Cîmpeanu, & Galben, 1983, p. 164);
- More starch accumulates in grain seeds (60-80%) or potato tubers (12-20%). Multiple utilizations:
- Raw matter to obtain products resulting from fermentative processes (ethyl alcohol, acetone and so on),
- As culture medium for microorganisms growth in antibiotics and vitamins industry,
- Glue or glue in textile and paper industry,
- Reagent for chemistry and biochemistry works,
- In food; lipids are more found in oil seeds (20-60%) and less in cereal seeds (2-5%). They are nutritional reserves and have the role of regulating cell permeability to substances that enter the cell;
- Most of the antella are found in legume seeds (23-40%) and less in vegetables and fruits (0.5-6.5%). They are *first importance* substances – *there can be no life without albuminoid compounds* (Neamţu, Popescu, Lazăr, Burnea, Brad, Cîmpeanu, & Galben, 1983, p. 190).

In order to supplement the structural biochemistry of the plant, it is important to mention the nucleic acids, vitamins, enzymes and phytohormones, without minimizing the role of water in plant life.

Muntean et al. (2003) exemplify the influence of climatic factors on current Romanian crops, which have a low efficiency in capitalizing on light. Winter crops, spending much of the vegetation period under less favorable conditions (6 months of long nights, temperatures and low light), have lower biomass growth. Thus, winter wheat (almost 100% in culture) develops about 3% of the total dry substance, during October-March (180 days), 97% being accumulated in 90 days, respectively April-June. Unlike winter crops, spring trees go through vegetation under more favorable conditions for photosynthesis. Out of the immense energy of the sun (1.360 W / square meter), only about a quarter of a billion can be absorbed by the green leaves and only 1-2% (on average) is fixed in shape of chemical energy in phytomass, only in crops very well managed from the phyto-technological point of view, the energy efficiency reaching 4-7% (maximum 10%; agrimedia.ro).

Enhanced solar energy intensifies the photosynthetic yield, but it is obvious to use varieties or hybrids with high yielding ability and the technology elements to be at high level (Figure 7).

Under agroecosystems, the photosynthetic yield is influenced, in addition to climatic conditions, by the nutrition conditions created by the farmer, the cultivated species, the plant age and the duration of the vegetation period. Species with a longer vegetation period achieve large photosynthetic yields, as a sum of yields depending on the vegetation period; the photosynthetic yield increases until the blooming phase, then decreasing to maturity (Muntean, Roman, Borcean, & Axinte, 2003). The growth of photosynthetic yield is achieved by the use of varieties and hybrids with high yielding ability and the use of technology that maintains photosynthesis, respectively the date of setting up the crop (sowing/planting), ensuring optimal plant density, with no stresses and so on. After all, agriculture means protecting crops and keeping undesirable organisms, such as weeds, diseases and insects, under control. The fertility and a good soil structure are also essential for a good harvest. This is why the soil is fertilized and mechanically tilled. As a result, the mulching of vegetal residues is worthy of promotion as an agricultural measure for increasing the carbon absorption and the production of grain crops in the field thus managed. All these processes have an impact on the environment.

Figure 7. Biological material role and selection for yield
Source: Personal interpretation

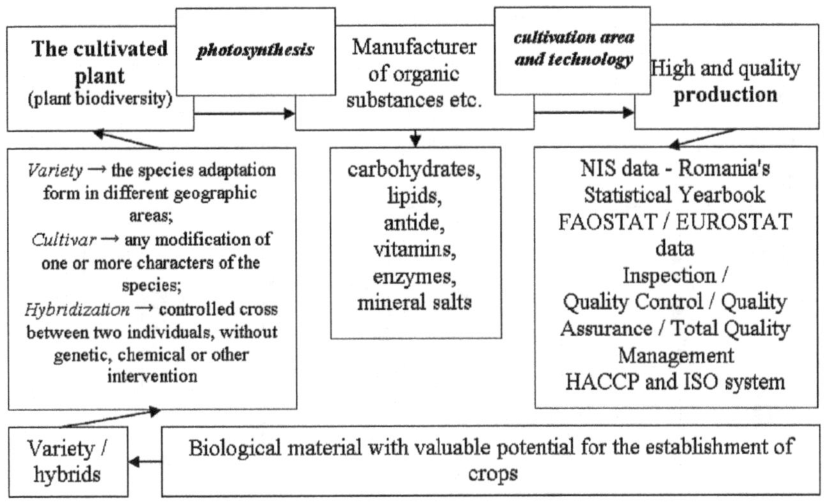

Referring to the fertilization of the land exploited on farms, it is noted that, in time, the amount of fertilizer was reduced (Figures 8 and 9). If, in 1999, 260kg NPK/ha were provided, in the year 2016 we used about 155kg NPK/ha (71.18kg N + 47.01kg P + 36.96kg K), ensuring a N: P: K ratio of 1: 0.77: 0.3. Compared to fertilizer provided in arable crop in 2016, the recommendation of 135 kg/ha of nitrogen, 70 kg / ha of phosphorus and 80 kg/ha of potassium for the production of wheat of 6-7 tons of wheat/ ha is not respected (www.agrointel.ro).

The manure has also been reduced, which in 2016 made arable land to benefit only from 17.31 tons.

If the 155kg NPK + 17.31t of manure were assured and administration would have taken into account the agricultural good practice, it would certainly result in superior yields to any field species and the area exploited would maintain its productivity.

Figure 8. Use of fertilizing products on Romanian arable: (a) Chemical fertilizers; (b) organic fertilizers
Source: Author own processing based on data from http://www.insse.ro/

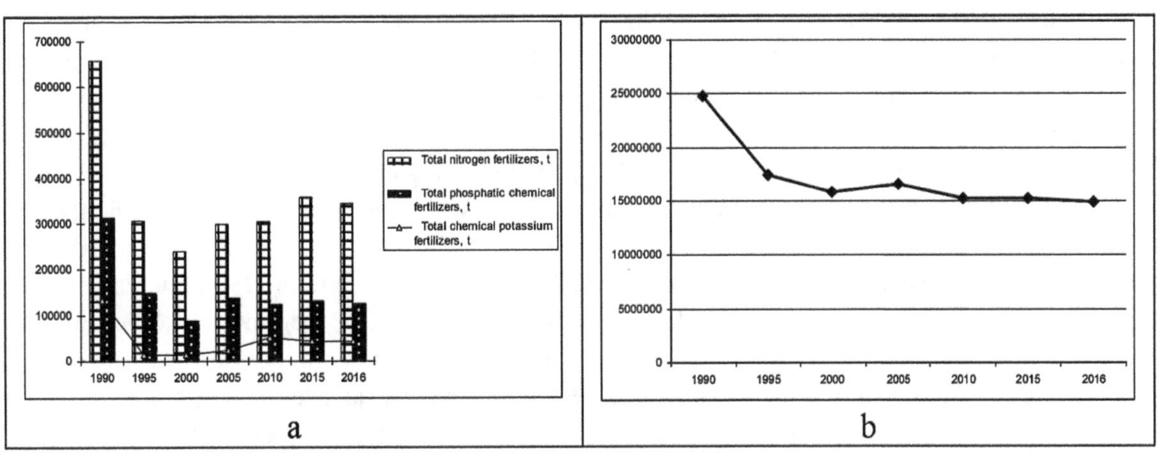

Figure 9. NPK active ingredient average amount into arable land
Source: Author own processing based on data from http://www.insse.ro/

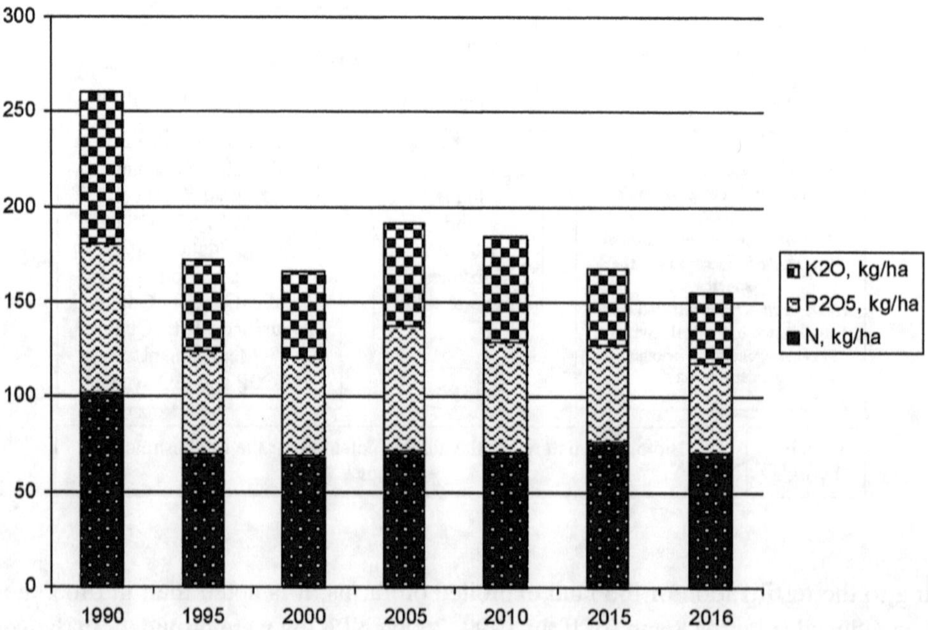

Part of the fertilizer must be found in the soil at the date of sowing. For this reason, fertilizers are applied prior to soil tillage, whether they are conventional tillage system, or we apply the *no-tillage method.* Soil tillage regulates the regime of chemicals by favoring the activity of microorganisms, but also by incorporating fertilizers. The 8-yr (2007-2015) average of winter wheat grain yield was 1.37% higher than that recorded for conventional cultivation, plowing tillage (5.91 vs. 5.83 t ha-1). The differences between soil treatment systems were significant in 4 of the eight seasons of the experiment (grain maize and soybean grain) and were reduced when the rainfall deficit occurred during the grain filling period (Cociu, 2016). Managing fertilizers in soil is the responsibility of water accumulated in capillary pores, so, the activity in the field must be coordinated specifically to the area and in accordance with planned economic results. The lack of soil water is supplemented by irrigation using economical watering methods. In a few arid locations in Romania it is irrigated. The arable area actually irrigated with, at least one watering in 2016 was of 152598 ha, less by 20 thousand ha compared to the previous year (164535 ha irrigated in 2012, 180126 ha in 2013, 145249 ha in 2014, 172624 ha in 2015). Although the negative effect of the drought has been found every year, because of the fragmentation of the properties, it is very difficult to mobilize the irrigation systems. Irrigation systems and changing the sowing interval as ways to adapt to these trends can lead to a 10-15% increase in global agricultural yield.

During crop growing, some undesirable organisms occur, with specific damage and, implicitly, lowering yields. As a result, input allocations and technological effort are required. The Figure 10 shows these allocations, which, being toxic and expensive, decreased compared to the 1990 reference year.

All human actions have as targets specific purposes and strictly pursue their interests. Thus, man has created an environment of great complexity, but, ignoring the natural environment, he senses the hostility of nature. The best-known example is pollution, the long-term consequences of which are manifested by global warming and the multiplication of climatic accidents. Pollution sources are multiple, and their

Figure 10. Arable land that (a) used pesticides; (b) and average amounts
Source: Author own processing based on data from http://www.insse.ro/

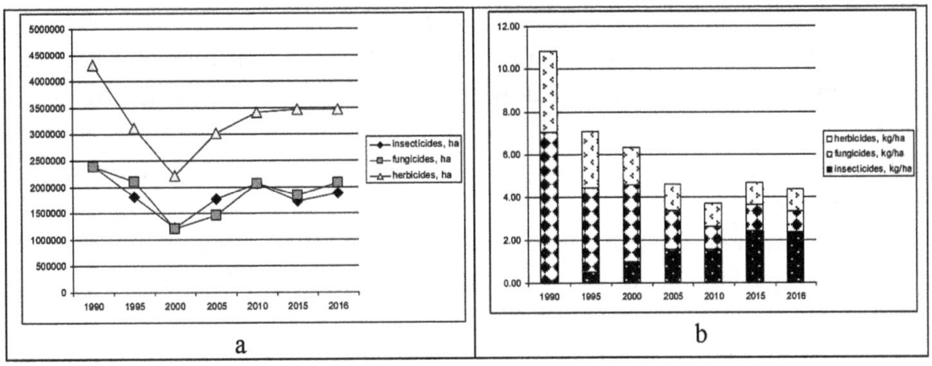

amplification has begun together with industrialization. Such consequences are nominated and analyzed in this paper based on the study of some bibliographic sources. Bioeconomics, as a science, is meant to integrate humanity into the environment. And today, health and sustainability are conditioned by the proper management of everything that surrounds us.

STRESS FACTORS IN THE FIELD VEGETATION

The field crops placed under natural environment are constantly under the natural risk factors (weather, producing floods, droughts and so on), with unpredictable repercussions, especially under the current climate and climate conditions.

Structural changes over time, as a result not only of the diversification and increase of anthropogenic pressure, are reflected in the current configuration of the ecological structure of our country's natural capital. The situation has led to the diminution of the bio-productive ability and the support of the requirements for resources and services from the national socio-economic system. This has increased the vulnerability of Romania's territory to geomorphological, hydrological and climatic hazards. In this sense, the Common Agricultural Policy (CAP) allocates incentives to farmers, in return for their compliance with management practices in land use: *best practices* (with the name of cross-compliance) to prevent soil erosion, preserve organic carbon in soil and maintains soil structure; the farmer's obligation is to keep his country under Good Agricultural and Environmental Condition (GAEC). Erosion occurs permanently and everywhere. The European Commission (Joint Research Center, 2010) has published centralizing materials for this purpose - erosion ratio of soil with water; our country ranks 9th in the EU-28. Romania has exceeded the average erosion value in the EU and requires compliance with good agricultural and environmental good practice.

The factors with influence on soil erosion may be natural (climatic, relief, lithological, edaphic) and anthropic or socio-economical (Dîrja, 2000; Toy, Foster, & Renard, 2002; Gobin, Jones, Kirkby, Campling, Govers, & Kosmas, 2004). The main consequences are the loss of fertile soil due to perturbed nutritional cycles, organic carbon losses and biodiversity. Measures for the judicious preservation and management of land and crops (egg: vegetal residues management and the introduction of crop-rotations, the keeping of slopes) have had a significant impact on the reduction of soil losses (9.5% on average) in EU during last decade.

Conventional technique has been the most common agricultural practice in most areas in Romania. This system requires high energy consumption and, through repeated soil tillage, water losses are recorded, and physical degradation of the soil is also observed on long-time. Loss of soil leads to a decline in organic matter and nutrient content, deterioration of soil structure, reduction of water in the soil tilled. Nutrients and carbon reserves have significantly changed their ratio (Quinton, Govers, Van, Oost, & Bardgett, 2010), as eroded soil can lose 75-80% of its carbon content, releasing it into the atmosphere (Morgan, 2005). The impact of soil erosion on the environment involves high economic costs of conservation and restoration.

For agricultural areas, loss of organic matter and nutrients, degradation of soil structure, uprooting of plants, reduction of available water and so on are ways to reduce soil fertility. The situation is even more alert as it is known that the surface layer of the soil is a fertile-carrier and it is first exposed to erosion. Soil erosion mechanisms preferentially remove organic matter, clay, sludge and fine materials. Soil erosion reduces the amount of soil needed to plant the plant and degrades its physical properties.

In most cases, additional fertilizer can offset the impact of erosion on soil fertility, but it is an additional cost to farmers and does not fully restore soil productivity.

The natural and economic conditions in which agricultural holdings operate and which do not always appear as expected by producers, almost naturally imply inconveniences, whether they involve production or sale of products or other aspects (Dobre, 2011). As a result, the following types of risk were differentiated in agriculture: yield (weather, disease, genetics, feed, other), price (input prices go up/down) liabilities, taxes) and human resources (dependability, tillage quality).

Under risk management process, after the detection of the factors that generate it, the risk reaction is manifested (Cornescu, Marinescu, Curteanu, & Toma, 2004).

The actual and necessary framework for agricultural activities is governed by the *state of nature*, the determinant factor of risk, which directly influences yield outputs and indirectly on economic ones. Extreme manifestations of natural phenomena (drought, late frost, storms) have negative effects on the yield and quality level. Only knowing the moment of their occurrence does not guarantee the achievement of favorable results. Natural conditions are therefore an unlikely adversary for the farmer. Climate fluctuations influence quantitative, qualitative and structural crop production, leaving a shadow of uncertainty over agricultural production. Therefore, agriculture is often characterized by the high variability of production outcomes or risk production. Unlike most other entrepreneurs, farmers are not able to predict the quantity resulting from the production process as a result of external factors such as weather, pests and diseases. Farmers may also be hindered by adverse events during harvesting or threshing that may result in production losses.

Economic causes are components of the macro-environment, which can generate the risk; has a decisive influence on the functioning and development of the agricultural holding through the market and the economic and financial policies. The market, by checking the consistency between supply and demand, is a confirmation and regulator of economic decisions on agricultural holdings.

The risk may also arise as a result of technologies (Zahiu, 1999). During current period, the technological factor influences the development of the agricultural holding through the technical level of the machinery. It lays the mark on the degree of technical endowment, the pace of modernization, the quality of agricultural products (for example, the accumulation of pesticide residues detected by Ministry of Agriculture and Rural Development (MARD) on 18.6% of the grain samples analyzed in 2015 - even if the recorded values were below the maximum admissible limit, however, the land exploited for the production of these cereals has accumulated some of the 7 pesticides used; madr.ro/en). The dynamics of applied technologies substantially alters the competitive relationship between two or more holdings.

In relation to risk situations, the behavior or risk response of the farmer is different depending on the type of holdings. In the case of family farms (households), due to insufficient financial capability (poor capitalization) and therefore the application of technologies based on the use of tradition, on the physical work of their members and less on modern elements, the possibilities of defense against some risks, such as natural ones, are limited. Agricultural holdings, especially legal entities are much larger in size, their management is in many cases carried out by specialists with university education and are connected to market relations. As a result, there is a different perception of the risks and they are much more active in the fight against them, the managers being favored by the fact that they have technological and economic knowledge and pay great attention to the information that circulates in the environment, from the ecological to the economic ones specific to the business world.

The general scheme of the factors and the types of risk they generate is shown in the Figure 11.

A method of reducing risk is found in the programming of the activity, which is used by agricultural companies and agricultural companies which, in order to survive the competitive environment, have to anticipate their production and economic results. Programming not only has the role of directing the holding in terms of resources, allocation of resources, output forecasts, revenues and expenditures, but also of predicting the size of risk by including it in the program. Therefore, any economic risks may also be foreseen in the program taking into account the price and cost changes in the course of an agricultural year. Thus, starting from a given forecast level of production and assuming a certain level of the risk coefficient (risk probability), loss of production can be determined. It is also possible to establish the loss value, including in relation the sales price of the product. The calculation algorithm is presented in two forms as shown in Table 3.

Scheduling production and economic risk will allow to better assess the farmer activity and, at the same time, the use the resources in such a way that even under risky conditions he can exceed the level of the profitability threshold.

Figure 11. The risk determinant scheme into agricultural exploitations

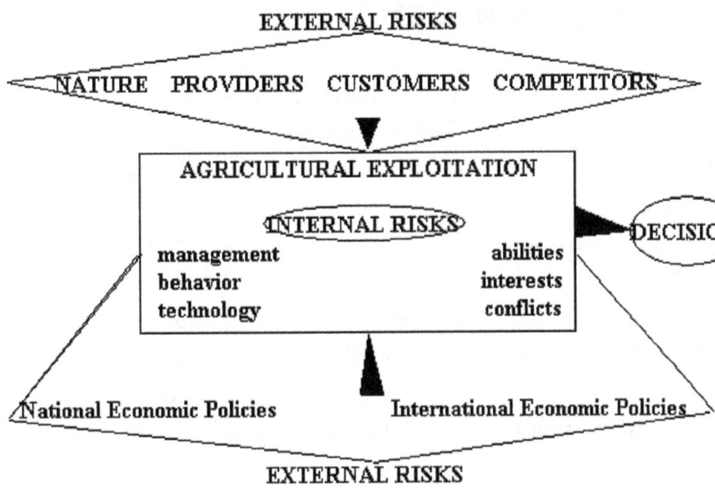

Table 3.

Case	Formula	In Which:
A: lowest selling price	Q = Q estimated – Q lost Vmep = Q lost * pmv Vme = Q * pmv	Q = total estimated yield; pmv = selling minimum price; Vmep = estimated minimum income as follows of yield loosing; Vme = estimated minimum income
B: the highest price	VMep = Q lost *pMv	VMep = estimated maximum income as follows of yield loosing; pMv = selling maximum price

In order to stabilize agricultural income, basic sources of agricultural risk, management tools and strategies must be known.

Risks can be overcome by production risk management strategies. Thus, action should be taken by:

- Follow-up of recommended production practices;
- Diversification of production by sowing as many species as possible, including new species. Diversification, as a form of production structure, is one of the factors that influence the eco-economic performance of the farm;
- Increasing production through intensive cultivation practices or by expanding surfaces;
- Crop insurance to stabilize revenue in loss periods;
- Adoption of risk reduction practices such as irrigation, drainage, location of resistant varieties and so on;
- Different field use and crop rotation. Having the optimal crop structure, you can move on to the arable land arrangements;
- Endowment under working order.

The manager of the agricultural society must recognize and manage the marketing, financial, legal and environmental risks and, last but not least, those related to the labor resource.

RESOURCE SUSTAINABLE MANAGEMENT

The concept of sustainable development, based on interference between human and ecological development, also includes influence factors (Figure 12); for example, the interest in the progress of humanity but also the effect of the technological impact on products and the environment (the eco-cyclical approach of products and technologies).

The agriculture, as ecosystem manager, assumes the increase in soil productivity and the preservation of a healthy environment. As a result, in order to promote a sustainable economy, the balance between inputs and outputs, investments and benefits must be maintained under agricultural yielding systems.

The choice of the agricultural system is conditioned, in addition to the previously noticed, by the level of technical endowment, professional knowledge, but also of mentality and education. It is also worth mentioning the criteria characterizing the different systems used in agriculture:

Figure 12. Sustainable development, resources, influence factors, and management
Source: Interpreting the authors

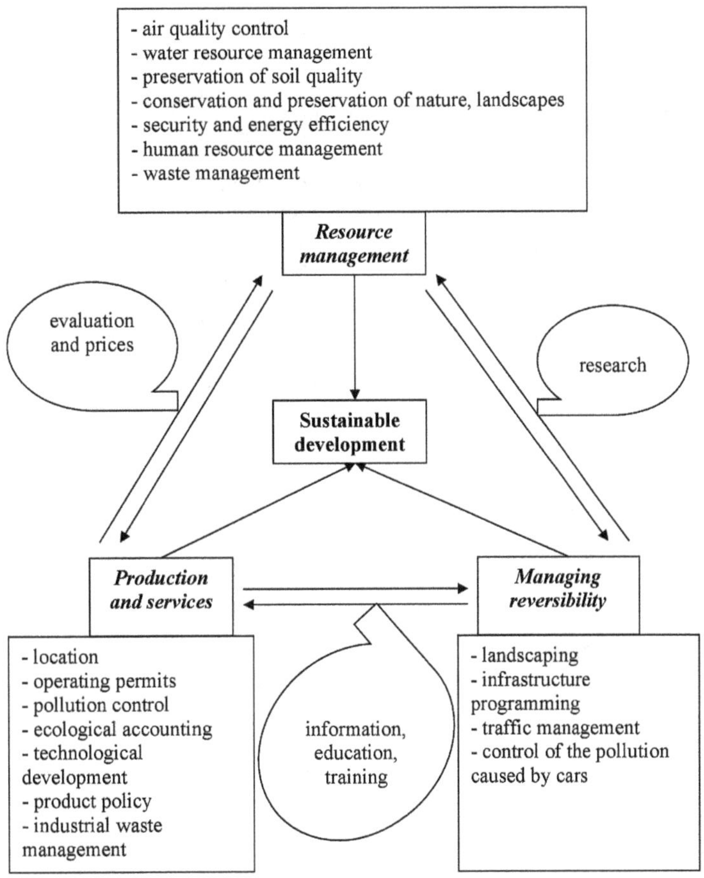

- Yield (quantity and quality);
- Competitive products (reasonable costs of production);
- Yielding stability (by sector, farm and farmland from year to year);
- Harmonious relations with the main natural resources (soil, water, fauna, flora, relief), their improvement, improvement and consistency for future generations;
- Flexibility in specialization and structure of agricultural production, according to supply and demand;
- Long-term balance between economic, environmental and social requirements.

The economic activity into agroecosystems is managed by the farmer (farm manager) applying technology (Figure 13); this is due to the natural succession of some links indispensable to production and sustainability. The complexity of the decision-making process is generated by the multitude and nature of the resources, which are mutually dependent, on how they are used.

The major objectives of sustainable farming are to optimize productivity, conserve basic natural resources, minimize and even eliminate the negative effects of intensifying farming.

Figure 13. Technological flow in crop yield
Source: Interpreting the authors

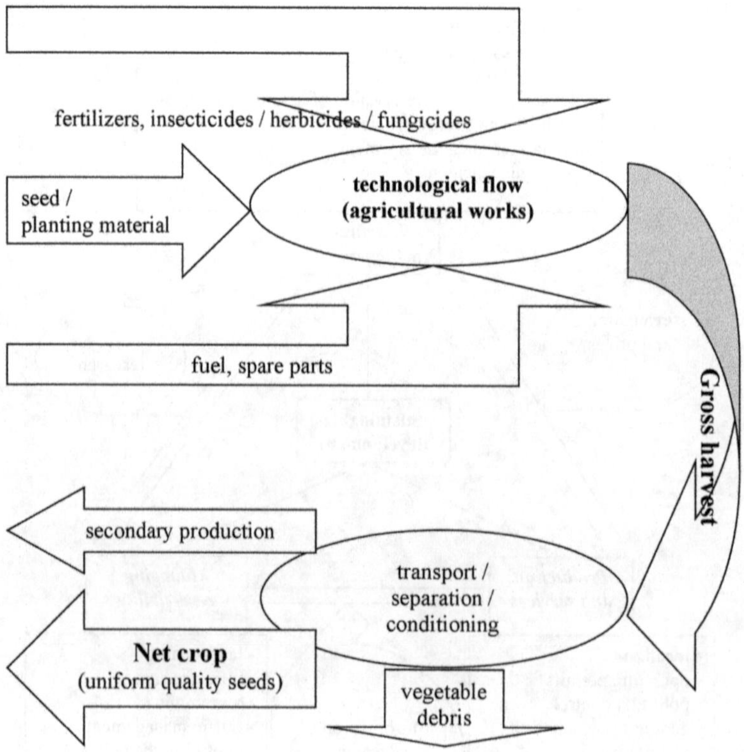

One can increase the productivity through sustainable methods such as organic farming and *high-level* (smart) irrigation systems (Figure 14). The crops, breeded by traditional and genetically modified methods, could stop the crop decline by increasing drought tolerance, nitrogen efficiency, pest resistance, and photosynthesis rates.

The fact that the intensity of human activity increases on the environment either through uncontrolled consumption of resources and space, or by the production of waste that nature cannot absorb without suffering, has led the international community to initiate and support concrete actions to prevent, counteracting and eliminate the repercussions of ecological balance disturbing factors (Brown, 2001). Under the same context, Brown L. highlights for the salvation of civilization and has the following advices:

- Preventing and combating the irrational use of natural heritage/natural capital, especially biotic one;
- Replacing the current consumption economy with one based on ecology and environmental protection;
- Man (constructor, regulator and consumer into artificial ecosystems) to become a wise partner of nature;
- Greening the anthropogenic polluting activities and expanding the use of green energy;
- Redesigning the world economic system → *bio-economic globalization based on bioethics through ecological and humanist education* and so on.

Figure 14. Water management through plant specificity
Source: Interpretation of the studied bibliography

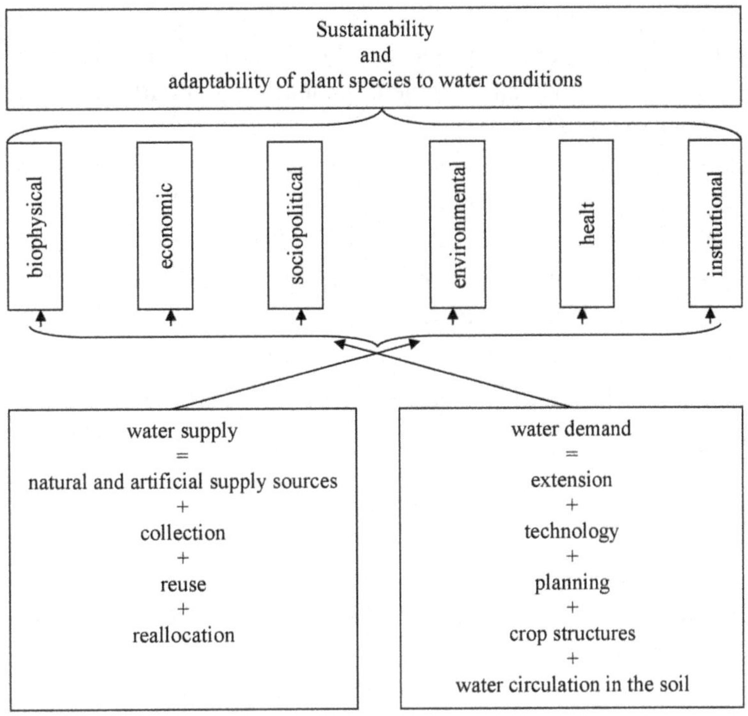

The farming exploits a pool of resources; those used in intensification have enriched energy and contributed to global warming.

The cumulative greenhouse gas emissions (of the sectors energy, agriculture, industrial processes, waste) in the year 2015 amounted to approx. 100,000 thousand tons, down from the peak registered in 2003 (about 130,000 thousand tons); in CO_2 equivalent, the emissions registered in 2015 were 6,587 million tons (EPA).

The farming releases a significant amount of greenhouse gases (as shown), a situation that can be regulated by the scientific/innovative management of the farm, generally speaking. Regarding the field crop, it is necessary to adjust the methods of land and crop management. Thus: judicious placement of plant species in the field and cropping; reducing soil aggregate crossing and adaptation for soil water conservation; fertilization should be balanced at optimum times (observance of the *closed period*); fight with biological stressors (weeds, diseases and pests) to use reduced inputs; proper management of plant debris resulting from harvesting the main product (in dry areas, to be kept in the field).

Versus the stated ones, it is clear that one can utilize the clean technology systems, but conventional ones are not excluded. As a result, for the Southern Muntenia Development Region, a comparative analysis of the two systems presents the following conclusions on sustainability (Bran, 2012):

- **Ecological Aspect:** Organic farming has a major contribution to sustainable development, increased economic activity with significant added value and increased interest in rural areas;

- **Technological Aspect:** To the same investment, to the same amount of work, on the same soil, the results obtained (quantity-quality) largely depend on the natural factors, which cannot be controlled precisely, always. Organic technologies require adherence to certain rules (planting crops, use of fertilizers and phytosanitary products permitted in organic farming only). Conventional technologies, although economically efficient, should be gradually replaced by the sustainable, conservative (the scheme in Figure 15);
- **The Economic Aspect:** The application based on the assessment data (the total existing / arable area, the crops available to the agricultural use category, the available financial resources, the estimated expenditures on crop types, the crop inputs and the availability of these resources) achieve maximum profit. The prerequisites are: full cultivation of arable land; cultivating field species in a balanced structure, according to existing natural conditions and the market; the possibility to achieve crop rotation and rotation according to scientific criteria;
- **The Social Aspect:** The agricultural production of 86.07% of the Arable Region of South Muntenia Region meets the consumption needs of the local population.

The concept of *sustainable development* is the unanimous response of mankind to the development of human society during next millennium.

An important condition for achieving the objectives of sustainable development is the simultaneity of progress in several dimensions. For this purpose, economic policies, environmental policy, investment, R & D, labor policy, education, health, and so on, are assigned to correlate their objections and actions according to certain priorities. All of this is found in the Europe 2020 Strategy: Promoting smart growth (knowledge and innovation economy) and sustainable (resource efficiency: greener and more competitive), both of which are at the same priority.

Thus, in terms of sustainable growth, the EU has promoted two initiatives:

- **A Resource Efficient Europe:** Climate, energy and mobility, which are essential resources for the production process in agriculture;
- **An Industrial Policy Adapted to the Globalization Era:** Turns to competitiveness.

Figure 15. Vegetable technological principles in conservative agriculture

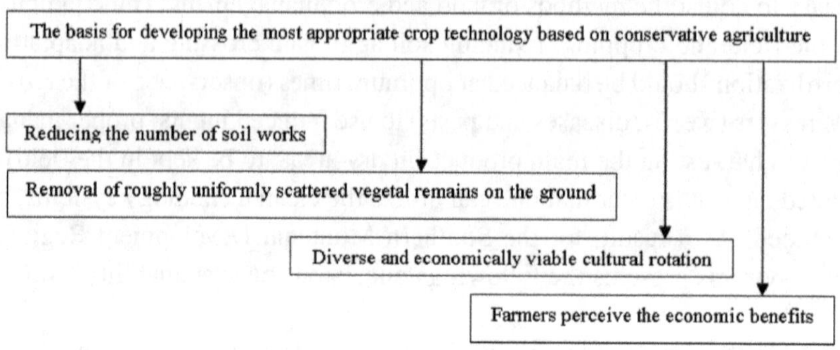

Renewable Energy

The consequence of intensification, the shortage of fossil fuels and their emanations in their CO_2 increase the level of use of alternative energy resources that can be fed by agriculture through the resulting products, by-products and waste. Vasile and Bran (2017), through the food grain's importance, orient the technologies of producing biofuels to non-food raw materials. It was found that vegetable waste can produce significant amounts of ethanol (225.7 l ethanol/ton of straw or coconut), low-emission fuel. As a result, the development of the biofuel industry has to be turned to the transformation of lignocellulosic materials (second-generation biofuels), unused residues (agricultural or forestry), but with renewable potential.

As we deepen the understanding of sustainability, it is clear that the idea of circulation of matter in nature and the management of natural resources must be updated judiciously, especially the fossil energy resources. The latter have, over the years, been the driving force behind industrial processes, which have proved to be the highest consumers, followed by decreasing transport, consumption by the population, other consumers in the economy and, last but not least, by the lowest consumption, agriculture and forestry industry (2015). Being limited and finding the direct polluting effect, energy alternatives, namely the use of renewable resources, were used.

Thus, Food and Agriculture Organization (FAO) statistics recorded biofuel production (Figure 16a) and global CO_2 accumulations (Figure 16b). It is noted that, as biofuel production increases and they are consumed, greenhouse gas emissions are lower.

Biofuels have increased their use (Figure 17) due to the contribution of specific raw materials generated by agriculture (agricultural products to the extent that they do not endanger food, plant debris and biomass). By exploiting agricultural space for this purpose, producer's income is stabilized. As a result, ensuring decent living, coupled with the tendency to preserve material, natural and spiritual assets, are responses to the implementation of sustainable agriculture.

The need to save the environment and maintain fundamental ecological balances appears as a reflection of human survival as a species between species.

Figure 16. The correlation between renewable energy production: (a) ethanol + biodiesel and; (b) CO_2 emissions
Source: FAO statistics; http://www.fao.org/faostat/en

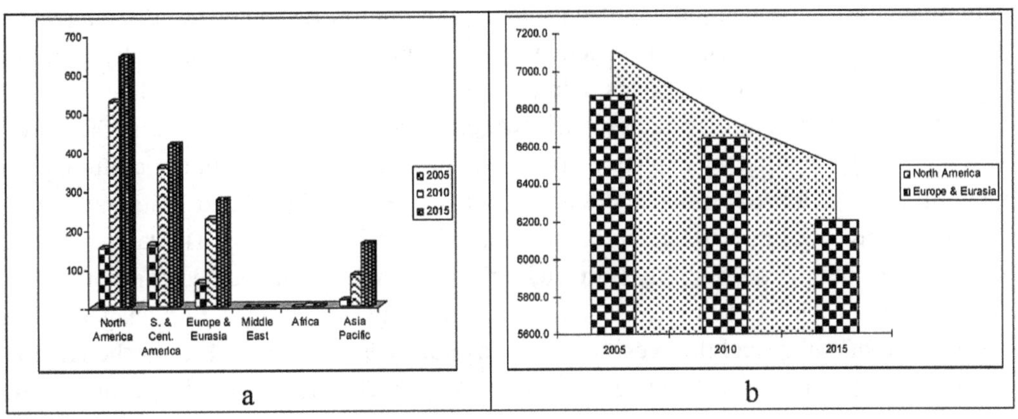

Figure 17. Global final energy consumption in 2013
Source: WBA Global Bioenergy Statistics 2016; http://worldbioenergy.org/uploads/WBA%20Global%20Bioenergy%20Statistics%202016.pdf

Sustainable development achieves reconciliation of agro-food production with conservation of non-renewable resources and environmental protection. It is therefore up to farmers to take account of the effects of their activities on the environment.

CONCLUSION

The chapter shown that concept of sustainable development of vegetal farming has an ecological, economic and social dimension. In this order the concept is the unanimous response of mankind to the development of human society during the next, also achieves reconciliation of agro-food production with conservation of non-renewable resources and environmental protection. It is therefore up to farmers to take account of the effects of their activities on the environment. It noticed that is over the time, there have been significant changes in the structure and functioning of ecosystems, both qualitatively and quantitatively. So, is needed judicious management of natural resources by resorting to the latest scientific results in the field because the vegetal products obtained in the field have diversified both by the number of species and by the cultivars, namely varieties and hybrids. It should be stressed that in the temperate climate characteristic of Romania, two productions can be obtained on the same area of land in one year. It appears that the exploited arable provides large quantities of agricultural products to the population. There is a predominance of cereals, which provides bread and pasta, but also concentrated fodder and industrial raw materials (bioethanol, oil, starch and so on). The land and the climate provide the support, but the profitability is given by the input allocation, the necessity of spending, the higher the intensity of the technological system. The agriculture, as ecosystem manager, assumes the increase in soil productivity and the preservation of a healthy environment. Part of the fertilizer must be found in the soil at the date of sowing. For this reason, fertilizers are applied prior to soil tillage, whether they are conventional tillage system, or we apply the *no-tillage* method. Soil tillage regulates the regime of chemicals by favoring the activity of microorganisms, but also by incorporating fertilizers. Conventional technique has been the most common agricultural practice in most areas in Romania. This system requires high energy consumption and, through repeated soil tillage, water losses are recorded, and physical degradation of the soil is also observed on long-time. For agricultural areas, loss of organic matter and nutrients, degradation of soil structure, uprooting of plants, reduction of available water are ways to reduce soil fertility. As a result, in order to promote a sustainable economy, the balance between inputs and outputs, investments and benefits must be maintained under agricultural yielding systems. Conclude that the chapter can be useful for specialists and farmers. Also, it improves documentation in the vegetal field and contributes to expanding knowledge to users.

REFERENCES

A.N.P.M. (2010). Raport anual – Starea factorilor de mediu în România. In *Cap. 5. Protecţia Naturii şi Biodiversitatea*. Retrieved from http://www.anpm.ro/anpm_resources/migrated_content/uploads/48584_5.Cap.5.Protectia%20Naturii%20si%20Biodiversitatea.2010.pdf

A.N.P.M. (2010). Raport anual – Starea factorilor de mediu în România. In *Cap.4. Utilizarea terenurilor.* Retrieved from http://www.anpm.ro/anpm_resources/migrated_content/uploads/48599_4.Cap.4.Utilizarea%20Terenurilor.pdf

Bogdan, A. T., Pop, V., Diaconescu, D. N., Chelmu, S. S., & Străteanu, A. (2009). *Despre eco-economie şi economie verde în România*. Referat ştiinţific, Academia Română, I.N.C.E. Retrieved from http://www.enciclopedia-economica.ro/upload/files/AT_2_DIMECOECONOMICE_2-5-1-prezentareprofBogdan.pdf

Bran, M. (2005). *Ecotehnica exploatatiei vegetale*. Editura Printech.

Bran, M. (2012). *Cercetări privind dimensiunea eco-economică a biodiversităţii vegetale din Regiunea de Dezvoltare Sud Muntenia, în condiţiile schimbărilor climatice şi de ecotehnică, în concordanţă cu dezvoltarea durabilă (Lucrare de cercetare ştiinţifică pentru finalizarea Programului Postdoctoral, Contract POSDRU/89/1.5/S/63258)*. INCE.

Brown, L. (2001). Eco-economie. Crearea unei economii pentru planeta noastră. Editura tehnică.

Cociu, A. (2016). Long-term effects of tillage systems on winter wheat, maize and soybean grain yield and yield stability under rainfed conditions in the eastern Romanian Danube plain. Romanian Agricultural Research, 33, 97-109.

Cociu, Al., & Alionte, E. (2017). The effect of different tillage systems on grain yield and its quality of winter wheat, maize and soybean under different weather conditions. *Romanian Agricultural Research*, (34): 59–67. Retrieved from http://www.incda-fundulea.ro/rar/nr34/rar34.8.pdf

Cornescu, V., Marinescu, P., Curteanu, D., & Toma, S. (2004). *Management de la teorie la practică.* Bucureşti, Romania: Editura Universităţii.

Dincu, I., & Bran, M. (2002). *AgroEcoFitotehnie* (Vol. I). Bucureşti, Romania: Editura Meteor Press.

Dîrja, M. (2000). *Combaterea eroziunii solului*. Cluj-Napoca, Romania: Editura Risoprint.

Dobre, I. (2011). Methodologies for analysis and evaluation of risk in agricultural exploitations. Internal Auditing & Risk Management, 4(24), 89-97.

E.P.A. (n.d.). *Greenhouse Gas Emissions*. Retrieved from https://www.epa.gov/ghgemissions/sources-greenhouse-gas-emissions#agriculture

Europeană, C. (2010), *EUROPA 2020 - O strategie europeană pentru o creştere inteligentă, ecologică şi favorabilă incluziunii*. Retrieved from http://www.fonduri-structurale.ro/Document_Files//fs2014-2020/00010115/ohsc1_Strategia%20Europa%202020.pdf

Florea, N. (2003). *Degradarea, protecţia şi ameliorarea solurilor şi a terenurilor*. Bucureşti, Romania: Editura Universitară.

Food and Agriculture Organization (FAO). (n.d.). Retrieved from http://www.fao.org/faostat/en

Gobin, A., Jones, R., Kirkby, M., Campling, P., Govers, G., Kosmas, C., & Gentile, A. R. (2004). Indicators for pan-European assessment and monitoring of soil erosion by water. *Environmental Science & Policy*, *7*(1), 25–38. doi:10.1016/j.envsci.2003.09.004

Guvernul României & Ministerul Mediului şi Dezvoltării Durabile. (2008). Strategia Naţională pentru Dezvoltare Durabilă a României. *Orizonturi 2013-2020-2030*, 19-22. Retrieved from http://www.insse.ro/cms/files/IDDT2012/index_IDDT.htm

Ianoş, I., & Heller, W. (2006). *Spaţiu, economie şi sisteme de aşezări*. Bucureşti, Romania: Editura Tehnică.

Ianoş, I., & Humeau, J. B. (2000). *Teoria sistemelor de aşezări umane*. Bucureşti, Romania: Editura Tehnică.

Kummamuru, B. (2016). *WBA Global Bioenergy Statistics*. Retrieved from http://worldbioenergy.org/uploads/WBA%20Global%20Bioenergy%20Statistics%202016.pdf

M.A.D.R. (2016). *Raport de monitorizare a reziduurilor de pesticide*. Retrieved from http://www.madr.ro/reziduuri-de-pesticide-in-plante-si-produse-vegetale/raport-de-monitorizare-a-reziduurilor-de-pesticide.html

Magurran, A. E. (1998). *Ecological Diversity and Its Measurement*. Princeton, NJ: Princeton University Press.

Morgan, R. P. C. (2005). Soil Erosion and Conservation (3rd ed.). Blackwell Publ.

Muntean, L. S., Roman, G. V., Borcean, I., & Axinte, M. (2003). Fitotehnie. Editura Ion Ionescu de la Brad.

Musila, W., Todt, H., Uster, D., & Dalitz, H. (2005). Is Geodiversity Correlated to Biodiversity? A Case Study of the Relationship Between Spatial Heterogeneity of Soil Resources and Tree Diversity in a Western Kenyan Rainforest. In B. A. Huber, B. J. Sinclair, & K. Lampe (Eds.), *H. African Biodiversity*. New York: Springer-Verlag. doi:10.1007/0-387-24320-8_40

Negulescu, M., Vaicum, L., & Pătru, C. (1995). *Protecţia mediului înconjurător*. Bucureşti: Editura Tehnică.

N.I.S. (2012-2016). Retrieved from http://www.insse.ro/cms/

Petrişor, A. I. (2008). Ecologie urbană, dezvoltare spaţială durabilă şi legislaţie. O abordare ecologică a relaţiilor dintre om, spaţiul construit şi mediu. Editura Fundaţiei România de mâine.

Quinton, J. N., Govers, G. V., Oost, K., & Bardgett, R.D. (2010). The impact of agricultural soil erosion on biogeochemical cycling. Nature Geoscience, 3, 311 – 314.

Rolland-May, C. (2000). *Evaluation des territoires. Concepts, modèle, méthodes*. Editura Hermes Science Publications.

României, P. (2004). Legea nr. 58 din 13 iulie 1994 pentru ratificarea Convenţiei privind diversitatea biologică, semnată la Rio de Janeiro la 5 iunie 1992. Monitorul Oficial, 199(1).

Sachs, J. D. (2015). *Age of Sustainable Development*. New York: Columbia University Press. doi:10.7312ach17314

Sârbu, C. N. (2006). *Locuirea în România: o abordare-cadru*. Bucureşti, Romania: Editura Universitară Ion Mincu.

Toy, T. J., Foster, G. R., & Renard, K. G. (2002). *Soil Erosion: Processes, Prediction, Measurement, and Control*. John Wiley and Sons.

Ţuţuianu, O. (2011). *Evaluarea şi raportarea performanţei de mediu. Indicatori de mediu ediţia a III-a*. Bucureşti, Romania: Editura AGIR.

Vasile, E., & Bran, Ş. D. (2017). Lignocellulose bio resources and renewable energy. Internal Auditing & Risk Management, 46(S1), 23-30.

Vătămanu, V. (2012). *Acumularea productiei vegetale si factorii care o conditioneaza*. Retrieved from http://www.agrimedia.ro/articole/acumularea-productiei-vegetale-si-factorii-care-o-conditioneaza

V.I.B. (n.d.). *Effect of genetically modified crops on the environment*. Retrieved from http://www.vib.be/en/news/Documents/vib_fact_genetisch%20gewijzigde%20gewassen_ENG_2016_LR.pdf

Zahiu, L. (1999). *Management agricol*. Editura Economică, Bucureşti.

KEY TERMS AND DEFINITIONS

Agricultural Production: The quantity of plant products.

Biodiversity: Spontaneous and cultivated plant species.

Crops Technologies: The natural succession of agricultural works in accordance with crops and the environment.

Development: Growth, evolution, amplification.

Environmental Factors: Climate, water, soil.

Farms: The place where agricultural products are obtained.

Risk: The probability of manifestation of some uncontrollable phenomena.

Sustainability: Judicious use of resources for future generations.

Chapter 4
The Social, Economic, and Environmental Impact of Ecological Beekeeping in Romania

Cristina Raluca Gh. Popescu
University of Bucharest, Romania & The Bucharest University of Economic Studies, Romania

Gheorghe N. Popescu
The Bucharest University of Economic Studies, Romania

ABSTRACT

The beekeeping activity in Romania benefits from an exceptional melliferous base and favorable climatic conditions, especially for Apis Melifera Carpatica species adapted for millennia to the specific Romanian conditions. Beekeeping is a beneficial activity for beekeepers who either supplement their income, whether it is the main or sole source of income, as well as for agriculture, society, and the environment. After 1989, during the transition period to the market economy, beekeeping experienced a severe decline, but the good organization experience in the Association of Beekeepers in Romania dating from 1958 and the relatively large share of the private sector before 1989 made the transition period much diminished compared to other sectors of the economy, and to experience continuous development. The quantitative and qualitative indicators that have been analyzed highlight the main economic, social, and environmental effects of apiculture practices in Romania.

INTRODUCTION

This chapter focuses mainly on the authors' active vision concerning the social, economic and environmental impact of ecological beekeeping on both the environment and human life, with a particular interest on Romania's situation, making however numerous comparisons with international trends and evolutions in terms of beekeeping business and activity, as well as bee honey consumption. First of all, it should be mentioned that the work is structured as follows: the background section emphasizes the

DOI: 10.4018/978-1-5225-5739-5.ch004

Copyright © 2019, IGI Global. Copying or distributing in print or electronic forms without written permission of IGI Global is prohibited.

role and the importance of beekeeping worldwide as well as the main literature review studies which are related to the authors' filed of interest; in addition, the research contains methods of evaluation and measurement of the economic, environmental and social performances of apiculture in the context of sustainable development, which in turn stress aspects, such as the importance of the harmonious development – a desideratum of the present age, the necessity of presenting and focusing on the indicators of sustainable development for Romania with impact on beekeeping, and also the necessity of discovering and strengthening all the relevant indicators for measuring social, economic and environmental impacts in beekeeping; moreover, the work is keen on analyzing and discovering the aspects related to Romania's apiculture sector, with its present and perspectives, having in mind the Romanian beekeeping – opportunities and threats; furthermore, the study aims to reflect upon the indicators which show the economic impact of beekeeping in Romania, with a direct reference on both the indicators reflecting the environmental impact of apiculture activity in Romania and the synergistically impact on economic, social and environmental of attraction and use of grants; at the end of this work the authors present some important and relevant future research directions as well as numerous important conclusion which come to strengthen the importance of this subject as well as the economic, environmental and social that the beekeeping business and activity, as well as bee honey consumption opportunities and advantages.

The paper has specific objectives that come to emphasize the beekeeping business and activity, as well as bee honey consumption opportunities and advantages and to stress the necessity of understanding and encouraging the beekeeping business and activity, as well as bee honey consumption, in terms of financial, economic, social, demographical, and ecological benefits for both the humanity and environment. In this context, the first objective is to show the historical, archaeological, folkloric data revealed by the works of many authors that come to attest the fact that the Romanian territory has offered favorable conditions for bee breeding since ancient times as well as the millenary continuity of this occupation and also the significant honey resources of the Romanian territory, as well as favorable natural conditions, are essential factors for the development of apiculture regardless of the period. In addition, the second objective is to present the methods of evaluation and measurement of the economic, environmental and social performances of apiculture in the context of sustainable development, which come to stress the importance of harmonious development - a desideratum of the present age, the indicators of sustainable development for Romania with impact on beekeeping, and the relevant indicators for measuring social, economic and environmental impacts in beekeeping. Moreover, the third objective is to show the apiculture from Romania in terms of its present and perspectives, focusing on aspects such as the Romanian beekeeping - opportunities and threats, the indicators which reflect the economic impact of beekeeping in Romania. Furthermore, in terms of the indicators which reflect the economic impact of beekeeping in Romania, the fourth objective of the paper is to discover and present the indicators reflecting the social impact of apiculture in Romania, and the indicators reflecting the environmental impact of apiculture activity in Romania.

In this complex and new context, the study considers several questions as being extremely relevant for the authors' research process, such as:

- When the beekeeping activity was first mentioned and in what way does this particular line of work makes itself remarked at an international level?
- Is the social, economic and environmental impact of the beekeeping activity as well as the honey bee consumption notable and relevant for the human life and for the environment, and are the researchers able to measure with the aid of specific indicators both the impact and the consumption?

- Can the beekeeping activity and business be regarded, on the long term, as a profitable line of work and is the beekeeping activity and business able to bring new social, economic and environmental trends capable to ensure ecological development, sustainable growth and care for the planet's future generations?
- Will the younger generations find themselves capable to see new perspectives and seek new development opportunities in the beekeeping activity and business, and are the younger generations interested in offering the beekeeping activity and business the chance to show its grate potential in terms of the social, economic and environmental impact?

The life of bees is so closely related to people's lives that Albert Einstein estimated that within 4 years of the disappearance of bees on our planet, man will also disappear (Fetea, 2015). It is estimated that the production increases and the qualitative increase by bee pollination exceed at least 20 times the value of the direct beekeeping products obtained through the valorization of specific products of honey, wax, pollen, royal jelly, venom, and many others. Bees capitalize the nectar and pollen of the plants that would lose anyway, the results being found, on the one hand, in apiculture products of great economic and nutritional value, with energetic, revitalizing properties and, on the other hand, in beneficial dietary supplements for human health. Honey is the main product of beekeeping, appreciated for both its nutritional qualities and its therapeutic effects. The level of honey consumption is considered an indicator of the health of the population and a barometer of welfare. In addition, bees have an extremely important role in maintaining ecological balance and perpetuating many species in the plant kingdom. Worldwide, specialized bodies of the World Organization for Agriculture and Food (FAO) have placed the bees on the 4th place in a classification of environmental pollution sensors, since beekeepers are the main pollutants: chemical; industrial and biological noxes; radioactive substances; powders loaded with heavy metals; toxic gases; noise pollution; and so on. With favorable natural conditions, Romania has a long tradition of bee growing and beekeeping products. Beekeeping, as a branch of agricultural production, has since ancient times constituted an activity appreciated by the society. Romania through: the abundance and variety of honey resources from the spontaneous and cultivated flora that provides bees, early spring, until late autumn; the number of bee families it has; the amount of honey obtained; the diversity of bee products; the results of scientific research and specialist training - is among the countries with a well-developed beekeeping.

The world beekeeping market has been marked by the major changes in recent decades. Both changes in the structure of food consumption and developments in the intensity and extent of trade between producers' beekeeping and consumers of honey products contributed to this. Honey with other bee products consumed worldwide developments recorded average consumption of which are different from one country to another, from one area to another, from one period to another. Some countries, such as China or Argentina, are large producers of honey, although they have a low domestic consumption and consequently take advantage of surplus, exporting a significant part of their production. High consumption of bee products shall also be recorded in the USA, but also in European Union countries. Although the average honey consumption in Romania is below the European level, our country is recognized not only by the quantity of honey exported but also by its special quality. Nowadays, honey production is on the rise worldwide due not only to the concern to increase the number of hives, increase average honey production, but also to better know the benefits not only of honey, but also of derived products, used not only in the food industry but also in the pharmaceutical and cosmetic industry, in various forms of therapy.

According to FAO statistics, world honey production has evolved as shown in Table 1.

Table 1. Evolution of honey production worldwide during 1961-2015

Specification	1961	1970	1980	1990	2000	2005	2010	2015
Production of honey (thousand tons)	670	803	975	1,180	1,252	1,414	1,542	1,826

Source: (http://www.fao.org/faostat/en/#data/QL)

These changes, as can be seen from international literature (Gu & Zhang, 2002), are also reflected in the honey production of the top ten producing countries in the world as shown in Table 2, between 1970 and 2000.

BACKGROUND

Numerous historical, archaeological, folkloric data revealed by the works of many authors (Antonescu, 1979; Avetisian, 1978; Hristea, 1943) attests the fact that the Romanian territory has offered favorable conditions for bee breeding since ancient times as well as the millenary continuity of this occupation. The significant honey resources of the Romanian territory, as well as favorable natural conditions, are essential factors for the development of apiculture regardless of the period (Cîrnu & Roman, 1986). Romanian beekeeping in the past 200 years has experienced both momentums, such as the 1870s when the production of honey not only covered domestic consumption but also generated surpluses that could be used for export as well as highlights such as the end of the Second World War, during which time the number of bee families in our country was only 460,000, the average annual production reaching about 3-5 Kg for a family. In socialism, apiculture enjoyed support, and individual beekeepers carried on their activity almost unhindered, even in the difficult period of a super-centralized economy. Romania had 1,418,000 families of bees in 1989 and honey production per bee family was 8.5 kg, according to the Romanian Statistical Yearbook of 1990. A decisive role for the organization and evolution of the Romanian beekeeping activity (Giogia, 2001) was: the reorganization of the Beekeepers Association in 1957,

Table 2. Honey production of top ten producing countries in the world

Rank	1970		1980		1990		2000	
	Country	Yield	Country	Yield	Country	Yield	Country	Yield
1	USSR	210,000	USSR	189,000	USSR	270,000	China	253,691
2	USA	106,499	China	96,300	China	193,000	USA	101,000
3	Mexico	38,984	USA	90,530	USA	84,000	Argentina	91,000
4	China	37,000	Mexico	65,245	Mexico	71,114	Turkey	63,500
5	India	35,000	India	45,000	India	50,500	Mexico	56,844
6	Argentina	25,000	Argentina	37,600	Turkey	40,000	Ukraine	52,000
7	Germany	23,829	Canada	29,235	Argentina	39,000	India	51,000
8	Canada	23,152	Turkey	25,170	Canada	33,000	Russia	50,000
9	Australia	22,258	Australia	2,4954	Germany	32,000	Canada	32,000
10	Turkey	14,889	Germany	14,907	Australia	28,200	Spain	32,000

Source: (FAO Production yearbook, www.fao.org)

which, besides the technical guidance, training and apicultural propaganda activity, was also directed towards the production of economic activities through the Beekeeping Association, which facilitated the endowment and supply of beekeepers and the valorization of the products obtained by them; the establishment in 1971 of the Centre for Beekeeping Studies, Design and Education, subordinated to the Association of Beekeepers Association in Romania; the establishment in 1974 of the Research Institute for Beekeeping, within the Association of Beekeepers Association in Romania.

In 1958, the Beekeepers Association of Romania became a member of the International Federation of Beekeeping Associations - APIMONDIA.

The most resounding event for Romanian beekeepers for that period was the organization in Bucharest, in 1965, of the 20th International Jubilee Congress on Apiculture. Thanks to the international recognition of the results of the beekeeping activity in Romania, Prof. Ph. Veceslav Harnaj, president of the International Federation of Beekeeping Associations - APIMONDIA was elected as President. She held this position for 20 years. In the same time, as a result of the confidence he had in the forces and capacity of the Romanian association, the Romanian government approved, at the request of the International Federation of Beekeeping Associations, the authorization to establish in Romania the International Institute of Apiculture, Technology and Apiculture and the APIMONDIA Publishing House.

In 1965, according to FAO statistics, Romania, with a honey production of 7,718 tons, produced over 1.08% of world honey production and ranks fourth in Europe, located after the USSR, Germany and Spain. In 2015 with a production of 27,893 tons of honey, Romania held over 1.52% of world production, ranking fifth place in Europe after the Russian Federation, Ukraine, Spain and Hungary.

The transition to a post-December 1989 market economy has affected beekeeping in a significant way, but fortunately, the impasse has been overtaken in a short while, covering step by step, the differences.

The apiculture sector (state, about 20 beekeeping and cooperative farms, 15% of the national beekeeping patrimony), similar to all sectors of economic activity, has disorganized, which has resulted in the abolition of many apicultural farms and the reduction and loss of flocks. In addition, the mirage of making immediate profits from other activities has led in many cases small beekeepers to abandon the practice of beekeeping. These are the main causes that have led to a fall in the number of bee families, as shown in Table 3.

The evolution of the beekeeping patrimony of Romania recorded, during 1990-2003, three distinct stages (Chirilă & Patruică, 2005), namely:

- **Stage 1990-1993:** When the apicultural heritage almost halved, with 638 thousand bee families disappearing, especially from the former state and cooperative beekeeping units;
- **Stage 1993-2000:** When more than 166,000 bee families entirely disappeared from the apicultural heritage, entirely from the private sector. Summarized, the total losses in the two stages amounted to 804 thousand bee families;
- **Stage 2000-2003:** When there was an increase in beekeeping stock with 225 thousand families of bees, with an average annual growth rate of 75 thousand families.

In the difficult period of transition, the Association of Beef Breeders in Romania, already reorganized into a territorial structure on counties, incorporating the vast majority of bee breeders from all over the country, played a decisive role. At the level of each county branch a commercial company operates, which takes the apiculture products made by the producers into its own apiaries and ensures the supply

Table 3. Evolution of bee herds and honey production during the transition period

Years	Families of bees (thousands)	Honey production (tons)	Average production for a family (Kg)
1989	1,418	12,124	8.50
1990	1,201	10,579	8.80
1991	1,091	8,279	7.60
1992	1,207	10,410	8.60
1993	780	9,936	12.70
1994	759	9,820	12.90
1995	747	10,434	13.90
1996	696	11,157	16.00
1997	656	10,543	16.10
1998	626	10,199	16.30
1999	620	11,153	18.00
2000	614	11,746	19.10
2001	745	12,598	17.00
2002	781	13,434	17.20
2003	849	19,007	22,40

Source: (Association of Beekeepers)

of the beekeepers with equipment, plants and apiculture materials made by S.C. Beekeeping complex S.A. and S.C. Institute for Development of Beekeeping S.A.

Romania's accession to the grand family of the European Union (EU) was a test of success for the apiculture sector in Romania (European Union (EU): Council Regulation (EC) No. 834/2007, 28 June 2007). The entrepreneurial spirit (Popescu, Popescu, & Popescu, 2014; Popescu, Popescu, & Popescu, 2017) of beekeepers, cultivated and transmitted from generations, has allowed the appropriation and capitalization of the opportunities offered by the Common Agricultural Policy (CAP) of the European Union. EU directly supports the beekeeping sector since 1997 (Regulation (EC), 1997). The successive rules for supporting beekeeping have allowed Member States to develop their own national programs. In this favorable context, they sought to capitalize on the national beekeeping potential for the production and marketing of apiculture products. Programs co-financed by the European Union in a percentage of 50% for successive three-year periods have also allowed the diversification of apiculture products. Besides honey production, they have increased significantly the quantities of royal jelly, pollen, propolis, venom or beeswax. The three-year plans of the EU have imposed a greater transparency on the financing, production and marketing of apiculture products. According to the 2016 Report (COM, 2016) with a production of around 250,000 tons / year, the EU is the second largest producer of honey after China. However, the EU does not produce enough honey to cover its own consumption. In 2015, the percentage of self-supply was around 60%. In this context, through its policies, the European Union imposed an almost continuous increase of its production of honey, with small deviations of the very unfavorable crop years. However, for beekeepers, maintaining this level of production can not only become much more difficult, but even impossible. The explanation is found in the new challenges that the beekeepers have to face, both in terms of the health of bees and the loss of their habitat due to agricultural intensification, which does not always comply with environmental standards. The immediate consequences are found primarily in higher production costs, which favor imports of cheaper honey from third countries, but often much weaker in terms of quality. This process inevitably leads to increased competition, otherwise positively, if it is not detrimental to the quality of apiculture products.

In Romania the beekeepers acceded to the quality requirements imposed by the European Union producing quality organic honey. Regarding the quality of honey, the European Commission adopted a series of measures to control severe content of antibiotics, pesticide, insecticide and fungicides, heavy metals, residues of veterinary medicines and generally full range of residues which can really put in danger the consumer health (Bura, 2010). With the 2013 CAP reform, the legal basis for supporting beekeeping has improved. The eligible measures have been adapted to the changing needs of the sector and the methods of allocating EU funds to the Member States, based on the number of hives, have been optimized. Out of the eight eligible beekeeping measures of the reformed Common Agricultural Policy, five of them directly or indirectly fund the quality and the attestation of quality in the production and marketing of apiculture products. These programs concern: technical assistance to beekeepers and beekeepers' organizations; combat aggressors and hive-related diseases, especially varroosis by ecological methods; measures to assist beekeepers' analysis laboratories to help beekeepers to harness and to market their products; extensive measures to include analysis of other bee products, such as royal jelly, pollen, propolis or beeswax ; cooperation with specialized bodies for the implementation of applied research programs in the beekeeping and bee products for conservation and the best use of bee products; improving the quality of products in order to better market their products. A segment currently insufficiently exploited in the European market is the production and marketing of organic honey. "Eco-friendly" honey must comply with Regulation (EC) no. 834/2007. In accordance with the requirements of this Regulation, organic honey must be obtained strictly under environmental conditions, both from the point of view of the environment and production. The requirements of that Regulation provide for the following rules: bees must be treated exclusively with veterinary medicinal products containing only approved organic substances; hives must be placed exclusively in a clean area where, within a radius of 3 km, there is no contamination with chemicals from industrial complexes, airports or high-traffic roads; hives must be built only from natural materials; the chemical treatment of agricultural holdings in the area is strictly forbidden; the artificial feed provided to bees must be certified as ecologically organic.

Some unethical trade mechanisms, including honey washing or re-labeling, have imposed new labeling rules on honey marketed in the European Union. For more certainty regarding compliance with the European Commission Regulation, European buyers set up a working group within the International Federation of Apiculture Associations (Apimondia, 2016) to monitor and prevent unfair trade.

It should be noted that apiculture, as an economic sub-branch of agriculture (Bradbear, 2009), represents on one hand a very practical and extremely useful occupation for the rural population, and on the other hand a form of increasing people's income or even of generating a constant profit if transform in some sort of business (Hall, & Vredenburg, 2003; Hall, Daneke, & Lenox, 2010).

Moreover, in this particular context, the strategy of combining several types of interests, such as the economic and environmental ones, prove to be extremely important (Cohen, & Winn, 2007): first of all, by improving beekeepers' strategies by helping them to find new ways to sell their healthy products on different markets their income will grow (Dean, & McMullen, 2007), and, second of all, by encouraging this type of entrepreneurial behavior sustainable (Gibb, 1996) and ecological development will be generated on the long run (Brugmann, & Prahalad, 2007).

Furthermore, weather the focus is on bee honey production in Romania (Antonie, 2016; International Federation of Beekeepers' Associations (Apimondia): Apimondia, 2016) or somewhere else in the world (Ene, 2015), in order to develop an entrepreneurial behavior in the beekeeping sector by taking into consideration elements such as international prosperity, sustainable development and ecological growth by meeting human development goals, the managerial aspects should be taken into account as well (Regulamentul (CE) nr. 1221/97 al Consiliului, 1997). In this context, sustainable development, regarded as a universal concept, which has currently become the primary goal of all individuals worldwide as well as the fundamental issue on the agenda of all governments on the planet, implicates at its managerial level the following key aspects which are interrelated: first of all, sustainable development relies today on the countries' economic development, and second of all, the process of economic development can be considered as an advantage for the human race in terms of sustainable development only if it implicates also social development, emphasized in its turn by the constant concern for environmental protection, cultural development, planet's preservation and biodiversity, national, regional and international good quality in terms of security development and assurance (Raport COM(2016) 776 al Comisiei către Parlamentul European şi Consiliu cu privire la punerea în aplicare a măsurilor privind sectorul apicol prevăzute de Regulamentul (UE) nr. 1308/2013, 2013).

However, the struggle for sustainable development needs to be seen far more than the simple transition process from the industrial society to the new economy, focused on cultural and ecological development as well as on information as part of the knowledge-based society. The managerial implications generated by the struggle for sustainable development refer to finding solutions in order to eliminate financial, economic, social and demographic crisis, poverty, pollution, migration and work-uncertainty, with a clear aim of returning to traditional values, which in this particular context are related to apiculture, bee honey production, beekeepers' strategies.

METHODS OF EVALUATION AND MEASUREMENT OF THE ECONOMIC, ENVIRONMENTAL AND SOCIAL PERFORMANCES OF APICULTURE IN THE CONTEXT OF SUSTAINABLE DEVELOPMENT

Harmonious Development: A Desideratum of the Present Age

About the need for harmonious development there have been many concerns over the last decades. These were in various forms and were meant to shoot more or less alarmed signals. Thus: In 1962 in the "Silent Spring" work, the author Rachel Carson, biologist, describes, in an imaginary way, a possible ecological catastrophe of massive proportions. (Carson, 1962)

The "Limits to Growth" report, published by the Club of Rome in 1972, predicts an apocalyptic decline as a result of the depletion of non-renewable resources; the first non-governmental ecological organizations such as the Earth Friends (1971) and Greenpeace (1971) appear to be involved in the fight for environmental protection; There is a visible similarity between the concept of Eco development, a term widely used during and especially after the 1972 United Nations Conference on Environment held in Stockholm and those of Sustainable Development; Although the United Nations Environment Program in 1975, which debated the report - What to do - another way of development - had a weak echo, The United Nations has given a mission to the World Commission on Environment and Development, led by Mrs. Harlem Brundtland, to conceives a "global change program". In 1987 was discussed and published the Brundtland Report, "Our Common Future", during which the concept of "zero growth" was replaced with "sustainable development". Although there are currently more than 1,000 definitions of this concept, the one that enjoys the widest international acceptance is the one in the Brundtland Report, which defines sustainable development as the development that ... *meets the needs of the present without compromising the ability of the next generation to meet their own needs*. Sustainable development includes at least two important ideas: Development has an economic dimension, a social dimension and an environmental dimension; the current generation has an obligation for the next generation to leave enough "stocks" of economic, social and environmental resources to enjoy levels of welfare at least as high as they are today.

Development will be sustainable only if there is a balance between these factors, which contribute to the overall quality of life, in the sense of changing the quality of economic growth in order to make it more bearable for both man and the environment.

In the Tokyo Final Statement of the United Nations Commission on the Environment and Development of 27 February 1987, all the nations of the world were invited to integrate sustainable development into their objectives by: revitalizing economic growth; conserving and improving the resource base; ensuring a sustainable population level; strengthening international cooperation. After 1987, progress in the development of sustainable development concepts was visible, continuous and rapid.

Indicators of Sustainable Development for Romania With Impact on Beekeeping

Defined, available and capitalized to assess the achievement of the targets set by the National Sustainable Development Strategy for the years 2013, 2020, 2030, the 103 Sustainable Development Indicators are structured on the Strategy's objectives and are hosted in: 19 level 1 indicators, as indicators main (basic); 37 Level 2 indicators, as complementary indicators for monitoring and reviewing sustainable development programs and 47 level 3 indicators, as progress indicators.

Among them, we have presented and analyzed those whom we considered to be able to directly or indirectly influence apicultural act in general and ecological beekeeping in particular. The dynamics of these indicators is presented in Table 4.

A brief presentation of the impact of these indicators on beekeeping activity:

Table 4. Indicators of sustainable development of Romania with impact on beekeeping activity

	2004	2005	2006	2007	2008	2009	2010	2011	2012	2013	2014
O1-1	9.00	4.80	8.70	8.50	10.30	-6.30	-0.20	1.60	1.10	3.90	3.50
O2-1	132,803.51	129,128.11	130,839.92	130,322.44	125,297.14	107,085.23	100,559.05	105,358.64	104,611.40	93,604.31	93,249.00
O3-9		4.29	4.23	3.79	4.07	3.95	3.73	3.87	4.20	3.78	3.50
O4-10			0.80	1.00	1.00	1.20	1.40	1.70	2.20	2.30	2.20
O4-14		4.29	4.23	3.79	4.07	3.95	3.73	3.87	4.20	3.78	3.50
O5-3	4.38	4.40	4.46	4.51	4.54	4.59	4.69	4.77	4.78	4.82	4.82
O6-10				34.50	31.30	34.90	31.50	28.00	27.10	26.50	

Source: (National Institute of Statistics on-line: http://www.insse.ro/cms/ro/content/indicatori-de-dezvoltare-durabila -22.09.2017)

- **O1_1. GDP Growth Rate per Capita:** Growth may favor honey consumption;
- **O2_1. Total Greenhouse Gas Emissions:** Growth can favor unfavorable climate change;
- **O3_9. Emissions of Suspended Particulate Matter from Shipment:** Growth is not favorable to bees;
- **O4_10. Share of the Area Grown Ecologically in the Agricultural Area Used:** Growth can help increase organic honey production;
- **O4_14. Total Emissions of Particulate Matter:** Growth is not favorable to bees;
- **O5_3. Surface of Artificial Space as% of Total Area:** Growth is not favorable to bees;
- **O6_10. Proportion of the Population that Believes that Households Suffer from Noise:** Growth is not favorable to bees.

Analyzing the dynamics of these indicators over the last decade, we find that their evolution generally favors the growth of bees, with the exception of the unfavorable increase in the share of artificial space (O5-3 indicator).

Relevant Indicators for Measuring Social, Economic, and Environmental Impacts in Beekeeping

In order to assess the social economic and environmental impacts, it is necessary to use a system of characteristic indicators. Defining and using this indicator system is not an easy operation, as it is a complex and lasting process involving efforts from several interested parties: businesses or civil society organizations, expert groups, research centers, international financial institutions, etc.

Although at the 1992 World Summit of the Earth Summit in Rio de Janeiro, it has come to the conclusion that the development of unitary means of measuring the level of sustainable development is vital to the success of its implementation. The most difficult issue was the establishment of a set of indicators of sustainable development, with general applicability.

A meeting of experts was set up in Bellagio, Italy in November 1996, in order to develop and solve this problem, which set out a set of principles on the selection, integration and interpretation of sustainable development indicators. These principles, guidelines, which apply uniformly, and represent a pragmatic expression of the most important features of the concept of sustainable development, refer to: Vision and guiding objectives; Holistic approach; Essential elements; Pragmatic orientation; Efficient communication; Continuous evaluation.

The purpose of these indicators is that through representativeness and replicability it allows to monitor and evaluate the different aspects of sustainable (ecological, social and economic) development and to provide concrete information from which economic policies can be used to self-check and correct.

In connection with the above principles to identify specific and relevant indicators that reflect the social, economic and environmental impact of bee activity in Romania we study several sources. The most important ones are:

- **Romania's Sustainable Development Indicators:** Managed by the National Institute of Statistics: we did not identify any indicators containing analytical data on the economic, social or environmental impact of beekeeping activity and its results; but many indicators have been identified that can dynamically influence them for good or bad beekeeping activity; they were previously presented in the material;

- **In the Literature (Daly, 1990; Rennings & Wiggering, 1997):** It is appreciated that the rules linking economic and environmental concepts in the context of sustainable development refer to: the rate of exploitation of renewable resources should be equal to that of their regeneration; the waste generation rate must not exceed the absorption capacity of the ecosystem in which the waste is stored; non-renewable resources must be exploited in a sustainable way, i.e. their consumption rate must not exceed their substitution rate through renewable resources. And the social dimension of sustainable development requires a fair distribution of opportunities between generations, which can be reflected both by GDP and employment, as primary macroeconomic indicators but also by the population health status index;
- **From the Performance Report List, Developed by the Global Reporting Initiative (GRI) in the G4 Sustainability Reporting Guidelines:** (https://www.globalreporting.org/information/g4/Pages/default.aspx; https://www.globalreporting.org/standards), which although has a somewhat different approach to corporate social responsibility, these issues are reflected by economic, legal, ethical and philanthropic indicators, relevant indicators can be identified for the economic, social and environmental impacts, even if they are presented as indicators reflecting legal, ethical or philanthropic aspects according to GRI classification.

In line with the above, we have identified as relevant for reflecting the economic, social and environmental impacts the following indicators:

1. Economic impact, reflected by:
    a. Generated and distributed direct economic value, including revenue;
    b. Significant financial assistance received from the Government;
2. The social impact, reflected by:
    a. Education, counseling training;
    b. Addressing public policies and participating in their development (lobbying);
    c. Lifecycle stages in which they are being evaluated to improve the impact of products and services on the health and safety of citizens;
    d. Significant financial assistance received from the Government;
3. Environmental impact, reflected by:
    a. Land owned, leased or administered;
    b. Describe the significant impact of activities, products or services on protected areas or other biodiversity rich areas;
    c. Programs for adhering to laws, standards and voluntary codes;
    d. Significant financial assistance received from the Government;

These indicators will be presented and further analyzed.

APICULTURE FROM ROMANIA: PREZENT AND PERSPECTIVES

Romanian Beekeeping: Opportunities and Threats

Beekeeping as a branch of animal husbandry in terms of absolute value has small compared to other sub-branches of livestock production in the national economy. However, the economic importance of bee products is much higher if we consider increasing their value through industrialization, use as a raw material in the manufacture of many food products, medicines, energetic vitalizing, cosmetics, and many others. Let's not forget that one-third of the food we eat is obtained from bee pollinating. It's about vegetables, fruits, field crops, with which we feed ourselves (Mateescu, 2015).

Since apiculture is far from reaching the maximum level of development, we will continue to analyze the types of beekeeping holdings and then make a SWOT assessment to identify sustainable development mechanisms.

If the small beekeepers who practice a degree of pleasure as mentioned in the literature (Bura & Patruică, 2005), the types of apicultural holdings and their characteristics in Romania are presented in Table 5.

The SWOT analysis method (Strengths, Weaknesses, Opportunities and Threats) is designed to determine the current situation of beekeeping in Romania. The aim is to identify the strengths and disadvantages of Romanian beekeeping, in order to determine the actions that should be taken and the measures to support the development that can be taken, taking advantage of the opportunities.

The results of the SWOT analysis are presented below.

- **Strong Points:**
    - Variety of natural conditions, with relief, which is part of three different categories: plain, hill and mountain, with micro-zones specific to the Danube Delta and meadows;
    - Substantial, insufficiently capitalized apical potential;
    - The existence of a long tradition;
    - Attracting young people to beekeeping;
    - Increase honey production on the bee family and reduce costs per unit of product;

Table 5. Types of beekeeping holdings in Romania

The type of the apiary	Small	Middle	Large
Category of beekeeper	Amateur	Semi-professional	Professional
Type of holding	Family management.	Farm.	Farm or commercial company.
Families of bees in maintenance	Minimum 50.	51-150 in Romania.	Over 600.
How to ensure reproductive material	Through own growth or acquisition activities.	By purchase from specialized units.	By purchase from specialized units.
Workforce	Its own.	Its own / Employed.	Employed.
Hours of work	Usually on weekends.	Under 2000.	Between 2000-2400.
The purpose of production	Completing the family budget, occasional spare time.	Profit, services (pollination).	Subsistence, services, profit.
The legal regime	Physical person.	Natural or legal person.	Legal person.

Source: (Processing after Bura & Patruică, 2005)

- Switching from amateur to professional apiculture;
- Combining tradition with elements of modernity.

- **Weaknesses:**
  - Existence of a high percentage of the aging population;
  - Reducing the employment of the population and shifting the population to other countries, especially to get jobs;
  - Using an insufficient number of modern technologies both in terms of increased production and recovery of by-products;
  - Generating a low-value production in general through the disappearance of honey-producing factories, which is exported in a natural state even to large honey producing countries that mix it with their own to increase quality;
  - The low economic power of the rural population, which also has low mobility and low productivity of work;
  - Dependence on manual labor and the whims of weather.
- **Opportunities:**
  - Access to the internal market of the European Union;
  - Increasing the use of computer communication technologies (ICT);
  - Transfer of technology and know-how;
  - Developing entrepreneurship;
  - Orientation of educational programs, training and training to specializations in the fields of beekeeping, especially in terms of preparing mature persons;
  - Developing civil society and civic spirit to protect the environment;
  - The development of organic farming, based on the existence of favorable land in our country, for which no chemical fertilizers were generally used;
  - Regeneration of rural communities;
  - Facilities for agricultural associations;
  - Increasing the number of environmental protection investments;
  - High potential for tourism development in the conditions of increased demand in the field of mountain, ecological, cultural and balneoclimat tourism;
  - Increasing interest in the concept of sustainable development that assures development at all three levels: economic, social and environmental;
  - Accessing non-reimbursable funds;
  - Apply new methods of growing and maintaining bee families;
  - Rational widespread practice of pastoral graduation;
  - Genetic improvement of Romanian bees;
  - Intensification and diversification of production;
  - Ensuring the quality of production and harmonization of the local legislation with the
  - European Union regulations;
  - Acceleration of the process of concentrating bee-breeding activities and capitalizing on apiculture products.
- **Threats:**
  - Number of population is decreasing;
  - Migration of young people to urban areas;

- Increasing the imbalance between supply and demand on the labor market and the unequal distribution of income by category and gender;
- Insufficient use of existing potential;
- Lack of entrepreneurial culture;
- Lack of capital to support investments in the economy;
- Slowing the development and diversification of the rural economy;
- The low process of modernization and restructuring of agriculture;
- Fragmentation of agricultural holdings;
- Import of unsuitable bees for the conditions in our country and the loss or weakening of the Romanian bee;
- The emergence of new diseases and pests of bees brought from other areas;
- The import of surrogate surplus honey, which is incorrectly labeled, which is used by hypermarkets.

As it emerges from the SWOT analysis, the melliferous base of Romania is unique and as such the growth and developmental strengths of the beekeeping activity will allow its sustainable development in the future.

Moreover, by being able to manage the sustainable development impact approach of beekeeping in Romania, certain elements will prove to be of extreme necessity in order to ensure this type of business' success on the long run, especially because this type of business should be regarded from this point on as extremely innovative, timely and with a simple and practical style:

- First of all, the sustainable development impact approach of beekeeping in Romania should be regarded and treated as an integrated management approach, which can be used across a wide range of sectors and domains, in a variety of contexts, such as in the medical field, food industry, and biomass technology production;
- Secondly, the sustainable development impact approach of beekeeping in Romania should be regarded and treated as a results-oriented management approach due to the fact that it addresses most pressing concerns such as pollution, poverty, and resources' scarcity;
- Thirdly, the sustainable development impact approach of beekeeping in Romania should be regarded and treated as a possible mean to contribute towards the sustainable development goals due to the fact that it has a people-oriented approach which is capable to integrate and interconnect specific managerial processes such as planning, strategizing, monitoring, and evaluating.

INDICATORS WHICH REFLECTS THE ECONOMIC IMPACT OF BEEKEEPING IN ROMANIA

As mentioned above, the main indicators reflecting the economic impact of beekeeping activity are:

- **MexT: Honey Extracted in Tons:** The dynamics of this indicator is reflected in Table 6.
- **VpaM: Income Obtained by Beekeepers at Purchase Price Thousand Lei:** The dynamics of this indicator is reflected in Table 7.

Table 6. Honey extracted in tons

Years	2003	2004	2005	2006	2007	2008	2009	2010	2011	2012	2013
MexT	17,409	19,150	17,704	12,576	15,106	16,250	15,202	15,184	11,593	13,443	14,861

Source: Processing by the National Institute of Statistics: Statistical Yearbook of Romania 2009 and 2014

Table 7. Income from the sale of honey (thousand)

Years	2003	2004	2005	2006	2007	2008	2009	2010	2011	2012	2013
VpaM	130.219,3	128.496,5	72.586,4	54.076,8	68.732,3	99.937,5	117.359,4	133.467,4	115.814,10	134.967,7	170.307,1

Source: Processing by the National Institute of Statistics: Statistical Yearbook of Romania 2009 and 2014

INDICATORS REFLECTING THE SOCIAL IMPACT OF APICULTURE IN ROMANIA

The main indicators reflecting the social impact of apiculture activity are:

- **AlMf: Bees (Thou Families):** The dynamics of this indicator is reflected in Table 8.
- **Education, Training:** Beekeepers who are part of professional associations have beekeepers certificate and participate in continuous professional training.
- **Approaching Public Policies and Participation in Their Development (Lobbying):** Out of about 40,000 beekeepers over 50% are registered with the Association of Romanian Beef Breeders (ACA) founded in 1958, they own over 75% of the bee families. The other members belong to several hundreds of beekeepers' associations; this division makes it difficult to define coherent policies for the common interest of beekeepers, as decisions are not made according to the number of bee-keepers or bee families (Fetea, 2015).
- **Life Cycle Stages Where they are Evaluated to Improve the Impact of Products and Services on the Health and Safety of Citizens:** Products derived from honey but also from propolis by-products, bee venom, pollen, royal jelly, and many others, carried out by the Institute for Research and Development for Beekeeping have the necessary attestations obtained, as the case may be, from the National Medicines Agency or the National Technical Committee for Medicinal, Aromatic and Stupin Products (Mateescu, 2015).

Indicators Reflecting the Environmental Impact of Apiculture Activity in Romania

- **EaHa: Agricultural Holdings in Hectares**: The dynamics of this indicator is reflected in Table 9.

Table 8. Bees (thousands of families)

Years	2003	2004	2005	2006	2007	2008	2009	2010	2011	2012	2013
AlMf	840	888	888	891	892	998	1,057	1,275	1,250	1,254	1,354

Source: Processing by the National Institute of Statistics: Statistical Yearbook of Romania 2009 and 2014

Table 9. Size class of agricultural area in use (hectares)

Size class of agricultural area in use (hectares)	Bees		
	2005	2007	2010
under 0.1	9,368	4,547	6,296
0.1-0.3	4,081	3,566	5,452
0.3-0.5	1,850	1,328	2,605
0.5-1	5,267	3,840	4,420
1-2	7,157	6,985	6,456
2-5	14,208	10,809	10,677
5-10	7,081	6,036	4,640
10-20	2,504	1,804	1,490
20-30	374	348	315
30-50	224	287	228
50-100	189	108	156
over 100	94	82	129

Source: Processing by the National Institute of Statistics: Statistical Yearbook of Romania 2009 and 2014

- **Adherence to Laws, Voluntary Standards and Codes:** The vast majority of beekeepers voluntarily accepted the quality control of honey delivered to the association. They have also aligned themselves with the voluntary requirements of EU environmental standards in the field. Romanian honey enjoys a high appreciation in the EU.
- **The Description of the Significant Impact of Activities, Products or Services on Protected Areas or Other Areas Rich in Biodiversity:** Bees by pollination contribute to the maintenance and development of biodiversity. Nectar that is picking bees would be lost. So the bees do not affect the environment in which they act.
- **Education, Training:** Beekeepers who are part of professional associations have beekeepers certificate and participate in continuous professional training.

The Synergistically Impact on Economic, Social and Environmental of Attraction and Use of Grants

Beekeepers in their professional associations have attracted and capitalized over 99% of European funds allocated from their grant to date. They have been used according to the eligibility requirements for: increasing the number of bee families, verifying and improving quality; procurement of equipment.

The social, economic and environmental impact of ecological beekeeping in Romania represents an innovative as well as a timely activity, which is seen both as a people-centered approach due to the fact that it has a profound impact in the sustainable development era and as a simple and practical business due to the fact that it has the potential to attract the younger generation support and involvement. In addition, only by being able to understand the social, economic and environmental impact of ecological beekeeping in Romania, the sustainable development practitioners will be helped to address the complexity of planning, strategizing, monitoring, and evaluating the development initiatives that this business requires. Moreover, the synergistically impact on economic, social and environmental of attraction and use of grants has the power to enrich the management process as well as the management process approach in order to increase in time the importance of this dynamic process. Furthermore, the synergistically impact on economic, social and environmental of attraction and use of grants has the power to guide and improve practice as well as build learning tools and strategies for the near future, which could prove to be the basic steps to manage the need for the sustainable impact in a correct manner.

However, the synergistically impact on economic, social and environmental of attraction and use of grants should inspire every person interested in how to manage a business in a practical, useful and effective way to think and work differently, by taking into consideration key aspects, such as: good and efficient management mainly oriented on the development impact and on the effective practice; independent international business based on ecological principals, focused on attracting wise investments, capable to generate effective practice flows which are founded on a solid ground of constructive use of information and knowledge; adaptability from understanding and engaging in complex line of businesses in more dynamic and competitive international economic, financial and social systems; integration elements, which are part of a comprehensive framework, having a complex strategic design, delivering not only profit but also respect and genuine concern to nature itself and people worldwide.

FUTURE RESEARCH DIRECTIONS

By analyzing the distribution of beekeepers and bee families on the eight development areas of the country, it is possible to identify an inverse correlation between employment and apiculture development in those areas. Beekeeping can be a real lifeline for those who cannot find a job. Although specialists estimate that beekeeping is only profitable after 3-5 years, it does not involve very large investments, and who is attracted to this activity can learn relatively quickly. In addition, there is a very good training infrastructure and the lack of workforce allows for an effective apprentice. It is also preferable that the proportion of bee families and beekeepers be lower in highly industrialized and polluted areas.

But, perhaps the most important aspect is represented by capitalizing on new sustainable energy resources, even finding new energy resources. We can think of fuel obtained in the form of peat, wood or non woody biomass, water from rivers, lakes, and groundwater (Xiangzheng, Zhihui, Jikun, Qingling, & Yanfei, 2013). Biofuels must be in the future very useful for food, energy, and, generally speaking for entire environment. This is a pressing necessity for a global review, influencing subsequently, of the impact of so called "Land Use Changes" on a series of biodiversity and ecosystem services, that will help avoid environmental harm, maintaining with a lot of care the biodiversity, referring especially to the impacts of the wellbeing human, major changes who alter already the entire ecosystem by his effects of negative effects in climate, human population, land use, water use, and who must be managed with a lot of attention by researches. It is why they have to create and implemented a set of thorough based on normative method, in order to estimate ecosystem services values and improve the accuracy of assessment results. On the other hand they need to begin an in-depth process-based analysis of the relationship between human activities and ecosystem services function. And, in the same time, they have to consider as an urgent requirement the promoted and the application of ecosystem services values in various aspects of production, livelihood, and government decision-making all of these measures serving eventually the human wellbeing (Euliss, Jr., Smith, Liu, Feng, Mushet, Auch, & Loveland, 2010; Jordan & Warner, 2010).

In addition, in terms of future research directions, other numerous elements need to be taken into account and should be mentioned in the lines below: first of all, one focus could be on the entrepreneurial methods concerning beekeeping activity and bee honey production business that should be analyzed in terms of profit growing by focusing on today's the new trends based on the increasing compliance with the requirements of ecology, sustainable growth, and sustainable development by enhancing with priority bio-diversity and bio-productivity; second of all, the beekeeping activity and business could be analyzed by focusing on its natural resources, means and current solutions able to generate economic efficiency and social equality in turbulent times such as the ones that humanity are facing nowadays; third of all, the beekeeping activity and bee honey production business could be analyzed by focusing on facilitating individuals' access to beekeeping education in terms of becoming more and more aware of the benefits of this line of work as well as of this type of business, by attracting in this way people interested in investing in this type of process which can offer, in time, effective means capable to generate money in a sustainable way, focusing on the young generation's participation as well as on the social, ecological and cultural aspects of life (Popescu, 2017).

Therefore, the specialists from all fields of activity should join their forces in order to further analyze, develop and improve the research on the social, economic and environmental impact of ecological beekeeping in Romania as well as at an international level. As a general idea, it should be mentioned that by emphasizing the social, economic and environmental impact of ecological beekeeping in Romania as well as at an international level in other works, new trends and new ways of managing the sustainable development impact approaches will be born as well as new and improved integrated, results-oriented approaches will emerge. So, it should clearly be stressed the fact that the research on the social, economic and environmental impact of ecological beekeeping in Romania as well as at an international level should represent the specialists' collaborative and participatory engagement to useful means of generating new information and knowledge capable to bring the latest research and insides concerning the subject of sustainable and ecological growth and development, focusing on bringing benefits to both people and nature as a whole.

CONCLUSION

The beekeeping activity in Romania, which has continued for millennia, has experienced both moments of glory when it was able to satisfy its internal requirements but also to capitalize on its export products, as well as the turning points such as those at the end of the Second World War as well as the end of the current transition. There have always been the resources to overcome these more difficult moments.

Being recognized as a safe and healthy food, having a very long history of consumption worldwide, due to its excellent conservation quality, honey can certainly be considered as a regular and necessary food for families, as well as a household drug. In addition, a real advantage is that beekeepers do not require their own land, so even a landless farmer can practice it, thus benefiting from its multiple properties and uses. Since apiculture exploits the available resources of any area cultivated a few kilometers of hives, the most important aspect for a beekeeper is to have access to a space large enough to place its own hives. Even in ours days, both traditional and modern approaches to beekeeping are used everywhere is possible in our country. Being crucial to the conservation and sustainability of the ecosystem, beekeeping offers natural pollination, future food security, high income generation, medicinal products, and many valuable new research opportunities. Due to pollination by bees, the increase and improvement of the quality of Romanian agricultural products, as well as the essential effects generated by the pollination process in ecosystems, are constantly increased, aiming not only at preserving them but also searching for new ways of expanding them. Honey is a product easy to sell as long as it has a very good and stable reputation along the years. If its reputation become doubtful due to an irresponsible way of being produced or even preserved, selling and capitalizing on them will generate great difficulties, which will be very hard to be hard to overcome by their producers and even harder to forget in time by their buyers. In order to avoid such possibility, it is very important that the participant farmers to be informed and educated to be aware such that the signification and the importance of their product's quality.

Today the Romanian beekeeping by the quality of the honey and the by-products obtained has a good reputation in Europe and the world and is fully adapted to market requirements and conditions. The activity of bees and beekeepers is beneficial to agriculture, the environment and society. The economic, social and environmental impact was highlighted by the set of financial and non-financial indicators that were presented and analyzed. The socio-economic restructuring known by Romania over the past 27 years has created particularly favorable conditions for the development of ecological beekeeping, unfortunately not enough exploited on the European Union market.

Moreover, the beekeeping activity and bee honey production business represent the key to develop new socio-economic strategies, with the aid of effective consumption and effective means of transportation, which could lead to saving people from the following problems: first of all, hunger and poverty due to the process of implementing undeniable methods to increase the importance of this business especially with the young generation's participation; second of all, certain illnesses due to the undeniable properties of honey discovered a long time ago by specialists in the medical field, which strongly recommend the use of honey for the health services; third of all, pollution due to the undeniable methods of honey production which first and foremost do not harm the environment and second support natural improvement, biotechnology, sustainable growth and development.

REFERENCES

Antonescu, C. (1979). *Albinele şi noi.* Bucureşti, România: Redacţia publicaţiilor apicole.

Antonie, I. (2016). Honey resources of Avrig City (Sibiu county) and economic relevance. *Scientific Papers. Series Management, Economic, Engineering in Agriculture and Rural Development, 16*(4), 37–41.

Avetisian, G. A. (1978). *Apicultura.* Bucureşti, România: Editura Apimondia.

Avetisian, G. A. (1978). *Apicultura.* Bucureşti, România: Editura Apimondia.

Bradbear, N. (2009). *Bees and their role in forest livelihoods.* Rome: Food and Agriculture Organization of the United Nations.

Brugmann, J., & Prahalad, C. (2007). Cocreating business's new social compact. *Harvard Business Review, 85*(2), 80–90. PMID:17345682

Bura, M. (2010). Pe piaţa europeană se poate comercializa doar miere de calitate, *Revista Ferma,* 5. Retrieved from https://www.revista-ferma.ro/articole/apicultura/pe-piata-europeana-se-poate-comercializa-doar-miere-de-calitate

Bura, M., & Patruică, S. (2005). *Tehnologie apicolă*. Timişoara, România: Editura Solness.

Carson, R. (1962). Silent Spring. *The New Yorker Magazine*.

Chirilă, A., & Patruică, S. (2004). *Tehnologii apicole moderne. Stupăritul pastoral*. Bucureşti, România: Agenţia Naţională de Consultanţă Agricolă.

Cîrnu, L., & Roman, Gh. (1986). *Din viaţa albinelor*. Bucureşti, România: Editura Ceres.

Cohen, B., & Winn, M. I. (2007). Market imperfections, opportunity and sustainable entrepreneurship. *Journal of Business Venturing*, *22*(1), 29–49. doi:10.1016/j.jbusvent.2004.12.001

Daly, H. E. (1990). *Toward Some Operational Principles of Sustainable Development* (Vol. 2). Ecological Economics.

Dean, T. J., & McMullen, J. S. (2007). Toward a theory of sustainable entrepreneurship: Reducing environmental degradation through entrepreneurial action. *Journal of Business Venturing*, *22*(1), 50–76. doi:10.1016/j.jbusvent.2005.09.003

Ene, A. E. (2015). *EU Funds for beekeepers*. Retrieved December 28, 2017, http://www.informativ.info/economic/237-fonduri-nerambursabile-pentru-apicultori

Euliss, N. H. Jr, Smith, L. M., Liu, S., Feng, M., Mushet, D. M., Auch, R. F., & Loveland, T. R. (2010). The need for simultaneous evaluation of ecosystem services and land use change. *Environmental Science & Technology*, *44*(20), 7761–7763. doi:10.1021/es102761c PMID:20809588

European Union (EU). Council Regulation (EC) No. 834/2007 of 28 June 2007 on organic production and labeling of organic products and repealing Regulation (EEC) No 2092/91. (2007). Retrieved December 28, 2017, from: http://eur-lex.europa.eu/oj/direct-access.html

Fetea, I. (2015). România, un rai al albinelor. *România rurală*, *19*(3), 10-12.

Gibb, A. A. (1996). Entrepreneurship and small business management: Can we afford to neglect them in the twenty-first century business school? *British Journal of Management*, *7*(4), 309–322. doi:10.1111/j.1467-8551.1996.tb00121.x

Giogia, M. (in press). Adevărate nestemate în coroana regală a timpului. *Ziarul Timpul, 34*. Retrieved August 10, 2017, from http://www.fiitea.org/foundation/files/martag/34Rmg.pdf

Gu, G., & Zhang, Ch. (2002). Analysis on the Structure of Honey Production and Trade in the World. *Apiacta*, *2*, 2002. Retrieved from http://www.fiitea.org/cgi-bin/index.cgi

Hall, J., & Vredenburg, H. (2003). The challenges of innovating for sustainable development. *Sloan Management Review*, *45*(1), 61–69.

Hall, J. K., Daneke, G. A., & Lenox, M. J. (2010). Sustainable development and entrepreneurship: Past contributions and future directions. *Journal of Business Venturing*, *25*(5), 439–448. doi:10.1016/j.jbusvent.2010.01.002

Hristea, C. L. (1943). *Stupăritul*. Bucureşti, România: Cartea Românească.

International Federation of Beekeepers' Associations (Apimondia). Apimondia. (2016). Retrieved December 28, 2017, from: https://www.apimondia.com/en

Jordan, N., & Warner, K. D. (2010). Enhancing the multifunctionality of US Agriculture. *Bioscience, 60*(1), 60–66. doi:10.1525/bio.2010.60.1.10

Mateescu, C. (2015). Cercetarea, baza dezvoltării apiculturii româneşti. *România rurală, 19*(3), 13-23.

National Institute of Statistics. (1990). *Romanian Statistical Yearbook*. Bucharest, Romania: Author.

National Institute of Statistics. (2009). *Romanian Statistical Yearbook*. Bucharest, Romania: Author.

National Institute of Statistics. (2014). *Romanian Statistical Yearbook*. Bucharest, Romania: Author.

Popescu, C. R. (2017). The Role of Total Quality Management in Developing the Concept of Social Responsibility to Protect Public Interest in Associations of Liberal Professions. *Amfiteatru Economic, 19*(Special No. 11), 1091-1106. Retrieved from http://www.amfiteatrueconomic.ro/temp/Article_2685.pdf

Popescu, C. R., Popescu, G. N., & Popescu, V. A. (2017). Assessment of the State of Implementation of Excellence Model Common Assessment Framework (CAF) 2013 by the National Institutes of Research – Development – Innovation in Romania. *Amfiteatru Economic, 19*(44), 41-60. Retrieved from http://www.amfiteatrueconomic.ro/temp/Articol_2593.pdf

Popescu, C. R., Popescu, V. A., & Popescu, G. N. (2014). The entrepreneur's role in the performance growth of the financial audit activity in Romania. *Amfiteatru Economic, 17*(38), 232-251. Retrieved from http://www.amfiteatrueconomic.ro/temp/Article_2382.pdf

Raport, C. O. M. (2016). *776 al Comisiei către Parlamentul European şi Consiliu cu privire la punerea în aplicare a măsurilor privind sectorul apicol prevăzute de Regulamentul (UE) nr. 1308/2013 al Parlamentului European şi al Consiliului de instituire a unei organizări comune a pieţelor produselor agricole*. Retrieved from http://ec.europa.eu/transparency/regdoc/rep/1/2016/RO/COM-2016-776-F1-RO-MAIN-PART-1.PDF

Regulamentul (CE) nr. 1221/97 al Consiliului din 25 iunie 1997 de stabilire a normelor generale de punere în aplicare a măsurilor de îmbunătăţire a producerii şi comercializării mierii, JO L 173, 1.7.1997, pp. 1.

Rennings, K., & Wiggering, H. (1997). *Steps towards indicators of sustainable development - linking economic and ecological concepts* (Vol. 20). Ecological Economics.

World Commission on Environment and Development (WCED). (1987). *Our Common Future*. Oxford, UK: Oxford University Press.

Xiangzheng, D., Zhihui, L., Jikun, H., Qingling, S., & Yanfei, L. (2013). A Revisit to the Impacts of Land Use Changes on the Human Wellbeing via Altering the Ecosystem Provisioning Services. *Hindawi Publishing Corporation Advances in Meteorology, 2013*, 907367. doi:10.1155/2013/907367

KEY TERMS AND DEFINITIONS

Bee Families: Represent number of bee families for honey production.

Beekeeper: Is the amateur or professional who is involved in raising and nurturing bees. Beekeepers enrolled in the professional associations of beekeepers must be certified by graduating courses organized by professional associations and carrying out a practical apprenticeship in an apiary under the guidance of an experienced beekeeper.

Beekeeping: It is a branch of animal breeding that cares for raising and nurturing bees.

Economic Welfare: Aims to generate a maximum revenue stream through maintaining in time the capital that has produced these benefits.

Environmental Protection: From the point of view of sustainable development, it aims to preserve the biological and physical stability of natural systems.

Honey: Is the main product obtained from the growth of bees. Its quality is dependent on the quality of the environment from which it was harvested and the health of the bees. Honey can be classified into multiple varieties depending on the single or predominant plant from which it originates.

Honey Base: Is the total number of honey plants within the useful activity range (3 km or 2,800 hectares) of bees and provides the raw material necessary for their survival and development.

Self-Supply: Is the ratio of domestic honey production to total consumption.

Chapter 5
Sustainable Rural Livelihood:
A Case of Hiware Bazaar, Maharashtra, India

Sushma Nayak
Symbiosis International (Deemed University), India

Shashikant Nehul
Symbiosis International (Deemed University), India

Aishwarya Narayan
Symbiosis International (Deemed University), India

ABSTRACT

In the present day, when human beings have excessively exploited the natural resources, sustainable development has become a catchword for countries around the world. With respect to the rural scenario, diversion of watercourses and settled agriculture are deemed as tactical imperatives for sustainable livelihood. The present study focuses on a village, Hiware Bazar, situated in the Ahmednagar district of Maharashtra, India. The village is known for its livelihood status, which can be distinctly categorized into two phases, the period before 1991 and the period after 1991. In the first phase (before 1991), the village was marked by severe livelihood crisis owing to acute water scarcity. The second phase (after 1991) is known for livelihood improvement realized through community engagement and planned investment initiatives for watershed development and organic farming.

INTRODUCTION

The principle of 'sustainable development' has been assuming significance in recent years, ever since its conception in the Brundtland Report titled "Our Common Future" released in 1987 by the World Commission for Environment and Development (WCED) (Vezzoli & Manzini, 2008). Sustainable development appreciates growth that is all-encompassing and environmentally nourishing to inhibit poverty and foster collective prosperity for the present-day population as well as future generations (Spring, 2015; Rigg, 2016). Correspondingly, countries across the globe are directing their efforts towards the end of

DOI: 10.4018/978-1-5225-5739-5.ch005

Copyright © 2019, IGI Global. Copying or distributing in print or electronic forms without written permission of IGI Global is prohibited.

judiciously managing resources for meeting the needs of the present by preserving the interests of posterity. For instance, water has a critical role in sustainable development and is allied to umpteen essential global problems, over and above a vastly diverse range of economic and human activities. Therefore, water security—access to sufficient quantities of water of suitable quality for sustaining human and ecosystem salubrity over a longer time horizon—is a sought-after goal for nations today.

Sustainable development can be followed only if demographic advancements are in conformity with the uneven productive potential of the ecosystem. Such endeavors are attempted in urban as well as rural settings. In fact, 'sustainable rural livelihood' is gradually gaining center place in the discussions about rural development and environmental management on account of its implications for the socioeconomic and ecological viability of nations (Scoones, 1998). Such a livelihood can adapt to and recoup from stresses and shocks, uphold or augment its capacities and resources, without depleting the natural reserves (Chambers & Conway, 1992; Scoones, 1998).

The following are the major objectives of the present chapter:

1. To examine and explore the concept of sustainable rural livelihood;
2. To substantiate the linkages between water conservation and environmental sustainability in rural settings;
3. To expound the potential and promise of organic farming systems for attaining ecologically sustainable pay-off;
4. To consider the case of a rural commune—Hiware Bazar, situated in Ahmednagar district of Maharashtra, India—a model village known for its custom-built development program and sustainable livelihood that is committed to environmental conscientiousness and social cohesion through community participation and engagement;
5. To assess how the village under consideration, once drought-prone and poverty-stricken with high unemployment rates, transformed itself to be recognized as a miracle village, through progressive initiatives such as adopting watershed development program for water conservation, and shifting to organic practices for retention and preservation of soil fertility;
6. To evaluate the outcomes of sustainability initiatives in connection with the case considered and identify future pathways for further development.

BACKGROUND

Sustainable Rural Livelihood

Rural poverty persists and continues to remain a concern for numerous developing countries notwithstanding the remarkable progress and innovations in agricultural technology. Several structural factors may be attributed to rural distress and deprivation. Some of these include inequitable allocation of land ownership, urban orientation of development policies, worldwide recurring fluctuations in prices of essential commodities, external shocks such as natural disasters and the ongoing climate change (Potter, Conway, Evans & Lloyd-Evans, 2012). Heightened priority to production of export crops over those meant for domestic production has worsened the agrarian crisis thus exacerbating rural poverty. For instance, noted Indian journalist P. Sainath reports how the state of Kerala in India, deemed to be the most globalized one in the country post-neo-liberal reforms, primarily produces cash crops much of which are not consumed by the local populace (Kumar, 2012). Likewise, Sri Lanka has emerged as a leading producer of gherkins, which is highly demanded in the European countries (Gopalakrishnan, 2007). Export-oriented farm production is likely to have feeble linkages with the native economy if the former fails to generate domestic demand for such products, either by means of development of agroprocessing or via consumer demand (Diao, Nwafor, Alpuerto, Akramov, Rhoe, & Salau, 2012). Rural livelihoods are ever more under stress due to paucity of land, erratic weather, substandard soil fertility, pests, pathogens and plant diseases (Mundle, 2017); in view of that, the rustic poor are distinctly identified as 'chronically poor' (Zoomers, 2013). An assessment of social protection programmes—interventions implemented to mitigate poverty, vulnerability and risk—across the world, brings to light the fact that regardless of the steady emergence of new models for driving rural transformation, only a handful are solely dedicated to or adapted towards rural people. In most countries, social protection policies are neither intended to assist the rural poor, nor do they attend to rural vulnerabilities and limitations.

An obvious outcome of this scenario is 'footloose migration,' whereby lack of stable and steady employment underpins massive influx of people from rural to urban areas in search of livelihood alternatives. Therefore, as pointed by Gill (2008), the disparity between rural and urban areas has to be narrowed with respect to human development and quality of life. Rural areas must specialize and diversify in economically viable and lucrative activities relating to production, processing and marketing. Simultaneously, the social aspects of general well-being such as health, hygiene, water supply, education, and other facilities must catch up with the pace that complements with the urban agglomerations.

Thus emerges the significance of the concept of 'sustainable rural livelihood' which connotes the ability and capacity of a rural commune to withstand crises, adapt to changing conditions, sustain, survive and thrive by adopting practices that support a decent life without hampering the natural resource endowments. Sustainable development and mitigation of poverty in rural areas depend, to a great extent, on the efficient management of common-pool resources and local leadership, which pivots around adaptation to native agro-ecological and social conditions. Popularly known as Participatory Assessment and Planning for Sustainable Livelihoods (PAPSL), the approach recognizes the involvement of local village people in gauging the existing conditions and drawing up action plans by embracing three tactical imperatives: training and upskilling for positive change that aims to kindle and empower the poor to assume control of their own growth and progress; participatory rural assessment, a scheme of investigation, exploration and data collection featuring local people in an exhaustive and meticulous inquiry of their community and livelihood practices; and a premeditated plan to realize the strategies outlined for development (Bansil, 2002). Thus, people are made to believe that they are a 'part' of the community who can change things for their better through self-help and mutual help, and work towards harmonizing projected resource availability with estimated demand, while also allowing for unforeseen shortfalls in resource supply or rise in demand (Dorward, Anderson, Clark, Keane & Moguel, 2001).

Various dimensions of sustainable livelihood are considered in Table 1.

Linkages Between Water Conservation and Sustainability: A Community Engagement Approach

Water is a vital natural asset that sustains every social and economic pursuit, and is elemental for human survival and well-being. Access to water is crucial for accomplishing livelihood potentialities. Correspondingly, sustainable water management is a critical global concern intricately allied to several livelihoods worldwide. With rapid urbanization fuelled by migration to the existing cities, and implementation of

Table 1. Dimensions of sustainable livelihood

Dimension of Sustainable Livelihood	Elucidation
Mitigation of poverty	An improvement in income levels, enhanced food security, access to basic necessities and public goods/services, expansion of income sources, equitable distribution of income, enhancement in quality of life (Neely, Sutherland & Johnson, 2004).
Stable employment	• Certainty of income, generation of productive output, recognition of providing valuable service (Sen, 1975). • 200 days of productive employment in a year (Lipton, 1994). • Availability of assured gainful employment for a definite period of the year, with due consideration to the concepts of 'work' and 'leisure' (Scoones, 1998).
Asset building	Undertaking projects that encourage creation of assets and building human as well as social capital with the objective of prevention from falling back into poverty trap yet again (Neely, Sutherland & Johnson, 2004).
Livelihood adjustment, acclimatization and resilience	• Ability to endure and recuperate from stresses and shocks (Scoones, 1998). • Capacity of socio-ecological systems (SES) to sop up disruption without spinning into another similar phase (Gunderson, 2000). • Resistance to vulnerability (Scoones, 1998; Neely, Sutherland & Johnson, 2004). • Enhanced risk preparedness (Neely, Sutherland & Johnson, 2004).
Well-being	Self-respect, self-worth, security, hedonism, power, inclusion, in addition to conventionally measured material concerns (Chambers, 1989).
Environmental protection	Avoiding exhaustion of stocks of natural resources (Scoones, 1998).
Community participation and social inclusivity	• Expansion of inventories and information systems where native and technical/ methodical/ systematic knowledge are integrated into specific high-quality knowledge systems to facilitate the connection of science and community action; awareness, learning and capacity building of farmers and conservationists (Kellner & Bosch, 2003). • Increased involvement of the people in the development processes – problem detection, planning, preparation, execution, supervision and appraisal phases (Bansil, 2002; Neely, Sutherland & Johnson, 2004).
Empowerment	Ability to voice, freedom of choice, and availability of growth opportunities (Carney, 2003).
Good governance	Leadership that is upright, scrupulous, competent, effectual, accountable and accessible (Neely, Sutherland & Johnson, 2004).

Source: Various sources

smart cities mission, there is increasing pressure on water resources in the urban settlements for diverse uses—public, domestic, commercial and industrial. Any shortfall in water availability has detrimental effects on the livability in these cities. By 2050, water demand across the globe is estimated to rise by 55%, primarily because of mounting demand from manufacturing, thermal electricity production and household use, all owing to burgeoning urbanization in developing countries (OECD, 2012; WWAP, 2015). In a typical rural setup, consistency in water supply ascertains the reliability of ecosystem in supporting poor people through fishing, grazing of cattle, etc (WWAP, 2015). Water scarcity can inflict constraints on farm productivity; aggravate the problems of rural unemployment; and adversely affect the financial status of rural households (Udmale, Ichikawa, Manandhar, Ishidaira, Kiem, Shaowei, & Panda, 2015).

The reality of water crisis may be best illustrated with the observation of George (1999) when he states, how, by the end of the second millennium, most countries arrived at a stage of rapid and extensive water depletion owing to misuse, contamination, drying up of wells and springs, and the strangling of rivers by way of myopic exploitation. Nearly everywhere, water is exhausted, and so far as people do not face paucity of water, they deem access to water as an apparent and natural phenomenon. Table 2 exhibits the enormity of water crisis by exposing the harsh realities in urban and rural settings. By demonstrating

Table 2. Water crisis: an urban-rural realism

Exhibit 1	Exhibit 2
Urban Scene	**Rural Scene**
Jagriti (name changed), the 47 year old sagacious homemaker, realizes her obligations towards her family. She now appreciates the interpretation of the dictum: "Early to bed and early to rise makes a man healthy, wealthy and wise." This, nevertheless, has been an upshot of chance, not choice. If Jagriti is late to rise by even 5 minutes in the morning, she may have to see dry taps for an entire day, till the next day dawns. The apartment where she resides has revised norms for water supply, in conformity with the restricted water provision by the city's municipal corporation. As a result, the duration of 30 minutes from 6:00 am to 6:30 am is crucial to store up enough water for an entire day's need, which may not necessarily suffice a family of four. On some occasions, her residential apartment is bound to meet the water requirement through private water tankers, who make the most of the opportunity by charging three times more than what they charge in normal situations. There is also an upsurge in the number of illegal water tanker operators in the city. Finally, the city residents have realized how water conservation is the panacea to water distress. This is the scene from Pune city in the Indian state of Maharashtra, deemed to be one of the fastest growing cities in the Asia-Pacific region; the experience narrated is one that occurred in August 2015, an unforgettable year in recent times with respect to water crisis.	Young Sakshi (name changed), along with three of her sisters, wakes up enthusiastically early in the morning to have another engaging day at play. She is too naïve to understand that situation is not all well at home. Her father, Ganesh (also name changed), does not own a farmland and makes a living through daily wages by serving as a laborer in other people's farms in the village. Things are getting worse as water scarcity has compelled farm owners in the village to undertake farming on tiny plots of land on a smaller scale that obviates the need of daily laborers like Ganesh. The village gets water from two wells, one of which is used for farming activities and another for drinking water and daily chores. For women in the village, the day begins even before the sun rises as they have to walk nearly a kilometer to reach the well and fetch water. At times, they are assisted in this activity by their young children. A major concern is that the wells are drying fast. Access to potable water is only a castle in the sky. Diminishing rains, rising temperatures and lack of alternative employment in the backdrop of seasonal nature of agriculture, has further intensified the crisis. This is the scene from Pimpaldhara, a far-flung village in the Yavatmal district of Maharashtra in India, reeling under the impact of acute water crisis.

Source: Authors

experiences—one from a city and another from a village—within the same state of Maharashtra, India, the table exemplifies that water scarcity is a pervasive problem that extends its ramifications far and wide.

Sustainable water security—universal access to fresh and pure water——is at the core of all developmental efforts. Correspondingly, in July 2014, the Open Working Group (OWG) on Sustainable Development Goals (SDGs) considered a worldwide objective of explicit targets connected to water resources management, as discussed in Table 3.

The world now realizes that droughts and other forms of water scarcity shall be a reality of the future, and, if the current patterns of water consumption continue, it is likely that a major chunk of global population shall be required to live in water stressed conditions. Accordingly, efforts are being initiated by countries to augment the water use efficiency through effective water governance (*Implementing Water Resources Management: Information Brief*, n.d.). The decade spanning from 2005 to 2015 was observed as the International Decade for Action 'Water for Life,' whereby efforts were made to promote initiatives to fulfill global commitments on water and connected issues.

Table 3. Targets set by the Open Working Group (OWG) on Sustainable Development Goals (SDGs) relating to water resources management

By 2030, bring about improvements in water utilization in a sustainable manner to ensure regular and unimpeded access to fresh water across all sectors; also bring down the number of people suffering from dearth of water
By 2030, put into action integrated water resources management at every level, also via transboundary cooperation as may be pertinent

Source: Open Working Group on Sustainable Development Goals (SDGs) (2014)

However, a good deal of expected outcomes can also be achieved through community participation in water resource management. The significance of community participation in environmental decision-making has been acknowledged and acclaimed at both national and global levels with a series of multifaceted proclamations and public appeals to encourage the participation of affected parties in decision-making (Boakye & Akpor, 2012). According to Dungumaro and Madulu (2003), the evolution of participatory approaches reveals the merit of local communities' assent in chipping in decision-making processes, particularly on concerns directly affecting their wellbeing; local communities in many parts of Tanzania have designed coping strategies to ensure preservation of water resources. Nongtraw and Khweng are two such villages from the state of Meghalaya in India, which, through participatory rural appraisal, have stood as exemplars of resilient local communes; a collaborative effort of the native populace, government and self-help groups, has helped the community to be insightful and responsive to their extensive, undocumented biodiversity heritage, and the need for conserving it (Rametra, 2016). In this milieu, the local community involvement offers indigenous knowledge, fundamental evidence, experience and ideas, which can bring about realistic, suitable, viable and agreeable solutions to environmental woes. Given the magnitude and intricate nature of international water challenges, there has been a tendency in moving towards a comprehensive bottom-up approach that encourages larger commitment of stakeholders and develops connections between government leaders and the community, propelled by previous breakdown of top-down approaches (Rolston, Jennings & Linnane, 2017).

Linkages Between Organic Farming Practices and Sustainability

Organic farming, which has moved beyond being a niche, specialized market to being a worthy substitute for commercial, conventional farming, has seen an upward trajectory in the last couple of decades. Organic farming, as defined by the National Organic Standards Board of the US Department of Agriculture is:

an ecological production management system that promotes and enhances biodiversity, biological cycles and soil biological activity; it is based on the minimal use of off-farm inputs and on management practices that restore, maintain and enhance ecological harmony. (Bennet, 2010)

It places special regard for the environment, human values, and nature, which is not seen in other agricultural systems.

Our world is moving toward urgently demanding agricultural techniques which can ensure availability to our expanding population, while securing sustainability of resources. Conventional agricultural systems have proven to be highly detrimental to the environment, and pose a threat to sustainability. The high dependence on chemical fertilizers and reckless crop rotation (which causes decline in soil productivity) are poor long term solutions for global food requirements.

Organic farming uses 45% less energy than its commercial counterpart, and replenishes soil nutrients, rather than depleting them. Organic farming, as a concept, can solve most of the above issues, by ensuring crop production without the risk of environmental harm. It completely matches the needs of sustainable agriculture due to its relationship with ecosystems. However, the question remains: can organic farming ensure equivalent yields to be able to replace commercial farming? Here, it is advisable to use a case by case approach, in evaluating which system provides the best results for a particular area and its people.

The desirable outcome of sustainability may be viewed from two perspectives—long term sustainability of the farming system and contribution of the farming system to the local community/ecology. Organic farming seems to fulfill both these perspectives. The four tenets of sustainability, as provided by the US National Academy of Social Sciences, are: be productive, economically profitable, environmentally sound and socially just. Organic farming does satisfy each one of these, making it an apt solution to be adopted as a sustainable method toward development. Organic farms have in the past proven more economically profitable than conventional farms (Seufurt, 2012). A possible reason for this is the market for organic products, which is worldwide, assigned some level of luxury or status. As a result, farms producing organic crops are likely to reap high rewards. Further, the high input costs of fertilizers and/or machinery are eliminated, thus expanding the bottom line for the farmer (Kobets, 2004).

An important feature of organic farming is the contextualization of local communities in the system. Organic farming places due emphasis on local farmers, providing local services, creating jobs and generating supply chains within communities. A rural farm going organic would create value additions from within the community; for example, manure for natural fertilizer could be procured from neighboring husbandries. Since organic farming is particularly labor intensive (Walsh, 2012), it is likely to absorb local workers. The implementation and subsequent success of organic farming systems, depends to a great extent, on the level of pollution caused by pesticides, heavy metals, etc (Kobets, 2004). This increases the reliance on locally available resources, rather than on external factor inputs. The same point, however, is also a deterrent to the progress of organic farming, as only those soils which can support the process can ever be used for its cultivation, limiting the land resource.

Satish, Bhaveshananda and Sengupta (2012) provide multiple studies which evince the success of organic farming. Experiments across USA, UK, Ukraine and other developed Western nations reveal that organic farming has outputs which are equal to or up to 20% less than that of conventional agriculture. A common argument against organic farming is the problem of 'conventionalizing' (Gafsi, Le Tron & Mouchet, 2010). This theory propounds that as organic farming attempts to scale up and fulfill market requirements; it will tend to lose its identity, and will blur the line between itself and conventional form (Best, 2008). However, in reality, large organic farms have managed to stay true to their principles. Comparative results of conventional agriculture and organic agricultural methods reveal that, even in those cases where organic yields fall short, the economy provided by lower costs, lower energy inputs and higher organic matter in soil compensates for the same.

Organic farming has also proven to improve farmer livelihood, which is inherently linked with rural development capabilities, especially in developing countries (Seufert, 2012). Organic farming reduces the exposure of farmers to the price uncertainties of commercial produce markets. Since organic products are less likely to be impacted by weather extremities, a certain amount of crop insurance is afforded to farmers. This is yet another benefit that organic farming provides. In Ghana, farmers have been advised by the state-owned marketing board to increase the production of organic cocoa, as the organic method of production is less affected by weather fluctuations. Today, Ghana is a world leader in organic cocoa production (Laven & Boomsma, 2012).

In yield comparison studies, it is observed that organic farming delivers a lower yield for every crop type, though the disparity varies widely. The disparity is greatest in major cereal crops such as wheat, and most vegetables (Walsh, 2012). Perhaps, an approach which focuses on certain crops alone in organic methods of production, while allowing others to capitalize on the large quantity provided by commercial methods of production is in order. This would allow a significant improvement in terms of environmental impacts, and ensure worldwide food security too.

The use of organic farming to promote sustainable development is prevented in developing countries by limited access to international markets. Unfortunately, the market for organic produce is limited greatly to US and UK markets. Local rural farmers find it increasingly difficult to reach these markets, due to the lack of infrastructure to create supply chains. Further, the certification processes and stringent international regulations also pose substantial obstacles to rural farmers (Seufert, 2012).

In developing countries, substantial policy efforts would be required to support the rise of organic farming as a mainstream activity. For example, certification procedures in these countries for organic products can lend validity to their quality, and possibly create local markets. There is also a need for informational support system for organic farmers, to educate them regarding the correct ways to maintain sustainable methods of farming.

MAIN FOCUS

The main focus of the present chapter is to bring out the vulnerabilities of rural livelihoods and methods that can be adopted to build resilience in the face of such vulnerabilities. This is done by considering the case of Hiware Bazar, a drought prone village in the state of Maharashtra, India, which was reeling under the impact of acute water crisis and was pulled out of this misfortune due to community participation and leadership offered by local citizens. As Sarojam (2016) rightly points:

Once impoverished and drought prone, Hiware Bazar in Maharashtra is a shining example of how a visionary leader can use good governance to make degraded areas resource-rich and transform the future of its people through empowerment and inspiration.

Although nature may present harsh realities, it is possible to deal with the same through prevention, mitigation, preparedness, response and recovery, deemed to be the vital constituents for building robust resilience.

HIWARE BAZAR: A MODEL OF SUSTAINABLE RURAL LIVELIHOOD

Background of the Village

Hiware Bazar is a village spread over 977 hectares in the Ahmednagar district of Maharashtra in India. The village, located in the rain-shadow zone of the Sahayadri range, is known to receive scanty rainfall, around 400-500 mm annually, and has been identified as a drought-prone area. An important characteristic of the village has been its livelihood status, which can be distinctly categorized into two phases, viz. the period before 1991 and the period after 1991. In the first phase (before 1991), the village was marked by severe livelihood crisis owing to acute water scarcity. Barring a few families, the rest of the village lived below the poverty line. Drought relief measures had failed to deliver expected outcomes. Access to water greatly depended on digging deep into the earth, which wasn't a lasting solution to the long-standing problem. Native women walked nearly 2 km to fetch water. Since agriculture, primarily rain-fed and seasonal, was the main occupation of the inhabitants, cultivation practices were limited to few months in the year. Consequently, a vast majority of the population was rendered unemployed, poverty-stricken and helpless during the off-season; farmers and farm laborers had to seek alternative forms of employment, often necessitating them to migrate to cities. Those, who resisted migration and preferred staying back in the village, engaged themselves in cutting trees for firewood and survival. Deforestation, poor soil water retention, misuse of groundwater resources and improper management of available water, further worsened the water crisis. Manufacturing and vending of illicit alcohol became a source of living for many due to lack of stable sources of income. Alcohol abuse and incidences of crime became increasingly prominent. Basic medical facilities were grossly inadequate.The student-teacher ratio in schools was abysmally low, which was also the reason why literacy remained at a miserable 30%, way below the national average. Thus, the village was distressed with respect to four essential parameters: economic development, environmental protection, social concerns of equity and an overall measure of livability from the standpoint of sustainability.

Local Leadership

The success or otherwise at community development endeavors, depend to a great extent, on the degree of initiative and leadership offered by the local citizens. The adversity and harsh conditions plaguing Hiware Bazar could have ended only with the active involvement of every member of the commune. In this regard, the village got timely guidance from a native resident, Mr. Popatrao Pawar. Mr. Pawar, after completing his post graduation from Pune, decided to move back to the village, instead of continuing his stay in the city for furthering personal aspirations. He was unanimously elected as the village *Sarpanch* (local leader), with soaring hopes and tall expectations of liberating the community from the clutches of poverty and distress. The situation then, required a radical action plan that could encompass every aspect of desired progress to ensure holistic development. Thus, in conformity with the Participatory Assessment and Planning for Sustainable Livelihoods (PAPSL) approach, the village designed a plan with the following main objectives:

- Ensure water security to all stakeholders (drinking; irrigation; cultivation) through water conservation practices;
- Improve water use efficiency and crop yields, by offering the appropriate quantity of water to the right farm at the apt time;
- Work towards soil management;
- Provide stable employment opportunities to the residents from the viewpoint of retaining them in the village and inhibiting migration to cities and metropolises;
- Improve living conditions in the village;
- Offer basic healthcare facilities;
- Increase literacy rate through improvement in enrollment ratio and student-teacher ratio.

Of all the aforementioned, water security was accorded foremost priority and a major step was undertaken through the initiation of watershed development.

Water Resource Management

In 1992, Hiware Bazar led the way to Watershed Development Program, whereby efforts were made to adopt distinctive methods for conservation of water. Through collective community participation, the local inhabitants adopted a series of water harvesting measures to facilitate water retention in the bounds of the watershed, a much-needed initiative for fighting persistent failure of monsoons and chronic droughts. A typical watershed refers to a host of conventional water structures that help in seizing and amassing rainwater. Accordingly, there were attempts at restoring the green cover by growing more trees in the forest area. The villagers developed contour trenches i.e. dug ditches along the hillside, in a manner that they ran perpendicular to the flow of rainwater. When it rained, water ran into these trenches and seeped into the soil. Continuous Contour Trenching (CCT), as exhibited in Figure 1, was taken up for arresting the surface run-off, thus recharging ground water and reducing soil erosion. Earthen bunds were constructed to allow water percolation. This was also meant to increase water availability and boost agricultural productivity.

Dams and percolation tanks were constructed to hoard water. Open wells were dug to access groundwater in underground aquifers. There was a stringent ban on the digging of bore wells to extract water for irrigation purposes. Flood irrigation, sprinkler irrigation and drip irrigation were deployed effectively.

Table 4 indicates the trends in water resource management of Hiware Bazar, the situation as it was in 1991, and the way it stands in recent years. The area under all forms of irrigation has increased appreciably in the year 2016, from what it used to be in 1991, further indicated in Table 4 and Figure 2. There is also an increase in the number of wells unearthed during this period, reflected in Table 4 and Figure 3.

Wherever possible, low-lying structures have been erected for the holding and retention of rainwater. Figure 4 depicts typical water resource management practices in Hiware Bazar. The water availability in monsoon, when used for recharge of underground water through watershed development, has proven

Figure 1. Continuous Contour Trenching (CCT) in a rugged topography at Hiware Bazar
Source: Authors

Table 4. Trends in water resource management of Hiware Bazar (1991 -2016)

Particulars	1991	2016
Area under irrigation (ha)	125	750
Area under summer irrigation (ha)	1-2	80
Area under drip irrigation (ha)	0	250
Number of dug wells	97	348
Groundwater level (feet)	90-120	15-40

Source: Hiware Bazar Panchayat Records (2016)

Figure 2. Area under irrigation at Hiware Bazar (1991- 2016)
Source: Hiware Bazar Panchayat Records (2016)

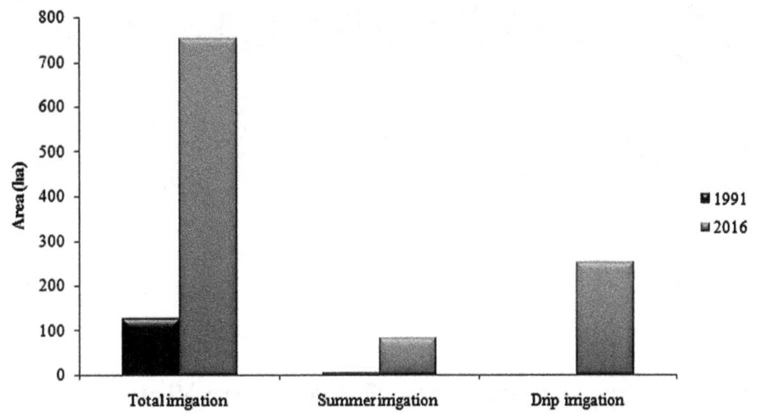

Figure 3. Number of wells excavated at Hiware Bazar (1991-2016)
Source: Hiware Bazar Panchayat Records (2016)

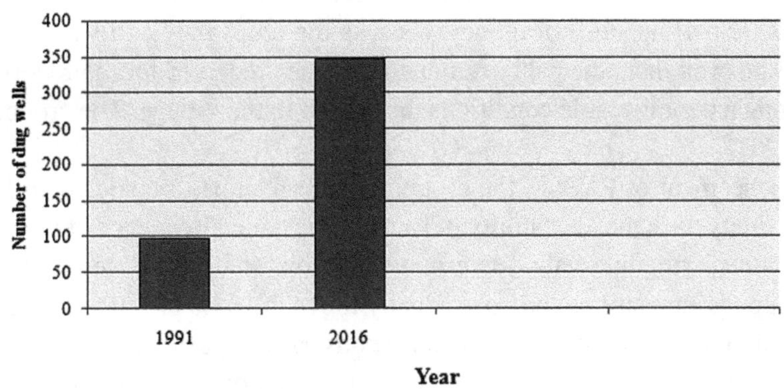

Figure 4. Water Resource Management at Hiware Bazar
Source: Authors

to be beneficial during summers. It also extends its cooperation in arresting the surface run-off thereby leading to efficient use of water.

Farming Practices: Cropping Intensity, Pattern, and Yield

The village experienced a gargantuan change in cropping intensity, cropping pattern and crop yield, ever since it embarked upon planned development in 1991. Prior to 1991, agriculture relied heavily on monsoon. Even those farmers who owned large pieces of land had to migrate to cities and work as daily wage laborers for trifling sums of money, because the crops they cultivated withered recurrently due to poor rains. And even then, the paltry remuneration they received for their work away from home, appeared far better than the miserable conditions they faced in the village. The cropping intensity stood at a mere 94% in 1991.

After the commencement of PAPSL, the cropping pattern came to be decided through community consensus, in conformity with the availability of water, ascertained by means of water budgeting. Water levels—measured at recurring intervals—determine the crops to be cultivated in the following season. As a result, the cropping intensity scaled up gradually to reach 164% in 2016. This is mainly attributed to the regular availability of water ensured by water conservation, selective cultivation of crops (those that are less water-intensive in nature) and enhanced soil fertility owing to organic farming methods. The cultivation of banana and sugarcane is strictly forbidden on account of their excessive water consumption. In the present day, only 20% of chemical fertilizers are used for farming practices (Kunke, 2012); there is heavy reliance on organic way of farming. The patterns of cultivation and crop yields exhibit an upward trend, as depicted in Table 5 and Figure 5.

Table 5. Transformation in crop yield capacity at Hiware Bazar (expressed in quintals per hectare)

Name of the Crop	Before 1995	2005	2016
Sorghum	8 to 10 q/hr	14 q/hr	25 q/hr
Gram	8 to 10 q/hr	12 q/hr	18 to 22 q/hr
Vigna radiata (Green Gram)	4 to 5 q/hr	7 q/hr	10 to 12 q/hr
Lentils	5 to 10 q/hr	6 q/hr	25 to 28 q/hr
Groundnuts	5 to 6 q/hr	6 q/hr	8 to 10 q/hr
Wheat	8 to 10 q/hr	15 to 20 q/hr	35 to 40 q/hr
Millets	5 to 7 q/hr	10 to 12 q/hr	20 to 22 q/hr

Source: Hiware Bazar Panchayat Records (2016)

Figure 5. Transformation in crop yield capacity at Hiware Bazar
Source: Hiware Bazar Panchayat Records (2016)

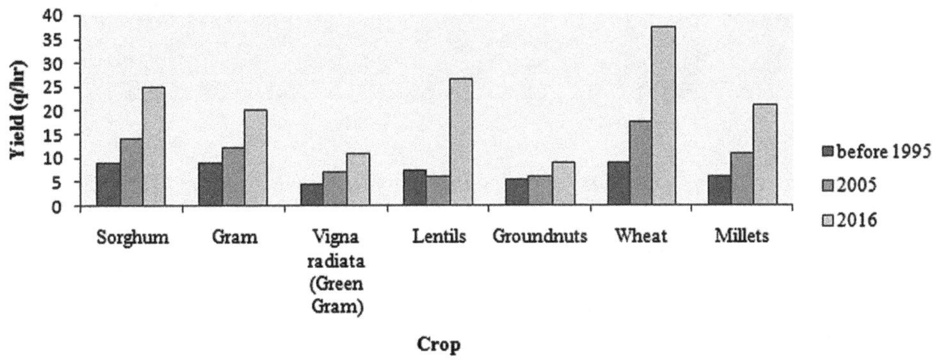

The yields of sorghum, gram, vigna radiata (green gram), lentils, groundnuts, wheat and millets have increased significantly during the period under consideration. The village also attained success with respect to the cultivation of onions, custard apples and tomatoes. The yield of custard apples increased from 10 kg/plant in 1995 to 25-30 kg/plant in 2016; likewise, the yield of tomatoes increased from 10 metric tons/hectare to 35 metric tons/hectare during the same period.

The village, in the coming years, endeavors to take on organic farming completely, and establish an independent market for organic food (Kunke, 2012).

Dairy Industry

The watershed development program has stimulated the progress of agriculture and allied activities in numerous ways. The increase in floral diversity, evident in Figure 6, is a byproduct of watershed development. Correspondingly, a ban on free grazing has resulted in an exponential increase in the production of grass. It signifies the availability of fodder for domesticated animals, which has contributed to increased milk yield; as an outcome, dairying constitutes an important activity of the economy (refer Table 6). The growth of dairy industry has resulted in generation of employment opportunities for a vast majority of the village population. This has not only brought down the rate of migration to urban areas, but has

Figure 6. Floral diversity at Hiware Bazar
Source: Authors

Table 6. Contributing factors for stimulating dairy industry in Hiware Bazar (1991-2016)

Particulars	1991	2016
Fodder availability (tons)	1500	6000
Number of milk producing cattle	<100	650
Milk production (liters per day)	150	3500

Source: Hiware Bazar Panchayat Records (2016)

also encouraged reverse migration, whereby former migrants have flooded back to the village. Nearly 93 families are known to have come back home, since 1989 (Byatnal, 2011).

Table 6 demonstrates how increases in the availability of fodder, rise in the number of milk producing cattle, and surge in daily milk production have given a lift to the dairy industry in Hiware Bazar. The marketing of dairy products, alongside cultivation of cash crops, has enabled the village attain self-sufficiency in terms of employment and income.

Improvement in Livelihood

An arid village with extremely low rainfall each year, Hiware Bazar has witnessed a far-reaching resurgence in the span of last two decades. A good deal of progress has also come, thanks to drawing on government programs that go well with the village requirements. In the initial years of planned development, in order to provide employment to jobless villagers, the local leader got in touch with the forest planning department and instigated a tree-planting program in the hills around the village. In 1994-95, the village was roped in the State's Adarsh Gaon Yojana (AGY), a 'model village' plan, which involves shared investment by the village and the government for the development of the former. An essential prerequisite of AGY is the contribution of funds by the village, without which, it does not receive any subvention. Upon succeeding atattainment of desired success in the projects undertaken, the village is expected to give back the funds to the patron. An 'Adarsh Gaon' is expected to abide by a set of laws and conventions to enjoy the patronage and backing of the state. This involves ecological awareness and adoption of sustainable methods such as effective water conservation by means of watershed management and water auditing, assuming accountability of the village's natural endowments—afforestation and ban on grazing, voluntarily offering labor for the development of the village, and endeavoring to induce lifestyle makeover within the local residents for bringing about social transformation.

Hiware Bazar has been drafting a triumphant narrative ever since its inclusion in the AGY. The model necessitates collective decision making by the entire village; there is no coercion or intimidation with respect to decisions made and solutions sought. An important objective has been to improve the livability in the village and do well in terms of resilience.

The village is devoid of addiction and does not host any shops that offer addictive substances for sale. Health centres have been set up across the village to perk up access to primary healthcare facilities. Anganwadis have been established for maternal and child care. The message of limiting the size of families is loud and clear; towards the realization of this objective, vasectomy has been advocated in recent years. Every prospective couple is expected to undergo an HIV test before getting into nuptial commitment.

Financial inclusion and economic empowerment have also been a part of the plan. There is increased awareness to access banking facilities for depositing savings and procuring loans. As a result, the dependence on local moneylenders has declined significantly. The villagers realize their credit obligations and have been prompt in paying installments and repaying loans.

As regards improving literacy, schools with adequate infrastructure—provision of electricity and toilets—have been instituted. These schools offer education up to class 10 and decently meet the requirements of student-classroom ratio and student-teacher ratio.

Table 7 exhibits the progress of the village with respect to essential livelihood parameters.

Table 7. Progress of Hiware Bazar since 1991

Particulars	1991	2016
Number of families	180	286
Number of families Below Poverty Line (BPL)	168	0
Number of landless families	22	3
Annual per capita income (Rs.)	832	32,000
Medical facilities	None	PHC*
Education facilities (Std.)	Up to IV	Up to X
Literacy (%)	30	96
Main workers (as % of total population)	42%	51%
Main workers in cultivation (%)	76%	76%
Main workers as agricultural laborers (%)	19%	10%
Number of tractors owned	1	30

Source: Hiware Bazar Panchayat Records (2016)
*Primary Health Care

Considering the socio-economic status of the village, the number of families below poverty line (BPL) has fallen from 168 to 'nil,' between 1991 and 2016. This picture is reflected through various observations namely, reduction in the number of landless families, increased per capita income, improved primary healthcare facilities, increase in literacy rate, and growth in agriculture.

The laudable progress of the village has been further acclaimed through its recognition as a 'model village' for Saansad Adarsh Gram Yojana, a rural development initiative by Prime Minister Mr. Narendra Modi, in the year 2014. Hiware Bazar has been chosen as one of the three villages as a benchmark for rural development through collective action (Fareed, 2015).

Despite the advancement of the village with respect to livelihood improvement, there are several key areas where the village has scope for further improvement.

SOLUTIONS AND RECOMMENDATIONS

Hiware Bazar is an exemplar of development made possible through interplay of knowledge with collective action. Nevertheless, the village has a lot to achieve in years to come. One of the areas where the rural commune requires careful deliberation and forethought is the adoption of organic farming. While farmers at Hiware Bazar intend to adopt extensive organic farming in future, a major challenge would be to do so without causing any reduction in the volume of total yield. While conventional farming offers higher productivity but is less sustainable, organic farming has comparably lower productivity but effectively stands the test of sustainability. Therefore, farmers at Hiware Bazar should take up farm budgeting for a fitting cost-benefit analysis of premeditated farming strategies. This shall aid in striking a right balance of productivity and ecologically desirable outcomes.

It is also observed that the village, after achieving a set of preferred targets for planned development, has settled at a point where no new targets are being identified with respect to those parameters in which progress is realized, but there is potential for further advancement. For instance, even though the level of literacy has improved, efforts should be directed toward setting up of centres of learning for education beyond class X. These institutions should be centred on imparting vocational education, which will give further impetus to village development.

The villagers, who have benefitted considerably from planned development and experienced significant increase in income, should work as angel funders for those who have innovative ideas in farming business and allied activities, but may be constricted due to lack of adequate funds.

FUTURE RESEARCH DIRECTIONS

While most studies so far have focused on the achievements of village development through community participation, there is a distinct twist in recent research which says that villages such as Hiware Bazar may not be the best in terms of altruism, as they have been known for years together.

A study by Hatekar, Kulkarni and Mehta (2015) suggests that residents of such 'so-called' model villages are equivalent to those of other ordinary villages with regard to voluntary or deliberate offerings for community causes, particularly when they are kept anonymous. The study aimed at identifying whether communities, with confirmed and deep-rooted social values of cooperation and mutual aid, make anonymous persons act more considerately, even when there is no penalty for self-seeking conduct.

Results suggested that the principles of cooperation and collaboration are highly feeble; their endurance relies significantly on supervision and means of punishment. It, therefore, shows that the time-honored chronicle of 'norms' might not always instill the steady and unwavering inclination for 'unconditional cooperation' in a person's value system.

It is essential to test the outcomes of the aforementioned study with respect to a set of different villages; whether the norm of collaboration and teamwork can instinctively assist in effective implementation of various government programs for local development, or whether enforcement is needed to attain desired results.

CONCLUSION

Sustainable rural development can become a reality only when the vision of the leadership body is complemented with the active involvement and support of every member of the commune. Hiware Bazar has transformed itself from a drought-prone village to a miracle village through Participatory Assessment and Planning for Sustainable Livelihoods (PAPSL) approach. The village had been reeling under the impact of acute water scarcity due to poor water management, forest degradation and overexploitation of groundwater resources. Crops and lives had come under threat and there was massive influx of people to cities in search for gainful employment. People who stayed back in the village were involved in the manufacturing of illicit alcohol for making a living. This was the trend till 1991, but things changed for better when a visionary named Mr. Popatrao Pawar assumed the reigns as the *Sarpanch* (local leader) of the village. Of the several progressive initiatives, the village embarked upon a watershed development programme for water conservation that has not only contributed to rising water levels, but has also led to reversal in cropping patterns and rise in cropping intensity. With the implementation of Continuous Contour Trenching method, it was possible to increase the ground water level in the village. Although the village typifies an arid topography, through corrective human intervention, it has attained success in soil and water conservation, afforestation, grassland development and protection of biomass. Agricultural laborers, who struggled to make ends meet, have improved conditions of living and own pieces of land. Farmers are gradually shifting to organic practices for retention and preservation of soil fertility, which is not the case with conventional farming. The village has attained self-sufficiency and is known as a 'model village' that boasts of several millionaires. The watershed development in the village not only provides solution to the water needs of the people, but also extends its cooperation in strengthening the rural economy sustainably. Thus, the village serves as a classic example of sustainable rural livelihood through planned investments. Whilst there are several ways of attaining a sustainable rural livelihood, diversion of watercourses and settled agriculture are deemed as tactical imperatives. Environmental stewardship has been a key factor for the progress of the village.

ACKNOWLEDGMENT

The authors express gratitude to the officials of Hiware Bazar Village Panchayat for sharing the necessary data and information. The authors further express their gratitude to reviewers and editors who have helped in improving the standard of writing and communication of the ideas systematically.

REFERENCES

Bansil, P. (2002). *Economic Problems of Indian Agriculture* (6th ed.). New Delhi: Techno Economic Research Institute Publication.

Bennet, G. (2010). *Food identity preservation and traceability*. Boca Raton, FL: CRC Press.

Best, H. (2008). Organic agriculture and the conventionalization hypothesis: A case study from West Germany. *Agriculture and Human Values*, *25*(1), 95–106. doi:10.100710460-007-9073-1

Boakye, M., & Akpor, O. (2012). Community Participation in Water Resources Management in South Africa. *International Journal of Environmental Sciences and Development*, 511–516. doi:10.7763/IJESD.2012.V3.277

Byatnal, A. (2011). 38 more 'ideal' villages to come up in Maharashtra by June. *The Hindu*. Retrieved from http://www.thehindu.com/news/national/other-states/38-more-ideal-villages-to-come-up-in-maharashtra-by-june/article2006895.ece

Carney, D. (2003). *Sustainable livelihoods approaches: progress and possibilities for change*. London: Department for International Development.

Chambers, R. (1989). Editorial introduction: Vulnerability, coping and policy. *IDS Bulletin*, *20*(2), 1–7. doi:10.1111/j.1759-5436.1989.mp20002001.x

Chambers, R., & Conway, G. (1992). *Sustainable rural livelihoods*. Brighton, UK: Institute of Development Studies.

Diao, X., Nwafor, M., Alpuerto, V., Akramov, K., Rhoe, V., & Salau, S. (2012). Nigeria. In X. Diao, J. Thurlow, S. Benin, & S. Fan (Eds.), *Strategies and Priorities for African Agriculture: Economywide Perspectives from Country Studies* (p. 231). Washington, DC: International Food Policy Research Institute.

Dorward, A., Anderson, S., Clark, S., Keane, B., & Moguel, J. (2001). *Asset functions and livelihood strategies: a framework for pro-poor analysis, policy and practice*. Academic Press.

Dungumaro, E., & Madulu, N. (2003). Public participation in integrated water resources management: the case of Tanzania. *Physics and Chemistry of the Earth, Parts A/B/C, 28*(20-27), 1009-1014.

Fareed, F. (2015). To develop Saansad Adarsh Gram Yojana villages, Centre cites three models. *The Indian Express*. Retrieved from http://indianexpress.com/article/cities/lucknow/to-develop-saansad-adarsh-gram-yojana-villages-centre-cites-three-models/

Gafsi, M., Le Tron, S., & Mouchet, C. (2010). Organic farming is it a sustainable agriculture. Development issues for sustainable organic farming in Midi-Pyrénées Région. In' Innovation and Sustainable Development in Agriculture and Food, June, 28.

George, T. (Ed.). (1999). India 1000 to 2000 (1st ed.). Chennai: Express Publications (Madurai) Ltd.

Gill, S. (2008). When Rural Development Succeeds? Experience from the Two Punjabs and West Bengal. In S. Bhaumik (Ed.), *Reforming Indian Agriculture: Towards Employment Generation and Poverty Reduction: Essays in Honour of G.K. Chadha* (pp. 293–294). New Delhi: SAGE Publications India Pvt Ltd.

Gopalakrishnan, T. (2007). *Vegetable crops. Pitam Pura*. New Delhi: New India Pub. Agency.

Gunderson, L. H. (2000). Ecological resilience—in theory and application. *Annual Review of Ecology and Systematics*, *31*(1), 425–439. doi:10.1146/annurev.ecolsys.31.1.425

Hatekar, N., Kulkarni, S., & Mehta, P. (2015). Culture, Community and Institutions Voluntary Provision of Public Goods in Maharashtra. *Economic and Political Weekly*, *L*(46-47).

Hiware Bazar Panchayat Records. (2016). *Hiware Bazar. Implementing Water Resources Management: Information brief.* United Nations. Retrieved 1 October 2017, from http://www.un.org/waterforlifedecade/waterandsustainabledevelopment2015/images/implementing_water_eng.pdf

Kellner, K., & Bosch, O. J. H. (2003). The value of local and indigenous knowledge for the development of information systems for conservation management. In Conserving Biodiversity in Arid Regions (pp. 105-118). Springer US. doi:10.1007/978-1-4615-0375-0_8

Kobets, M. I. (2004). Organic farming in the context of sustainable development. Kyiv: Agrarian Policy for Human Development, 22.

Kumar, A. (2012). P. Sainath on "Inequality, livelihood and agrarian crisis" in India. *New Socialist Alternative*. Retrieved 11 September 2017, from http://www.socialism.in/index.php/p-sainath-on-inequality-livelihood-and-agrarian-crisis-in-india/

Kunke, R. (2012). One village. 60 millionaires. The miracle of Hiware Bazar. *Development Dialogue*. Retrieved 17 October 2017, from http://agrariancrisis.in/2013/08/17/one-village-60-millionaires-the-miracle-of-hiware-bazar/

Laven, A., & Boomsma, M. (2012). *Incentives for sustainable cocoa production in Ghana*. Amsterdam: Royal Tropical Institute.

Lipton, M. (1994). *Growing points in poverty research: labour issues*. International Institute for Labour Studies.

Mundle, S. (2017). Agrarian crisis: the challenge of a small farmer economy. *Livemint*. Retrieved from http://www.livemint.com/Opinion/Y3Fp6CcumJhRIEwl2WeUMM/Agrarian-crisis-the-challenge-of-a-small-farmer-economy.html

Neely, C., Sutherland, K., & Johnson, J. (2004). *Do sustainable livelihood approaches have a positive impact on the rural poor? A look at twelve case studies*. Academic Press.

OECD. (2012). Environmental Outlook to 2050: The Consequences of Inaction. Paris: OECD.

Open Working Group. (2014). *Open Working Group proposal for sustainable development goals*. New York: Author.

Potter, R., Conway, D., Evans, R., & Lloyd-Evans, S. (2012). *Key concepts in development geography*. London: SAGE. doi:10.4135/9781473914834

Rametra, A. (2016). *India Together: Rays of hope for the 'local' in Meghalaya*. Retrieved 13 October 2017, from http://www.indiatogether.org/rays-of-hope-for-the-local-in-meghalaya-agriculture

Rigg, J. (2016). *Challenging Southeast Asian Development: The Shadows of Success*. New York: Routledge.

Rolston, A., Jennings, E., & Linnane, S. (2017). Water matters: An assessment of opinion on water management and community engagement in the Republic of Ireland and the United Kingdom. *PLoS One*, *12*(4), e0174957. doi:10.1371/journal.pone.0174957 PMID:28369136

Sarojam. (2016). *The Village Of Millionaires - Hiware Bazar - Spiritual Discussion*. Retrieved 3 February 2018, from https://www.speakingtree.in/discussion/the-village-of-millionaires-hiware-bazar

Satish, A. B., Bhaveshananda, S., & Sengupta, B. (2012). Organic agriculture: Way towards sustainable development. *International Journal of Environmental Sciences*, *3*(1), 209.

Scoones, I. (1998). *Sustainable rural livelihoods: a framework for analysis*. Academic Press.

Sen, A. (1975). *Employment, Technology and Development*. Oxford, UK: Clarendon Press.

Seufert, V. (2012). Organic agriculture as an opportunity for sustainable agricultural development. *Research to Practice Policy Briefs. Policy Brief*, 13.

Spring, J. (2015). *Economization of Education: Human Capital, Global Corporations, Skills-Based Schooling*. New York: Routledge.

Udmale, P., Ichikawa, Y., Manandhar, S., Ishidaira, H., Kiem, A., Shaowei, N., & Panda, S. (2015). How did the 2012 drought affect rural livelihoods in vulnerable areas? Empirical evidence from India. *International Journal of Disaster Risk Reduction*, *13*, 454–469. doi:10.1016/j.ijdrr.2015.08.002

Vezzoli, C., & Manzini, E. (2008). *Design for Environmental Sustainability*. London: Springer Science & Business Media.

Walsh, B. (2012). WholeFood Blues: Why Organic Agriculture May Not Be So Sustainable. *TIME.com*. Retrieved 26 January 2018, from http://science.time.com/2012/04/26/whole-food-blues-why-organic-agriculture-may-not-be-so-sustainable/

WWAP (United Nations World Water Assessment Programme). (2015). *The United Nations World Water Development Report 2015: Water for a Sustainable World*. Paris: UNESCO.

Zoomers, A. (2013). Rural livelihoods. In V. Desai & R. Potter (Eds.), *The Companion to Development Studies* (2nd ed.). New York: Routledge.

ADDITIONAL READING

Belay, M., & Bewket, W. (2015). Enhancing rural livelihoods through sustainable land and water management in Northwest Ethiopia. *Geography, Environment. Sustainability*, *8*(2), 79–100.

Biggs, E. M., Bruce, E., Boruff, B., Duncan, J. M., Horsley, J., Pauli, N., & Haworth, B. (2015). Sustainable development and the water–energy–food nexus: A perspective on livelihoods. *Environmental Science & Policy*, *54*, 389–397. doi:10.1016/j.envsci.2015.08.002

Brock, K., & Coulibaly, N. (1999). *Sustainable rural livelihoods in Mali.* Brighton: Inst. of Development Studies.

Carney, D. (1998). Implementing the sustainable rural livelihoods approach. *Sustainable rural livelihoods: What contribution can we make, 3*, 27.

Gautam, Y., & Andersen, P. (2016). Rural livelihood diversification and household well-being: Insights from Humla, Nepal. *Journal of Rural Studies, 44*, 239–249. doi:10.1016/j.jrurstud.2016.02.001

Jonah, C., Maitho, T., & Omware, Q. (2015). Water Access and Sustainable Rural Livelihoods: A Case of Elementaita Division in Nakuru County, Kenya. *International Journal of Science. Technology and Society, 3*(1), 9.

Kanel, K., & Niraula, D. (2017). Can rural livelihood be improved in Nepal through community forestry? *Banko Janakari, 14*(1), 19. doi:10.3126/banko.v14i1.17055

Martin, S. M., & Lorenzen, K. A. I. (2016). Livelihood diversification in rural Laos. *World Development, 83*, 231–243. doi:10.1016/j.worlddev.2016.01.018

Pelletier, B., Hickey, G. M., Bothi, K. L., & Mude, A. (2016). Linking rural livelihood resilience and food security: An international challenge. *Food Security, 8*(3), 469–476. doi:10.100712571-016-0576-8

KEY TERMS AND DEFINITIONS

Adarsh Gaon: An ideal village from the standpoint of rural development that encompasses stable employment, income, social development, cultural development, as well as overall wellbeing for its residents.

Anganwadi: A form of rural mother and childcare center that provides basic health care in Indian villages.

Footloose Migration: Migration that is underscored by the need for better employment and livelihood opportunities.

Organic Farming: A method of crop production that does not use pesticides, synthetically compounded fertilizers and chemicals but employs crop residues, animal manures, green manures, off farm organic wastes, and biofertilizers in the interests of maintaining soil health and fertility.

Participatory Assessment and Planning for Sustainable Livelihoods (PAPSL): An approach that focuses on community participation and involvement at every step in assessment and planning for sustainable livelihood.

Resilience: The ability to recover and bounce back quickly from stresses and shocks.

Sustainability: Judiciously managing resources for meeting the needs of the present by preserving the interests of posterity.

Vulnerability: Reduced capacity of an individual or group to foresee, withstand, endure and recover from the impact of a natural or man-made calamity.

Watershed Development: Conservation, regeneration, and the judicious use of water within the watershed area.

Chapter 6
Sustainable Development From a "Nature vs. Nurture" Perspective

Lucia Ovidia Vreja
University of Economic Studies, Romania

Sergiu Bălan
University of Economic Studies, Romania

ABSTRACT

This chapter presents the role of nature and nurture in shaping the behavior of human beings toward sustainability identifying instances of both dramatic extinctions of species and collapse of entire societies, as well as successful, peaceful, and healthy adaptation of human communities to their environment, in an attempt to presents the imperative conditions necessary for attaining sustainable development. A very long and intriguing history reveals that from the nature's point of view humans are rather destructive, interested in their own short-term survival. Nevertheless, the same long history of human species bears valuable lessons and examples of adaptive behaviors grounded by nurture, and based on these examples, the chapter aims at advancing a new perspective of thinking sustainable development that could lay the foundation of a new education curriculum.

INTRODUCTION

We have entered the Anthropocene, without question. Some of the changes, for example those brought about by large reservoirs and megacities, are here to stay for hundreds if not thousands or even millions of years. We will simply have to get used to and find a way of living with such changes. But there are things we can and should change to keep improving human wellbeing and to avoid crossing potentially dangerous thresholds. Our strength as humans is the capacity to recognize problems, to understand them and to develop solutions. The final chapter of the Anthropocene story is yet to be written: the narrative will depend on our collective self-awareness and the capacity to correct our course, for the relentless pressure on our planet portends unprecedented destabilization. (Syvitski, 2012)

DOI: 10.4018/978-1-5225-5739-5.ch006

Copyright © 2019, IGI Global. Copying or distributing in print or electronic forms without written permission of IGI Global is prohibited.

Attaining sustainable development in our contemporary increasingly complex societies is of utmost importance for global entities, national-level political decision makers, and ordinary individuals, equally. Sustainability is simply defined as the capacity of an entity to ensure its survival without endangering the survival of future generations, thus touching upon the well-known "triple bottom line" of sustainability (Elkington, 1997). The definition points out to the three main interdependent domains of action that are necessary to ensure sustainability translated into economic growth, social development and environmental protection (Hitchcock & Willard, 2009). Finding global scale or locally applicable solutions to problems associated with economic development, environment protection and responsible exploitation of resources has become a critical goal of every development policy, pertaining to the very endurance of the human species on the long run. Nevertheless, national states or local communities continue to have severe difficulties in promoting both economic development and environmental protection, in spite of carefully planned national policies and strategies.

From nature's perspective, human beings do not necessarily manifest a sustainable behavior, and examples of species going extinct for not being able to adapt to their environment are informative in this regard. At the same time, various societies have succeeded in identifying, promoting and manifesting adaptive behaviors, ensuring the peaceful and successful acclimatization to a specific environment. What these latter societies seem to have and do differently refer to a set of cultural values and norms reflected in a respected and promoted "traditional knowledge", defined as valuable information resulted from a long process of thinking and *tested in the rigorous laboratory of survival* (Hunn, 1993, p. 13). Traditional societies are collectivist societies, primarily interested in the survival of the community rather than the survival of the individual, which use culture and its carriers (among which, traditions, moral and social norms, religion, etc.) as an instrument of social control.

Numerous examples of traditional knowledge used to ensure the long-term survival of a society's members (Berkes, 2008, p. 28) are still instructive for any policy maker. For instance, populations living in tropical forest and in other habitats practice crop rotation to maintain the qualities of the soil; traditional pastoral societies use a complex system of alternation or transhumance for protecting the grazing areas; mountain people have long discovered terracing as a method to conserve soil and water; almost all societies have discovered and instituted taboos and various utilization procedures to protect animals, plants and places. In such societies, the traditional knowledge is transmitted from generation to generation through socialization, rule observance is ensured by interiorization and group monitoring, and the breaking of the rules is informally sanctioned at the level of the primary group, mainly the extended family.

Nevertheless, in current modern individualistic societies traditions, religion and customary rules are considered obsolete and harshly debated, primary groups, especially the family and the elders, have lost their authority over the younger generations, and the social control is exerted by secondary, formal contractually based groups, which watch over law observance and apply formal sanctions.

The main objective of the chapter is to elaborate on examples of both adaptive and maladaptive behaviors of individuals and societies that led to either their self-destruction or their beneficial adaptation to the environment in an attempt to identify the elements that were essential for survival and endurance. At the same time, the chapter presents the role and importance of both nature and nurture in directing the behavior of human beings towards sustainability, with a particular focus on culture and education as valuable and yet insufficiently exploited sources of adaptive behaviors, in an attempt to promote a new perspective of thinking sustainable development, that could provide the foundation of a more efficient informal and formal education.

ARE HUMANS NATURAL ECOLOGISTS?

Homo Sapiens as the Dominant Species

According to the authors of *The Geologic Time Scale 2012*, we are still officially living in the geological epoch known as *Holocene*, which is the final epoch of the *Quaternary* period, which also is the final period of the *Cenozoic* era (Gradstein & Ogg, 2012). Nevertheless, in the 32nd chapter of the book, named *The Anthropocene*, we are told that an International Commission on Stratigraphy (ICS) working group is studying the potential formal introduction of this denomination in the Geological Time Scale, formally admitting the fact that we are no more living during the Holocene, but the Anthropocene, which would mean an official recognition of the significant geological impact of human activities worldwide. Whether the introduction of a new denomination will happen or not depends on certain factors, such as the estimated usefulness of the new terminology for the natural scientists, the theoretic value of scientific arguments for and against it, and the ability of geologists to find an undisputed effective boundary marking the beginning of this new geological epoch, either a Global Boundary Stratotype Section and Point (GSSP) or a Global Standard Stratigraphic Age (GSSA) (Zalasiewicz, Crutzen & Steffen, 2012). As any geological period is defined by its beginning or lower boundary, it is necessary to identify a GSSP, i.e. an event in the stratigraphic material or a marker, such as a sediment layer, together with some auxiliary markers that indicate some substantial changes in the Earth system. Such an identification would mean the beginning of a new geological era, such as the peak in iridium concentration dated 66 million years ago that marks the Cretaceous-Paleogene boundary and the start of the Cenozoic Era (the extinction of dinosaurs and the beginning of the mammal dominance era, presumably caused by the impact of a meteorite). If the boundary marker cannot be identified, then a GSSA, i.e. a conventional date, is agreed upon as is the case with periods situated further back in time, such as Precambrian (more than 540 million years ago), when precise events are more difficult to locate and markers not so easy to find (Lewis & Maslin, 2015, pp. 172-173).

Up to now, the final verdict of the ICS working group is still awaited, but a whole series of candidate events and markers were proposed, such as the first nuclear weapon detonations, the presence in the soil of the persistent industrial chemical products, the Industrial Revolution, the creation of anthropogenic soils, extensive farming, the invention of agriculture and so on. The two scientists that formally proposed the introduction of the term *Anthropocene* in 2000, the Nobel laureate for chemistry Paul Crutzen and the biologist Eugene Stoermer, did choose as a possible GSSA the end of the 18th century because they believe that since then the ecological consequences of human activities have become clearly and undeniably evident (Crutzen & Stoermer, 2000, p. 17). As they argue, our species' environmental footprint is ever increasing as time goes by, and the impact our activities have on the ecosystem is almost always a negative one. For instance, during the last 300 years Earth's human population increased ten times, from circa 6-700 million to 7 billion people, so together with our domesticated animals, we account for more than 90% of terrestrial vertebrate biomass, i.e. approximately 300 million tons (humans) plus 700 million tons (livestock) versus 100 million tons of wild vertebrates (Harari, 2016, Ch. 2).

Already twenty years ago Vitousek, Mooney, Lubchenco and Melillo (1997) presented data from different domains, such as land and ocean transformation, alteration of geological, chemical and biological cycles, biotic changes, which unequivocally prove that the alteration of Earth's ecosystem due to human action is serious and continues to increase, having among its consequences the global warming and irreversible loss of biodiversity. As regarding the transformation of land by human action, they maintain that the fraction of land affected or degraded through urbanization, industry, agriculture, forestry, and animal grazing can be estimated at between 39% and 50% of the total Earth surface. The alteration of oceans, though more difficult to quantify, comes from the fact that about 60% of human population is concentrated within 100 km limit from the coastline, so human action is more intensive here. For instance, as a consequence, approximatively 50% of mangrove ecosystems, situated in these coastal areas, have already been destroyed or irremediably transformed by human action. Even if humans do not use more than 8% of the production of the oceans, many commercial fisheries are not sustainable for a long time, and already in 1995 22% of the marine fisheries were depleted and another 44% were at their maximal limit of exploitation. Moreover, fishing is not selective, so one third of the total capture (around 27 million tons of non-target animals caught by mistake in 1995) is thrown back into the oceans, substantially increasing the damages (Vitousek, Mooney, Lubchenco & Melillo, 1997, p. 495).

Both these kinds of transformations caused by humans, together with other actions and activities have serious consequences concerning the bio-geo-chemical cycles of life on Earth. The most obvious and important example is the carbon cycle, since life on our planet is carbon-based. Here, the results of human activities are first and foremost related to the CO_2 emissions resulted from the burning of fossil fuels, the best documented sign of human alteration of the environment. The analysis of polar ice samples shows that the CO_2 concentration in atmosphere before the year 1800 was about 0.028% and relatively constant, but has increased exponentially since that period to this day, and at the end of the twentieth century the burning of fossil fuels added around 5.5 billion tons of CO_2 in the atmosphere each year. The main consequence of the increasing level of CO_2 is the greenhouse effect, conducing to a serious climate change that is already happening, yet there are also perturbations in the growth rhythm of plants that furthermore decrease the animal food quality (Vitousek, Mooney, Lubchenco, & Melillo, 1997, p. 496). Another bio-geo-chemical cycle seriously perturbed by our activities is the water cycle, given that more than 50% of the total fresh water quantity is used by humans, mostly – approximately 70% – in agriculture. In order to be able to satisfy this huge demand, humans were forced to alter the courses of about two thirds of all world's rivers in such a measure that today very little water from major rivers, such as the Nile, the Ganges or the Colorado can finally reach the sea, while the surface of major inland lakes, such as Aral Sea and Lake Chad was seriously reduced (Vitousek, Mooney, Lubchenco, & Melillo, 1997, p. 497).

All these environmental changes have important consequences on the biosphere. While a certain rate of species extinction is natural, the current rate of genetic diversity loss is far above the normal, and is, of course, irreversible. The estimates are that the rates of species disappearance are up to 1000 times higher than those happening before the human conquest of the planet (Vitousek, Mooney, Lubchenco, & Melillo, 1997, p. 498). Besides extinction, the destruction of ecosystems has caused also invasions of new species in some areas, leading to serious changes of biotic systems, by the mixing of species of plants and animals that were for long time geographically isolated. For instance, on many islands, more than half of species are not indigenous. The problem is that most invasions are irreversible, and the introduction of new species has negative effects both on the local species, driving them to extinction, as well as on human health, since, for example, most infectious diseases, such as plague or cholera, are caused by invading organisms.

The conclusion drawn by Vitousek and his colleagues is a radical one: the global negative effects of human activity are not a problem of the future but of the present, as they are already happening here and now at a scale of magnitude and a speed that must raise the most serious concern. There is only one cause to be singled out behind all these consequences, namely *the growing scale of the human enterprise*

(Vitousek, Mooney, Lubchenco, & Melillo, 1997, p. 498). This conclusion seems to be in accord with Crutzen's and Stoermer's suggestion (2000) that the beginning of the Industrial Revolution should also be considered the starting point of the Anthropocene. In other words, the "human enterprise" means technology used on an ever-increasing scale, therefore we should conclude that the human alteration of Earth's environment represents a side-effect of our contemporary technological culture. Humans as technological and cultural beings have become more and more destructive, while back then, when they were only natural beings among all others, endowed with only a very primitive technology, they lived in complete harmony with the rest of the world.

But is our destructive attitude towards the environment an essentially cultural one, driven and reinforced by our species' scientific and technological progress, i.e. by our ever-increasing ability to further alter and modify the natural world? In other worlds, has the individual from contemporary large-scale societies become more and more destructive, while the humans from traditional societies had an essentially benign relationship with their world and lived in harmony and equilibrium with their environment? Is technology the real and ultimate "Original Sin"?

The Myth of the Ecologically Noble Savage

When it comes to the question of the ecological knowledge and practices of people in traditional societies, there are at least three different perspectives or, as the eminent ecologist Fikret Berkes called them, three contradictory myths that sometimes overlap, about people in these societies and their relation with the environment (Berkes, 2008, p. 225). The first, called the myth of "the Exotic Other", is the conception according to which people in traditional societies live in harmony with nature because they are closer to their environment and well attuned to it, so they can do no harm to nature whatsoever. In that, they are very different from us, the people of modern technological societies. They possess a certain ecological knowledge lost forever for us, which makes them superior to us in this respect, therefore we see them with reverence, as the *ecologically noble savages*. The second myth sees both traditional and modern individual as "the intruding wastrel" and advances the opposite view, that humans are always alien intruders and destroyers of natural environments. Individuals from traditional societies are not in possession of special ecological knowledge, but are superstitious, ignorant, careless and destructive, and apparently never lived in harmony and equilibrium with nature. Their existence was always on the brink, at the discretion of natural forces, and so long as their numbers were small and the technology simple enough, the impact they had on the environment was minor, although they always had an inherent destructive tendency. Finally, the third myth about people in traditional societies is somewhere in between the first two: traditional individual is seen in a sort of duality, as *noble savage/fallen angel*. From this point of view, as long as they practice their traditional, primitive way of living, humans are in harmony with the environment, but when these societies are contaminated by the contact with modern world, they become a threat to themselves and to the ecosystem in which they live (Berkes, 2008, p. 226).

Of these three ideal images of traditional societies, the most interesting seems to be the first one, namely the myth of "the Exotic Other". It seems to be born from the various adaptations of the "noble savage" myth, originated in Rousseau's view on the human life in the ideal, free and uncorrupted "state of nature" (as opposed to the "state of society") and from its adaptation to the environmentalist context, where it generated the "ecologically noble savage" paradigm (Wheelan, 1999). But exactly how did scientists come to adopt this perspective *that seized such a strong hold on the anthropological imagination in the last decade of the twentieth century* (Ellingson, 2001, p. 345)?

As Redford (1991) suggested, the original myth of the noble savage originated in the idealized European image of the natives of the New World, seen by John Dryden, Jean-Jacques Rousseau or Thomas More as innocent dwellers of the Garden of Eden, beyond all the sins of the modern individual, egalitarian and living according to the laws of nature, in harmony with each other and the environment. As he continues,

it is the latter idea, that Indians lived in conformity with nature, that inspired this century's reincarnation of the noble savage. Writings of several scientists and indigenous rights advocates echo the early chroniclers' assumption that indigenous people lived in 'balance' with their environment. Prominent conservationists have stated that in the past, indigenous people 'lived in close harmony with their local environment.' [...] The idealized figure of centuries past had been reborn, as the ecologically noble savage. (Redford, 1991, p. 46)

However, the argument goes on, the more recent gathered evidence shows that pre-Columbian natives were not such perfect environmentalists or just another species of big animals with limited capacity of altering the natural environment, but just the opposite as their actions severely altered their surroundings. Therefore, the people in traditional societies were just as destructive as the modern people are.

The same kind of idealization can be found in the rhetoric of the ecologist political movements all over the world. As Wheelan (1999) argues, the emergence of Green or environmental political movement brought a new kind of worldview into the political arena, adopted by its followers with the fervor of a religious faith. This view is founded on the idea that due to the ever-increasing demands of the modern, industrial way of life, human species is now essentially in conflict with nature and on the way of completely destroying it as our disruptive actions will eventually lead to a major environmental catastrophe. The Greens argued that the entire modern lifestyle, based on science, technology, consumerism, industrialism and an arrogance towards nature that has its sources in the Judeo-Christian tradition of conceiving man as the pinnacle of creation and master of the world, was fundamentally wrong. Therefore, we must reconsider our relationship with the planet, which means a radical change of our lifestyle by a re-definition of our relationship with nature. The Green militants believed that they have found some convenient role-models, namely people from traditional societies, who were seen as living in a perfectly harmonious relationship with their world, inflicting minimal damages on it. Therefore, the Green movement activists more and more confidently promoted the idea of emulating the ecological wisdom of primitives:

The wisdom of the tribal peoples, and its favorable contrast with the cruel rapacity of the white man's approach to the environment, has become one of the mantras of the Green movement. Primitive people, supposedly living in harmony with nature, have become, in Wallace Kaufman's phrase: 'the gold standard of the environmental movement. Against this standard, it measures the values and achievements of our society'. The environmental holocaust will only be averted, according to this argument, when we are humble enough to sit at the feet of tribal peoples and absorb their wisdom. (Wheelan, 1999, p. 17)

All this exacerbated ideological hubris was based on the postulate of the natural environmentalist attitude of people in traditional societies. In order to assess the correctness of this assumption, human behavioral ecologist Bobbi S. Low (1996) used ethnographic information from the greatest anthropological database, the Human Relations Area Files, and compared the data gathered from 168 traditional societies concerning their natural resource attitudes and practices. She wanted to find answers to questions such as if there is evidence of environmental degradation in traditional societies, if this is connected to ecological conditions, if people in these societies have a conservation ethic or if the societies that are more individualistic suffer more or less environmental degradation than those that are more altruistic. Her conclusions derived from the statistical data analysis are drastic: people in traditional societies do not have a general conservation ethic, but act rather according to individual needs than to some long term common interest. Traditional societies can cause great degradation of the environment, especially when there are significant and fast changes in population or technology (Low, 1996, p. 368). In addition to the ethnographical data analysis, Low uses in support of her conclusions some examples of ecological disasters from the past, such as the Pleistocene extinction of many species of large and small birds and mammals from Europe, America, Asia and Australia, probably caused by human actions, or the collapse of Mayan empire and civilization due to failures of the agricultural economic system, and the Polynesian and Hawaiian extinctions of native species caused by the first human colonization of Pacific islands (Low, 1996, pp. 361-362).

The first one of these examples, the Pleistocene extinction, is of special interest for the current discussion, because it concerns pre-historical events that did happen before human species discovered agriculture and thus, before the beginning of the journey towards civilization and modernity. Those events happened even before the birth of traditional societies, in the days of the great migrations that brought *Homo sapiens* from their original areal in Africa to the other continents and made humans a global species. So, if it results that humans in this stage of evolution, before the advent of civilization or "in state of nature", did live in harmony with their environment, then the "ecologically noble savage" myth will not be contradicted for now. Therefore, it can be assumed that humans are natural environmentalist, and only the birth of culture, large scale societies and technological development made them become a menace for the planetary ecosystem. If, on the other hand, it will be discovered that even in their natural state, at the dawn of their existence as a new hominid species, their actions had destructive consequences on the environment and if, in other words, the Pleistocene extinctions can be connected with human activities, then the myth of the "ecologically noble savage" is falsified. In such a case, humans are far from being natural environmentalists, the Anthropocene has begun much earlier than Crutzen and Stoermer believed, and *Homo sapiens*, by their very nature, must thus be seen as a menace for the environment. Accordingly, solutions to this problem must be found not in human nature, but in the cultural capabilities of our species.

The Pleistocene Extinction of Megafauna

Our species, *Homo sapiens*, is just one of the many hominids that evolved from an earlier genus of apes, *Australopithecus*, starting from about 2.5 million years ago: *Homo rudolfensis, Homo erectus, Homo ergaster, Homo neanderthalensis, Homo soloensis* and so on, that inhabited the Earth side by side until around 10.000 years ago, when archaeological evidence shows that humans remained the sole hominid species on the planet (Harari, 2015, Ch. 1). The earliest evidence found in fossil record shows that *Homo sapiens* appeared in East Africa (South-West Ethiopia) about 195,000 years ago and then, after a long period of more than 100,000 years of sedentary life, migrated and dispersed out of Africa in the same manner that *Homo erectus* and *Homo heidelbergensis* did before him. An intriguing feature of these early *Homo sapiens* is that for a very long period of time, they had a very similar behavior with their ancestors, and only about 70,000 to 50,000 years ago they gradually became an altogether different kind of hominid, destined to rule the world (Maslin, 2017, p. 38). This means that in the first period of their existence, *Homo sapiens*, similar to their ancestors, did have stone tools and discovered the use of fire, but nevertheless subsisted mainly by gathering fruits and digging roots, hunting small animals, such as rodents and almost never large game, and eating carrion left by the more powerful predators (Harari, 2015, Ch. 1). Then, something happened that helped humans to "jump to the top of the food chain" in a blink of an eye. The problem is that the other animals at the top of the pyramid, such as lions and sharks, reached that position by evolving in a very long period of time, over millions of years. This fact granted the ecosystem the time needed to develop mechanisms of balance that maintained the ecological equilibrium, preventing the predators to inflict too much damage on other species. On the contrary, humans ascended so quickly in the top position of the food chain that the environment did not have time to adjust, as neither did humans. Suggestively expressed by Harari,

most top predators of the planet are majestic creatures. Millions of years of dominion have filled them with self-confidence. Sapiens by contrast is more like a banana republic dictator. Having so recently been one of the underdogs of the savannah, we are full of fears and anxieties over our position, which makes us doubly cruel and dangerous. Many historical calamities, from deadly wars to ecological catastrophes, have resulted from this over-hasty jump. (Harari, 2015, Ch. 1)

Starting from about 70,000 years ago, humans from eastern Africa begun to spread, at first into the Arabian Peninsula, then in the Middle East, Asia (60,000 years ago), Europe (45,000 years ago), Australia (45,000 years ago), North America (16,000 years ago) and South America (12,000 years ago). At the very same time, more than 97 of 150 then existing genera of large mammals (over 44 kg) went extinct immediately after the arrival of humans in their habitat: 8 species in Sub-Saharan Africa, 9 in Europe, 21 in Australia, 33 in North America, and 50 in South America. For these dramatic events, three types of causes were proposed as explanations: the impact of human activities (especially in Australia, and North and South America), climate changes (especially in Africa, Europe and Asia) and a combination of the climate changes and human actions (Barnosky, Koch, Feranec, Wing, & Shabel, 2004, pp. 70-71). Barnosky and his colleagues analyzed the latest evidence from simulations, chronology, archaeology, paleontology, climatology, and ecology in order to find arguments for or against the three possible explanations and discussed different anthropogenic extinction models, such as *overkill* (excessive hunting), *blitzkrieg* (rapid overkill) and *sitzkrieg* (extinction due to use of fire, habitat destruction and the introduction of new species and diseases). Their conclusion is that *the case for humans contributing to extinction is now much stronger than it was in the early days, with support from recent work on chronology, simulations, paleoclimatology, paleontology, and archaeology* (Barnosky, Koch, Feranec, Wing, & Shabel, 2004, p. 73). However, they believe that it would be an oversimplification to say that the arrival of humans meant everywhere the beginning of a process of hunting to extinction of all indigenous large mammals. The *blitzkrieg* hypothesis can be rejected for the case of Europe, Siberia and Alaska, where climate change had a role to play too, together with human activities, so:

the recent information now points toward humans precipitating the extinction, but also to an instrumental role for late Pleistocene climatic change in controlling its timing, geographic details, and perhaps magnitude. (Barnosky, Koch, Feranec, Wing, & Shabel, 2004, p. 74)

The *blitzkrieg* hypothesis was discussed and tested time and again by scientists. For instance, ecologists Barry W. Brook and David M. J. S. Bowman (2004), from the Key Centre for Tropical Wildlife Management of the Charles Darwin University, used population modelling and meta-analysis of empirical data to investigate the possible causes and scenarios of the Pleistocene extinctions. They concentrated their attentions on the *blitzkrieg* hypothesis, analyzing the rate at which the extinctions may have occurred, in connection with the empirically documented relation between animal's body mass and its proneness to go extinct due to human overkill. Their conclusion is that, given the incomplete and imperfect information at our disposal, we cannot single out the most plausible scenario of the observed extinctions, but however,

> *one universal prediction, which applied in all scenarios in which the empirical distribution was correctly predicted, was for the extinctions to be rapid following human arrival and for surviving fauna to be suppressed below their pre-'blitzkrieg' densities. In sum, human colonization in the late Pleistocene almost certainly triggered a 'blitzkrieg' of the 'megafauna', but the operational details remain elusive. (Brook & Bowman, 2004, p. 517)*

Apparently, the conclusion must be that all these studies should lead to the rejection of the "ecologically noble savage" hypothesis and to the conclusion that, by their nature, humans are not inborn environmentalists but, *au contraire*, their actions have had destructive effects on the environment even before the advent of civilization and sophisticated technology. But, in these circumstances, if humans are so destructive, one cannot but wonder how come they didn't cause even more damages, and what kept them from bringing the whole natural world to extinction? After all, certain human societies, either primitive or even some more advanced and sophisticated, such as the Maya civilization, the Pueblo Amerindians, the Vikings, the Norse in Greenland or the people from Easter Islands, did exactly that: brought an end to their own existence by way of exhausting or destroying their natural environment (Diamond, 2005). Nevertheless, humans not only did not bring themselves to extinction, but in the end, they became the most successful predator on Earth, occupying virtually every ecological niche, from the Equator to the Poles, and did that in a very sudden manner. How did this happen?

An answer to this question can be found in the evolutionary theories that explain human behavior, and especially in the most sophisticated one, the theory of gene-culture coevolution (Laland & Brown, 2002). Charles Darwin believed that the distinctive note that differentiated us from the other primates is our big brain. However, this cannot be the answer, because the Neanderthals had an even bigger brain, and yet they went extinct, most probably exterminated by the humans that invaded their territories when they arrived in Europe from Africa (Harari, 2015, Ch. 1).

The secret of human species' success is neither their native intelligence, nor some innate instincts that prevent humans from inflicting self-destruction, but the fact that they gradually became a cultural species. This means that humans began to learn from each other's experience in such a way that culture became cumulative, so as each generation could use the knowledge accumulated before, and at the same time could built up on it and improve it. Human individuals no longer used first and foremost the personal experience to guide their behavior, but rather the cumulative experience of their society, transmitted as cultural knowledge, through learning. Once this process started, it snowballed, and no one could survive without learning all this complex toolkit of practices and techniques: the hunting secrets, the tool making procedures, the way to distinguish between edible and non-edible plants, cooking methods, and so on. In these new conditions, natural selection began to favor those individuals who were better cultural learners, who could better appropriate and exploit the body of accumulated knowledge in a society. Thus, as anthropologist Joseph Henrich (2015) puts it, *cultural evolution initiated a process of self-domestication, driving genetic evolution to make us prosocial, docile, rule followers who expect a world governed by social norms monitored and enforced by communities* (Ch. 1). A new form of status appeared in human societies, alongside with dominance status that we inherited from our ape ancestors, namely the status of prestige. Humans begun to grant respect to those individuals who were wiser, that is who had more cultural knowledge than others, and who can thus be imitated and emulated and be sources of learned information. Of equal importance were the social norms that influence all human actions, from the relations between the members of a society to relations to strangers and, last but not least, relations to the natural environment. The norms were formalized as rules, permissions, interdictions and taboos, and norm following was of paramount importance in traditional societies, where a rule violation could bring loss of status, punishment, ostracism or even the death penalty. It is in these rules governing the relations with the environment that humans could find those cultural safety nets that prevented their ancestors to inflict irreparable damages to their environment and ensured the survival of their societies: only those societies that developed sophisticated and efficient norms that severely limited the native tendency to destroy the ecosystems did eventually survive and prosper.

An interesting feature of these norms is that they need not be explicit, which means that in order for them to effectively function, people don't have to understand their true signification and ultimate motivation, but only to obey them. In other words, one could expect to see people strictly following rules without really knowing why they are doing that. As the American philosopher Daniel Dennett (2007) argued, there can be the so-called "free floating rationales", or reasons without them necessarily being represented explicitly in the mind of human individuals:

Free-floating rationales emerge as the reasons why some features exist; they do not presuppose intelligent designers, even though the designs that emerge are extraordinarily good. For instance, there are reasons why termite colonies have the features they do, but the termites, unlike Gaudí, do not have or represent reasons, and their excellent designs are not products of an intelligent designer. (Ch. 3)

So, one could expect to find in the mass of ethnographical data gathered from various human societies that did eventually succeed to avoid fatally damaging their environment, and implicitly their own extinction, some examples of social norms that made this possible, yet without the people from those societies to understand that explicitly. Therefore, as long as social norms were enforced and tradition followed, people were not inflicting damages on their environment, although they did not explicitly understand what were they doing except that they were obeying to some "free floating rationale". Such instances constitute sufficient evidence to prove that social norms and traditions are the cultural constructs that did prevent human beings to inflict damage on their environment, whether they did understand and desire it or not.

HUMANS AS CULTURAL ECOLOGISTS

Naturally, all human beings are interested in their immediate survival, without concern for the long-term effects of their actions, thus being potentially destructive. Nevertheless, during their long history, humans succeeded in finding solutions to environmental problems and in adapting to their environment, especially through culture. Even though the distinction between "nature" and "nurture" is not quite accurate given the fact that no individual is the exclusive result of either one, the separation is necessary to better understand that culture plays an essential role in the history of humankind, in the survival of humans and nature equally.

Although not as spectacular and well-known as cases where individuals or entire societies failed to protect and conserve their resources and their physical and social environment, the instances of successful adaptation to environment and nature conservation are quite diversified and informative. The so-called *species-specific taboos* and *area-specific taboos* (Berkes, 2008, p. 40), as well as distinctive practices instituted by traditional societies, apparently irrational from an outsider's point of view, make perfect sense when analyzed from the perspective of the long-term survival of those communities. For instance, food taboos, defined as interdictions to consume certain types of food, are found in virtually every traditional society all over the world, although the rationale for establishing and enforcing such prohibitions varies from community to community. Meyer-Rochow (2009) has identified diverse reasons for instituting a food taboo, such as the preservation of a certain resource or the identification of a more efficient or optimal use of that resource, the protection of people's health, especially of certain categories of persons (such as pregnant women or children), the accentuation of certain events significant either for the individual (such as passage rites or initiation rituals) or for the community (climatic events such as droughts or floods), the desire to monopolize the access to and use of a resource. Nevertheless, the management of a valuable and scarce resource seems to be one of the most frequent motives for imposing a taboo or a certain practice, as resulted from a research of Colding and Folke (1997), who found out that approximately 30% of identified prohibitions referred to food taboos instituted on endangered species. Similarly, area taboos or so-called sacred groves, which are interdictions to enter, visit or exploit a certain region, mainly a forest area, are also found all over the world, especially in traditional societies dependent on the produce of the forest, such as India, Japan, Morocco, Ethiopia and Ghana (Kandari, Bisht, Bhardwaj & Thakur, 2014). Such sacred forests, protected through a complex system of religious practices, rules and obligations, play an essential social role for the community (Ramakrishnan, Saxena & Chandrashekara, 1998), by ensuring an unpolluted environment and a viable ecosystem, equally for humans, as well as for flora and fauna.

In traditional societies, the protectionist or conservationist practices are almost never enforced by law, but by a system of less formal social norms, such as traditions and mores, of which taboos and religious practices and constraints play a very important role (Berkes, 2008). It is worth mentioning here the numerous examples of food taboos promoted by various communities as religious norms, such as the interdiction to consume beef meat for the Hindu population or the ban on pork meat consumption in the Middle East. While it might seem irrational or, at least, idiosyncratic and even maladaptive nowadays, the ban on beef consumption should be understood as an unconscious cost-benefit analysis related to the environmental, demographic, technological and economic conditions of the population. According to Marvin Harris (1974), worshipping the cow, seen in the Hindu religion as the "mother of all things", is perfectly reasonable and explainable when understood in terms of a rational response to the ecological and economic conditions of rural life in India over millennia ago, manifested by the depletion of natural resources doubled by an increase of population density. For many individuals living in rural areas, where the subsistence of the population was ensured by the produce of the fields, which were traditionally ploughed by using working animals, the cattle as draught animals became an indispensable and scarce resource that needed protection against consumption, especially in times of famine and extreme drought. As a result, the Hindus designed a complex mechanism promoted in the form of a religious taboo that banned beef consumption and made it sacrilegious. A religious interdiction to eat pork is also found in Judaism and Islam and, in spite of numerous alternative or complementary explanations for the establishment and perpetuation of such a taboo (Simoons, 1994), converging arguments indicate that raising pigs in the Middle East, a region that suffered severe environmental degradation and climatic changes, would have been both "ecologically maladaptive" (Harris, 1974, p. 44) and even hazardous, given the fact that pigs and humans might exploit the same ecological niche.

The "pig hate" encountered in Judaic and Islamic societies is nevertheless opposed by an intricate and debatable "pig love", especially manifested by populations from New Guinea and islands of the South Pacific (Harris, 1974). The *kaiko* pig festival, extensively described by Rappaport (1984), refers to an extremely elaborate array of events involving massive pig ritual slaughter and culminating in a war. The "pig lovers" of New Guinea raise pigs for years, without sacrificing and using them as a permanent source of meat. On the contrary, pork meat is rarely consumed except when ritually feasting during a *kaiko* festival, which takes place at large time intervals. The festival seems at least whimsical if not completely irrational for an outsider, especially the preliminary activities leading to the ceremonial slaughtering of pigs and war preparations. In reality, the love of pigs and the war it involves are adaptive behaviors, aimed at regulating the density and structure of the population in accordance with the characteristics and components of the environment (Rappaport, 1984, p. 4). As Harris (1974, p. 48) consistently demonstrated, the festival has a very practical grounding although not consciously grasped by the actors, being a complex and efficient mechanism for adapting to the environment by controlling the number of human and animal population in accordance with existent resources and subsistence opportunities.

Regardless of their particularities related to form or content, all traditional cultures include social norms, especially taboos imposing restrictions on either "sacred" or "unsacred" animals, plants, places, etc. Such restrictions are primarily promoted as religious norms or passed from generation to generation as founding myths or moral parables, without being critically examined by the "believers". Although they do not usually know the real causes of such interdictions, members of traditional societies are complying with norms without questioning them, invoking either the tradition and stories circulating in society about the danger of violating the ban or the benefits associated with observing the rule for the health or wellbeing of the individual or the entire community (Quiroz & van Andel, 2015). In some situations, not only are members of traditional societies clueless about the logical or scientific explanations of the taboos, but they are even illiterate, yet with a rich and intense body of oral stories, folkways, mores, rituals and ceremonial procedures that ensure compliance with the norm (Kandari, Bisht, Bhardwaj, & Thakur, 2014).

When explaining a phenomenon or a behavior one should take into account both its *proximate* and *ultimate* causes. For instance, the particular behavior of a person can be interpreted as a result of the cultural norms existent in a society, this being the proximate explanation, while the ultimate explanation of that behavior should deal with why that specific behavioral pattern emerged and what was its origin (Sanderson, 2001). Using this line of reasoning, although the proximate causes of taboos are related to traditional practices or religious norms orally transmitted from older to younger generations, the ultimate causes seem to be related to more utilitarian and ecological reasons, such as preserving a resource that would go extinct because of overexploitation in the absence of protectionist measures (Colding, 1998, p. 14) or taking care of the people's health. Without being even aware, various populations have removed from their diet those species that pose the greatest risk of intoxication and even poisoning for some categories of individuals or those that, if driven to extinction, endanger the entire ecosystem and thus the very subsistence of the society.

In fact, behavioral constraints and "supernatural sanctions" for disobeying the rule emerge in societies when the use or consumption of a resource becomes too costly compared to the benefits of the community. Abundant resources, either animals or plants, that can be obtained without high costs and can be consumed without putting in danger the ecosystem were rarely, if ever, the target of supernatural prohibitions (Harris, 1978, p. 146). On the contrary, such severe prohibitions refer only to highly nutritional or valuable resources whose continued use would become not only more expensive, but even threatening to the very subsistence or endurance of a community.

Social norms, including folkways, mores or taboos as part of a society's culture are destined to limit the destructive force of humans, or so were they in traditional communities. Although such rules or taboos may seem irrational when considered from an outsider's perspective, they are a thorough cost-benefit analysis, not necessarily conscious, of humans trying to adapt to their environment and survive on a long run. In all traditional societies, cumulative cultural evolution has generated adaptive responses to the environment, allowing human populations to accumulate traits, techniques, practices, norms, institutions that have allowed them to survive and thrive. Such complex adaptations arose precisely as a result of common wisdom accumulated over generations, under the form of culture, through individuals who have not questioned cultural heritage and existing knowledge of society. Cumulative culture, through its carriers such as shared norms, ensured the social control of the population and, ultimately, the very survival of the society. Most of the taboos, in fact environmentally protective measures, were transmitted as religious norms in traditional societies because in such societies religion was not an individual matter, but a *social fact*, a common way of acting, doing things or feeling in a society, which exerts an external pressure or constraint on every member of the society (Durkheim, 1982, p. 59). For such societies, religion is part of their everyday life, whose roots are to be found in the real experiences and whose rites and myths, no matter how peculiar, are transcriptions of human needs or of various important aspects of life (Durkheim, 1995, p. 2).

However, in our contemporary societies both tradition and religion tend to become at most secondary in people's lives if not completely thrown away. Similarly, the authority that was once exerted by family or prestigious individuals within the community is lost in today's impersonal world. Modern societies have rendered tradition and religion as being groundless and even irrational and have adopted the so-called rational approach for explaining the social phenomena and social life. Societies have suffered what Weber (2015, p. 28) called the "disenchantment of the world", meaning that people replaced the "old culture", based on traditions, folkways, taboos, religion, primary groups, informal institutions and informal social control with a new one, based on impersonal laws, formal institutions for exerting the social control, secondary groups for carrying out the formerly fundamental roles of the family. The compliance with the norms would not be ensured by the external constraint exerted by culture, through its vehicles such as traditions or religion, but by a complete rationalization manifested by every individual. In a blunter utterance, the well-known philosopher and environmentalist Edward Goldsmith (1974) indicated that in current modern societies the lack of constraints became synonymous with liberty, although such an association would be nonsensical. Giving the example of the family, Goldsmith reasonably argues that constraints, understood as limitations of behavioral choices within a range of commonly accepted behaviors, are essential for the very survival of any group of individuals, ensuring order, cohesion and solidarity. Moreover, he stresses the role of culture, with religion as part of it, as a "control mechanism" in the absence of which no social system would be able to provide its stability (Goldsmith, 1974). In traditional societies or present-day collectivist ones, focused on the welfare of the group, the social control or, in simpler words, the compliance with the group norms, is ensured by informal, small entities, usually primary groups such as the extended family or the elders. In modern societies, especially the Western ones, fundamentally centered on the individual, impersonal, big, secondary groups tend to replace the primary groups and to take over their functions within society. The essential functions of a society become formalized, therefore norm observance is ensured by formal institutions, which enforce the laws appealing to the rational nature of every individual.

In societies in which the informal control of the primary groups, especially family is efficient, the level of social order and conformity to group norms is higher, yet in our modern, industrial, individualistic societies the primary groups tend to be replaced by large, contractually-based secondary groups even for the fulfilment of expressive need. In individualistic societies everything is questioned, the traditional knowledge is considered obsolete, children are thought that the welfare of the individual is more important than the welfare of the group and that the independence from the group is more important than the interdependence among the members of a group. At the same time, the informal pressure of the group is replaced by the formal social control which involves explicit and legal sanctions.

In a suggestively entitled book, namely *Order by Accident*, Miller and Kanazawa (2000) argue that collectivist societies, in which the informal social control is efficient or the social pressure of the primary groups works well, are better at ensuring social order. This happens because, through the process of primary socialization, the individuals are internalizing the social norms and, as a result, the rules are observed rather out of conviction and not necessarily because of the fear of being punished. Moreover, primary groups, such as family, peers or work groups are better at monitoring the observance of the rule and at punishing the transgressors (Miller & Kanazawa, 2000, p. 9). In countries where the informal social control is efficient, such as Japan, a country studied in detail by the mentioned authors, order at macro level is an "unintentional by-product" of the informal pressure of the group on the individual (p. 8). Simply expressed, in collectivistic societies rule observance becomes a habit through primary socialization, mainly exerted by and within the family, the latter being responsible for monitoring and sanctioning the appropriate or inappropriate behavior. The unintended yet beneficial outcome of this process at the level of the entire society is a higher degree of order and conformity. Talking about collectivist countries, which are still governed to a considerable degree by the force of habits, tradition or taboos and by informal social control exerted by the family or the elders of the society, the ecologically protective measures are promoted by appeal to traditional wisdom and enforcement mechanisms. For instance, in many Asian countries littering on the streets is still considered a taboo (Schweitzer & Alexander, 2015), and not littering has clearly become a cultural value (Edmond, 2017), acquired by every individual through the process of socialization and enforced by the pressure of informal social control.

Nevertheless, individualism, both in terms of practices and values, is on the rise all over the world, including in collectivist societies (Santos, Varnum & Grossmann, 2017), meaning that independence and uniqueness are gaining ground as cultural values in front of other social values such as interdependence, family relationships and conformity to group norms. The main factor driving the spreading of individualism is the socioeconomic development, including the access to educational formal institutions. In individualist societies, where the role of the family decreases significantly and the traditional knowledge is put aside as elders are becoming marginalized, other state institutions, such as the educational system should take over the role of educating individuals in observing rules that are beneficial for their future and, more importantly, for the future of their offspring. Formal education must include measures for teaching children that environmentally protective rules of behavior are essential for the long-term survival of the humankind. The "holders" of the knowledge, which are nowadays agents of secondary socialization, most especially formal educators and schools, are responsible for taking over the former role of the family and the elders and promote behaviors that would ensure sustainability of the society. Educational curricula all over the world must include compulsory disciplines that would promote a conservationist ethic, exactly as they include arithmetic or biology. Learning to protect the environment has become as vital as learning to count, write, or speak, therefore the formal educational system and explicit knowledge should replace the former informal groups and implicit knowledge.

CONCLUSION

Numerous examples demonstrate that the hypothesis of the "ecologically noble savage" should be clearly rejected, as by their nature, humans are rather careless about their environment, the destructive effects of their actions being registered even before the onset of civilization. Nevertheless, although they are not natural ecologists, humans seem to be cultural ecologists, as they managed to create a vast body of knowledge, transmitted as culture through learning, from generation to generation. Norms were primarily the carriers of culture that prevented the members of numerous societies to inflict irreparable damages to their environment and ensured the survival of their societies. The transmission, enforcement and observance of such norms were ensured by traditional societies through an efficient system of informal social control, that guaranteed the promotion of protectionist or conservationist practices and limited the destructive force of humans. Most of the taboos and interdictions were, in fact, environmentally protective measures, transmitted as religious norms or traditions that exerted an external pressure or constraint on every individual.

In our contemporary societies, however, two of the most important carriers of culture, namely tradition and religion, tend to lose their importance in people's lives. Similarly, the role of the family or prestigious individuals within the community, entities that once ensured the compliance of every member with the norms of the group, is lost in today's impersonal world. As the role of the family is constantly decreasing, traditional knowledge is rejected as obsolete and elders are becoming marginalized in modern individualistic societies, the formal educational system should take over the role of educating individuals in complying with rules that are directly related to sustainability.

Although naturally humans might be unable to reach sustainability by themselves, formal education, whose main role is to promote the values of the dominant culture, provides an invaluable vehicle for promoting a new, sustainable way of living. The state has a principal role in the process of secondary socialization (formal education), and it should use this role by capitalizing on humans' cumulative cultural evolution and integrating historical knowledge that generated adaptive responses to the environment. Since education can make humans individualistic, it results that it can also make them ecologically responsible, by appealing to a well-designed and targeted curriculum.

REFERENCES

Barnosky, A. D., Koch, P. L., Feranec, R. S., Wing, S. L., & Shabel, A. B. (2004). Assessing the Causes of Late Pleistocene Extinctions on the Continents. *Science*, *306*(5693), 70–75. doi:10.1126cience.1101476 PMID:15459379

Berkes, F. (2008). *Sacred Ecology* (2nd ed.). New York, NY: Routledge.

Brook, B., & Bowman, D. M. J. S. (2004). The uncertain blitzkrieg of Pleistocene megafauna. *Journal of Biogeography*, *31*(4), 517–523. doi:10.1046/j.1365-2699.2003.01028.x

Colding, J. (1998). Analysis of hunting options by the use of general food taboos. *Ecological Modelling*, *110*(1), 5–17. doi:10.1016/S0304-3800(98)00038-6

Colding, J., & Folke, C. (1997). The relations among threatened species, their protection, and taboos. *Conservation Ecology*, *1*(1), 6. doi:10.5751/ES-00018-010106

Crutzen, P. J., & Stoermer, E. F. (2000). The 'Anthropocene'. *Global Change Newsletter*, *41*, 17–18.

Dennett, D. C. (2017). *From Bacteria to Bach and Back: The Evolution of Minds*. New York, NY: W. W. Norton & Company.

Diamond, J. M. (2005). *Collapse: How Societies Choose to Fail or Succeed*. New York, NY: Wiking Penguin.

Durkheim, E. (1982). *The Rules of Sociological Method*. The Free Press.

Durkheim, E. (1995). *The Elementary Forms of Religious Life*. New York, NY: The Free Press.

Edmond, C. (2017, July 18). *Why Japan's streets are spotless*. Retrieved September 4, 2017, from https://www.weforum.org/agenda/2017/07/why-japanese-dont-litter/?utm_content=buffera55a5&utm_medium=social&utm_source=facebook.com&utm_campaign=buffer

Elkington, J. (1997). *Cannibals with Forks. The Triple Bottom Line of 21st Century Business*. Oxford, UK: Capstone Publishing Limited.

Ellingson, T. (2001). *The Myth of the Noble Savage*. Berkeley, CA: University of California Press. doi:10.1525/california/9780520222687.001.0001

Goldsmith, E. (1974). *Religion - the control mechanism of a stable society, a talk given before the Farmington Trust, 1974*. Retrieved September 4, 2017, from http://www.edwardgoldsmith.org/695/religion-the-control-mechanism-of-a-stable-society/

Gradstein, F. M., & Ogg, J. G. (2012). The Chronostratigraphic Scale. In F. M. Gradstein, J. G. Ogg, M. D. Schmitz, & G. M. Ogg (Eds.), *The Geologic Time Scale 2012* (pp. 31–42). Oxford, UK: Elsevier BV. doi:10.1016/B978-0-444-59425-9.00002-0

Gradstein, F. M., Ogg, J. G., Schmitz, M. D., & Ogg, G. M. (Eds.). (2012). *The Geologic Time Scale 2012*. Oxford, UK: Elsevier BV.

Harari, Y. N. (2015). *Sapiens: A Brief History of Humankind*. New York, NY: Harper & Collins.

Harari, Y. N. (2016). *Homo Deus: A Brief History of Tomorrow*. New York, NY: Harper & Collins.

Harris, M. (1974). *Cows, Pigs, Wars, and Witches: The Riddles of Culture*. New York, NY: Random House.

Harris, M. (1978). *Cannibals and Kings. The Origins of Culture*. Fontana, CA: Collins.

Henrich, J. (2015). *The Secret of Our Success: How Culture Is Driving Human Evolution, Domesticating Our Species, and Making Us Smarter*. Princeton, NJ: Princeton University Press.

Hitchcock, D., & Willard, M. (2009). *The Business Guide to Sustainability. Practical Strategies and Tools for Organizations.* Earthscan.

Hunn, E. (1993). What is traditional ecological knowledge? In N. M. Williams & G. Baines (Eds.), *Traditional Ecological Knowledge: Wisdom for Sustainable Development* (pp. 13–15). Canberra: Centre for Resource and Environmental Studies, Australian National University.

Kandari, L. S., Bisht, V. K., Bhardwaj, M., & Thakur, A. K. (2014). Conservation and management of sacred groves, myths and beliefs of tribal communities: A case study from north-India. *Environmental Systems Research, 3*(1), 16. doi:10.118640068-014-0016-8

Laland, K. N., & Brown, G. R. (2002). *Sense and Nonsense: Evolutionary Perspectives on Human Behaviour*. Oxford, UK: Oxford University Press.

Lewis, S. L., & Maslin, M. A. (2015). Defining the Anthropocene. *Nature, 519*(7542), 171–180. doi:10.1038/nature14258 PMID:25762280

Low, B. S. (1996). Behavioral Ecology of Conservation in Traditional Societies. *Human Nature - An Interdisciplinary Biosocial Perspective, 7*(4), 353-379.

Maslin, M. (2017). *The Cradle of Humanity: How the changing landscape of Africa made us so smart*. Oxford, UK: Oxford University Press.

Meyer-Rochow, V. B. (2009). Food taboos: Their origins and purposes. *Journal of Ethnobiology and Ethnomedicine, 5*(1), 18. doi:10.1186/1746-4269-5-18 PMID:19563636

Miller, A. S., & Kanazawa, S. (2000). *Order by Accident. The Origins and Consequences of Conformity in Contemporary Japan*. Boulder, CO: Westview Press.

Quiroz, D., & van Andel, T. (2015). Evidence of a link between taboos and sacrifices and resource scarcity of ritual plants. *Journal of Ethnobiology and Ethnomedicine, 11*(1), 5. doi:10.1186/1746-4269-11-5 PMID:25573058

Ramakrishnan, P. S., Saxena, K. G., & Chandrashekara, U. M. (Eds.). (1998). *Conserving the Sacred for Biodiversity Management.* New Delhi: Oxford and IBH Publishing.

Rappaport, R. A. (1984). *Pigs for the Ancestors. Ritual in the Ecology of a New Guinea People*. New Haven, CT: Yale University Press.

Redford, K. (1991). The Ecologically Noble Savage. *Cultural Survival Quarterly, 15*(1), 46–48.

Sanderson, S. K. (2001). *The Evolution of Human Sociality. A Darwinian Conflict Perspective*. Lanham, MD: Rowman & Littlefield Publishers, Inc.

Santos, H. C., Varnum, M. E. W., & Grossmann, I. (2017). Global Increases in Individualism. *Psychological Science*, *28*(9), 1228–1239. doi:10.1177/0956797617700622 PMID:28703638

Schweitzer, S., & Alexander, L. (2015). *Access to Asia*. Hoboken, NJ: John Wiley & Sons, Inc.

Simoons, J. F. (1994). *Eat Not This Flesh. Food Avoidances from Prehistory to the Present* (2nd ed.). Madison, WI: The University of Wisconsin Press.

Syvitski, J. (2012). Anthropocene: An epoch of our making. *Global Change*, *8*(78), 12–15.

Vitousek, P. M., Mooney, H. A., Lubchenco, J., & Melillo, J. M. (1997). Human domination of Earth's ecosystems. *Science. New Series*, *277*(5325), 494–499.

Weber, M. (2015). Science as a vocation. In P. Lassman & I. Velody (Eds.), *Max Weber's 'Science as a Vocation'* (pp. 3–32). New York, NY: Routledge.

Wheelan, R. (1999). *Wild in Woods: The Myth of the Noble Eco-Savage*. London: IEA Environment Unit.

Zalasiewicz, J., Crutzen, P. J., & Steffen, W. (2012). The Anthropocene. In F. M. Gradstein, J. G. Ogg, M. D. Schmitz & G. M. Ogg (Eds.), The Geologic Time Scale 2012 (pp. 1033-1040). Oxford, UK: Elsevier BV.

KEY TERMS AND DEFINITIONS

Adaptive Behavior: A behavior that helps an organism survive and reproduce in a certain environment.

Anthropocene: Concept introduced by Paul Crutzen and Eugene Stoermer to designate the current geological epoch, defined by the increased magnitude of the effects of human activity on environment.

Ecologically Noble Savage: Concept according to which people in traditional societies are living in perfect harmony with their natural environment.

Free-Floating Rationale: The rational motivation of an agent's adaptive behavior, that can be unconscious (i.e., not explicitly known to that agent).

Social Control: The pressure exerted by the group or society over their members in order to ensure the compliance with shared social norms.

Social Norm: A common rule of conduct existent in a society, transmitted from generation to generation and from individual to individual, that is assimilated and observed by members of that society.

Taboo: A strong norm imposing a proscriptive behavior or action, whose break is reprehensible.

Chapter 7
Green Public Procurement:
Instrument of Sustainable Development

Carmen Lenuta Trica
Bucharest University of Economic Studies, Romania

Luminita Ghita
Ministry of Environment, Romania

ABSTRACT

At present, legal provisions and environmental policy regulate the possibilities of using environmental considerations in the development of award criteria, as well as in the performance clauses of procurement contracts. The first part of the chapter analyzes the concept of green procurement and product categories for which green procurement can be used. The second part of the chapter presents the benefits of using green procurement. The third part of the chapter will include assessing the progress and impact of using green procurement. The fourth part of the chapter analyzes the legal framework for public procurement in Romania, as well as the capacity of the market to offer and develop products and services that include minimum environmental requirements and criteria. In the fifth part of the chapter, the authors analyze the possibility of implementing a mechanism for the operation and implementation of the legal provisions in Romania in order to improve the quality of the services and optimize the costs of the public procurement.

GREEN PUBLIC PROCUREMENT CONCEPT

Sustainable Public Procurement (SPP) is the broader concept, including also social consideration among economic and environment ones. Relying on Agenda 21 (the outcome document of the Earth Summit in 1992 – UN Conference on Environment and Development) in the 4th chapter on Changing Consumption Patterns, mentioned 10 years after Rio Summit, in Agenda 21's operationalizing plan: The Johannesburg Plan of Implementation JPOI (the outcome of World Summit on Sustainable Development in Johannesburg in 2002), as,

DOI: 10.4018/978-1-5225-5739-5.ch007

Copyright © 2019, IGI Global. Copying or distributing in print or electronic forms without written permission of IGI Global is prohibited.

encourage relevant authorities at all levels to take sustainable development considerations into account in decision-making, including on national and local development planning, investment in infrastructure, business development and public procurement,

Sustainable Public Procurement has gained international recognition in 2012 as part of the outcome of Rio+20 Conference (UN Conference on Sustainable Development) being one of the five programmes in The 10-Year Framework of Programmes on Sustainable Consumption and Production patterns (10YFP) is a global framework of action to enhance international cooperation to accelerate the shift towards sustainable consumption and production (SCP) in both developed and developing countries.

At global level, the agreed concept is Sustainable Public Procurement, including environmental, social and economic considerations, based on national circumstances. Sometimes "green" is associated only with environment and economic criteria, missing one of the three dimensions of sustainable development – social dimension.

At European level, more often used is Green Public Procurement. Public authorities represent high consumers in Europe, spending about 16% of the EU's Gross Domestic Product (an amount equivalent to half of Germany's GDP).

Using their purchasing power to achieve environmentally friendly goods and services, they can make an important contribution to sustainable development. Green public procurement covers areas such as the purchase of computers and energy-efficient buildings, office equipment made of sustainable wood-based wood products, recyclable paper, electric cars, ecological public transport, organic food canteens, electricity from renewable energy sources and air conditioning systems that correspond to the most advanced technical solutions for environmental protection.

Green procurement also refers on influencing the market by the power of the example. By promoting green procurement, public authorities can really stimulate industry to develop green technologies. For certain production, works and services, the impact can be particularly significant, as public institutions command a large part of the market (computers and buildings that are energy efficient, public transport, and so on).

Finally, by taking into account the costs of the entire life cycle of a contract, green public procurement gives the opportunity to save money while protecting the environment. By buying wisely, you can save materials and energy, reduces waste and pollution, and encourages sustainable behavioral patterns.

The legal framework for public procurement is defined by the provisions of the Treaty on the Functioning of the European Union and by the EU Public Procurement Directives, as interpreted by the European Court of Justice. From an international perspective, the EU must comply with the terms of the World Trade Organization (WTO) Agreement on Procurement (WTO) and bilateral trade agreements. In practice, compliance with these instruments is, as a general rule, achieved through the extension of the same rights applying to economic operators from the EU to agencies established in third countries.

Specific EU legislation imposes obligations on the procurement of certain goods and services, for example by setting minimum energy efficiency standards to be applied. Such mandatory obligations are currently applied in the following sectors, among others:

- **IT and Office Equipment:** IT products purchased by central government authorities must meet the latest minimum energy efficiency requirements set out in the EU Energy Star Regulation (Regulation No 106/2008 on a Community program of Labeling energy efficiency of office equipment).

- **Road Transport Vehicles:** All contracting authorities must take into account the environmental impact of the vehicle's energy and environmental performance in procurement procedures. A common methodology for the calculation of operational lifetime costs (Directive 2009/33 / EC on the promotion of clean and energy-efficient road transport vehicles).
- **Buildings:** Minimum energy performance standards for public buildings are defined at national level, based on a common EU methodology. From January 1, 2019, all new buildings occupied and owned by public authorities must be "nearly zero-energy buildings" (Directive 2010/31 / EU on the energy performance of buildings (recast)). The Energy Efficiency Directive9 also provides for mandatory renovation requirements for public buildings and for purchasing or re-renting contracts that meet minimum energy performance standards.

GREEN PROCUREMENT BENEFITS

The benefits of green procurement include: savings in long-term resource efficiency, elimination or reduction of the use of toxic materials that affect the environment, and implicitly mitigation of the impact of hazardous substances on human health and the environment, encouragement of innovation, reduction of landfills through the purchase of products recycling and creating less waste, promoting the organization's commitment to sustainable development, saving money through reuse and recycling materials, saving water, reducing greenhouse gas emissions, preserving biodiversity.

The European Commission co-founded a research project - called Relief (1) - to scientifically assess the potential environmental benefits that would be achieved if green procurement were widely adopted in the EU. The results produced the following conclusions: *If all public authorities would buy electricity from renewable sources, they would save 61,350,363 tons of CO2 eq (equivalent) each year. This figure is equal to 18% of the EU's Kyoto commitments to reduce CO2 emissions by 340 million tons.*

If all public authorities switch from purchasing food and beverages produced in conventional farming to those obtained from the organic farming system, 41,560 tons of phosphates (equivalent to PO4)

Figure 1. CO2 eq (equivalent) impact

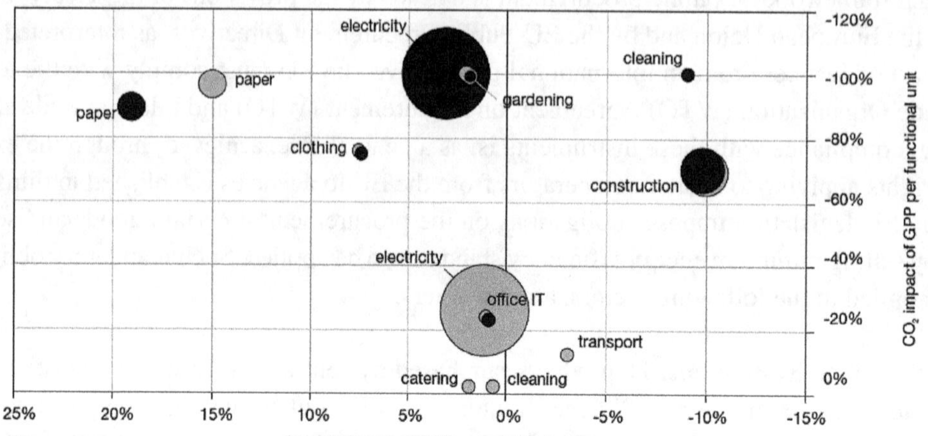

would be saved. This figure is equal to the amount of phosphates used each year for about 3.5 million Europeans. "

Chemical synthesis chemicals such as fertilizers, soil improvers, pesticides, and advanced chemical chemistry specific to conventional agriculture are the causes of groundwater pollution, exhaustion of natural soil fertility, rapid advancement of desertification and eutrophication.

Wider acquisition of more energy-efficient computers - meeting standards higher than the EnergyStar criteria - would reduce electricity bills and greenhouse gas emissions in Europe by up to 8 million tons of CO2 eq. (The Figure 1 applies to the entire market not only to public purchasers).

PROGRESS ASSESSMENT AND THE IMPACT OF GREEN PROCUREMENT

In order to stakeholders, the "Green Procurement Manual" has been developed, which presents how to introduce environmental considerations into public procurement procedures.

According to this, the contracting authorities introduce environmental criteria in different phases of the public procurement process as follows: the object of the contract (contract name), the qualification and selection criteria; the technical specifications of the specification; evaluation criteria (where the award criterion is the most economically advantageous tender); the performance clauses of the contract.

At present, the measures proposed through re-regulation are:

- The establishment of an appropriate policy framework to incorporate environmental criteria into procurement policies for goods and services;
- Using financial, budgetary and accounting measures to ensure that procurement policies take account of environmental costs of products and services;
- Providing information, training, and technical assistance to those involved in the procurement process, including those setting the performance criteria, as well as those using the products and services;
- Developing tools to facilitate green public procurement at all levels;
- Information dissemination needed to encourage green public procurement decisions, as well as the results and benefits of adopting them;
- Setting procedures to identify products and services that meet the objectives of green procurement policies;
- Encouraging the development of indicators for measuring and monitoring progress in using green public procurement;
- Analyzing and assess public procurement policies to implement cost-effective and environmentally friendly criteria.

The procurement criteria are based on data from a variety of sources and a general technical report is drawn up for each product group to signal the considerations considered. The evidence base uses available scientific information and data, adopts a lifecycle approach and involves a wide range of stakeholders.

The EU GPP criteria include two "levels" for each sector concerned:

1. Basic criteria are designed to allow simple GPP implementation, focusing on the key area (key areas) of the environmental performance of the product or service, and aim to keep administrative costs for companies at a minimum.
2. Full criteria take into account several aspects of the environmental performance or higher levels of it and are intended to be used by authorities wishing to go further in supporting environmental and innovation objectives.

In addition to the EU GPP criteria, many international, national and regional bodies have developed sets of criteria for a wide range of product and service groups. Processes for adopting these criteria vary, although many of them are similar to those of the EU GPP scheme. The EU GPP website contains links to some of the main sets of criteria.

There are many eco-labels designed to help buyers on identifying sustainable products or services. The most valuable labels from the GPP perspective are those that are based on objective and transparent criteria and are awarded by an independent third party. These labels can play a special role in designing

Table 1. Green public procurement criteria

	Product group	Product	Core criteria	Comprehensive criteria
1	Cleaning products & services	Cleaning services (including cleaning products)	• Use of cleaning products without hazardous substances	• Training of employees • Use of reusable microfiber cloths and/or dry-cleaning techniques
2	Construction	New buildings & offices	• Consideration of energy-saving measures in design and usage phase of building • Water-saving technologies in kitchen and sanitary facilities • Use of materials without hazardous substances • Use of timber from legal sources	• Use of localized renewable energy sources
3	Electricity	Electricity	• 50% or higher electricity from renewable energy sources	• 100% electricity from renewable energy sources
4	Catering & food	Catering services (including food)	• organic production of food products • Use of seasonal fruit, vegetables and fish	
5	Gardening	Gardening services and machinery	• Fuel type use of gardening machines • Use of soil improvers without peat and sewage sludge	•
6	Office IT equipment	Computers (desktops & laptops) and monitors	• Energy star standards • Accessibility and changeability of memory, hard disks and/or CD/DVD drives	•
7	Paper	Copying & graphic paper	• Production from recovered paper fibres • Use of ECF/TCF paper • Pulp production from sustainably managed forests for paper based on virgin fibres	•
8	Textiles	Clothing	• Öko-Tex Standard 100	•
9	Transport	Passenger cars and light duty vehicles	• Maximum CO_2-emissions per vehicle segment • Euro 5 standard	
10	Furniture	Office furniture	• Use of wood from legally sourced timber and sustainably managed forests	

Source: EC, 2016

technical specifications and award criteria, as well as in checking compliance. The specific rules that apply to the use of labels in public procurement are analyzed in the appropriate sections of this handbook.

The EU Ecolabel was created in 1992 to encourage businesses to bring less polluting products and services to the market. Products and services that have been awarded the EU Ecolabel bear the flower logo, allowing consumers - including public and private purchasers - to identify them easily. Currently, the EU Ecolabel has been awarded to 34 products and services, including paper, textiles, cleaning products, lubricants, household appliances, home and garden products and accommodation for tourists.

To this, new groups of products and services are added. The different types of eco-labels are presented below:

- **Multiple Criteria Labels:** These are the most known types of eco-labels and are also the most commonly used for GPPs. Multi-criteria labels are based on scientific information on the impact of a product or service on the environment throughout its lifecycle, from raw material extraction, through production and distribution stages, to use and final disposal. They apply a set of criteria that set the standard for the label in question.
- **Single Domain Labels:** These are based on one or more acceptance / rejection criteria related to a specific field, for example energy efficiency. If a product meets those criteria then it can bear the label. Examples of such labels include the EU Ecolabel or the Energy Star label for office equipment.
- **Industry-Specific Labels:** Industry-specific labels include forest certification schemes run by organizations.
- **Product Classification Labels:** Classify products or services according to environmental performance in the targeted area and do not use acceptance / rejection criteria.

Examples include the EU Energy Label, which classifies energy-using products according to their energy efficiency.

In July 2015, the Commission proposed revising the Energy Efficiency Label to reintroduce a labeling scale from A to G, A being the most efficient class, and G, the least efficient class.

The 2014 Directives provide for a transition period for exclusively electronic tender by 2018 (2017 in the case of central procurement bodies)

E-Procurement systems are at an advanced stage of development in most Member States and can be a valuable tool to support the implementation of GPPs. In particular, e-Procurement systems can allow you to track the use of GPP criteria and verify that suppliers have provided the necessary information to demonstrate their compliance.

Assessing the progress made towards achieving the proposed objectives requires the creation of an effective monitoring system. Figure 1 and Figure 2 sows products groups reported to compliance with green public procurement criteria. It should keep a record of the tenders and /or contracts awarded that have included GPP criteria. Ideally, the monitoring system adopted should also include information on the environmental impact of procurement decisions.

A number of EU Member States have introduced or are in the process of introducing monitoring systems for the implementation of GPPs at national level that can establish specific procedures to follow for the collection of information.

In addition to quantitative monitoring of progress, it is also advisable to periodically assess GPP activities from a qualitative point of view, analyzing the obstacles encountered, the corrective actions and improvements needed, as well as the future objectives.

Many of the problems encountered in applying GPP are common to all public authorities, being beneficial to engage in networking and cooperative activities with others. Sharing information, for example on the environmental criteria used in procurement procedures or on the availability of organic or green products on the market, can save time and effort.

Several networks centered on green and sustainable procurement have been created at national or regional level.

GREEN PROCUREMENT LEGAL FRAMEWORK IN ROMANIA

The public procurement system in Romania is similar to that of other EU Member States, as the relevant EU legislation (Directives, 2004/18 / EC and 2004/17 / EC of the European Parliament and of the Council of the European Union) has been transposed into Romanian legislation.

In Romania, the Law on Green Public Procurement (Law No. 69, 2016) represents the legal framework on GPP, created aims to: promote environmental protection and sustainable development; promoting sustainable consumption and production patterns, as well as resource efficiency; encouraging the development and application of clean and environmentally friendly technologies; promoting social progress that enable economic development; ensuring efficient use of funds by promoting products, services and works with minimal environmental impact; improving the quality of benefits and optimizing costs for short, medium and long term public procurement; the development of the internal market for green products, services, public works and technologies.

Figure 2. Overall level of green of total products

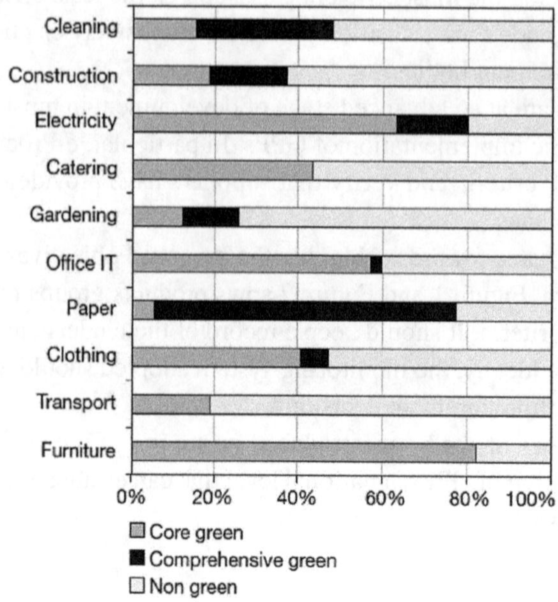

Figure 3. Overall level of green of contracts

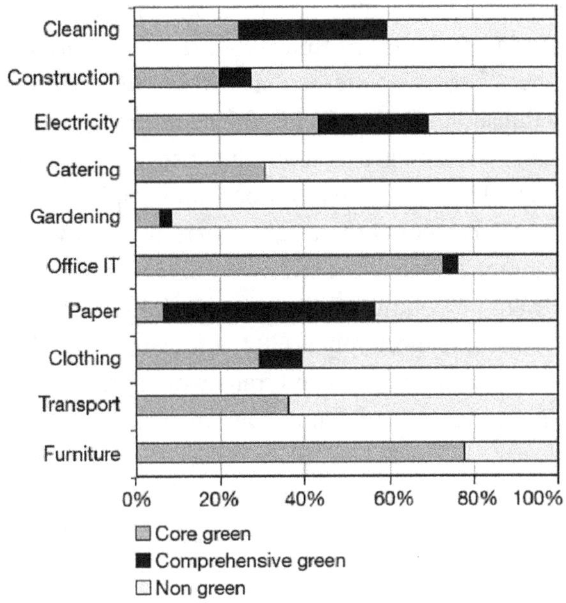

The law established the obligation for the contracting authority to award green public procurement contracts by establishing in the environmental protection awarding documentation the following forms:

- At least one qualification or selection criterion of the nature to the extent that that criterion relates to the approach taken by the economic operator in relation to environmental protection and is relevant to the possibility of fulfilling the future contract.
- Technical specifications describing the subject matter of the contract or parts of it, specifications relating to compliance with systems, standards and / or performance levels with an impact on environmental protection, favoring sustainable development by improving the quality of the benefits, their efficiency and cost optimization in the short, medium and long term.

Where a criterion of the kind set out above is established, the contracting authority seeks to verify the technical and/or professional potential of the economic operators participating in the green public procurement, so it is bound to correlate the requirements with the specificity, volume and complexity of the products, services or works to be provided, performed or executed.

The contracting authority shall define and apply the technical specifications of the kind referred to above, by reference to systems, standards and technical prescriptions which refer either to the European Union Ecolabel scheme; or to the specific European Commission green procurement criteria; or to the European Union criteria for green/ecological production and labeling of ecological/organic products in the field of agriculture and food.

The purpose of Law 98/2016 and Law 99/2016 is to provide the legal framework necessary to achieve the purchase of goods, services and works in terms of economic and social efficiency.

At the same time, the principles underlying the award of the public/sectoral procurement contract are non-discrimination, equal treatment, mutual recognition, transparency, proportionality and accountability.

For economic operators operating in the sphere of social economy, it is important to note that, for public procurement contracts/framework agreements on services which have as their object social services and other specific services, set out in Annex no. 2 of Law 98/2016, the award procedures apply to the award of public procurement contracts/framework agreements whose estimated value net of VAT is equal to or higher than the following threshold: ROL 3,334,050 (according to Article 7 (1) (c) of Law 98/2016).

For sectoral service contracts which have as their object social services and other specific services, set out in Annex no. 2 of Law 99/2016, the award procedures apply to the award of sectoral contracts whose estimated value net of VAT is equal to or higher than the following threshold: RON 4,445,400 (according to article 12, paragraph (1), letter c) of Law 99/2016).

For social services, the award procedure applicable to social services and other specific services as specified in art. 68 of Law 98/2016, respectively art. 82 of Law 99/2016. The award criteria used for the award of public / sectoral procurement contracts and framework agreements dealing with social services and other specific services, as set out in Annex no. 2, are the best value for money or the best value for money according to art. 111, par. (4) of Law 98/2016, respectively art. 124, par. (4) of Law 99/2016.

According on the article no. 56, paragraph (1) of Law 98/2016, respectively art. 69, par. (1) of Law 99/2016 the contracting authority / entity may reserve the right to participate in the award procedure only to protected units authorized under Law no. 448/2006 on the protection and promotion of the rights of persons with disabilities, republished, as subsequently amended and supplemented, and social insertion undertakings provided by Law no. 219/2015 on the social economy.

Moreover, art. 112, para. (1) of Law 98/2016 provides that the contracting authority has the right to determine that the participation in procedures for the award of public contracts exclusively covering health, social and cultural services is reserved to economic operators such as non-profit legal persons, social enterprises and protected entities, accredited as social service providers, public service providers.

According to art. 125, para. (1) of Law 99/2016, the contracting entity has the right to determine that participation in the award procedures of sectoral contracts exclusively for health, social and cultural services is reserved to economic operators, such as legal entities without lucrative purpose, social enterprises and protected entities, accredited as social service providers, public service providers.

In the new public procurement legislation there are public procurement procedures that are scheduled to take place exclusively online, at least for classical and sectoral procurements, and this will be possible after the new SICAP (Intelligent Collaborative Information System for Performance Environment of public procurement), which will replace the SEAP.

According to article No. 141 of Law 98/2016:

The contracting authority shall have the right to use public procurement and framework agreements to award bids and, in this case, to determine the size and subject matter of the lots, provided that this information is included in the procurement documents.

The contracting authority will determine the subject of each batch on a quantitative basis, adjusting the size of the individual contracts so as to better fit the capacity of small and medium-sized enterprises, or on a qualitative basis, according to the different trades and specializations involved, individual contracts more closely to the specialized sectors of SMEs or in accordance with the different phases of the project.

In the event that the contracting authority does not use the award of the contract on batch, it has the obligation to justify the decision not to award the contract on batch.

Also, information on the batch distribution of contracts is also available in art. 150 of Law 99/2016 on Sector Procurement.

IMPLEMENTING A MECHANISM FOR GREEN PUBLIC PROCUREMENT IN ROMANIA

The success of a national or regional green procurement strategy is a progressive approach. While there are successful models in the European Union that have left the bottom-up, other models, laws and targets imposed by the state are at least cumbersome. The second option is mainly found in the new Member States, while the local approach has become a common practice in the Scandinavian states. This has evolved into national norms that have transformed habit into imperative norms.

The progressive approach we consider to be most useful for Romania is to start with a limited range of products and services whose environmental impact is evident or for which greener alternatives are available and do not involve additional costs (eg paper recycled, energy efficient office equipment). Alternatively, we need to make sure that the requirements of the specification do not have a negative environmental impact, for example by excluding the use of recycled components7.

It should be noted, however, that in 2003 the European Commission in its Communication on Integrated Product Policy (PIP) encouraged Member States to draw up their National Action Plans (NAPs) for greening public procurement, accessible to the public.

The NAP should contain an assessment of the existing situation and ambitious targets for the next three years, specifying what measures will be taken to achieve them. NAP is not legal, but provides a political impetus to the process of implementation and desensitization of green public procurement. They allow Member States to choose the options that best fit the political framework and the level they reached.

Only Romania, Hungary, Greece, Estonia, Luxembourg and Croatia have not adopted the NAP.

In Romania, Ministry of Environment and National Authority on Public Procurement has shared responsibility on Green Public Procurement policies and implementation of Law on GPP. At present, both insitutions are working close on finalizing the common order of the Minister of Environment and President of the National Authority on Public Procurement, which sets the criteria for some specific group of products as well as guidelines on drafting the tenders dossiers for Green Public Procurement. Moreover Ministry of Environment works with Italian authorities within a LIFE+ project on developing the National Plan for Green Public Procurement in Romania.

Romanian environmental authorities need to be more active in public procurement development initiatives and also reach other local or regional networks than those listed in GPP lists or acts. purchase, a grouping including a certain number of sales units, whether that group is sold to the end user or end user or whether it serves only as a means of filling shelves at the point of sale; these packages may be separated from the product without affecting its characteristics; transport packaging or tertiary packaging, e.g. packaging designed to facilitate the handling and transport of a number of packed sales units or packs in order to prevent physical damage during handling and transport.

Contracting authorities will need to indicate in the contract notice and in the tender documents how many additional points will be awarded for each award criterion. The award criteria relating to the environment should together account for at least 10-15% of the total points available.

If the award criterion is expressed as "superior performance compared to the minimum requirements included in the technical specifications", the points will be allocated in proportion to the performance improvement.

Specific Sector of Public Procurement

Traditional Products

An alternative legal path to organic/ ecological registration and eco-labeling is the guarantee of traditional specialty. Government Decision no. No 134/2008 on agricultural products and foodstuffs as traditional specialties guaranteed implements Regulation (EC) No. 509/2006 on agricultural and food products as traditional specialties guaranteed.

To qualify as a Traditional Specialty Guaranteed (STG), an agricultural or food product must comply with certain specifications.

The specification includes the registration with or without reservation of the name by a production group from the Ministry of Agriculture. The specification must include, (33)also a description of the agricultural or food product, including its main physical, chemical, microbiological or organoleptic attributes, a description of the traditional production method to be followed by producers, including the nature and characteristics of the raw material or ingredients used and the method preparation of the agricultural or food product. The product also has key elements that define specific character requirements and procedures to verify specific character.

The National Authority for Consumer Protection has the obligation to check the compliance and the specific character of the products.

Moreover, in order to qualify as a traditional guaranteed specialty, an agricultural or food product must comply with certain product specifications. It must bear:

- The name that indicates if the group requests registration with or without reservation of the name;
- Description of the agricultural or food product, including its main characteristics;
- A description of the production method to be followed by producers, including, where appropriate, the nature and characteristics of the raw material or ingredients used and the method of preparation of the agricultural or food product, and;
- The key elements defining the specific character of the product and, where applicable, the reference basis used.

Protection of the traditional product applies without prejudice to national or EU rules or governing intellectual property, and in particular those relating to geographical indications and trademarks.

We have some terminology notes to operate here. The term "traditional" means proven use on the market for a period of time and shows intergenerational transmission. This period of time is generally attributed to a human generation, that is, not less than 25 years. "Traditional guaranteed specialty" means a traditional agricultural or food product recognized by the EU under Regulation 509/2006 on agricultural and food products as traditional guaranteed specialties.

Finally, and very importantly, through the "group" the law accepts any association, irrespective of its legal form or composition, of producers or processors operating with the same agricultural or food product. Therefore, any type of manufacturer is covered by this regulation.

Sustainable agriculture refers to methods of producing food that are healthy, do not harm the environment, respect the workers' rights, are animal friendly, offer fair income to farmers and support farming communities. Sustainability includes purchasing food as locally as possible.

However, buying local food is not a guarantee that it is produced in a sustainable way. Pesticides, chemical fertilizers, industrial agriculture and non-therapeutic use of antibiotics may be involved in local food production, so it is important to ensure that local food is from farmers or gardeners who use sustainable methods.

At its roots, sustainable agriculture is beneficial to the local community and the local economy, helps to enrich soil and agricultural land, protects air and water quality, and minimizes energy consumption. Industrial food production, on the other hand, is totally dependent on fossil fuels and chemicals used in fertilizers and pesticides.

Small local farms are being exploited by farmers who live on their land and work to preserve it. They protect natural areas by maintaining land for agricultural purposes and preserving natural habitats by maintaining forests, pastures or areas of land. A local distribution also provides a reduction in the materials used to package the products.

And finally, local products keep secondary culture locally and possibly traditional means of production. They also open the way for secondary business for farmers such as ecotourism.

There is no legal basis for the term "local product", but the last "agricultural reform" in the EU has prepared the ground for good practice. A principle set out in Agenda 2000 and the Common Agricultural Policy (CAP) mention "Good Agricultural Practices" (GFP). Member States can define GFP codes at local or national level. GFP is the basic requirement for farmers wishing to adhere to agri-environment schemes. These can undoubtedly be broad criteria for GPPs.

Catering Services

The specific element in catering is transportation. Vehicles to be used for the provision of services shall at least meet the EURO 5 or V exhaust emission requirements. Suppliers shall provide a list of the vehicles to be used for service provision and their technical datasheets specifying the respective emission levels.

At the same time, as the food is prepared, evidence of its technical and professional capacity to comply with the environmental requirements of the contract appears also by: an environmental management system (EMS) for catering services (eg EMAS, ISO 14001 or equivalent, national or regional official systems are inserted) or an environmental policy for catering activities and working instructions and procedures for the execution of services in an environmentally sound way or previous experience in the application of environmental management measures in similar contracts.

Paper

First of all, we must emphasize that it is generally agreed that the recycling of a ton of paper would avoid the emission of:

- 1.4 tons of CO2 equivalents compared to simply converting paper to waste at the landfill, and;
- 0.6 tons of CO2 equivalents compared to incineration.

The mere fact of selective collection and paper recycling succeeds in making the above emission savings.

Recycling is not the only solution on "greening" a project on paper. Using chlorine for bleaching paper increases the complexity of the problem. The process of paper making and chlorine bleaching increases CO_2 equivalent emissions. These are almost impossible to calculate over the entire production chain.

The impact on ecosystems is one of the issues that further complicate the calculation because ecosystem services are still quite difficult to assess. As a result of the generation of liquid waste or wastewater containing chlorine, the aquatic fauna and plants around water or aquatic ecosystems suffer. The recommended choice in projects (in case the paper is not recycled) is chlorine free - actually means supporting ecosystem services in the paper production process and limiting the unfriendly paper market to the environment.

In the case of recycled paper, the "bleachless chlorine" formula can by no means be used as it is impossible to trace the origin of all types of paper used in the recycling process (and thus the total absence of chlorine cannot be guaranteed). In fact, the recycling process is the one that can be done without chlorine. In this problem of chlorinated paper we will have to train ourselves to reconsider the aesthetic sense and prefer a less white paper.

Certification standards generally include:

- Prohibiting the conversion of natural forests and wild habitats;
- Prohibiting the use of hazardous pesticides;
- Banning the cultivation of genetically modified trees;
- Respect for the rights of indigenous peoples or green belt around communities;
- Annual auditing.

Reducing emissions through forest certification cannot be scientifically proven with accuracy, but certainly provides many benefits for the environment and communities. In the case of paper, the optimal preference is for purchasing recycled paper for the simple reason that unlike wood or lumber, paper is an easily consumable item.

In many cases, paper available in the market cannot be 100% recycled or 100% unprocessed, but may contain a mixture of sources. Therefore, the object does not refer to unprocessed fiber paper, but to unprocessed-fiber paper, allowing the use of recycled fiber for paper production as long as the specifications defined above are met.

- **Percentage Increase:** According to the EU Ecolabel, at least 10% of unprocessed fiber must be produced from sustainable managed forests and at least 20% according to the Nordic Swan label.

Contracting authorities may wish to attach greater importance, at the award stage, to paper with a higher percentage of certified unprocessed fibers coming from a resistive and / or recovered paper fiber.

I.T. and Birotics

The product category includes a wide range of products and services:

- Standard hardware (computers, printers, copiers, etc.)
- Servers
- Software

- IT solutions (usually a mix of hardware and software to solve a specific problem).

Electonic and Electrical Products

The EU has introduced a series of directives aimed at reducing the environmental impact of electronic and electrical products. The RoHS Directive restricts the use of hazardous substances in electrical and electronic equipment and the WEEE Directive promotes the collection and recycling of equipment.

In addition to EU legislation, no legal instruments of particular importance for IT products have been introduced in Romania. Depending on key market players in the IT sector, national legislation will probably have a very limited effect on large multinational producers.

Economic instruments are not used for IT equipment directly, but electricity costs can influence purchases because operating costs are directly related to electricity consumption. This is especially relevant for server systems where power is used for both equipment operation and cooling.

There is an agreement at EU level that central government departments purchase IT office equipment that meets Energy efficiency requirements. This document presents the EU's green procurement processes developed for the office IT equipment product group. The Explanatory accompanying technical report provides full details of the reasons why these criteria have been selected, as well as references for additional information.

Contracting authorities will need to indicate in the contract notice and in the tender documents the number of additional points awarded for each award criterion. Environmental award criteria should represent at least 15% of the total points available.

Ecolabels type I or ISO 14024: Ecolabels type I or ISO 14024 are labels based on criteria established by an independent body and monitored in a certification and audit process. They therefore have a high degree of transparency, are reliable and constitute an independent source of information. The eco-labels must meet the following conditions:

- Requirements for labels based on scientific evidence;
- Are adopted with the participation of all stakeholders such as government bodies, consumers, producers, distributors and environmental organizations, are accessible to all stakeholders.

On public procurement, authorities involved in public procurement may require that the criteria underlying a particular eco-label are respected and that the eco-label is using a form of proof of compliance. However, they cannot require the product to be eco-labeled. Moreover, authorities involved in public procurement may use only the Ecolabel criteria that refer to the characteristics of the product or service itself or to the production processes rather than the criteria relating to the general management of the company.

Declaration of Conformity

If the criteria check specifies the fact that other appropriate means of proof may be used, they may include a technical dossier from the manufacturer, a test report from a recognized body or other relevant evidence. The contracting authority will have to decide on each case if the submitted application can be considered technically or legally appropriate.

Considerations on Costs

It is recommended to apply a "total cost of ownership methodology" when awarding a contract. This means instead of taking into account only the purchase price of the product when assessing the bidder offering the best value-for-money ratio, the contracting authority took into account the life cycle cost of the product for the estimated period of ownership of the device. This would cover the purchase price, maintenance costs and other services, the cost of energy consumption and other consumables (such as paper or ink) per device, as well as all disposal costs. This would allow the Authority to take into account environmental aspects both in quality assessment (through environmental technical specifications and/or award criteria) and in price setting (including life cycle cost).

As with any electricity-consuming product, the purchase of energy-efficient models is generally a beneficial option in both points of view - reducing operating costs and environmental impacts.

Transport

Types of vehicles purchased by public administrations vary widely from current ones (e.g. official vehicles, control vehicles, vans or gardening equipment) to emergency vehicles (ambulances, fire engines, cars and police cars, etc.) and special vehicles (sweepers, bins, buses, etc.).

Criteria have been developed for the following three product groups:

- Cars purchased directly or in a leasing / leasing system;
- Vehicles and public transport services, and;
- Waste collection machines and services.

The criteria and contracting procedures defined in this document may also be used as guidelines for the definition of specifications for the purchase of vehicle types and service contracts not explicitly mentioned in this Schedule.

In addition, the proposed criteria need to be interpreted in conjunction with Directive 2009/33/ EC on the promotion of clean and energy efficient road transport vehicles and national legislation implementing the directive. The Directive obliges public authorities and operators contracted under a public service contract to purchase road transport vehicles to take into account the energy and environmental impacts during the lifetime, which at least low energy consumption, CO_2 emissions and pollutant emissions, including NOx, NMHC and particulate matter. This objective can be achieved either by including energy and environmental performance requirements for each of the types of impacts considered (as technical specifications or minimum allocation criteria), or by calculating the monetary value of these types of impact when taking the purchase decision, according to a calculation methodology provided for in the Directive. The recommended criteria included in this product sheet may guide public authorities wishing to implement the Directive using the first or second option, i.e. by including energy and environmental performance requirements as technical specifications and / or award criteria. The criteria may also be used together with a life cycle cost assessment using the methodology set out in Article 6 of the Directive or an equivalent instrument such as the one on the website www.cleanvehicle. I for making a purchasing decision based on lifetime costs as well as minimum environmental criteria. Since not all Member States authorized the use of both of the above options when implementing the Directive on the promotion of clean road transport vehicles, it is imperative that public authorities comply with the obligations laid down by national law before choosing the GPP approach for this product group.

Until a few years ago, most public transport services were managed by public authorities (mainly local and regional administrations) either directly by civil servants or through a public company responsible for providing the service. However, in recent years, the number of tender procedures for public bus services has increased significantly. Consequently, criteria are laid down both for the direct purchase of buses and for the purchase of public transport services.

Buses

Bus services in Romania are generally a public responsibility. These are usually organized by a local or regional public company responsible for traffic and offer plans. All operating costs are managed by the operator following a tender procedure. The bus operator is traditionally paid a fixed amount for each bus line plus running costs related to driving hours.

Legal instruments related to the purchase of new buses are at EU level. This means that new buses purchased in the EU must comply with certain requirements regarding the emission of particulate matter to protect the local environment. CO_2 is not yet a factor, but it should be - as far as electric or alternative vehicles are to be promoted. These requirements are renewed once every three or five years, but today EURO IV, EURO V and EURO VI are relevant.

From the environmental point of view, it is interesting that no economic instruments have been introduced to influence the sector in the form of lower taxation on alternative and less polluting fuel or bus types. On the existing market, alternative fuels are more expensive and so are the vehicles that use them.

One could argue that fuel tax is an incentive to use electric vehicles, and so it is in theory. However, at the current technological and economic stage, electric buses are not competitive in terms of distance, reliability and price.

Cleaning Services

Cleaning products are regulated extensively in the European and Romanian legal frameworks. The following European regulations are transposed into national legislation, focusing on health and environmental issues at work in relation to cleaning products:

- REACH;
- Detergents;
- Classification, packaging and labeling of dangerous preparations.

Regulation (EC) No 1907/2006 of the European Parliament and of the Council of 18 December 2006 on the Registration, Evaluation, Restriction and Authorization of Chemicals - REACH is a European Union regulation designed to ensure a high level of protection of human health and manage and control the potential risk to human health and the environment from the use of chemicals in the European Union, given the free movement of substances as such, in mixtures or in articles42

Therefore, legal instruments are very much in use, focusing on occupational health. This has a certain effect on the environment as well as, in particular, on recipients and the requirements of non-toxicity and degradation of air.

Voluntary Eco-labeling of cleaning products and environmental information for consumers could be used for differentiated taxation or for different treatment within the GPP.

Ecolabels for Type I cleaners or ISO 14024 are labels whose basic criteria are set by an independent body and are monitored through a certification and audit process. They are therefore a source of extremely transparent, reliable and independent information. Eco labels must meet the following conditions:

- The requirements for obtaining the eco-label are based on scientific evidence;
- Eco-labels are adopted with the participation of all stakeholders, such as government bodies, consumers, manufacturers, distributors and environmental organizations;
- Eco-labels are accessible to all stakeholders.

In the framework of public procurement, purchasers may require that certain criteria underlying eco-labeling be met, as well as the use of the eco-label as evidence of compliance. However, purchasers may not require a product to be labeled as organic/ green. Purchasers may also use only eco-label criteria that refer to the characteristics of the product or service or production processes rather than to the general management of the enterprise.

In Romania, the first steps for the introduction of environmental criteria in public procurements procedures were made in 2006, with the adoption of the Emergency Ordinance no. 34/2006 regarding the award of public procurement contracts, public works concession and services, which allows contracting authorities to impose special conditions for fulfilling the contract, aiming to achieve social or environmental effects and promotion of sustainable development. However, most purchasers do not have information about green products and services. Moreover, they are confronted with legal and technical issues related to the introduction of environmental criteria in the awarding documentation.

Environmental Issues in the Awarding Documentation

An in-depth need analysis will help define the scope for green procurement, and avoid unnecessary purchases.

If relevant, materials, production methods or environmental performance levels can be specified.

Various offers help the contracting authority in selecting the most economically advantageous, but also environmental performance.

The contracting authority shall establish through the tender documentation, the rules for participation, the requirements to be fulfilled by the economic operators, the conditions that must be fulfilled by the tender, the formalities for submission, the terms of the contract, so that, in the way it is elaborated, assumes responsibility for conducting the award process, the outcome and the achievement of the established objective.

Awarding documentation includes:

- Formal requirements;
- Technical requirements;
- Financial requirements, which allow an objective description of what is to be provided, performed, executed, on the basis of which the economic operator will develop its offer.

The structure of the documentation adapts to the procedure to be applied.

The contracting authority will include in the notice / invitation to participate information on:

- The formal rules for the submission and submission of tenders / applications;
- The conditions of participation, eligibility and qualification / selection of bidders / candidates;
- The award criterion of the contract and, where applicable, the relative weightings of the valuation factors or the calculation logarithm.

Tender Specifications/ Tender Dossier

Tender dossier comprising tender specifications represents the starting point for the preparation of the awarding documentation, as it highlights the subject of the future public procurement contract as well as the necessary quantity of products/services/works. The drafting of the specifications is based on the requests made by the departments of the contracting authority. When preparing the specifications, the department responsible for awarding public procurement contracts within the contracting authority shall work with the institution's specialists, or possibly hired consultants. In addition to the technical specifications, mandatory in the content of the specification, it may also contain information on the quantities of products/services/works subject to the contract, the time and place of supply/performance and other relevant information that the contracting authority considers them necessary.

The technical standards, eco-labels and criteria for green procurement established at European and national level are valuable sources of information when drafting the specifications and can be included directly in the tender documents without requiring the bidders a certain type of eco-label.

Technical Specifications

They describe the products, services or works that are wanted to be purchased, as well as being a determining factor for companies whether they are interested or not, thus determining the level of competition.

These offer measurable requirements on the basis of which tenders can be evaluated.

The technical specifications must allow free and equal access to the award procedure and shall be defined in such a way as to meet the needs or exigencies of any user. They define:

- Qualitative, technical and performance features:
- Environmental impact requirements;
- Dimensions, terminology, symbols, tests and test methods, packaging, labeling, marking and instructions for use of the product, technologies and production methods;
- The quality assurance system and the conditions for certifying the conformity of the product/service with relevant standards or the like.

The use of standards in defining technical specifications is useful because it provides extra transparency and a non-discriminatory approach to economic operators in that products/services/works fit in their accepted and recognized quality. Many standards include features such as energy consumption and waste management.

The tender dossier may contain the statement that the products or services offered that hold a particular "eco-label" are considered to meet the required technical specifications implicitly.

A technical proposal cannot be considered as inappropriate for the sole reason that the products or services offered are not "Ecolabel" if the tenderer demonstrates by any appropriate means that they meet the required technical specifications.

The formulation of environmental requirements must be in line with environmental and sustainable development regulations, so that they are not considered as "barriers" in the competitive environment.

To purchase green products, the technical specifications need to specify the environmental performance/characteristics of the product/service or work.

When developing technical specifications for products/ services/ works that aim to protect the environment, it is recommended to use the operational tools provided by the European Commission: http://ec.europa.eu/environment/gpp/toolkit_en.htm.

If certain environmental characteristics are required, full or partial specifications may be used, defined by:

- European "eco-labels";
- "Multi-national" "eco-labels";

Any other "certified eco-labels" certified if the following conditions are met:

These must be appropriate to define the characteristics of the products/services. This implies that only those eco-label specifications corresponding to the Public Procurement Contract can be used. The general criteria contained in the eco-labels that refer to the general management of the company producing a good or a service are ineligible as technical specifications.

The eco-label requirements should be developed on a scientific basis.

Requirements must have been adopted through a participatory approach and be accessible to all stakeholders.

Tenderers cannot be required to be registered in a specific Ecolabel scheme. Always accept equivalent means of verifying compliance, such as a technical dossier of the manufacturer or a test report issued by a recognized body.

It is not necessary for the contracting authority to stipulate in detail all the green features that the product/service/work should fulfill, but at least the positive effect each of them must produce.

Establishment of Contractual Provisions/Terms

When defining the subject of a contract, contracting authorities have the freedom to choose what they want to purchase. This allows the inclusion of environmental considerations, provided this is done without distorting the market by limiting access to it.

The structure of the public procurement contract is as follows:

- The preamble, the introductory part of the contract;
- Mandatory contractual clauses, those contractual provisions that any public procurement contract must contain;
- Specific contractual clauses, those provisions specific to each public procurement contract.

Specific contractual clauses may refer to:

- Subcontractors;
- The performance guarantee;
- Quality guarantee;
- Price adjustment;
- Deadlines for the execution of the parties' obligations;
- Reception, inspections, tests;
- Packaging, marking, transport;
- Insurance;
- Other conditions of performance of the contract;
- Environmental protection requirements.

Each public procurement contract involves a different set of potential environmental impacts to be taken into account.

The contractual clauses for a green procurement must take into account:

- Subject of the contract;
- The desired effect through the purchased procurement ;
- The ecological benefits from the final result;
- Impact on the environment;
- Sustainability of the procurement.

Examples of clauses in public works contracts for green public procurement.

Supply Contracts:

- The environmental impact of the materials from which the product is made and the production process;
- The use of recyclable materials in the production process;
- Water and energy consumption during product use - product life;
- Opportunities for recycling / reuse of the product at the end of its life;
- Packaging and shipping of the product.

Service Contracts:

- Technical expertise and staff qualifications allow performance (execution, execution) of the contract in an environmentally friendly manner;
- The products / materials used;
- The management procedures implemented to minimize the environmental impact of the service;
- Energy and water consumed, waste generation.

Works Contracts:

- Can have significant effects on the environment through land use or traffic;
- There are projects for which the EIA (Environmental Impact Assessment).

Mandatory Tests:
The contract carries out work that requires an appropriate level of operation:

- Water installations;
- Ventilation installations;
- Heating / air conditioning;
- Electrical installations.

Transport of materials and tools to the site:

- Delivery of products on site in concentrated form and dilution on site;
- The use of reusable containers for the transport of products to the site;
- Waste management system (eg packaging waste - recovery, recycling and reuse of packaging materials for construction materials).

How to execute the works:

- Use of dosage indicators to ensure that appropriate amounts of material are used.
- Removal of products used or product packaging.
- Products or packaging are lifted by a contractor for reuse, recycling or disposal.
- Training of the contractor's staff.S is trained on the environmental impact of the work and on the environmental policy of the contracting authority.

For example, in the award documentation of the project for building the construction of Environmental Protection Agency Bacau, environmental criteria have been included in Quality Requirements - Thermal, Waterproof and Energy Saving, and Installations - Performance Level of Works electrical installations.

Selection of Suppliers of Goods and Services

The selection criteria of tenderer must be objective, non-discriminatory and in accordance with the principle of proportionality.

When assessing the tenderer's ability to contract, the public authority may consider issues such as experience and competence to meet the environmental requirements relevant to the subject matter of the contract. Operators violating environmental legislation may be excluded if this constitutes a professional misconduct and, for service and works contracts only, may require evidence of the ability to enforce environmental management measures in the performance of the contract.

Exclusion Criteria

The exclusion criteria refer to the circumstances in which an operator can be found and which normally lead to disqualification.

The legislation in force provides for cases in which the contracting authority may reject tenders submitted by economic operators in a public procurement procedure.

Qualification and Selection Criteria

Qualification criteria are the minimum requirements imposed by the contracting authority that have to be fulfilled by an economic operator in order to qualify as qualified.

The selection/pre-selection criteria are the conditions that the qualified economic operator must fulfill to select/preselect.

The qualification and selection criteria refer to:

- Personal situation of the candidate or tenderer;
- The ability to exercise professional activity;
- The economic and financial situation;
- Technical and / or professional capacity;
- Quality assurance standards;
- Environmental standards.

Qualification and selection criteria are determined according to the specific requirements of the contract and always in accordance with the principle of proportionality.

Contracting authorities may require tenderers to demonstrate the technical and/ or professional capacity to meet the environmental requirements of the contract.

Environmental criteria could include:

- Minimization of the quantities of waste produced,
- Preventing the discharge / leakage of pollutants,
- Reducing the consumption of pollutants,
- Minimization of the fragmentation of natural habitats.

A useful tool in checking candidates is experience in similar contracts. It is therefore advisable to clearly define the types of information considered relevant and the supporting documents to be submitted.

In other cases, the environmental aspects can be verified by examining the educational and professional qualifications of the staff involved in the contract initiation, which are very important for achieving the environmental objectives.

For service and works contracts, evidence can be required to certify the company's ability to implement environmental management measures in the execution of the contract.

Environmental requirements are often complex, but there is means to verify compliance. Product environmental statements can be a useful tool in assessing compliance. They provide information on the environmental impact of the life cycle of a product or service. References to environmental statements can be made to ISO or EMAS environmental standards. Environmental management systems

If the economic operator does not hold an environmental certificate as required, then the contracting authority has the obligation to accept any other evidence or evidence submitted by it insofar as such evidence or evidence confirms that an adequate level of environmental protection is assured.

Examination and Evaluation of Tenders

The procedure for selecting tenders is based on the application of selection criteria as specified in the contract notice and which must be objective, non-discriminatory and in accordance with the principle of proportionality.

At the award stage, points can be allocated in addition to the minimum requirement set in the specification that recognizes environmental performance. There is no maximum that can be assigned to the weight of the environmental criteria.

The "lifecycle cost" approach reveals the real cost of a contract. An analysis of water, energy, maintenance and disposal costs during the assessment may indicate that the green option may be the cheapest.

General Rules for the Award of Contracts

Award Criteria

At the award stage, the contracting authority shall assess the quality of the tenders and compare the prices.

When assessing the quality of bids, the pre-defined award criteria, previously published, are used to decide the best bid. These criteria may be:

- The lowest price.
- The most economically advantageous tender, which implies that, in addition to the price, other award criteria will be taken into account, e.g.: quality, environmental characteristics, aesthetic and functional features, etc.

It is up to each contracting authority to determine the award criteria to be applied and the weighting assigned to each criterion.

CONCLUSION

The need to adopt a sustainable, eco-friendly way of making public procurement derives from the reality that communities across the planet are facing the dramatic consequences of climate change, excessive exploitation of natural resources and threats to biodiversity. Our habit of consuming unsustainably, without considering long-term effects, is the key issue of all the other issues mentioned above.

An important factor in any green procurement is to reduce the negative environmental impact of the product / service over the entire life cycle: production, use, waste and disposal.

The way we produce and acquires influences the environment, the effects include: ecosystems, climate change, global warming, air and water pollution, population growth, biodiversity impact, species and habitats destruction, general environmental degradation.

To the extent that only the adoption of a National Action Plan on GPP at central government level did not bring major positive changes to other EU Member States, imperative implementation of the Law on GPP (Law no. 69/2016) would be necessary in Romania. This would create a framework for progressively increasing the percentage of GPP in public procurement from year to year. Areas of implementation would remain in the area of good governance and good practice, and in the alternative, public institutions and public authorities (central and local) would set their own priorities. However, a minimum threshold will always exist, will increase from year to year and will be imposed by force of the law.

Each purchased product or service has an environmental impact throughout its lifecycle, from raw material extraction, product manufacture to its use and recycling / disposal. Green public procurement contributes to reducing these impacts, with benefits being felt both locally and globally.

Green public procurement is part of this sustainable development framework, but the reduction in environmental impact of the purchase of products, services and works can only be achieved if all public purchasers act together to send a clear and unified signal to the market.

Green Public Procurement (GPP) stimulates the industrial sector - especially small and medium-sized enterprises - in developing and promoting clean technologies on the market. Exploitation of this sector will create new jobs and growth.

Green public procurement leads to the development of environmentally friendly products, services and works that meet recognized quality standards.

Green public procurement improves quality of life, and organic/ecological farming practices create new jobs, especially in rural areas, and thus contribute to the revitalization of rural space. The implementation of the National Action Plan can influence consumers' attitudes by making them aware of the fact that by purchasing organic or green products, the quality of life can be improved.

As the share of public procurement accounts for about 16% of the European Union's GDP, the environmental impact generated by green public procurement is significantly reduced, with environmental objectives being met thanks to the public authorities' financial strength.

On concluded the chapter, relying on the words of a distinguished professor addressing the Open Working Group of elaborating the SDGs at UN auspices in New York at the debate on Sustainable Consumption and Production in spring of 2014 "when we are talking about life cycle we should mention that life is only a characteristic of living organisms and to be aware that the products have no life at all", the policies on Sustainable or Green Public Procurement represent an valuable instrument and an important vehicle of starting a global movement of mainstreaming or integration of human made products in Nature's cycles, being fully aware of our role as top living organism on this Planet and respect the ecosystems and Nature's balance as our unique support of life.

REFERENCES

Communication from the Commission to the Council and the European Parliament Stimulating Technologies for Sustainable Development. (2004). Bruxelles: An Environmental Technologies Action Plan for the European Union.

Eco-procurment, the path to a greener market place, European Project – Environmental Relief Potential of Urban Action on Avoidance and Detoxification of Waste Streams through Green Public Procurement Buying Green Handbook. (2007). *Collection of statistical information on Green Public Procurement in the EU*. Academic Press.

EKOenergy. (2013). Retrieved from www.ekoenergy.org/criteria/

No, R. 106/2008 on a Community energy-efficiency labelling programme for office equipment. (n.d.). Retrieved from RELIEF Project Information webpage: www.sustainable-procurement.org/about-us/past-projects/relief/

PriceWaterHouse Coopers. (2009). *Significant si Ecofys*. Author.

(2002). Recommendation of the Council on Improving the Environmental Performance of Public Procurement. *Improving the Environmental Performance of Public Procurement: Report on Implementation of the Council Recommendation of the Council on Improving the Environmental Performance of Public Procurement, C*, 3.

Chapter 8
Romanian Agriculture Funding:
Approaches Regarding the Funding in Romanian Agriculture After EU Integration

Ancuta Marin
Research Institute for Agricultural Economics and Rural Development, Romania

ABSTRACT

Romania's EU integration has determined major economic changes, including the transition to a functional market economy which meant the existence of a competitive environment, free commercial exchanges, and free movement of people, capital, and services. To solve the existent delays, Romania has benefit from the European Community's funds for investment in agriculture and rural areas through the financial mechanism which assures non-reimbursable funds for farmers, private societies, and local authorities. Without claiming that we are exhausting the issue of financing in Romanian agriculture, the topic addressed brings some issues regarding the financing under the National Rural Development Program (NRDP) and the possible explanations of the low absorption rate of these funds. NRDP is a program which allow people to access non-reimbursable funds from EU and Romanian Government for economic and social development of the rural area.

INTRODUCTION

Romania's geographical position, the soil type, the climate and the hydrographic network offer the necessary conditions for agricultural activity. The modern development of agriculture requires, besides land, also human and material resources, financial means for the implementation of new technologies and for improving its technical-material endowment.

Romania's EU integration has determined major economical changes, including the transition to a functional market economy which meant the existence of a competitive environment, free commercial exchanges and free movement of people, capital and services. In order to solve the existing delays, Romania benefits from European Community funds for investments in agriculture and rural areas through the financial mechanism which assures non-reimbursable funds for farmers, private societies and local authorities.

DOI: 10.4018/978-1-5225-5739-5.ch008

Copyright © 2019, IGI Global. Copying or distributing in print or electronic forms without written permission of IGI Global is prohibited.

Without claiming that we are exhausting the issue of financing in Romanian agriculture, the topic addressed brings some issues regarding the financing under the National Rural Development Program (NRDP) and the possible explanations of the low absorption rate of these funds. NRDP is a program which allow people involved in agricultural and social activities of rural development to access non-reimbursable funds from EU and Romanian Government.

The present paper analyses a method of funding agriculture in Romania by offering non-reimbursable funds from EU and Romanian Government in order to avoid delays and to develop a modern agriculture. Also, some possible causes related to the low rate of funds absorption (only 70% of them), were analysed.

MAIN FOCUS OF THE CHAPTER

When Romania joined the EU, the Romanian economy, especially agriculture, are, in terms of endowment, far behind EU member states. From the multiple ways of financing, this chapter analyzes the structural funds allocated to Romania by the European Union, through the European Fund for Agriculture and Rural Development. The purpose of these allocations was that Romanian agriculture could overcome the lag between Romanian agriculture and the agriculture practiced in other countries of the Union and to comply the modern standards of development. This paper outlines the structure of these funds and their degree of absorption offering some explanations for the amounts that have not been accessed.

Romania is located in Central Europe, nearly at the half of the distance between the Atlantic Ocean and Ural Mountains. It has a surface of 238.391 km^2 and is the 12th-largest European country. 45°N Parallel and 25°E Meridian cross Romania near its geometric centre, at 100 km NW from the capital, Bucharest. Romania has borders in East with Moldova and Ukraine, in North with Ukraine, in west with Hungary and Serbia and in south with Bulgaria. Population density is around 90 people per kilometer square, above the European average.

Romania is a Carpathian country because 2/3 of the Carpathian chain is located on its territory. The Carpathian Mountains determine bio-pedoclimatic flooring, representing a barrier to air masses. Romania has important natural resources that have been of particular importance in the development of economic life. Romania is a Danubian country because the whole lower sector of the Danube River (1075 km) is on the territory of Romania or represents parts of the border area with the states of Serbia, Bulgaria and Ukraine. The Danube is an important axis of navigation since antiquity. Currently, the Black Sea and the North Sea are connected through the Danube and the Rhine River, and also through the Main-Rhine and the Danube-Black Sea channels. Romania is a Pontic country because it shares the Eastern border with the Black Sea. The length of the Romanian shore is about 245 km, which offers many economic advantages.

The geographical location of Romania, its pedo-climatic conditions and the hydrographic network provide favourable conditions to agriculture. For the modern development of agriculture requires, besides land, also human and material resources, financial means for the implementation of new technologies and for improving its technical and material endowment, the judicious choice and substantiation of the ownership forms, holding and organizing, managerial training, a legal and legislative framework specific to the market economy and to ensure the protection of the environment.

Until December 1989, Romania was a communist country. Between 1965 and 1989, there was no private property, land, forests, factories, any other property being the property of the communist state.

In 1990, the new regime adopted a new law which established that all lands would be returned to their owners as it was before 1965. As we mentioned earlier, in 1965 a lot of lands were confiscated and consequently were managed as a whole. The heads of the newly created holdings were members of the Communist Party, most of them without economic or agronomic studies. They focused on increasing production and exporting unprocessed products (wheat, corn, meat, milk, live animals, etc.) by their own means, without loans. With the risk of starving the population, the authorities decided that Romania must be a zero debt state which also had effects in agriculture. In this domain as in others, loans were not allowed, so a barter system was created to supply all needs (corn for tractors, for example) (Marin, 2017a; Marin, 2017b).

Currently, Romania is a national, sovereign and independent state, unitary and indivisible, and the form of government is the presidential republic. Since 29th of March 2004, Romania is a member of NATO, and since 1st of January 2007 it is a member of the European Union. The European Union's membership required, among other things, the alignment of national legislation with the European one and the reduction of Romania's economic and social lag. In the last 27 years, a lot of transformations took place in the agricultural domain.

The change of the political regime and the accession of Romania to the European Union brought ample transformations into the country. After Romania's accession to the EU, a mandatory condition was the harmonization of national legislation with the EU legislation. In this regard, laws and programmes have been adopted so that they respect EU requirements and the national specificities.

Romania's accession to the EU has led to major transformations of the economy and society, by accelerating the transition to a functioning market economy, opening up to the competitive environment, liberalization of trade, movement of people, capital and services. Business has begun to develop in a legal environment close to the European one. In order to overcome the lag, Romania benefited from agricultural investment through the Community, in the rural areas, through financial instruments that provided non-reimbursable funds to farmers, private companies and local public authorities.

The National Rural Development Program is divided into two periods: 2007-2013 and 2014-2020. The first is already over and the second is still in progress. On the 26th of May.2015 the European Commission approved the last version of NRDP 2014-2020.

Important funds for agriculture have been made available for Romania by the EU. Thus, over the 2007-2013 period, the National Rural Development Program allocated 8.12 million euro from the budget of the European Fund for Agriculture and Rural Development (EAFRD) and 1,20 billion euro from the national budget, while for 2014-2020, 8.01 billion euro from the EAFRD budget and 1,32 million euro from the national budget (Figure 1).

The National Rural Development Program (NRDP) is a program of measures developed by the Ministry of Agriculture and Rural Development in collaboration with the European Community to help the development of the Romanian agriculture. NRDP is a program that grants funds from the European Union and the Government of Romania for the economic and social development of the Romanian rural area. These measures are aimed to develop and to enhance agriculture, and in particular, to make the transition from a fragile and unprofitable agriculture to a modern agriculture. Investors have new financial opportunities for investment projects in the development of agriculture and rural areas. On each axis, according to the objectives pursued, measures and sub-measures are designed in such a way that sums of money can be allocated if several criteria are cumulatively fulfilled.

The European Fund for Agriculture and Rural Development (EFARD) is a financing instrument created by the European Union to support EU Member States in implementing the Common Agricultural Policy. The Common Agricultural Policy is a set of rules and mechanisms that regulate the production, the processing and the marketing of agricultural products in the European Union and which pays great attention to rural development. The EFARD represents a financing opportunity for the Romanian rural area and it is based on the principle of co-financing private investment projects.

The development of the rural environment is the general objective of the NRDP, but it is also explicitly addressed by the market measures on agriculture and other measures foreseen in the operational programs:

1. Sectorial Operational Program for Increasing the Economic Competitiveness (SOP IEC)
2. Sectorial Operational Program – Transport and Infrastructure (SOPT)
3. Sectorial Operational Program for Environment (SOP ENV)
4. The Regional Operational Program (ROP)
5. Sectorial Operational Program for Human Resources Development (SOP HRD)
6. Operational Program for Administrative Capacity Development (OP ACD)
7. Operational Program for Technical Assistance (PO AT)
8. Operational Programs of Cross-border Cooperation (One Operational Program for each implied state).

Transforming the mentality of rural people has also led to the transformation of the rural environment. The gradual penetration of urban consumption patterns, increasing the rural population's aspirations, international labour migration are just a few factors that have an impact on the rural environment along with the impact of the planned measures.

The National Program for Rural Development was designed on 4 axes:

Priority Axis 1: Increasing the competitiveness of the agricultural and forestry sectors;
Priority Axis 2: Improving the environment and rural space;
Priority Axis 3: Quality of life in rural areas and diversification of the rural economy;
Priority Axis 4: LEADER;

For a better understanding of the National Program for Rural Development, the author has systematized from the NRDP the general objectives, the strategic objectives and the specific objectives, as well as the measures that come out of them and systematized them in the form of logical schemes that will be presented below (Figure 1, Figure 2, Figure 3, Figure 4 and Figure 5.).

Analysing the National Program for Rural Development in the 2007-2013 segment, the allocation of funds by measures (Table 1 and Figure 6) and their absorption rate (Table 3 and Figure 7) are presented.

The economic crisis from 2009 has produced visible negative effects in Romania. It has reduced the capacity of public and private co-funding sources of European programs as a result of the decline in state and economic revenues. Demand reduction in most products, uncertainty about the duration of the crisis has considerably reduced the absorption capacity of the available funds for market-oriented activities.

By making a brief analysis of the funds allocated to these activities, we find that for the period 2007-2013 from the total earmarked amount of 9.324,8 million euro the total payments were 6.613,3 million euro, which represents a 70.92% rate of absorption. From the total allocated, on the mentioned interval, the contribution from the national budget was 12.9%, the difference coming from the Union budget.

Figure 1. The NRDP Components

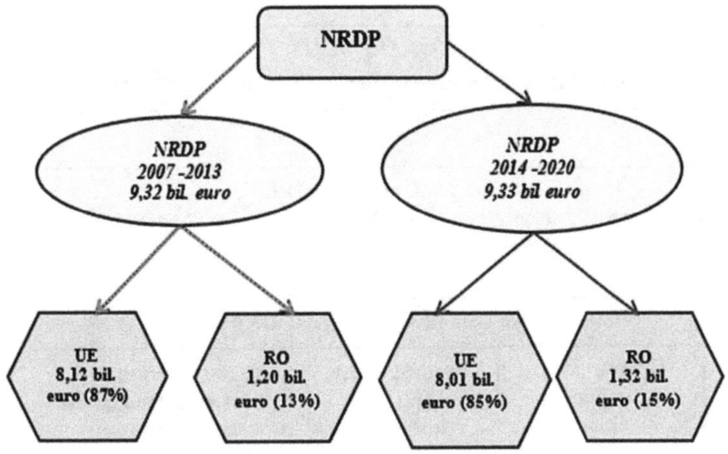

For the 2014-2020 period, the National Rural Development Program has several strategic objectives: restructuring and increasing the viability of agricultural holdings, diversifying economic activities, creating jobs, improving infrastructure, improving services, improving the quality of life, implementing a sustainable management of natural resources, combating climate change.

The NRDP 2014-2020 financial framework (presented in Table 3) has six priorities. These are:

P1: Encourage the transfer of knowledge and innovation in agriculture, forestry and rural areas;

P2: Increase the viability of farms and the competitiveness of all types of farming in all regions and promote innovative agricultural technologies and sustainable forest management;

P3: Promote the organization of the food chain, including the processing and marketing of agricultural products, animal welfare and risk management in agriculture;

P4: Restoration, conservation and consolidation of ecosystems that are related to agriculture and forestry;

P5: Promote effective resource use and support the transition to a low-carbon and climate-resilient economy in the agricultural, food and forestry sectors;

P6: Promoting social inclusion, poverty reduction and economic development in rural areas.

NRDP 2014-2020 contains measures whose aims are the rural development in Romania, taking into consideration both the socio-economic realities of our country, implicitly of the rural area, as well as the Common Agricultural Policy and the Europe 2020 strategy. The NRDP 2014-2020 will focus on a small number of measures that will address the needs identified on the basis of sectorial, socio-economic and environmental analysis for the Romanian rural area, reflecting the directions of action in the development of agriculture in Romania. NRDP 2014-2020 includes 14 measures divided into 29 sub-measures and 3 sub-programs.

The budget of the NRDP 2014-2020 has been reduced, so that the financial allocation Romania will benefit from in the next financial year is lower, and will decrease from 8.12 billion € in the period 2007-2013 to 8.015 billion € in 2014 -2020. The amounts allocated by the EAFRD for 2014-2020 period are thus divided (see Table 4 and Figure 8):

Figure 2. Axis 1 and measures for the allocation of funds

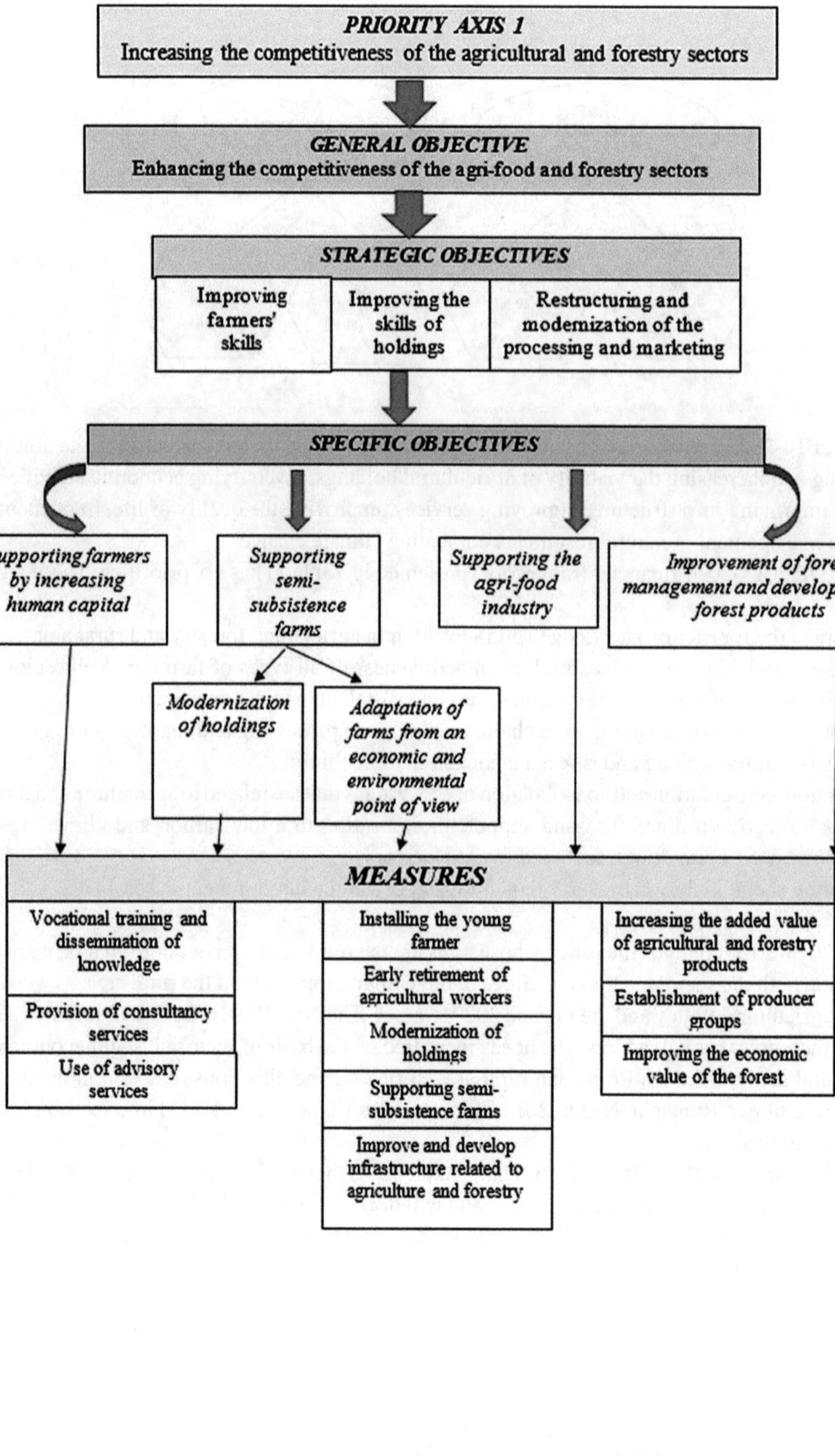

Figure 3. Axis 2 and the allocation of funds

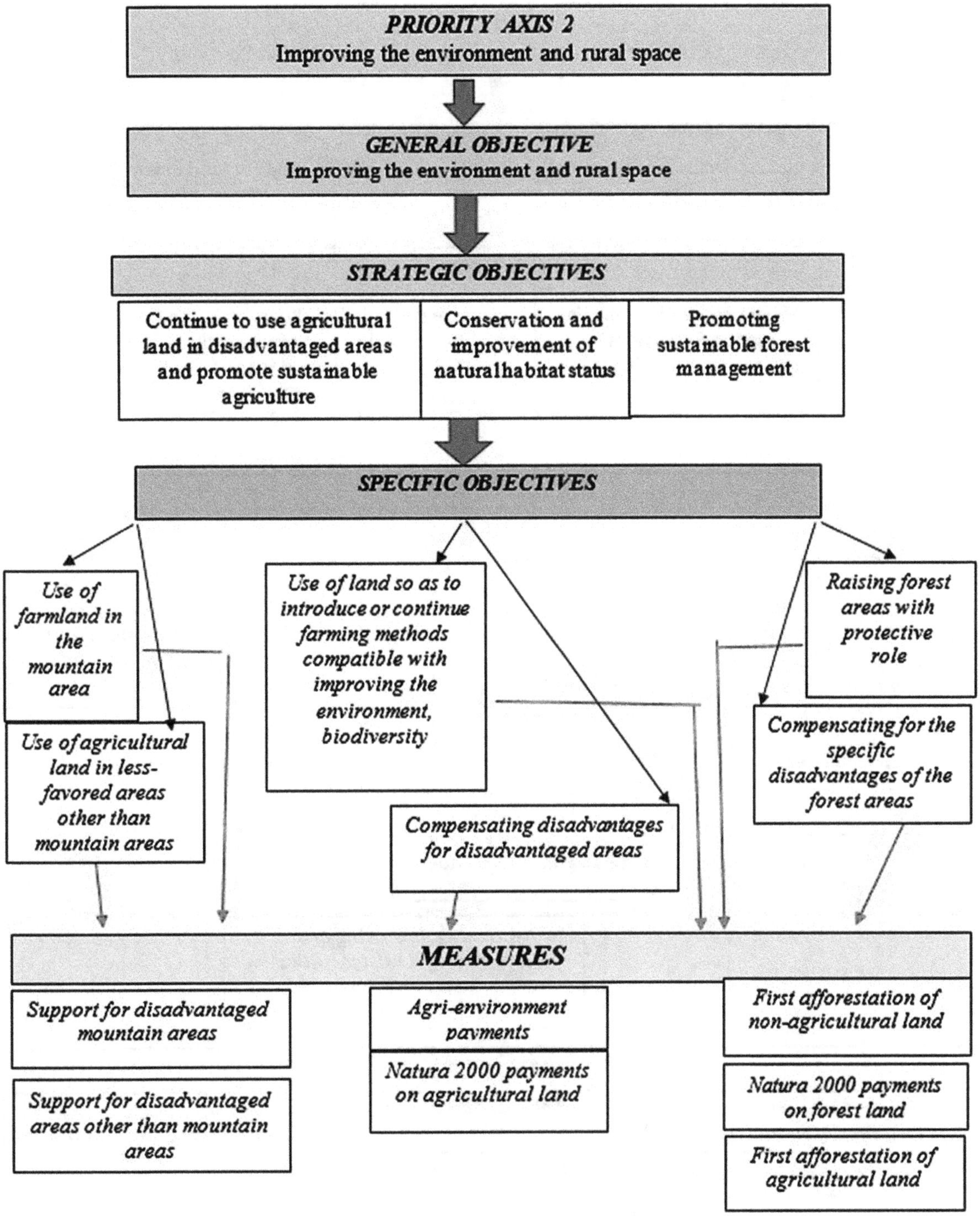

Figure 4. Axis 3 and measures to allocate funds

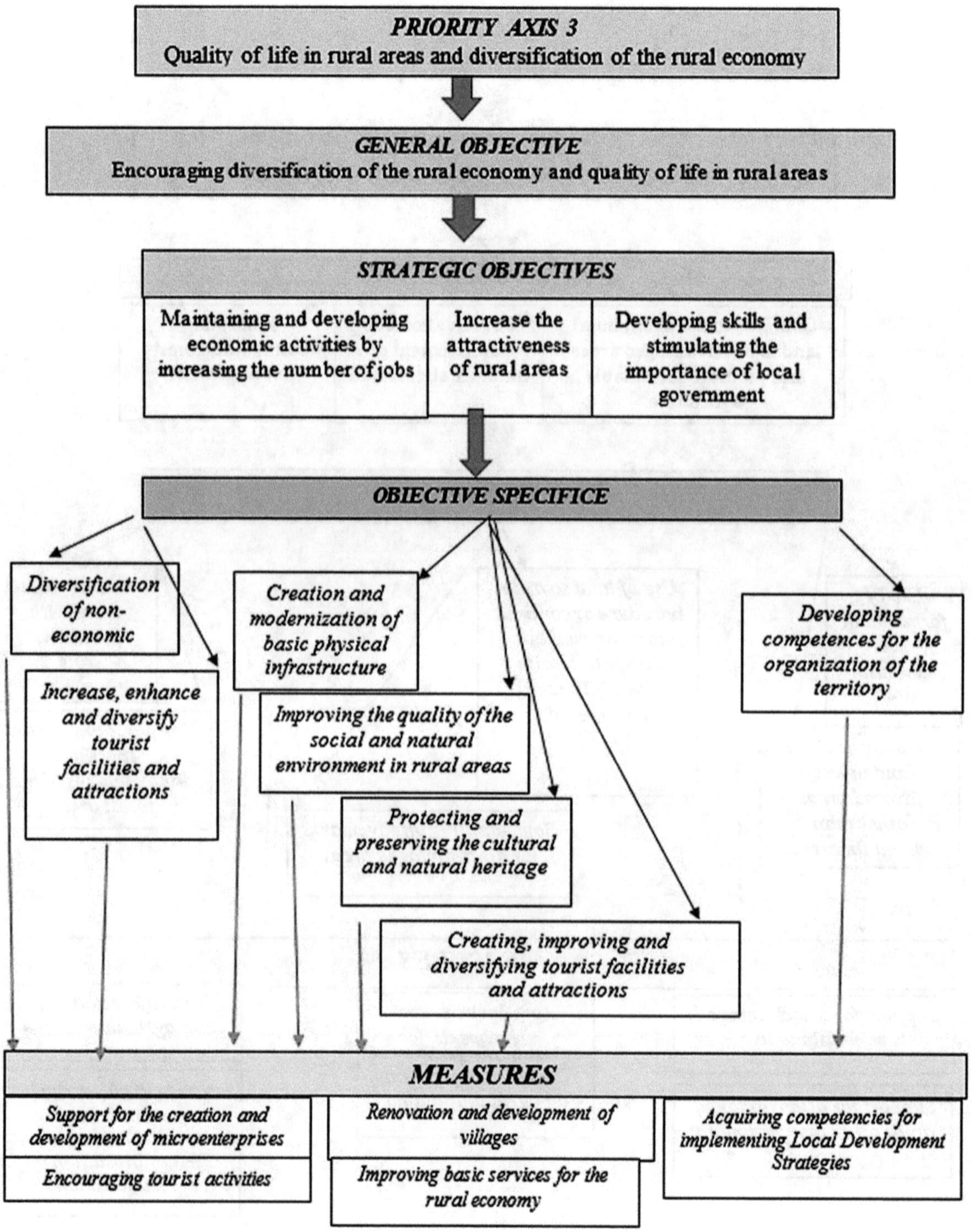

Figure 5. Axis 4 and measures to allocate funds

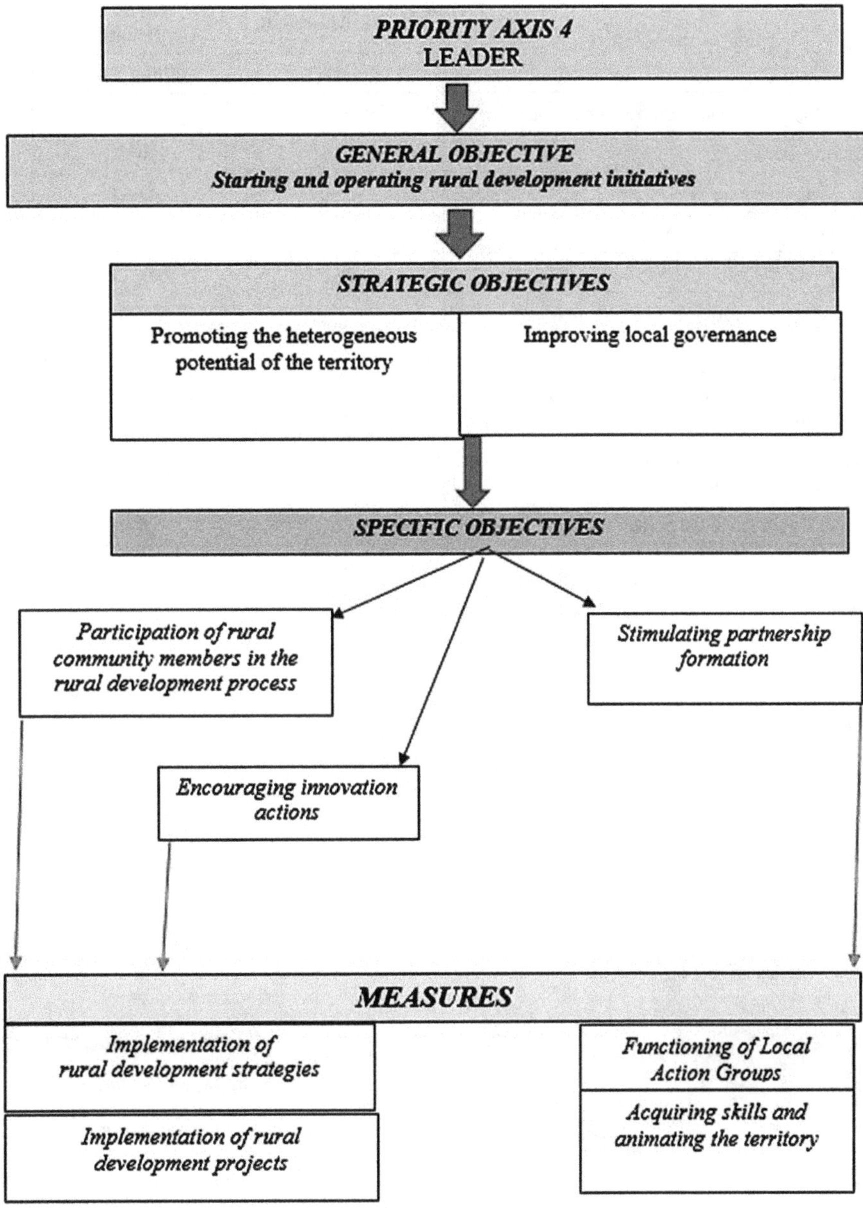

Regarding the absorption of structural funds through measures under the National Rural Development Program for 2014-2020 (Table 5), so far, out of 34141 submitted projects a number of 11995 projects are in progress, representing 35.14%. Projects completed since the opening of the measures are only 477, ie 1.39% of the number of submitted projects. Projects cancelled are in a number of 38, which means an insignificant percentage, namely 0.11% of the total.

Analysing the before mentioned information, we can explain the low degree of absorption of funds as having several causes. The first one is related to the size of holdings whose average size varied be-

Table 1. Allocation of funds by measures in NRDP 2007-2013

Code of Measure	Name of the Measure	FEADR Allocation	National Contribution	Total
		Millions EUR	Millions EUR	Millions EUR
M1	Transfer of knowledge and information actions	50,3861	5,5985	55,9846
M2	Counselling services	50,3861	8,8991	59,2852
M4	Investments in physical assets	2.502,04	444,9674	2.947,01
M6	Developing farms and businesses	866,3575	127,0992	993,4567
M7	Basic services and village renewal in rural areas	1.100,60	197,3303	1.297,93
M8	Afforestation and creation of forested areas and forest curtains	105,6952	18,6521	124,3472
M10	Agro-environment and climate	849,9649	149,9938	999,9587
M11	Organic farming	200,6862	35,4152	236,1014
M13	Payments for areas experiencing natural constraints or other specific constraints	1.130,80	199,5523	1330,3487
M15	Forestry and environmental protection of forests	100	17,6471	117,6471
M16	Cooperation	28,0211	3,1135	31,1346
M17	Risk management	200	35,2941	235,2941
M19	LEADER	652,3546	72,485	724,8495
M20	Technical support	178,368	31,4767	209,8447
	TOTAL	8.015,66	1.347,52	9.363,19

Source: Ministry of Agriculture and Rural Development (n.d.)

Figure 6. Allocation of funds in 2007-2013 period

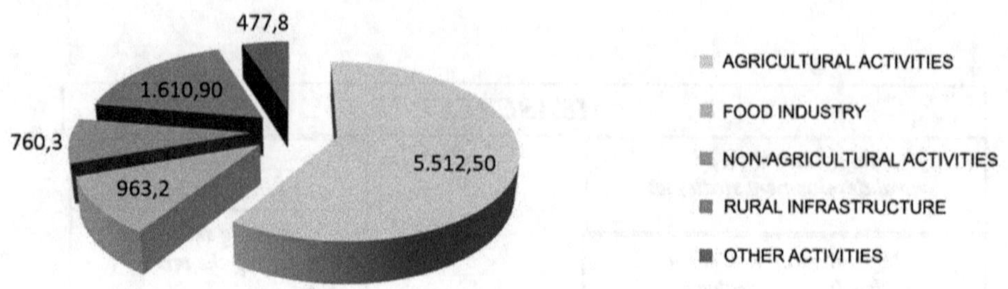

tween 1.73 and 2.29 ha per farm for individual property and between 190.78 and 274.43 ha per holding for those with legal personality (Table 6). In the guidelines about accessing the funds, there are special stipulations on the size of holdings, most of them that do not fulfill this condition.

Territorial fragmentation is characterized by a low yield, the cultivated areas cannot be worked mechanically. Farmers are unable to use efficient fertilizers and selected seeds or to find workers. It is necessary, therefore, to associate the owners of agricultural land in holdings with legal personality, to protect their interests through proper management, by using more efficiently the means of production: land, financial resources and labour force.

Table 2. The degree of funds absorption NRDP 2007-2013

Activity	Allocation Millions EUR	Payments Made Millions EUR	Absorption Rate %
I. Agricultural activities	5.512,50	4.218,70	74,71
II. Food industry	963,20	297,80	30,91
III. Non-agricultural activities	760,30	330,20	43,43
IV. Rural infrastructure	1.610,90	1.345,90	83,54
V. Other activities	477,80	284,30	77,48
Total general of which:	9.324,80	6.613,30	70,92
- EU budget	8.124,00	-	-
- national budget	1.200,80	-	-
% national contribution	12,90	-	-

Source: Ministry of Agriculture and Rural Development (n.d.)

Figure 7. Degree of absorption in 2007-2013 period

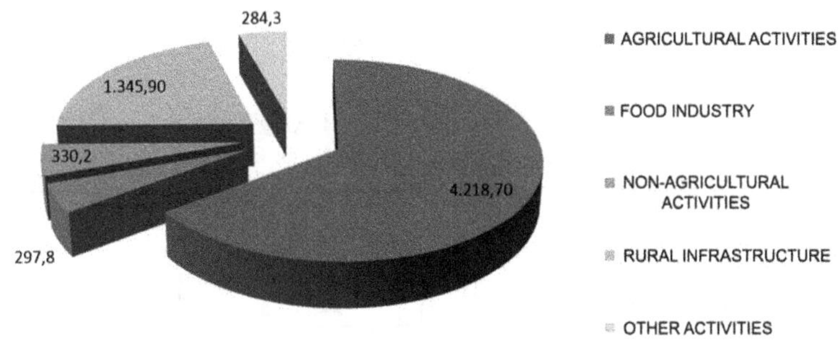

Table 3. Financial framework NRDP 2014-2020 by priorities

Priority	Financial Allocations					
	EU Budget (Mil. EUR)	%	National Budget (Mil. EUR)	%	Total (Mil. EUR)	% Structure
P1	178,4	98,8	1,7	1,2	180,1	1,9
P2	1.516,9	86,3	241,1	13,7	1.758,0	18,8
P3	867,6	85,1	151,6	14,9	1.019,2	10,9
P4	2.309,5	85,6	387,7	14,4	2.697,2	28,9
P5	856,2	83,6	167,5	16,4	1.023,7	11,0
P6	2.287,1	86,2	366,2	13,8	2.653,3	28,4
TOTAL	8.015,7	85,9	1.315,8	14,1	9.331,5	100,0

Source: Ministry of Agriculture and Rural Development (n.d.)

Figure 8. Allocation of funds for 2014-2020

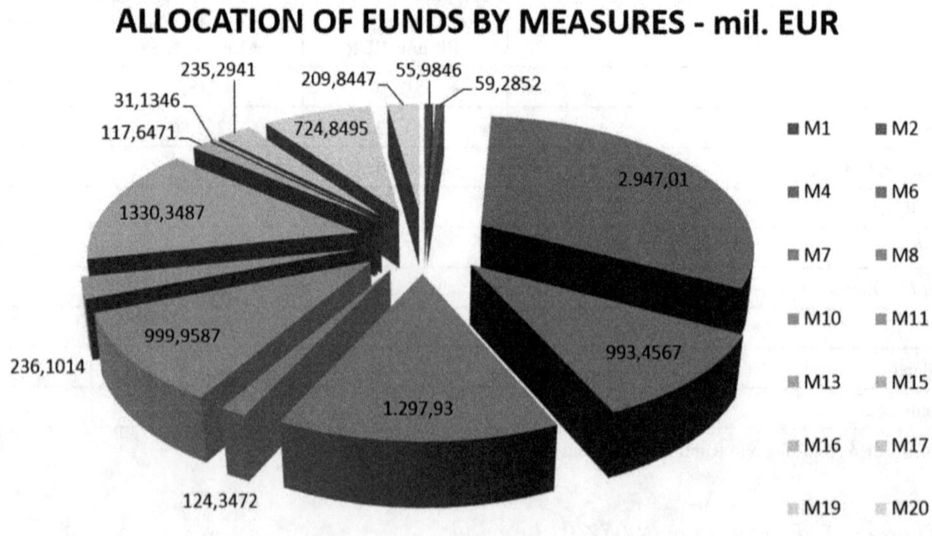

Table 4. Allocation of funds to the 2014-2020 NRDP measures

Measurement Code	Name of the Measure	FEADR Allocation Millions of EUR	National Contribution Millions of EUR	Total Mill. EUR
M1	Transfer of knowledge and information actions	50,38	5,60	55,98
M2	Counseling services	50,38	8,90	59,28
M4	Investments in physical assets	2.502,04	444,96	2.947,00
M6	Developing farms and businesses	866,35	127,11	993,46
M7	Basic services and village renewal in rural areas	1.100,60	197,33	1.297,93
M8	Afforestation and creation of forested areas and forest curtains	105,70	18,65	124,35
M10	Agro-environment and climate	849,97	149,99	999,96
M11	Organic farming	200,69	35,41	236,10
M13	Payments for areas experiencing natural constraints or other specific constraints	1.130,80	199,55	1.330,35
M15	Forestry and environmental protection of forests	100,00	17,65	117,65
M16	Cooperation	28,02	3,11	31,13
M17	Risk management	200,00	35,30	235,30
M19	LEADER	652,35	72,48	724,83
M20	Technical support	178,37	31,47	209,84
	TOTAL	8.015,66	1.347,52	9.363,18

Source: Ministry of Agriculture and Rural Development (n.d.)

Table 5. Degree of absorption for the period 2014-2017

Sub Measure	Projects Submitted		Contracted Projects					
			Contracted Projects (in Progress and Completed)		Completed Projects		Projects Cancelled	
	No.	Value Mil EUR	No.	Value Mil EUR	No.	Value Mil EUR	No.	Value Mil EUR
Sub-measure 1.1 "Support for Vocational Training and Acquisition of Competencies"	254	4,36	0	0	0	0	0	0
Sub-measure 4.1 "Investments in agricultural holdings"	3530	339,94	836	58,03	276	14,57	9	1,82
Sub-measure 4.1 "Investments in agricultural holdings" - ITI Danube Delta	126	11,24	0	0	0	0	0	0
Sub-measure 4.1a "Investments in fruit holdings"	260	25,31	20	1,87	0	-	2	0,33
Sub-measure 4.2 "Support for investment in the processing / marketing of agricultural products"	191	35,26	47	8,66	2	0,02	1	0,19
Sub-measure 4.2 "Support for investment in the processing / marketing of agricultural products" - ITI Danube Delta	3	1,18	0	0	0	0	0	0
Sub-measure 4.2 "GBER State Aid Scheme"	86	10,44	9	2,02	0	0	0	0
Sub-measure 4.2 "De minimis scheme"	3	0,01	0	0	0	0	0	0
Sub-measure 4.2a "Investments in the processing / marketing of fruit-growing products"	20	1,53	3	0,07	0	-	1	0,03
Sub-measure 4.3 "Investments for the development, modernization or adaptation of agricultural and forestry infrastructure - irrigation"	97	21,78	53	11,04	0	0	0	0
Sub-measure 4.3 "Investments for development, modernization or adaptation of agricultural and forestry infrastructure - agricultural access infrastructure"	444	92,61	45	9,49	0	0	0	0
Sub-measure 4.3 "Investments for development, modernization or adaptation of agricultural and forestry infrastructure - agricultural access infrastructure" - ITI Danube Delta	5	0,96	0	0	0	0	0	0
Sub-measure 4.3 "Investments for development, modernization and adaptation of agricultural and forestry infrastructure - forestry infrastructure"	96	29,73	0	0	0	0	0	0
Sub-measure 6.1 "Support for the Installation of Young Farmers"	11352	102,76	5441	49,38	4	0,04	6	0,05
Sub-measure 6.1 "Support for the Installation of Young Farmers" - ITI Danube Delta	163	1,46	0	0	0	0	0	0
Sub-measure 6.2 "Support for the establishment of non-agricultural activities in rural areas"	6175	76,79	1494	20,03	0	0	0	0
Sub-measure 6.2 "Support for the establishment of non-agricultural activities in rural areas" - ITI Danube Delta	181	2,57	0	0	0	0	0	0
Sub-measure 6.3 "Support for the development of small farms"	5953	19,62	2663	87,79	9	0,03	0	0
Sub-measure 6.3 "Support for the development of small farms" - ITI Danube Delta	13	0,04	0	0	0	0	0	0
Sub-measure 6.4 "Investing in the creation and development of non-agricultural activities"	2020	73,95	273	9,04	36	0,99	9	0,27
Sub-measure 6.4 "Investments in the creation and development of non-agricultural activities" - ITI Danube Delta	43	1,77	0	0	0	0	0	0
Sub-measure 6.5 "Scheme for Small Farmers"	15	0,01	0	0	0	0	0	0
Sub-measure 7.2 "Investments in the creation and upgrading of small scale basic infrastructure - water / waste water infrastructure"	502	159,22	105	31,78	0	0	0	0

continued on following page

Table 5. Continued

Sub Measure	Projects Submitted		Contracted Projects					
			Contracted Projects (in Progress and Completed)		Completed Projects		Projects Cancelled	
	No.	Value Mil EUR	No.	Value Mil EUR	No.	Value Mil EUR	No.	Value Mil EUR
Sub-measure 7.2 "Investments in the creation and modernization of small scale basic infrastructure - water / wastewater infrastructure" - ITI Danube Delta	5	1,29	0	0	0	0	0	0
Sub-measure 7.2 "Investing in the creation and upgrading of small-scale basic infrastructure - road infrastructure of local interest"	993	225,22	210	49,75	0	0	0	0
Sub-measure 7.2 "Investing in the creation and modernization of small-scale basic infrastructure - local infrastructure" - ITI Danube Delta	16	3,46	0	0	0	0	0	0
Sub-measure 7.2 "Investing in the creation and upgrading of small-scale basic infrastructure - educational and social infrastructure"	450	37,51	136	11,04	0	0	0	0
Sub-measure 7.2 "Investing in the creation and modernization of small-scale basic infrastructure - educational and social infrastructure" - ITI Danube Delta	4	0,33	0	0	0	0	0	0
Sub-measure 7.6 "Investments associated with the protection of cultural heritage"	769	57,49	225	14,97	0	-	1	0,03
Sub-measure 7.6 "Investments related to the protection of cultural heritage" - ITI Danube Delta	4	0,36	0	0	0	0	0	0
Sub-measure 8.1 "Afforestation and the creation of wooded areas"	37	1,59	0	0	0	0	0	0
Sub-measure 9.1 "Establishment of producer groups"	8	0,69	3	0,33	0	0	0	0
Sub-measure 15.1 "Payments for commitments in the field of forestry and climate"	28	0,61	0	0	0	0	0	0
Sub-measure 16.4 "Support for horizontal and vertical cooperation between actors in the supply chain"	92	1,93	0	0	0	0	0	0
Sub-measure 16.4a "Support for horizontal and vertical cooperation between actors in the supply chain"	23	0,48	2	0,04	0	0	0	0
Sub-measure 19.1 "Preparatory support for the development of local development strategies"	180	0,54	164	0,49	139	0,42	9	0,03
Sub-measure 19.4 "Support for running and animation expenses"			239	15,35	0	-	0	-
Measure 20 "Technical Assistance" **			27	0,81	11	0,04	0	-
TOTAL	34141	1.344,03	11995	302,95	477	16,10	38	2,75

Source: Ministry of Agriculture and Rural Development (n.d.)

Another explanation is linked to the immigration of the active population to other European countries, Canada and the USA as the immigration from Eastern Europe increased after the fall of communist regimes in this part of the continent in period 1989-1990. Today, the European Community counts more than 10 million foreigners, representing almost 3% of the EU population. There are also more than 3 million people still living illegally in the EU. Against this background, EU policymakers have reinforced measures to limit migratory flows with a policy of integrating aliens living in the EU, seeking solutions to the demographic crisis predicted in the nearest future of Europe.

Table 6. Situation of agricultural holdings

Situation of Agricultural Holdings					
Indicators	Years	Legal Status of Agricultural Holdings		
		Total	Individual Agricultural Holdings	Units With Legal Personality
Agricultural holdings (total number)	2002	4.484.893	4.462.221	22.672
	2005	4.256.152	4.237.889	18.263
	2007	3.931.350	3.913.651	17.699
	2010	3.859.043	3.823.130	30.698
	2013	3.615.604	3.587.724	27.880
Agricultural holdings that have used the agricultural area	2002	4.299.361	4.277.315	22.046
	2005	4.121.247	4.103.404	17.843
	2007	3.851.790	3.834.407	17.383
	2010	1.021.937	992.602	27.702
	2013	3.549.994	3.522.544	27.450
Used agricultural area (ha)	2002	13.930.710	7.708.758	6.221.952
	2005	13.906.701	9.102.018	4.804.683
	2007	13.753.046	8.966.309	4.786.738
	2010	13.306.128	7.151.186	5.856.507
	2013	12.531.367	6.746.367	5.784.842
Agricultural area used on an agricultural holding	2002	3,11	1,73	274,43
	2005	3,27	2,15	263,08
	2007	3,50	2,29	270,45
	2010	3,45	1,87	190,78
	2013	3,60	1,88	207,49

Source: NIS (n.d.)

At European level there is a 50% reduction in agricultural workforce from 12,730,000 in 1980 to only 6,527,000 in 2002. The labour force used in the total economy increases considerably in absolute figures in all countries, from 154,988. 000 people in 1999 to 162,974,000 people in 2002. There has been a reduction in the share of labour force in agriculture in the last quarter of a century, reaching only 1.4% in England, 2.5% in reunified Germany, 4.1% in France. The explanation for this could be the labour force absorption by the tertiary sector, that of the services, for example: 67.8% in the EU in 2002. The labour force in agriculture is relatively high (three times more employees than in Romania) with great variability, however, from one country to another (4.5% in Greece and 50% in Denmark) with the average for the EU-15 being 30.9% (Albu, Caraiani, & Iordan, 2012).

In 2012, as shown in table no. 7 only 42% of the country's population was employed while 58% were retired, students or unemployed. In 2015, the share of the employed population decreased to 38% while the share of the inactive population cumulated with the unemployed increased to 62% (Table no. 8). Another conclusion drawn from the two tables is that females have important shares in the category of inactive and unemployed people ranging from 66.83% to 78.08% (NIS, n.d.).

Table 7. Population structure of Romania in 2012

Population Structure in Romania in 2012						
	Total		Males		Females	
	Thousands of People	%	Thousands of People	%	Thousands of People	%
Active population:	9759	45,70%	5434	52,28%	4325	39,78%
*employed	9019	92,42%	4980	91,65%	4039	93,39%
*unemployed	740	7,58%	454	8,35%	286	6,61%
Inactive population (pupils, students, retirees):	11397	54,30%	4960	47,72%	6437	60,22%
Total population	21356	100,00%	10394	100,00%	10872	100,00%

Source: NIS (n.d.)

Table 8. Population structure of Romania in 2015

Population Structure in Romania in 2015						
	Total		Males		Females	
	Thousands of People	%	Thousands of People	%	Thousands of People	%
Active population:	9159	41,08%	6657	61,16%	2502	21,92%
*employed	8535	93,19%	6365	95,61%	2107	84,21%
*unemployed	624	6,81%	292	4,39%	332	15,79%
Inactive population (pupils, students, retirees):	13139	58,92%	4227	38,84%	8912	78,08%
Total population	22298	100,00%	10884	100,00%	11414	100,00%

Source: NIS (n.d.)

Of the total people employed, in 2015 25.6% worked in the agricultural sector, 28.4% in industry or construction and 46.0% in services. In non-agricultural activities, 6,352 thousand people were occupied, significant shares of which were owned by those operating in the manufacturing (24.4%), commerce (18.1%) and construction (10.0%).

Out of 178,000 agricultural workers, as many as they were in 2015, 15.9% were unqualified workers, 21.4% were specialists with higher education, and 62.7% had switched to other areas where they could earn constant income.

The net international migration rate is significant and is mainly related to the younger age group, and is the second leading cause of the drop in the number of graduates with higher education. External migration has become a significant issue for Romania, especially in recent years. Statistical estimates indicate that over 2 million Romanians - or about 10% of the total population - work abroad (Spain, Italy, Greece and Germany). Young people are not necessarily aware of the shortage of well-trained people in specialized fields, but rather focus on the faculties they consider to be good. That's because they think of the bigger chances they could get once they're out on the labour market.

As shown in Table 9, the importance of training active people from all EU countries has increased over the past 5 years, the share of the population with higher education in the total population in a country, varying slightly, but on an increasing trend (see Table 9) (Marin, 2014; Palicica, Gavrilă, & Ion, 2007).

At European level, Romania ranks the penultimate place in terms of the number of specialists trained in one year. Many of them are attracted by large wages in other countries and do not work for Romania.

In Romania, their degree of absorption on the labour market in the graduated field is low, except those who work in IT and financial-accounting. Analysing and processing the data published by the National Institute of Statistics (NIS), a situation of the graduates with higher education in Romania, by domains, was made during 2006-2013. As it can be seen in Table no. 10 in Romania, in 2008 there was a doubling of the number of graduates, followed by a slight decrease in the following years (see Table 10) (Marin, 2014).

Over the last decade, lifelong learning has taken a central stance in the Community's education and training policy. Adult learning has been recognized as an important component of lifelong learning, with adult learning opportunities being essential for ensuring social and economic progress, but also for individual personal development. Lifelong learning in Romania is not addressed in a coherent and comprehensive manner at the level of the educational system and policies. This limits the coherence and flexibility of individual lifelong learning paths. Despite the progress made in regulating past learning validation, insufficient use of the existing legal framework remains one of the weak points of the lifelong learning approach to education and training.

From the author's research on the basis of structural surveys among small and medium-sized fruit and vegetable producers in representative counties for these sectors of Romanian agriculture, there were other reasons why the absorption rate in Romania is so small compared to other states from the EU (Poland, Hungary). Advanced age of Romanian farmers influences the absorption of funds in two ways:

1. Guides on accessing European funds provide additional points for those who are under the age of 40; many farmers consider this an insurmountable threshold and no longer apply for the measure they would be interested in;
2. The fact that over 58% of the Romanian fruit and vegetable producers are over 40 years old and more than 10% over 60 years of age leads to a great confusion in the perception of the association / cooperation term, most of them putting the same sign with the former agricultural cooperatives (CAP); refusing the association, they lose important points when it comes to selecting files on accessing European funds;

In the books published in 2016 - "Design and experimentation of models of development of short chains for the capitalization of vegetable production", respectively in 2017 - "Design and experimentation of models for the development of short chains for the capitalization of fruit production", both in the ASE Publishing House, Bucharest, the authors conducted two studies, one among the vegetable producers (2016) and one among the fruit producers in Romania (2017) (Marin et al., 2016; Marin, Ion, Chetroiu, & Iurchevici, 2017).

Between March and April 2016, a survey was conducted among vegetable growers in Romania. The survey aimed to identify the problems they are facing and to highlight their views in order to develop solutions / models for the efficient use of vegetable production. A number of 180 questionnaires were applied in six representative counties in terms of vegetable production: Dâmboviţa, Călăraşi, Galati, Ilfov, Olt and Dolj (Marin, Chetroiu, & Iurchevici, 2016).

Table 9. Balance of population with higher studies in total population

Balance of Population With Higher Studies in Total Population											
		2009		2010		2011		2012		2013	
Country	Total Population	Persons With Higher Education	% From Total Population	Persons With Higher Education	% From Total Population	Persons With Higher Education	% From Total Population	Persons With Higher Education	% From Total Population	Persons With Higher Education	% From Total Population
Germany	81.802	11.940	14,60	12.060	14,74	12.610	15,42	12.961	15,84	13.154	16,08
France	65.822	10.359	15,74	10.501	15,95	10.730	16,30	11.118	16,89	11.499	17,47
United Kingdom	62.008	11.774	18,99	12.402	20,00	13.106	21,14	13.675	22,05	14.078	22,70
Italy	60.626	5.044	8,32	5.141	8,48	5.211	8,60	5.481	9,04	5.683	9,37
Spain	46.149	8.537	18,50	8.848	19,17	9.105	19,73	9.265	20,08	9.463	20,51
Poland	38.187	4.763	12,47	5.018	13,14	5.233	13,70	5.519	14,45	5.770	15,11
Romania	20.122	1.689	8,39	1.789	8,89	1.946	9,67	2.030	10,09	2.071	10,29
Holland	16.945	3.086	18,21	3.032	17,89	3.041	17,95	3.113	18,37	3.164	18,67
Hungary	10.014	1.145	11,43	1.166	11,64	1.227	12,25	1.277	12,75	1.304	13,02
Sweden	9.428	1.667	17,68	1.716	18,20	1.770	18,77	1.832	19,43	1.918	20,34
Austria	8.404	898	10,68	917	10,91	929	11,05	966	11,49	1.004	11,95
Denmark	5.561	950	17,09	969	17,43	977	17,56	998	17,94	1.012	18,19
Finland	5.461	1.090	19,96	1.117	20,46	1.144	20,96	1.149	21,05	1.171	21,44
Slovakia	5.435	526	9,68	591	10,87	636	11,70	660	12,14	687	12,63

Source: Marin (2014)

Table 10. Situation of graduates of Romanian higher education studies

Situation of Graduates of Romanian Higher Education Studies (State Institutions + Private Institutions)									
Field	2006	2007	2008	2009	2010	2011	2012	2013	Total Graduates per Fields
Technical	24.605	24.758	23,949	49,342	30,287	25,493	27,190	25,317	230,941
Agricultural	3.200	3.750	2,087	3,336	2,228	2,453	2,499	2,396	21,949
Medico-pharmaceutical	6.292	6.633	6,596	8,122	7,763	9,729	9,434	9,437	64,006
Economic sciences	32.098	37.211	91,884	67,420	72,641	62,685	34,415	25,724	424,078
Legal Sciences	10.175	12.568	14,458	21,418	17,954	26,404	19,215	12,521	134,713
Pedagogical University	36.502	42.093	92,118	66,140	60,198	57,507	41,514	33,430	429,502
Artistic and journalistic	2.572	2.236	3,880	2,384	2,448	2,629	2,404	2,203	20,756
Total graduates per year	112.244	125,499	232,885	214,826	191,291	186,900	136,671	111,028	1,311,344

Source: Marin (2014)

Of the interviewed persons, 72% started their activity in vegetable growing before the accession to the EU and only 21% after 2007. It is noted that 97% of the respondents are not involved in any form of association/cooperation, only 3% being associated (1% in the cooperative, 1% in the associations, 1% in the producer group). 109 persons (62%) of the 180 interviewed do not consider joining in one of the forms of association in the near future. The reasons for this reluctance include: first of all high taxes and charges (double taxation), as well as the lack of loans with preferential interest.

In terms of farm size, 37% are under 1 ha, 49% between 1 and 5 ha and only 13% over 5 ha. Of those interviewed, 85% had only field crops, which involves a major risk of calamity (frost, flood, drought).

From the analysis of the answers, it is noticed that 74% of the questioned vegetable growers apply the mechanized works with their own means, 18% resort to private persons and only 8% employ agricultural services companies. The labour force used in these holdings is 93% of the family members (2-4 persons), 5% employ skilled workers and 2% have unskilled / seasonal workers.

Regarding the capitalization of production, the results of the questionnaires confirm the problems raised by producers regarding the sale of fresh vegetables, namely: 95% of the interviewees have no contracts concluded for sale; 74% sell directly to the farm gate; over 80% of customers are at long distances of 10 to 50 km.

The bulk of the produced quantities (64%) is sold in retail, therefore in small quantities, leading to high quantitative and qualitative losses by increasing the capitalization time. State financial support is considered by most producers (67%) to be insufficient, even insignificant. However, only 26% of respondents accessed the funds allocated through the specific measures in the NRDP, the main reason for those who did not apply for European funding was the high degree of bureaucracy in the process of filing the documentation.

Of those who benefited from NRDP funds: 48% went to Sub-measure 1.4.1. Support for semi-subsistence agricultural farms in the NRDP 2007-2013, 21% Sub-measure 6.3 Support for the development of small farms in the NRDP 2014-2020, only 3% Sub-measure 6.1. Installing young farmers in the NRDP 2014-2020.

However, the prospect is gratifying, as 71% of the respondents to the questionnaire want to access NRDP funds in the near future. Between March and April 2017, a survey was conducted for fruit growers in Romania. The survey aimed at identifying the problems faced by small and medium-sized producers in the sector and highlighting their views in order to develop efficient solutions / models for fruit harvesting. A number of 153 questionnaires were applied in five representative counties in terms of fruit production: Dâmboviţa, Argeş, Valcea, Prahova and Constanţa (Marin, Chetroiu, Iurchevici, & Tudor, 2017)

Most respondents (32 respondents), representing 20.9%, started their activity in the period 1996-2000, followed by the period 2011-2015, when 20.3% (31 respondents) started their activity, and the third place is the period 2006-2010, when 16.3% of the people participating in the study started their activity. Only 20.9% of respondents started their activity before 1989, the remaining 79.1% after 1990. Most of them (63%) consider that it is not appropriate to join an association, while 26% are part of an association, 6% are part of a producer group, and 5% are organized in cooperatives.

Unfortunately, more than half (66%) of non-affiliated farmers are not interested in accessing such a form of organization either in the future. Regarding the motivation underlying the decision of non-involvement in an associative form, we note:

- On the first place is the lack of legislation to support small and medium-sized producers (27%);
- Followed, with a close percentage, by the difficulty in accessing European funds (22%);
- Lack of loans with preferential interest for farmers in an association (14% of respondents);
- The existence of high taxes and fees (12% of the total participants in this study).

The agricultural holdings owned by the farmers surveyed are allocated according to the size of the holding: under 1 ha (10%), 1 to 5 ha (75%), 5 to 10 ha (7%), 10 to 20 ha (6%) and over 20 hectares are owned by only three of the farmers (their share in the total area owned by the respondents being only 2%). 22% of the total of 153 farmers participating in the study set up orchards on an area of less than 1 hectare, 70% on surfaces between 1 and 5 ha and, in an approximately equal percent (4%), on surfaces between 5 and 10 ha and between 10 and 20 ha. In terms of crop diversity, more than half of the holdings are dedicated to apple and plum crops, with 29% and 26% respectively, followed by peach, apricot, hair or walnut, and strawberry crops occupy a percentage of 6% of the area under cultivation. Regarding the mechanization of the crop maintenance works, 76% of the cases use own means, existing on the farm, and rarely resort to companies specialized in the provision of services in agriculture (3%). Almost one quarter of respondents resort to carrying out the works with private individuals. The structure of human resources used reflects the fact that, in general, in small-scale farms, family members are those who work on the farm, occasionally also resorting to seasonal workers.

The market for the sale of production is another aspect that was taken into account in the realization of this study:

- It is noticed that only 15% of respondents plan their production structure on the basis of contracts concluded in advance, having a well-structured, predictable management plan aiming at achieving well-established indicators;
- Most of the production goes to sale, even if a percentage is usually kept for self-consumption.

Most farmers market their products in local markets. A very small number of producers have access to large supermarket or hypermarket networks, with one producer of the 153 respondents who sells exclusively in the supermarket (Marin, Chetroiu, & Iurchevici, 2017).

Farmers surveyed are relatively well connected to the current realities, almost 60% of them have already accessed EU funds. Also, the majority (72 out of 153 respondents) also accessed funds from the NRDP, the most accessible measure being measure 1.4.1 "Support for semi-subsistence farms" (57%) and the least approached (1%) – measure 6.1 "Support for the installation of young farmers".

Unfortunately, although a high percentage of farmers have already accessed European funds, they have become rather mistrustful and reluctant, basically 43% of them are not sure that they will access European funds in the future, 9% are even certain that they do not have that in plan. Thus, the percentage of those intending to access EU funds has fallen below 50%.

First in terms of obstacles in accessing funds is in the minds of farmers bureaucracy associated with this line of work (80% of respondents) (Marin, 2016).

Another impediment is co-financing, which in many cases accounts for 50% of the requested amount. Many producers do not have the amount required for co-financing and would favour the amendment of the law on access to funds for the purpose of granting a 50% advance and, as a result of doing business, be able to repay the advance.

Much of the respondents currently receiving subsidies from APIA. Asked whether current agricultural activity could be profitable if there were no subsidies, only a quarter of farmers claimed to be able to cope with market competition, with the rest responding negatively to this question. Moreover, more than half of the farmers consider the state support to be insufficient.

Many farmers consider bank lending to be the last solution for the development of the business they manage. Changing banking legislation would be of great help by granting loans with preferential interest to associate or non-affiliated agricultural producers.

Also, the fragmentation of holdings involves small and inconsistent production, which is reflected in the lack of contracts for the capitalization of production in supermarkets or hypermarkets, most of them spreading products on the local market or directly at the farm gate. Small and inconsistent production is reflected directly in the revenues from their sale, which are at a level of less than 4,500 lei per year, for half of the farmers.

Another problem is the small number of collection and storage facilities for small and medium-sized producers, to which only large producers have access, due to the high costs of conservation and storage.

Last but not least, poor information is a major impediment, 25-27% of the interviewed producers claiming that sites are extremely complicated and often give unclear information. The large number of requested documents and their complexity are the first amongst the obstacles to accessing funds.

CONCLUSION

The change of the political regime in Romania has led to extensive changes in the economic life. In agriculture, the "Land Fund Act" was adopted in 1990, whereby the lands nationalized by former Communists were returned. Thus, under 1% of the holding remained in the state's ownership, the rest being divided among farmers. Territorial fragmentation makes holdings averaging between 1,73 ha and 2,29 ha, which do not respect the selection criteria for accessing European funds for many farmers.

High unemployment and low wage levels determine many Romanians to leave their country looking for better paid jobs. This has led to the aging of the active population up to 65 years. The age of those working in the agricultural sphere determines two new issues: according to the guidelines for accessing funds, projects of those people who have not turned 40 until the date of filing the application obtain a higher score while, those who lived in the communist period do not want any debt or association / cooperation because they consider this a return in time.

In conclusion, changing people's mentality over the terms of cooperation and association, improving legislation to create conditions for young people, reducing unemployment and emigrating, are only a few aspects that would significantly improve agricultural activity. The new, innovative ideas acquired during the study years will be put into practice in Romania in the agricultural field, and young people will not have to leave or reconsider their career options. The remaining of young graduates with higher education in the country will be a prerequisite for increasing the absorption of European funds, because they fulfil all criteria in the age and studies guides.

The best solutions for improving the situation in Romanian agriculture come from those directly interested. Thus, over half of small and medium-sized producers admit that accessing structural funds is a way to improve their business. A percentage of over 50% is held by those who support the amendment of the legislation regarding both the ways of accessing and simplifying the applicants' guides, as well as by diminishing or even exempting the taxes and fees for small and medium-sized farmers who are in the first years of their activity or who are in a form of association. Farmers also advocate improving legislation on the marketing of Romanian products and eliminating unfair competition. Many appreciate state support as insignificant and want subsidies to support their work. Removing intermediaries from the production-collection-warehousing-processing-valorisation process by making as many operations as possible by the producers is another solution that could be viable under the conditions of changing the legislation so as to allow easier access to bank credits.

In conclusion, changing people's mentality over the terms of cooperation and association, improving legislation to create conditions for young people, reducing unemployment and emigrating, would be only a few aspects that would significantly improve agricultural activity. The new, innovative ideas acquired during the study years would be put into practice in Romania in the agricultural field, and young people would not have to leave or reorient to other fields. Remaining young graduates of higher education in the country would be a prerequisite for increasing the absorption of European funds, thus fulfilling the age criteria and studies required in the guidelines for accessing funds.

REFERENCES

Albu, L., Caraiani, P., & Iordan, M. (2012). *Perspectivele pieţei muncii din România în contextul Strategiei Europa 2020*. Report within the project Îmbunătăţirea capacităţii instituţionale, de evaluare şi formulare de politici macroeconomice în domeniul convergenţei economice cu Uniunea Europeană a Comisiei Naţionale de Prognoză, cod SMIS 27153. Retrieved from www.cnp.ro/inovatie/docs/seminar-studii-25-06-2012/Rezumat%20studiu%20Piata %20muncii.pdf

Marin, A. (2014). *Education and specialized training - Ways to increase performance in agriculture*. Paper presented at the 5th edition of the international symposium Agrarian Economy and Rural Development – Realities and Perspectives for Romania, Bucharest, Romania. Retrieved from http://mpra.ub.uni-muenchen.de/61725/

Marin, A. (2016). *Short chain - A solution for current capitalization of vegetable production*. Paper presented at the 7th edition of the international symposium Agrarian Economy and Rural Development – Realities and Perspectives for Romania, Bucharest, Romania.

Marin, A. (2017a). *Analysis of the association forms of small and medium producers in the fruit and vegetables sector*. Bucharest: ASE Publishing House.

Marin, A. (2017b). *Considerations regarding the association/cooperation of small and medium fruit producers*. Paper presented at the 8th edition of the international symposium Agrarian Economy and Rural Development – Realities and Perspectives for Romania, Bucharest, Romania.

Marin, A., Chetroiu, R., & Iurchevici, L. (2016). *Study regarding the situation of the romanian vegetables sector and the analysis of the branch of vegetables production valuation*. Bucharest: Romanian Academy Publishing House.

Marin, A., Chetroiu, R., & Iurchevici, L. (2017). *Researches on the growth of the production of fruit*. Bucharest: Romanian Academy Publishing House.

Marin, A., Chetroiu, R., Iurchevici, L., Ion, R. A., Dobre, I., Bran, M., & Costaiche, M. (2016). *Study on the situation of vegetables sector national and community*. Bucharest: ASE Publishing House.

Marin, A., Chetroiu, R., Iurchevici, L., & Tudor, V. (2017). *Study on the situation of the fruit sector at national and community level*. Bucharest: ASE Publishing House.

Marin, A., Ion, R. A., Chetroiu, R., & Iurchevici, L. (2017). *Designing and experimenting with the development models of short chains to capitalize on the production of vegetables*. Bucharest: ASE Publishing House.

Marin, A., Ion, R. A., Dobre, I., Chetroiu, R., Iurchevici, L., Pop, R. E., & Costaiche, M. (2016). *Design and experimentation of the development models of the short cutting values of the vegetables production*. Bucharest: ASE Publishing House.

Ministry of Agriculture and Rural Development. (n.d.). *The National Rural Development Program (NRDP)*. Retrieved from www.pndr.ro

National Institute of Statistics (NIS). (n.d.). *Data sets*. Retrieved from www.insse.ro

Palicica, M., Gavrilă, C., & Ion, L. (2007). *Pedagogy for agricultural education*. Timisoara, Romania: Mirton Publishing House.

KEY TERMS AND DEFINITIONS

Agency for Payments and Intervention in Agriculture (APIA): An institution subordinated to the Ministry of Agriculture and Rural Development, comprising in a main centre, 42 district centres and 266 local centres. APIA runs European funds to implement support measures financed from the European Agricultural Guarantee Fund (EAGF).

Common Agricultural Policy (CAP): A set of rules and mechanisms governing agricultural production chains-collection-storage-processing to exploit fresh and/or processed agricultural products. These rules have been adopted by all members of the Union and are applicable in all its member states.

European Agricultural Guarantee Fund (EAGF): A financing instrument created by the European Union to support agricultural markets.

European Fund for Agriculture and Rural Development (EAFRD): A financing instrument created by the European Union to support EU Member States in implementing the Common Agricultural Policy. The EAFRD is a financing opportunity for the Romanian rural area and is based on the principle of co-financing private investment projects.

Ministry of Agriculture and Rural Development (MADR): Specialized body of the central public administration, with legal personality, subordinated to the government.

National Institute of Statistics (INS): A specialized body of central public administration, with legal personality, subordinated to the Government and funded by the state budget.

National Rural Development Program (NRDP): A program of measures developed by the Ministry of Agriculture and Rural Development together with the European Community, meant to help the development of agriculture in Romania. NRDP is a program that grants non-reimbursable funds from the European Union and the Government of Romania for the economic and social development of the Romanian rural area. These measures aim at developing and improving agriculture and, above all, switching from a fragile, unprofitable agriculture to a modern agriculture. Investors have new financial opportunities for investment projects in the development of agriculture and rural areas. On each axis are designed, according to the objectives pursued, measures and sub-measures whereby sums of money can be allocated if several criteria are met cumulatively.

Rural Area (RA): A continental or coastal area with small towns and villages where the proportion of the population living in rural settlements is over 15%. Most of the area's land is used for agricultural purposes (agricultural crops, forestry, fishing), local economic and cultural activities (crafts, services). The area can also include leisure facilities and natural reservations.

Section 2

Institutional Structures

Chapter 9
Prerequisites for Relaunching Economic Growth in Romanian Agriculture by Promoting Associativity

Gabriel Popescu
Bucharest University of Economic Studies, Romania

ABSTRACT

The level of socio-economic development, the types of relations, and the existing legislation at some point in society represent basic elements in the operation of public-private partnership. In the post-revolution period, a series of difficulties and discontinuities in agricultural policies are manifested in the Romanian society, even after 27 years. It has affected the configuration and development of associative structures, including public-private partnerships from rural areas. Currently, the evolution of the representative associative structures in public-private partnership and the diversification of operation record more than the progressive process, a correlation with the local development plans. Also, concerns on behalf of the central public administration are recorded regarding the update of the legislative and institutional set-up of associative type, because it is one of the main tools for operating the common agricultural policy in accordance with the rules of the European Union.

INTRODUCTION

The realities of Romanian agriculture - even after 27 years after the removal of the communist regime - represented by high costs, lack of technology, the continued fragmentation of agricultural land, the three million of subsistence and semi-subsistence agricultural households, to which on added to aging and lack the experience of many new owners - some of them transferred from urban areas - are the main causes that affect the efficient functioning of the whole sector, production entities and services established in all these years.

DOI: 10.4018/978-1-5225-5739-5.ch009

Copyright © 2019, IGI Global. Copying or distributing in print or electronic forms without written permission of IGI Global is prohibited.

The main target of this analysis is the processes and phenomena of agriculture association and co-operation with extensions for the whole Romanian rural area.

In Romania, after 1989, the co-operative phenomenon, as well as the main public actions undertaken by the post-1989 governments, have been extensively analyzed in the book "Agricultural Co-Operation, from Historical Argument to the Transfer of Knowledge" (Popescu, 2014). The author of this book critically identifies the various processes and phenomena that have affected the functionality of agricultural co-operation and association, out of which: the absence of an integrated vision of the role, complexity and functionality of institutional structures in general; actions initiated by public decisions, which were sequential in character and strictly aimed at solving current needs; lack of experience of public decision-makers in the field of institutional building; the absence of a priorities agenda at the level of public decision-makers - on the one hand, at central and local level and, on the other on medium and long term; the poverty of the population and its main orientation towards immediate activities.

The evaluation of the Romanian rural state, namely the reconfiguration of the associative and cooperative structures specific to the new socio-economic system established after 1989, took into account the institutional instruments developed to support the public-private partnership through the two National Rural Development Plans from 2007-2013 and 2014- 2020, as well as the Local Development Plans initiated after Romania's accession to the European Union.

The re-launching of economic growth in Romanian agriculture through the promotion of associative structures is strongly pointed by the difficulties faced by the Romanian Agricultural Policy after 1990, out of which we mention: the less developed co-operative culture of the Romanian peasants; excessive fragmentation of agricultural properties; the attitude of rejecting the cooperative system and any form of partnership by the whole society, especially by the rural population; the low level of information, of confidence and the weak training of farmers; fears of farmers' participation within markets, etc. In order to highlight the role of the associative structures, including the public-private partnerships, this issue emphasis highlights, on the one hand, the institutional instruments for supporting the public-private partnership in Romania and, on the other hand, the drivers of economic recovery in the Romanian agriculture parallel with the special importance on the necessity of their correlation with the entire domestic and European legislative and institutional framework.

REDEFINING THE COMMON AGRICULTURAL POLICY BY TAKING INTO ACCOUNT THE COOPERATIVE SYSTEM

The Importance of the Cooperative Sector

After more than 150 years of the formation of the cooperative system, about one billion people are members of these entities. "The World Co-operative Monitor 2016 Database: Exploring The Co-Operative Economy – Report 2016" and of the European Research Institute on Cooperative and Social Enterprises - EURICSE - which has a database within 2,370 cooperatives and mutual organizations, on can found this structures in 63 countries, with a turnover of over 100 million US dollars (The World Co-operative Monitor 2016 Database, 2017).

According to the already mentioned study on large geographical areas, the situation is the following: Europe has 68.5% of total number of cooperatives and mutual organizations and 61.6% of cooperatives and mutual organizations with a turnover of over 100 million US dollars; in North and South America there are 21.1% of total cooperatives and 28.6% of cooperatives with a turnover of over 100 million US dollars; Asia - Pacific region has 9.9% of total cooperatives and 9.4% of units with a turnover of over 100 million US dollars; in Africa there are 0.5% of the total number of cooperative and 0.4% of cooperative with a turnover of more than 100 million US dollars. The same source points out that the first 300 cooperative and mutual organizations together have a turnover of 2,5 billion US dollars (Table 1).

On activity sectors, the mentioned document presents the following structure: 39% of total cooperatist units are activating in insurance; 32% in the field of agriculture and food industries; 19% in wholesale and retail business; 6% in banks and banking; 2% in industry and utilities; 1% in health and social care; 1% in other areas (Table 2).

On is noticed that Europe has several leading positions in terms of numbers and economic power of its cooperative system (Table 1).

Table 1. Number of cooperatives and mutual organizations from large geographical areas of the basis of The International Cooperative Alliance Report, 2016

	Cooperatives and Mutual Organizations		**Entity With a Turnover of More Than 100 Million US Dollars**	
	Number	**%**	**Number**	**%**
North and South America	501	21.1	40	28.6
Europa	1,623	68.5	875	61.6
Africa	12	0.5	5	0.4
Asia-Pacific	234	9.9	133	9.4
Total	**2,370**	**100**	**1,420**	**100**

Source: The World Co-operative Monitor 2016 Database, 2016

Table 2. Capital structure and net income of cooperatives and mutual organizations of International Cooperative Alliance. Percentages

	Total Liabilities	**Equity**	**Net Income**
Capital Structure by Activity Sectors, %			
Agriculture and food industries	63.1	33.5	3.4
Banks and banking	92.2	7.1	0.7
Insurance	80.0	18.3	1.7
Wholesale and retail business	65.0	31.9	3.2
Other sectors	67.0	30.1	1.0
Total	**72.2**	**25.3**	**2.5**
Capital Structure by Regions, %			
America	72.8	24.4	2.8
Asia and Pacific	67.9	29.8	2.3
Europe	72.9	25.0	2.1
Total	**72.2**	**25.3**	**2.5**

Source: The World Co-operative Monitor 2016 Database, 2017

In this context, on can mention several actions initiated by the European Commission in view to redefine the Common Agricultural Policy. Also, on can add that a development in world cooperative system was strongly influenced by the International Labor Office activity, which, in 2002, adopted the Recommendation Concerning the Promotion of Cooperatives.

EU ACTIONS AIMING TO REDEFINE THE COMMON AGRICULTURAL POLICY

The preoccupations of the European Commission to revive and strengthen the cooperative system in the economy, including the rural one, were characterized by continuity. In 2003, in Cork, Ireland, the process of reforming and redefining the CAP in the context of the next New Member States accession was debated and on was considered that "... the potential of cooperatives has not been fully exploited and the image of these entities should be improved both at national and European level ... ". Thus, the European Commission (EC) adopted, in 2003, the Regulation on the Statute of the Cooperative Society in Europe (SCE) - Regulation (EC) no. Council Regulation (EC) No 1435/2003 of 22 July 2003 on the Statute for a European Cooperative Society - the only legal form of the social economy valid at the EU level. The mentioned document adopted in Cork strengthens the role of cooperative societies, because them are structurally related to the territory within they function. Because of this particularity, these entities are also a key factor for accelerating local development, creating true social, economic and territorial cohesion. Also, by the co-operative process of continuous training regarding social responsibility and entrepreneurship, the system can become a true "vehicle" for the implementation of numerous Community objectives in the field of employment policy, social inclusion, regional and rural development, agriculture, development of local / regional markets, etc.

The public consultation carried out by the EC at the level of the European co-operative enterprises emphasis that ... *despite the extensive reforms initiated, the instrument that they represent (cooperatives) is not fully exploited* ...; Also, on this occasion was noticed that the dimensions of the sector are directly correlated with the general level of society development, with the level of social cohesion (income homogeneity), the state of the business environment, but also with the state of the general democracy. On can mention that in the EU, in 2012, were 160,000 co-operatives, which provided jobs for 5.4 million people. At the same time, they were an important pillar of the economy and a key driver of social innovation, especially in rural areas and urban agglomerations.

EXISTING RELATIONS IN ROMANIAN SOCIETY: A REAL PREMISE FOR EMERGING PUBLIC-PRIVATE PARTNERSHIP

In society, in general, the objective of association is not only to establish: (1) a better correlation between the actors with an economic role in the territory, including the rural areas and in local/ regional markets, but also; (2) to develop and strengthen a system of public-private partnerships, by: providing alternative sources of funding; providing to economic actors, individual and institutional, a wider range of options; providing data, information and knowledge of interest to both partners.

In this complex context, central government institutions - in Romania the Ministry of Agriculture and Rural Development (MARD) - must continuously monitor developments and identify actions required for the development and strengthening of the associative framework, in view to implementation of the

Figure 1. Existent relations in Romanian society: Premise for ensuing the public-private partnership

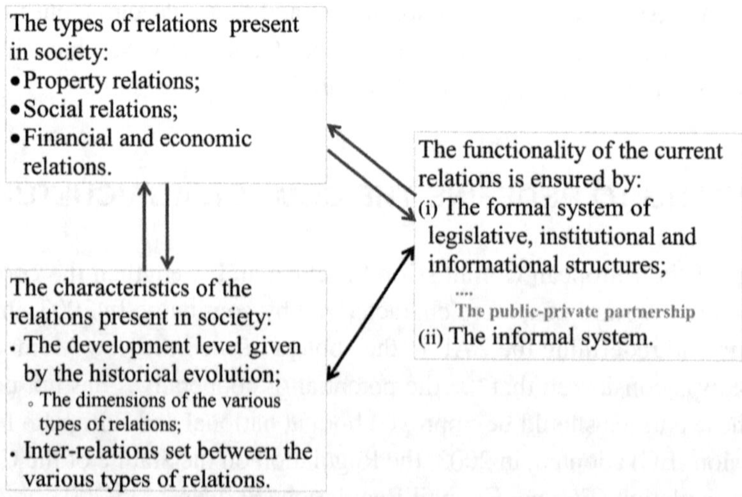

National Rural Development Program (NRDP) 2014-2020; these entities will act as "engines" of general progress. The materialization of these issues is reflected in the Regulation (EC) no. 1303/2013 on the implementation *of Community-Led Local Development* (CLLD), which is supported by the European Regional Development Fund - ERDF (by the Regional Operational Program) and the European Social Fund -ESF (by the Operational Program for Human Capital).

Between 2015-2017, in Romania, on were several steps in view to implement the procedural mechanism for CLLD, starting with bureaucratic framework (criteria and methodologies for assessing and selecting LSD) for Local Development Strategies (LSDs), to the establishment of the managing authorities, to stipulate the conditions that potential beneficiaries (Local Action Groups – LAGs) must meet until to launch projects. Those steps involved: setting up the Joint Steering Committee (JSC) - as decision forum for funding approval of the LDS.

Up to now, as a result of the first two calls for LDS projects, there have been created rural premises for the emergence of another 63 associative structures (were selected 23 preparatory support projects - 36.5% of total, 37 projects for less developed regions - 58.7% - and 3 projects related to the Bucharest-Ilfov region - 4.8%).

OVERVIEW OF THE PUBLIC-PRIVATE PARTNERSHIP

In Romania, developments of the *associative phenomenon post-1990* have recorded several stages. The first measures planned primarily the *rapid implementation of the Package of Principles for the Association and Cooperatives, adapting to the new realities for strengthen the local markets.* Next developments have been relatively slow phenomenon explained by: mentalities of people before to 1990; some inconsistencies of the legislative framework, even constructions assumed on how to build and implement The Land Law - and why not - insufficient general knowledge about new relations from the economy, including agriculture; lack of investment capital and one for current activities.

Difficulties and Errors of Romanian Agricultural Policy in the Configuration Public-Private Partnership

One of the *main structural problems* of Romanian agriculture was and still is the *excessive fragmentation of agricultural properties*. On came as a result of legislation (The Land Law no. 18/1991, The Law No.1 / 2000 regarding the Restoration of Property Rights over Agricultural and Forests Land demanded by the Law no. 18/1991), as at that time the main aim was to ensure "*acts of righteousness / social justice*" regarding the rights of former land owners by restoring land that have been belong to them before the command economy system.

In this context, the land restitution process generated some negative phenomena, such as: land fragmentation, demolition of the technical-material basis of agriculture and employment in agricultural activities of aging people without advanced knowledge and information. Following attempts of policy-makers to "correct" some of the effects of The Land Law no.18 / 1991 were found in *Law No.1 / 2000, which try to increase the size of the restituted land - from 10 to 50 ha.* Also, public measures regarding the financial incentives and support of landowners' associations were very much delayed – first of them appear only after ten years (in 2000) as a result of EU pre-accession funds and, later, by measures of CAP implementation.

Also, the historical events during the period 1945-1990 left a strong mistrust faced to the state authorities, of their decisions, a phenomenon which also explains the attitude of the whole society - especially of the rural inhabitants - after 1990 regarding the rejection of any forms of association, or partnership established in view to solve various issues - from the need for local land exchanges between landlords and until to the acquisition of technical knowledge and of agricultural management, participation of agricultural producers in local markets with homogeneous and representative goods (qualitative and quantitative). By Law no. 36/1991 on Agricultural Companies and Other Forms of Association in Agriculture (in force from 06.05.1991) was defined *simple forms of association and associations based on the agreement among two or more families, with the purpose for agricultural land use, livestock, supply, storage, conditioning, processing and sale of goods, provision of services and other activities.* Also, the Law mentioned specified modalities of association ... *through verbal or written agreement, without further legal formalities, those concerned establishing their own business object and conditions they intend to operate ....* However, the association was not welcomed by farmers because at that time (in 1991) the people were far too involved (especially emotionally) in the processes of establishing and rebuilding their land property rights.

Another effect of the laws of constitution and reconstitution of the land ownership rights was the establishment of many farms for subsistence and semi-subsistence - which still cause problems of a structural nature. In Romania, defining the role, size and importance of the subsistence and semi-subsistence farms has some delays also. Only in 2006 was defined the "agricultural farm" (it was a foreign influence) because it was asked by the process of establishing the eligibility criteria for the EU granting of payments per unit area (Emergency Governmental Ordinance no.125 for Approval of Direct Payment Schemes and Complementary National Direct Payments within Agriculture, started with 2007). After nearly 10 years, in 2015, the definition of agricultural farms has been upgraded due to the need for more details regarding the EU financial allocations. However, the imprecision in defining the agricultural households represented by the subsistence and semi-subsistence ones - which holds the largest share in the Romanian agriculture - is still lacking by its minimal attributes (which are also found in the EU Member States) related to: (1) the degree of market participation of agricultural farms and; (2) the surface of the land owned. Together, these issues have had their own influence in setting up the associative structures set-up by the public-private partnership formula. On can mention that regarding the economic size of the agricultural households on were adopted the *standard production coefficients* and *the limits* by which the size of farm influences their access to the European funds through the NRDP 2014-2020.

Lack of mechanization and of capacity for modernization of agricultural work of small farms that have added of the high inputs prices (fertilizer, seed and chemicals, etc.), but also the reservations of agricultural producers to enter on markets, *generally determined the individual producer to make products strictly for its and his family consumption and the sporadic access of local markets.* All these phenomena and processes that explain several delays in EU funds absorbing for the modernization of agricultural production activities and the Romanian rural development.

Beginning of Associative Structures Building in Accordance With European Legislation Through the National Rural Development Program

Historical facts have shown that, in general, association is search for and accepted only in the extreme economic situations and not in the middle conditions - as are in our present economy. For example, association was acceptable: (1) either when people poverty has reached serious levels or; (2) when the market has provided favorable conditions for achieving some attractive earnings. To stimulate the intensification of associative process, within the National Rural Development Program (NRDP) 2014-2020 *have been included five measures - supported by the European Fund for Agriculture and Rural Development (EFARD)*, which are covering:

1. "Actions for the Knowledge Transfer and Information Actions - M01"to ensure the provision of data, information and knowledge regarding the necessity of association and cooperation of actors from the rural area. Among the concrete expected effects of this measure on can mention: raising awareness and providing the knowledge and skills about the modern and innovative methods of processing and marketing of agricultural products in the context of the emergence and development of short supply chains; providing knowledge on environmental practices that sustain biodiversity, soil and water; deepen knowledge of agricultural practices that contributes to a better adapting of the activities with impacts of climate change in risk areas; providing knowledge on reducing greenhouse gas emissions (GHG).
2. *The establishment of groups and producer organizations in agriculture and forestry – M 09.* This measure was dedicated to association and cooperative processes. On should be noted that in the previous NRDP 2007-2013 was a similar measure - (M 142) *Support for Producer Groups.* "The Report from November 2013 on the Results Achieved within the M 142 "shows that set out 149 groups and producer organizations recognized and were selected for funding 72 entities (48.3% of all actors) representing Producer Groups (PG) for cereals and oilseed; PG for milk and dairy products; PG for pork; poultry and eggs; PG for lamb and goat meat; PG for honey and honey products, etc. The Report from November 2017 mentions some of the main reasons which are behind the relatively small number of recognized Producer Groups and organizations that have met the legal requirements to receive European funds, such as: low interest of agricultural producers in follow these entities; the poor awareness of producers; their financial problems; lack of experience of potential members for designing the development projects; the instability of the legislation; some legislative provisions emphasizing the contribution in kind - land and /or adult animals - of

the future members and this issue reminding them the period of forced co-operativisation (during 1949 to 1962), but also it induced to the agricultural producers the anxiety do not lose their land acquired by The Land Law no.18/1991, etc.

3. In NRDP 2014-2020 the main objective of the Producer Groups and of the producer organizations from agriculture and forestry is focused on commercialization of members' products, relieving thus the agricultural producer to sell their own production; on this way it is estimated that on will directly contribute to both a better integration of agro-food producers in the local and / or regional markets and to the development of short supply chains;

Measure Cooperation – M 16– provides the financial support for setting-up and functioning of Operational Groups (OG) which will be established by rural actors (farmers, researchers, advisors, businesses and NGOs) in order to jointly develop a project of development-innovation, to address specific problems and valorize some of the opportunities from the food industry and forestry. By initiating this measure, on is desirable to link the Romanian rural area to the European Innovation Partnership on Productivity and Sustainability of Agriculture (PEI-AGRI), which is a new approach which promot innovation in European agriculture of the EU Member States. On should be pointed out that innovation will look up to create links between science and practice, in particular through the development of projects with interconnected activities. Thus, the new products, represented by the practices and processes resulting from such actions can become the main engines for diversifying agricultural activities as well as for improving the competitiveness of the rural economy as a whole;

4. *Support for local development by the LEADER Programme - Community Led Local Development (CLLD)* in conformity with Art. 35 of EU Regulation no. 1303/2013 – Measure M19. The CLLD approach allow to local partners: (a) to draw up their own rural development strategy by analyzing the needs and priorities of each territory and; (b) to facilitate the implementation of projects with an innovative, multi-sectorial and transversal approaches in line with the Europe 2020 Strategy. By NRDP 2014-2020, Romania intends to cover the entire LEADER eligible territory to realize balanced regional development, in view to ensure: the "critical mass" needed: (a) for a correct and efficient implementation of the strategies proposed; (b) cost-effective operation of the communities involved.
5. When assessing investment measures for agricultural production are to be considered (additional points are added during projects selection) belonging of the future beneficiaries to an associative structure and / or cooperative.
   a. It should be noted that when the actions of communication and information about the NRDP 2014-2020 were realized, many rural residents were surprised by the significant Romanian financial allocations for semi-subsistence farms and their orientation, predominantly, to various associative and cooperatives in order to access European funds. For example, the situation about submission and contracting of projects from European funds from by the NRDP 2014 - 2020, on 14th September 2017, in the case of five sub –Measures (sM) - sM 9.1 "Setting Up Producer Groups", sM 16.4 "Support for Horizontal Cooperation and Vertically between Actors in the Supply Chain ", sM 19.1" Support preparation for the development of local development strategies ", sM 19.2" Support for the implementation of actions under the local development strategy ", sM 19.4" Support for running costs and animation "- is recorded the

following situation (Table 3):Projects submitted for above sub-Measures of NRDP 2014-2020 (including transition):

i. **Contracted Projects in Progress:** 414 (2.8% from total) - amounting 75,493,192 euro (4.0% from total);

ii. **Finalized Projects:** 143 (22.7%) - amounting 1,961,505 euro (2.1%);

iii. **Canceled Projects:** 9 (18.4% from total), amounting 122,482 euro (0.7%);

b. Projects transferred/ relocated through the transition procedure - it means the contracts uncompleted during the period 2007-2013 programming period which are paid from 2014-2020 programming period had the following features:

i. **Contracted Projects Which Are in Progress:** 2,237 (10.0% of the total number of projects from NRDP 2007-2013), amounting to 43,964,856 euros (10.0% of total value of the NRDP 2007-2013);

ii. **Finalized Projects:** 1,322 (9.5%) amounting of 31,587,301 euros (9.5%);

iii. **Canceled Projects:** 105 (3.3% of the total) amounting 640,426 euros (3.2%).

6. Another example is on the three sub-measures dedicated to "LEADER - Local Development Support- Community Led Local Development (CLLD) according to Art. 35 of EC Regulation no. 1303/2013- namely sM 19.1 "Preparatory support for local development strategies," sM 19.2 "Support for the implementation of actions under the local development strategy" and sM 19.4 "Support for running costs and animation" - which opened/continued and there have been the following situations:

a. Projects of sM 19:

i. Sub-measures with a total number of 403 contracted projects in progress, summing up 72,07405 million euros, out of which 40.7% this amount 3.1% of the sum of categories of projects (2,224,725 euros) are represented by the sM 19.1 "Preparatory support for local development strategies," and the remaining 59.3% (96.9% in value - 69,849,325 euros) for the sM 19.4 "Support for running costs and animation"

ii. A total number of projects completed in the number of 143, amounting to 1,961,505 euros - are all from sM 19.1 "Preparatory support for the development of local development strategies";

iii. Total number of projects canceled was 9, them worth 122,482 euros - all of which are from the sM 19.1 "Preparatory support for the development of local development strategies".

b. Projects from sub-measures transferred by transition procedures – it means unfinished projects in the period 2007-2014 which are paid from 2014-2020 funds – all those projects are under sM 19.2 " Support for the implementation of actions within the framework of the local development strategy ":

i. **Number of Contracted Projects, in Progress:** 2,204, summing up 39,899,679 euro;

ii. **Contracted Project in Progress:** 2,237 (10.0% from total), summing up g 43,964,856 euro (10.0%);

iii. **Finalized Projects:** 1,308, summing up 30,519,698 euro;

iv. **Canceled Projects:** 105, summing up 640,427 euro.

Table 3. Status of implementation of the NRDP 2014-2020 in the Year 1, the total and for sM 9.1., sM 16.4a, sM 19.1, sM 19.2 and sM 19.4, in September 2017. Euro

Sub-Measure			**sM 9.1 "Setting up producer groups"**	**sM 16.4a "Support for horizontal cooperation and vertically between actors in the supply chain."**	**Total Measure 19.*)**	**TOTAL NRDP year 1**	**Out of which for sM9.1+ sM16.4a+ M19**	**% of sM9.1 + sM16.4a + M19 from the Total**
Submitted projects		Nr.	10	23	180	42,309	213	0.50
		Val.	3,674,879	2,183,027	2,435,307	6,637,686,041	8,293,213	0.12
Selected projects		Nr.	8	6	175	19.567	189	0.97
		Val.	3,124,065	556,960	2,379,233	2,898,368,680	6,060,257	0.21
Contracted projects	Projects (in progress and finalized)	Nr.	8	3	403	14,803	**414**	2.80
		Value	3,124,064	295,078	72,074,050	1,876,899,225	75,493,192	4.02
	Finalized projects	No	0	0	143	629	**143**	22.73
		value	0	0	1,961,505	93,384,890	1,961,505	2.10
	Canceled projects	No	0	0	9	49	**9**	18.37
		Val.	0	0	122,482	17,007,580	122,482	0.72
Projects transferred by transition procedure**)	Contracted projects (in progress and finalized)	No.	33		2,204	22,654	**2,237**	9.87
		Val.	4,065,177		39,899,679	440,083,932	43,964,857	9.99
	Finalized projects	Nr.	14		1,308	13,940	**1,322**	9.48
		Val.	1,067,603		30,519,698	156,009,543	31,587,301	20.25
	Canceled projects	Nr.	0		105	3,217	**105**	3.26
		Val.	0		640,427	19,890,716	640,427	3.22
Payments			286,541	0	27,460,045	1,339,121,472	27,746,586	2.07
Payments (transition)			1,769,326		28,882,975	543,615,708	30,652,301	5.64

Source: Report: "Status of implementation of the NRDP 2014-2020 on 09.14.2017 "http://www.pndr.ro/situatia-proiectelor-depuse-2014-2020/item/279-situatia-proiectelor-depuse-in-data-de-14-septembrie-2017-pndr-2014-2020.html

*) sM19.1 "Support preparation for the development of local development strategies", sM 19.2 "Support for the implementation of actions under the local development strategy", sM 19.4 "Support for running costs and animation".

**) Contracts non-finalized during 2007-2013 programming period which are paid from funds for the 2014-2020 programming period.

Note: If the sM 1.1 "Support for training and skills acquisition" at the time of the report were 254 projects submitted equivalent with 19,816,177euros, out of which 41 projects were selected (16.1% of the category) amounting to 3,264,296 euros (16.5%), but none of these projects were contracted.

As a conclusion to the above examples mentioned - a significant number of canceled projects and contracts remaining unfinished during the programming period 2007-2013 (which will be financed by the current period), which reflects, first, the absence of knowledge management, absence of resource for co-financing from the beneficiaries and, on the other hand, poor functionality of partnership structures in Romanian rural areas, at more than 25 years after the change of the socio-economic and cultural system.

Table 4. Status of implementation of the NRDP 2014-2020, at 14.09.2017, for sM 19.2 and sM 19.4, in September 2017. Euros

Sub-Measure			sM 19.1 Preparatory Support for Development Local Development Strategies	sM 19.2 Support for the Implementation of Actions Within the Framework of the Local Development Strategy	sM 19.4 Support for Expenditures of Functioning and Animation	Total M 19.
Submitted projects		No.	180	0		180
		Val.	2,435,307	0		2,435,307
Selected projects		No.	175	0		175
		Val.	2,379,233	0		2,379,233
Contracted projects	Contracted projects -in progress and finalized	No.	164	0	239	403
		Val.	2,224,725	0	69,849,325	72,074,050
	Finalized projects	No.	143	0		143
		Val.	1,961,505	0		1,961,505
	Canceled projects	No.	9	0	0	9
		Val.	122,482	0	0	122,482
Projects transferred by transition procedure*)	Contracted projects -in progress and finalized	No.		2,204		2,204
		Val.		39,899,679		39,899,679
	Finalized projects	No.		1,308		1,308
		Val.		30,519,698		30,519,698
	Canceled projects	No.		105		105
		Val.		640,427		640,427
Payments			1,975,015	0	25,485,030	27,460,045
Payments (transition)*)				28,882,975		28,882,975

Source: Report: "Stage of implementation NRDP 2014-2020, 14.09.2017 http://www.pndr.ro/situatia-proiectelor-depuse-2014-2020/item/279-situatia-proiectelor-depuse-in-data-de-14-septembrie-2017-pndr-2014-2020.html.

*) Unfinished projects from the programming period 2007-2013 which are paid from current programming period 2014-2020 funds

INSTITUTIONAL INSTRUMENTS TO SUPPORT PUBLIC: PRIVATE PARTNERSHIP IN ROMANIA

Preparation and then adopting of the Acquis Communautaire to joining with the EU implied:

imposing requirements, technical standards and quality and ecological standards as well as introducing European monitoring mechanisms of market and prices, creating systems of registration and control of products and activities, development of mechanisms of control and access to local, regional, national and international markets, to ensure the protection and health of consumers, etc.

However, the transposition of the Acquis Communautaire to Romanian society was materialized also by *adapting to conditions/national situation* of a significant number of laws, rules, regulations to support various measures of harmonization and set-up compatibility of Romanian space with the European one, *creating new links between the structures of central government and regional authorities, to encourage the emergence of any kind of partnerships, including public-private ones,* as a condition of EU funds absorption - in this case those for materializing the objectives of Common Agricultural Policy.

Associative Forms

In conformity with the legislation in force, in the Romanian agriculture, there are the following forms of association:

- *Agricultural companies and other forms of association in agriculture* (Law no. 36/1991 on Agricultural Companies and Other Forms of Association in Agriculture, amended and supplemented for many times);
- *Associations* established under the Government Ordinance (GO) no. 26/2000 on Associations and Foundations, approved by the Law no. 246/2005, amended and supplemented;
- *Agricultural cooperatives* established under the Law no. 566/2004 on Agricultural Cooperatives, amended and supplemented;
- *Groups and Producer Organizations for the Marketing of Agricultural and Forestry Products* established under the provisions of GO no. 37/2005 on recognition and functioning of groups and producer organizations for the marketing of agricultural and forestry products. Producer groups have as main aims to: ensure production planned and adjusted to market demand; promoting the supply and allocating on the market the products get by their members; optimizing production costs, setting producer prices and promoting the use of cultivation practices, production techniques and waste management practices that do not harm the environment;
- *Agricultural Chambers* were established by the Law no. 283/2010 regarding the Chambers for Agriculture, Food Industry, Fisheries, Forestry and Rural Development. Their role is to represent, protect and support the interests of their members and the business community in relation with public authorities and bodies in the country and abroad. However, these institutional structures have the role to promote basic objectives in rural areas included in the Lisbon Agenda as follows: competitiveness; improve farmland and forestry; the primary role of agro-environment sustainable development of rural areas; increasing the quality of life in rural areas.

All these associative entities from Romanian agriculture have several advantages starting with the economic advantages for their members and the local communities in which they operate until to social, cultural and even political link to facilitate implementation of CAP and NRDP priorities, but also to boost the absorption of European funds.

Even if the legal framework gives the impression of a fertile field for asserting these structured in terms of concrete facts, results still remain at modest levels.

LEADER Program and the Implication of Local Communities

A major step in the association and cooperation Romanian of agriculture was to create the first *Local Action Groups* (LAGs) in order to design and implement, at local level, specific development strategies including institutional building. In this context it should be noted that: in Romania, by NRDP 2007-2013, *the public-private partnership* played a vital role in rural areas, this contributing to development of rural areas by creating specific institutional entities. *The Managing Authority for NRDP* - which has the function of management and administration of European Agricultural Fund for Rural Development (EAFRD) - is MARD by its Directorate General for Rural Development (DGRD), according to Government Decision no. 725/2010 on the Reorganization and Functioning of the Ministry and its Subordinate Structures. Managing Authority for the NRDP is the structure of central government, which authorize diverse types of partnerships that will be set up locally in order to access the measure 19 "Support for local development LEADER" according to the art. 35 of the EU Regulation no. 1303/2013.

Local Action Groups

In Romanian rural areas, Local Action Groups (LAGs) are the main expression of public-private partnerships; they are like a small management authority and consist of representatives of:

- **Public Sector Represented by:**
    - Public administration (local and county - mayors, local councils, county councils, etc.);
    - Public services (social services, health and transport, schools, universities, etc.);
- **Private Sector Represented by the Following Types of Entities:**
    - Commercial Sector (joint stock companies, limited liability companies, etc.);
    - Financial sector (banks, credit institutions, etc.);
    - Agricultural sector (agricultural cooperatives, producer groups, etc.); Organization of entrepreneurs;
    - Private companies for the provision of community services (cultural, radio, TV, non-cultural services, etc.).
- **Civil Society Represented by:**
    - Non-profit organizations, associations, federations (environmental associations, cultural associations, social, religious, chambers of commerce, religious entities, etc.);
    - Individuals, groups of people which are not officially registered.

In LAGs the *partnership in decision-making* is 49% represented by public sector and 51% by private sector. LAGs have as main task the elaboration of an "integrated local rural development strategy" and they are responsible for its implementation. In this respect LAGs have, at local level, a significant role in accessing European funds, as well as, in the choice of projects to be financed under the strategy proposed. Also, LAGs can also select potential cooperation projects with other entities.

In EU Member States, LAGs have different *legal structures for functioning*: non-profit associations; associations / foundations; local or regional authorities; companies; cooperative. The main tasks of LAG:

1. Elaboration of Local Development Strategy;

2. Implementation of their Local Development Strategies which has achieved 6-7 lines of priority funding for the area where it is located (for example, farm development, tourism investment, non-agricultural services, setting up of young farmers, development of associations and cooperatives). Please note that the establishment of these priority funding lines depends on the results of the public consultation realized by the LAG Coordination Committee;
3. Project selection;
4. Taking over by LAGs the main administrative functions, such as: preparation and publication of calls for project selection in accordance with LDS; animating the territory; analyzing, evaluating and selecting projects; monitoring and implementation of the LDS; compliance check of payment claims for projects selected (excepted the LAG where it is the beneficiary); project monitoring arrangements; public procurement procedures, running costs and animation; specific aspects: financial, accounting, legal, human resources, etc.

During the 2007-2013 programming each LAG had established a budget for launching its projects (approx. 2,850,000 euros by each LAG); the total public financial allocation for those structures was approx. 370 million euros. This measure aimed to encourage local communities to be actively involved in solving local problems of general interest. However, LAGs established in the first programming period projects - see Tables 3 and 4 - on the "Status of implementation of the NRDP 2014-2020", on 14.09.2017, but for five sub-measures has not fully succeeded to finalize the projects, which is affected the results of the current period – projects not completed in 2007-2013 are supported by NRDP funds for the period 2014-2020.

The NRDP 2014-2020 took on the positive experiences of LAGs from previous programming period on: (a) increasing the importance of LAGs and in this context them allocated nearly a double budget (around 650 million euro); (b) Differentiated budget allocation per LAG based on: population number and the size of territory covered by each such entity; the quality of the proposed strategy; (c) encouraging the increase of the number of LAGs (in the end, on estimates reaching about 200 entities); in this way the area covered by the activities of LAGs (which will cover about 80% of the eligible territory through the LEADER Program) with positive effects on local development, and; (d) offsetting development disparities at the level of rural communities through the objectives of the projects developed.

In 2015, 163 Local Action Groups were still active from the period programming period with the following features: a surface occupied of more than 140,000 km^2 (an average area of 859.0 km^2 per LAG); them represented about 63% of the LEADER eligible territory; them cover 58% of total eligible LEADER population. In 2016, for the currently programming period, the central authority MARD recognized 239 LAGs, an increase with 68.2% compared with the former period.

In 2017 the first 10 LAGs with a successful implementation of their strategies were: LAG South East Vrancea; LAG Association Vedea – Găvanu - Burdea; LAG High Podu Vaslui Association; Arges County Middle LAG Association; LAG Oltenia Hills; LAG the Carts' Road; LAG Ampoiului Valleys area and Mures; LAG Association Oltenia Hills; LAG Association of Ampoiului Valleys area and Mures; The Local Development Association Tecuci; LAG Association Plaiurile Olteţului; LAG Central Dobrogea.

All of them managed to contract over 27,35 million euros by European funds. The payment to these beneficiaries amounts to over 20 million euros. Also, in 2017 there were three LAGs which have developed cooperation projects. These associations set up at national and international level have started cooperation projects worth (public) of over 110,000 euros per each entity. The objectives of these LAGs concerned, especially promoting the potential of rural tourism and local traditional products. These LAGs are "Almaj Gate" Association (as coordinator) and the LAG Delta Dunarii (the public/project value were: 20,000 Euros each LAG); ENPORT - European Network for Promoting Rural Tourism Outdoor composed of: Association of Local Action Group Porolissum Napoca (as coordinator); Local Action Group Association Cluj; Pays Vichy-Auvergne. In each of these LAGs the public/ project value was 75,150 euro.

National Federation of Romanian LAGs

National Federation of Romanian LAGs (NFLAGs) is the institutional structure of unitary and coherent representation of LAGs interests at national level. So, the goals of NFLAG is the representation of LAGs in Romania in relation with the central government and / or local or any other institution, associations, individuals or legal entities operating in rural areas, to create and / or improve the necessary framework for sustainable development of territories included in the LAGs local development strategies. NFLAG is a private legal person with non-economic purpose, independent, apolitical, private which was organized according to Government Ordinance no. 26/2000 on associations and foundations and approved by Law no. 246/2005. Founding members of NFLAG are 19 LAGs, which, in 2015, were top entities in implementation of the project strategies.

NFLAG objectives pursue its declared goals such as:

- Promoting the rights and interests of its members in all of activities;
- Participation in the development and/or modification of normative acts issued by local and/or central authorities, which may have impact on rural development;
- Participation in the development, updating and implementation of local public policies, national and European ones concerning rural development;
- Support for increasing the absorption and efficient use of European funds;
- Raising the professional qualification of the staff working in LAGs;
- Improving the professional skills of potential applicants to EU-funds projects for LAGs.

The principles of operation and organization of NFLAG are: volunteering (people elected in the leadership of NFLAG are not paid for their performance); territorial representativeness in decision-making bodies of the organization; reasonable operating costs that are covered by membership fees and / or other sources, according to the statute of organization; ensuring cooperation between partners; transparency of decisions taken.

Among the main activities of Romania NFLAG were launching of several debates, such as:

- Complementarity of LEADER projects with Human Capital Operational Program;
- Promotion of quality schemes for agro-food products;
- Arrangements for access to environmental schemes;
- Compliance of payment applications for projects funded by LAGs;
- Association and cooperation, incentives for a better use of food.

NFLAG Romania also organized, at the request of its members, the training programs for occupations less common for the rural areas, but for which increasing in the demand on the local labor market. Among the occupations for which NFLAG had organized professional training programs are: risk manager, code of The Occupation Code of Romania (OCR) 241 240; credit analyst, OCR code 241 241; intermediary in commercial and financial activity (broker), OCR code 331 103; entrepreneurial skills, OCR code 241 205.

This professional training activities organized by NFLAG were made by Sectorial Operational Program of Human Resources Development (SOP HRD) /135/ 5.2/S/ ID 135 486 "Microcredit- fundamental component of entrepreneurship in the rural areas". The programs mentioned have been developed by NFLAG were completed by delivering certificates which recognized skilled achieved by the National Qualifications Authority - NQA. So far, NFLAG Romania organized for its members, 12 training sessions, with an attendance by 180 people.

VECTORS FOR REVIVAL OF ECONOMIC GROWTH IN ROMANIAN AGRICULTURE BY ENHANCING THE ROLE OF ASSOCIATIONS

Correlation of NRDP Measures With Existing Partnership Structures and With Provisions of Local Development Strategies

The correlation of the measures requirements in the two NRDPs with the priorities of the Local Development Plans through associative structures existing in the territory respects a series of common principles, such as:

- **The Principle of Decentralization of the Decision-Making Process:** Through the transfer of information at national, regional and local level;
- **The Partnership Principle:** Which consists of building partnerships with all the local actors involved;
- **Subsidiarity Principle:** Aims to establish the most appropriate level of intervention in the areas of shared competences amongst the Community, national and regional levels, respectively the local one so that the action taken to be necessary for the reference target group;
- **Planning Principle:** To achieve the objectives set-up;
- **The Principle of Co-Financing of the Projects Initiated:** Consists in the obligation to set-up the financial contribution for each actor involved in the Local Development Plans.

In Romania, on estimates to achieve significant and realistic results concerning the partnership entities, such as:

- A more rigorous use of European funds available;
- Valorizing the existing local resources - natural, human, material, financial, qualifications/skills, information;
- Creating preconditions for achieving integrated approaches at the level of micro-areas and according to their specific needs;
- Reducing imbalances between different communities;

- Ensuring premises for achieving economic growth at the micro-zones;
- Suitability of functionality of the institutional framework - represented by public-private partnerships established to design strategies, accessing to the European funds and, also, for implementation of projects by common interest - to the specific micro-areas included in LDS;
- Adapt and integrate certain provisions of the NRDP for various levels of decision making;
- Stimulate domestic and international cooperation at local level through existing partnerships for sustainable economic and social development.

Accelerating Suitability to Romanian Realities of Legislation Packages on Association and Cooperation in Agriculture

In terms of accelerating the implementation of legislative packages on association and cooperation in agriculture is estimated that is necessary to carry out actions based on proposals coming "bottom-up" regarding:

1. A unitary record of these entities in order to urgent involve them, in specific actions, at local level that are difficult to realize at central level - such as, for example, accelerating the knowledge the state of the land assets from agriculture;
2. Restructuring the size of agricultural holdings;
3. Increasing the role of associations and cooperation in local development;
4. Define further priorities designed to support:
   a. Development of associative structures configured in public - private partnership formula;
   b. Development of specific retail financial services;
   c. Establish of an expert group in view to develop institutional and regulatory structures systems for financing association and cooperation in agriculture and non-agricultural rural activities.

Table 5. Example of correlation of the NRDP 2007 -2013 measures with the Local Development Plans through existing partnership structures

Economic Agents	NRDP Measures			
	Measure 111 "Professional training, information and knowledge diffusion"	**Measure 112 "Installing of young farmers"**	**Measure 121 "Modernization of agricultural holdings"**	**Measure 141 "Support for semi-subsistence farms"**
Agricultural cooperative societies		✓	✓	✓
Cooperatives for Capitalizing of Agricultural Production		✓		
Agricultural cooperatives		✓	✓	
Producer Groups		✓	✓	

For helping the Romanian farmers to get European funds maybe it will be useful to set-up a fund for co-financing the NRDP projects. Such a fund could develop several types of guaranties for access to credits of potential beneficiaries for the European funds, because this is still a relatively complicated process, given the standards of collaterals put by banks.

The confirmation this fact mentioned above is found in the total number of decommitment of contracts from NRDP 2014-2020 (see the Report on September 14, 2017), which reached 36,898,296 euros, out of which 54.0% (19,890,296 euros) were represented by the expiated projects contracted during the 2007-2013 programming period and them will to be paid from the next funds (2014-2020 programming period). The resolutions established and related to a number of 3,266 projects, out of which 3,217 projects, mostly 98,5% of the projects were ones a subject to the "transition procedure", i.e. paid from the funds for the 2014-2020 programming period.

The five sub-measures on local partnership projects (M 9.1 "Establishment of Producer Groups", sM 16.4a "Support for Horizontal and Vertical Cooperation between Actors in the Supply Chain", sM 19.1 "Preparatory Support for the Development of Development Strategies Local Government ", sM19.2" Support for the Implementation of Actions Under the Local Development Strategy ", M 19.4." Support for Operating and Animation Expenses") reached 762,909 euro, out of which 640,427 euro, (84%) referred to project terminations from the previous programming period 2007-2013 (them will be financed by the funds allocated for the current programming period 2014-2020). These resolutions referred to 114 projects, out of which the bulk (105 projects, 92.1% of total) were ones undergoing the transition procedure. This situation is explained by the resistance of banks to co-finance EU projects, because the banks from Romania avoided lending as guarantee the agricultural and forest land, and the most of landowners still have unsolved problems with their ownership, respectively with the release of property documents. Against this background, the decision to accelerate the cadaster work for agricultural land is in line with the agreement between Romania and the European Commission - DG-AGRI - to finish this action until the end of 2019 (this is a priority objective for decision-makers).

The Government's assumption of acceleration the legalization of the public-private partnership is another direction for acceleration the correlation of the legal and institutional framework within the Romanian realities in order to encouragement the association and cooperation in agriculture and in the whole rural space.

Next, it is necessary to step up actions the creation of the real premises for the central public authority to be able to interfere in the establishment of partnerships based on private initiatives in which the state takes its "organizing role in strictly specialized areas", such as: organization of an agricultural stock exchange; assure the enough credit guarantee services needed to co-finance European projects - precisely to increase confidence in applicants; building partnerships with farmers' associative structures for storage facilities within the strategic points in order to abandon the practice of agricultural producers regarding the immediate capitalization of their output, especially of cereals (sodded directly from the plain); the organization of charging services for the transfer of agricultural production to foreign markets, as in other countries; the building of laboratories - with an international accreditation - to establish the quality of agricultural production, because without a such certification, the traders underestimate the quality of the goods - for example, wheat bread is bought at the price of forage wheat, etc.

Innovation: Vector of Revival the Public-Private Partnerships in Rural Areas

Innovation, which is one of the main objectives of the LEADER approach, is also reflected in how the LAGs manage the opportunities and challenges identified at local level, as well as how these entities / partnerships operate to grow the interest of local communities in which they operate to develop innovative public utility projects and / or with economic, social, cultural or local impact. Innovative measures developed by LAGs can target areas such as: energy efficiency and renewable energy; Information and Communication Technologies - ICT; material and immaterial patrimony - including natural heritage of local interest; organizing agri-food markets for local producers; supporting vulnerable groups and disadvantaged communities, etc.

In this context, the innovative measures developed by the LEADER Program by its local partnership structures - existing or to be established - will encourage:

- *Innovative projects developed in accordance with the objectives of Regulation (EC) No. 1305/2013, including the local development objectives of each community, also;*
- *Supporting the preservation of local heritage and traditions that contribute to: (a) enhancing the quality of life in the areas covered by the LEADER Program; (b) stimulating rural tourism activities; (c) the development of short supply chains made up for local products, and; (d) job creation.*

On this way, projects supported by the local partnerships in the framework of the community local development strategies will have a positive impact on the objectives financed by the EAFRD, respectively by the NRDP.

CONCLUSION

1. The important place held by the cooperative system in Europe in supporting the cooperative phenomenon at global level and the process of EU accession of Eastern and Central European States have imposed in 2003 in Cork, Ireland, by reconsidering of cooperatives place and its role in socio -economic development, for which on adopted the Regulation on the *Statute of the European Cooperative Society* in order to encourage the development of the internal market, to accelerate local development and to create of a real social, economic and territorial cohesion.
2. Institutionalization in Romania of public - private partnership in rural areas, is aimed to eradicate poverty, to defeat barriers to market access of agricultural producers, by providing information and knowledge of interest to support economic convergence and increasing resilience of Romanian economy within the European and international level. In this context the implementation of the NRDP 2014-2020 for the LEADER is done by LAGs which implement LDS, approved by the central authority represented by the MARD.
3. The creation, implementation and dynamism of a real and functional associative sector in the economy, including agriculture, as well as the initiation of a public-private partnership depends on the re-launch of the real economic growth in the industrial sector, which is the true deliverer of progress. It should be noted that while Romania recorded the highest economic growth among all

EU MS, on was mainly due, in 2016, to the increase in household final consumption expenditure (them increased by 7.5% compared to 2015), which contributed with 4.5% on overall growth. In addition, gross fixed capital formation declined by 3.3%, in 2016 compared to 2015, with a negative contribution of -0.8% to economic growth in 2016 (according to INS). Considering these figures, it is noted that the current economic situation is still deficient in terms of GDP.

These macroeconomic features add to a number of difficulties facing the Romanian agrarian policy after 1990, such as: excessive fragmentation of agricultural property; the attitude of rejecting the society, but above all the rural inhabitants, of any form of partnership in view to solve some of the severe problems faced to the population - from the necessity of local land exchanges between the owners (so as to favor the acquisition of technical knowledge of organization and management in agriculture, to the participation of farmers in homogeneous and representative commodity markets.

However, the process of developing associative structures and public-private partnerships should consider both the need for compliance with the requirements for an effective functionality and the need to create real premises for building institutional infrastructure, solid and specific environment in which they will exist. Among these requirements strictly mandatory, are the following:

- *Associative ownership, including the cooperative, must be accepted in active partnership rather than as an alternative and / or a substitute relationship to other types of relationship in the economy, such as individual private ownership and public ownership.*
- The principles of establishing and functioning of associative structures will need to be implemented in their entirety and unity - precisely to ensure both the stability of the system and its comparability with existing ones in other countries. On should be noted that although the principles recognized by the international cooperative doctrine are included in Romanian legislation (Law no. 1 of 21 February 2005 on the organization and functioning of co-operation at Art. 7), however these issues have been taken partial in organizational statutes of co-operatives.
- Adhesion of an associative / cooperative group is based on criteria and scientific procedures formulated for the selection and recruitment of potential members. It is appreciated that, in addition to the unit of profile, size and size of the future entities that will be constituted, it will be necessary to consider: (a) the attitude of the future members, an position in which the entrepreneurship, generosity, the desire for knowledge, the innovative aspirations, etc. are the skills needed to meet the specific objectives of future partnership structures which will be established, and; (b) the economic homogeneity of the group represented by the associate members or partners is a condition that guarantees the good functioning of these entities. For example, an association, or a cooperative which is intended to be functional will not be made up of a mixture of occupations (craftsmen / farmers and other occupations) and or legal structures of different dimensions; they will follow the principle: the entities will be only of large or small size actors
- The scope of the associative / partnership structures must be lied within the scope of the specific markets of the various products, financial markets, knowledge transfer.
- The power of associative structures is given by: (a) the intensity of the relationships between the members and the entity to which they belong, and; (b) the relationship between the association and its market partners.
- In fact, the "energy support" of these structures is given by the unity of interests of their members, and not by the "joint" work (as the Communists support).

- The activity of these structures must be primarily covered the local markets, as well as the area of collection, processing, storage, etc. In the case of the agricultural cooperative, its activities must not be linked to the earth, especially when them are addressing to people with small and very small properties, because there is a danger of repeating the mistake of the socialist period.
- Associative and cooperative structures should not be subordinated to the politics issues. Naturally, the associative system has less resistance to political order.

4. Under the 2014-2020 NRDP, measures - supported by the EAFRD - have been launched, targeting: actions for the transfer of knowledge and information actions - M 01; setting up producer groups and organizations in agriculture and forestry - M 09; cooperation - M 16; support for local development LEADER. In addition, the assessment of investment measures for agricultural production should consider (the projects received additional points) the future beneficiaries' membership in an associative and / or in a cooperative structure.

According to the submission and contracting of projects from European funds, through NRDP 2014-2020, on September 14, 2017, in the case of the sub-measures mentioned above, the following situation was found: 414 contracted projects (2.8% of the total), amounting to 75,493,192 euro (4.0%); 143 completed projects (22.7%), amounting to 1,961,505 euro (2.1%); 9 terminated projects (18.4% of the total), amounting to 122,482 euro (0.7%).

It is also noted that there are also projects from unfinished contracts coming from the programming period 2007-2013, which will be paid from the funds related to the current programming period (contracted projects in progress amounting to 2,237 - 10.0% of the total number of projects NRDP 2007/13, amounting to 43,964,856 euros, respectively 10.0%, completed projects in the amount of 1,322 and 9.5% respectively, amounting to 31,587,301 euros, respectively 9.5%, projects terminated in number of 105.33% of the total, amounting to 640,426 euro, respectively 3.2%).

5. The existence of a significant number of finish projects as well as of projects from previous contracts (but to be financed from current funds) reflects, on the one hand, the absence in the case of project beneficiaries of serious management knowledge, insufficient resources to co-finance some possible projects and, on the other hand, poor functionality of partner structures from the Romanian rural area. The institutional instruments which support the association and the public-private partnership in Romania are represented by various forms of association. The creation of the first Local Action Groups (LAGs), setting up to design and implement local specific development strategies and community participation in the LEDEAR Program, was a beginning to develop associative structures like those from other EU Member States.

Empowerment of different types of associative structures established in the rural area for the purpose of accessing the measure on "Support for local development LEADER - Local Development under Community Responsibility, Art. 35 of the EU Regulation no. 1303/2013, which is carried out under the Managing Authority for the NRDP of MADR, is, in fact, an example of a public-private partnership in Romania, but the examples of functionality are still relatively few.

6. Local Action Groups (LAGs) representing the interests of the inhabitants and the rural community, which have as main structure of public-private partnerships (49% private representatively and 51% private one), clearly play the role of small authorities (local) for: managing access to European funds; selecting potential projects for cooperation with other entities; developing and implementation of its LDS, etc.

In 2017, following the evaluations of the Managing Authority for the NRDP from MADR, it was found that ten LAGs worked best (they managed to contract more than 27.35 million euro in non-reimbursable European funds), also, were other three LAGs which developed cooperative projects (110,000 euros per each entity).

7. The institutional structure of unitary and coherent representation of the legitimate interests of LAGs at national level is the NFLAGs. This entity has the merit of organizing, at the request of its members, professional training programs (risk manager, OCR code 241 240, credit analyst, code OCR 241 241, intermediary in financial and commercial activity / broker, code OCR 331 103, entrepreneurial skills, OCR code 241 205), as well as to launch diverse topics of debates on the issues of interest of its members, etc.
8. Increasing the role of associative structures, including public-private partnerships, is likely to generate vectors to re-launch economic growth in Romanian agriculture, among which: a more rigorous correlation of NRDP measures with existing partnership structures, and with the provisions of the Local Development Plans adopted or updated; to accelerate the adjustment to the Romanian realities of the current legislative packages regarding the association and cooperation in agriculture, parallel to their correlation with the entire existing legislative and institutional framework; intensifying the support of innovation in the public-private partnerships set-up in the Romanian rural area.

The essential of the association and development of public-private partnerships in agriculture and in the Romanian rural area is that they permit and support the amplification of positive results and experiences through the diversification of opportunities, contributing to the increase of the general well-being.

REFERENCES

Alliance International of Cooperative Property. (1998). *Statistics and information on European cooperation*. Retrieved from http:/www.ica.coop/europe

Altman, M. (2010). History and Theories of Cooperatives. In H. K. Anheier, S. Toepler, & R. List (Eds.), *International Encyclopedia of Civil Society* (pp. 563–570). New York: Springer Science –Business Media.

Bădescu, I., Buruiană, C., Şişeştean, Gh., & Abraham, D. (Eds.). (2011). *The peasants and the New Europe*. Bucharest: Ed. MicaValahie.

European Commission. (2003). *Report from the Commission to the European Parliament, the Council, the European Economic and Social Committee and the Committee of the Regions on the implementation of Regulation (EC) No 1782/2003, Publication no. 1435/2003 of the Council of 22 July 2003 on the Statute of the European Cooperative Society*. ECS.

Federatia Nationala a Grupurilor de Actiune Locala (FNGAL). (2017). *Module cursuri formare*. Retrieved from http://www.fngal.ro/images/module/CURSURI_FNGAL.pdf

Ministry of Agriculture and Rural Development (MARD). (2015). *National Program for Rural Development 2014 – 2020*. General Direction for Rural Development (MA -NRDP), Program adopted by EC on 26.05.2015

Ministry of Agriculture and Rural Development (MARD). (2017). *Situatia proiectelor depuse in data de 14 septembrie 2017, PNDR 2014-2020*. Retrieved from http://www.pndr.ro/situatia-proiectelor-depuse-2014-2020/item/279-situatia-proiectelor-depuse-in-data-de-14-septembrie-2017-pndr-2014-2020.html on 20 sept.2017

Ministry of Agriculture and Rural Development (MARD). (2017). *The stage of implementation of NRDP 2014-2020 to date*. General direction for rural development (MA-NRDP), Report adopted in 14.09.2017

Petrescu, C. (Ed.). (2013). *Cooperatives from Romania. Actors of socio-economic development*. Bucharest: Polirom Publishing.

Popescu, G. (2013). *Old problems, new relation in agriculture*. Bucharest: Romanian Academy Publishing.

Popescu, G. (2014). *Cooperation in agriculture, from the historical argument to transfer of knowledge*. Bucharest: Romanian Academy Publishing.

Romanian Land Fund Law no 18/1991, published in Official Gazette from 20.02.1991

Romanian Land Fund Law no 18/1991, republished, with subsequent amendments and completions. (n.d.). Retrieved from http://www.dreptonline.ro/legislatie/legea_fondului_funciar.php

Romanian Law no. 1/2000 for restoring ownership of agricultural land and forestry

Romanian Law no 1/2005 on the organization and operation of the cooperation, published in Official Gazette of 03.03.2005.

Romanian Law no. 31/1990 on commercial society, publish in Official Gazette no.1066/17.11.2004

Romanian Law no. 338/2005 for approving OG 37/2005 Report on the recognition and operation of producer groups for the marketing of agricultural and forestry products, Official Gazette no.1098/12.05.2005.

Romanian Law no. 36/1991 concerning agricultural societies and other forms of Association in agriculture, published in the Official Gazette of 06/05.1991.

Romanian Law regarding the Organization and encouraging agriculture, published in Official Gazette no. 67 of 22.03. 1937. (n.d.). Retrieved from http://www.cdep.ro/pls/legis/legis_pck.htp_act_text?idt=27536

The Council of State. (1977). *Status of agricultural production cooperative of 28 September 1977.* Official Gazette no. 105 of 30 September 1977. Retrieved from http://www.monitoruljuridic.ro/act/statutul-din-28-septembrie-1977-cooperativei-agricole-de-productie-emitent-consiliul-de-stat-publicat-n-monitorul-oficial-20811.html

The World Co-operative Monitor. (2016). *The World Co-operative Monitor Database: Exploring the Co-Operative Economy.* Retrieved from http://old.ica.coop/sites/default/files/WCM_2016.pdf?_ga=2.176310509.978773152.1506020642-1176687090.1506020642

Chapter 10
The Role of Agricultural Cooperatives Models Among Europe

Rebega Elena Daniela
Academy of Economic Sciences, Romania

ABSTRACT

The chapter describes the situation from several EU countries on cooperation among farmers, with a focus on the approach of the cooperative concept related to legislation and function. The study comprises 10 member states from different parts of Europe: east, south, and north. The objective was to identify the differences and the existence of an integrative model for cooperative or producer organizations in agriculture. The member states included in the study were France, Italy, Spain, Denmark, Finland, Germany, Poland, Ireland, The Netherlands, and Romania. In order to find common features, the history and previous developing of cooperation were analyzed. Focused on the bibliographic research and comprising an analysis of the history and legislation, the author tried to underline some aspects that could facilitate the setting-up of new agricultural cooperatives and at the same time, a proper operation of the existing ones. The information gathered was presented and interpreted, in order to capture the situation of agricultural co-operative structures, legal type, and economic operation.

INTRODUCTION

"Cooperation can be defined as working or acting together for a common purpose or benefit (Webster's College Dictionary; 1991, p. 300).

In essence, cooperation represents the process of interaction between: (1) cooperatively committed members, employees and leaders and their expectations for the future; (2) cooperative values inherited from the past and expressed in principles, programmes, statues, books, education material, etc; (3) practical cooperative applications, structures, methods of activity, education, etc. also inherited from the past, and; (4) the environment of cooperatives, e.g. the government, the institutional structures of the society at large, the economic system, the values in the community, etc. (ICA; 1998).

DOI: 10.4018/978-1-5225-5739-5.ch010

Copyright © 2019, IGI Global. Copying or distributing in print or electronic forms without written permission of IGI Global is prohibited.

The overall aim of this chapter is to present the current situation of cooperation in the agricultural sector in several EU Member States. The objective is to identify an integrative economical, viable model which can be defined as proper, suitable and successful at European level.

In the study a number of 10th Member States from different parts of Europe are included: France, Italy, Spain, Denmark, Finland, Germany, Poland, Ireland, The Netherlands and Romania. The analysis is focused on several points:

- Short history, evolution of the cooperation system;
- Legal bases and recognition;
- Economic information.

BACKGROUND

As a concept, at European level, cooperation was found in different activity sectors as an element of social economy, starting from the18th century and with a significant development in the 19th century, mostly on the bases of association or mutual societies.

In 2002, the International Labour Organisation adopted the recommendation no. 193 on the promotion of cooperatives considering the cooperative as *an autonomous association of persons constituted on a voluntary basis for economic, social and cultural purposes, collectively owned and democratically controlled.*

A cooperative was created and still exists but not like a non-profit organization, like an entity aimed to obtain profits from an economic activity democratically controlled (one man, one vote) by its members and in their interest. While a commercial company has the objective to maximize profits, a cooperative exists to maximize the benefits of its members by participating in the cooperative activity, underline an economic and social nature of the structures (e.g. negotiating power, lower prices at inputs, access to equipment, but also to information and capital).

A cooperative has an economic and a social perspective. From the economic point of view in case of the cooperatives governed by democratic principles it is noted that they are constituted to assure support to their own members to access the market, strengthening the negotiating power for their interest against the retailers, using in common the assets/machinery, negotiation of the production and marketing costs.

From the social perspective, a cooperative constituted on democratic principles leads to cohesion, increasing social responsibility, helping each other, on the equality, equity and solidarity principles.

Apart of cooperatives, in European context there are many structures set up on association base as producer organisations that follow cooperative principles in their structure and functioning but are not cooperatives as defined by the general principles and common characteristics.

MAIN FOCUS OF THE CHAPTER

History, Legislation, and Organisational Models of the Cooperatives

1. Legal Framework and History

In the identification of an integrative, viable model for cooperation the starting point in the analysis is the legal framework of the Member States concerned.

In order to reach a pertinent conclusion and to focus on a targeted approach to fulfil the proposed objectives, some questions were raised:

- How rich is the history of cooperation?
- How many types of legal framework on cooperation in agriculture could be identified among the Member States?

In this respect the author analysed the context in ten different Member States and the situation is as presented below:

Agricultural Cooperation in France

Agricultural cooperation is defined and regulated by the Rural Code. The core part of the Code sets out most of the principles applicable to agricultural cooperatives. In addition to application norms, the regulatory part of the code established the condition for approval in each agricultural subsector. Agricultural cooperatives, producer organisations and their unions, subject to an approval or accreditation withdrawal during the previous year, are listed in a legal act issued by the Ministry of agriculture.

The legislation on agricultural cooperatives is laid down in Book 5 of the Rural Code. Agricultural cooperatives and SICAs are approved by the Hault Conseil de la Cooperation Agricole in conformity with the provision of art L 525-1 of the Rural Code.

French agricultural cooperatives have a long history started in the 12th century and followed by a modern movement in the 19th century.

A major influence on the cooperation development in France comes from the cooperatives in the wine sector (Gueslin, 1990). In the vineyards areas from southern France, at the beginning of the 20th century, it was often claimed that the cooperatives were "the daughters of poverty" (Rinaudo & Gavignaud, 1990).

Around 1890, the owners of the destroyed vineyards in Charentes started to breed dairy cows and thus laid the foundation for dairy cooperatives. In 1904, during the large wine crisis, several wine growers joined together to produce and market their wine. Until the First World War, agricultural techniques were improved, markets for agricultural products expanded, and the number of cooperatives increased. During the inter-war period, public authorities and farm organisations started to collaborate and it was the time for restructuration of cooperatives. The economic and wheat market crisis in the 1930s led to the development of cereal cooperatives and in 1936, the "Bureau du Blé" (Wheat Office) was established (Van Bekkum & Van Dijk, 1997).

The creation of national associations of cooperatives began in 1945 by the government order. After the Second World War, farmers increased the production and the number of cooperatives also increased. Nevertheless since the 1960s, the agricultural cooperatives have been in a process of transformation and concentration.

In general, French farmers are members as well as participants in cooperatives. An exception to this rule is the possibility of trading with non-members and the possibility of non-participants to become members. Businesses with non-members are legally allowed, but restricted to 20% of the cooperative turnover. Certain provisions can also stipulate that individuals and legal entities who will not use the cooperative services, but who intend to support the cooperative to achieve its objectives by providing capital, may become members.

In the case of investor-members, voting rights are proportional to their shares.

According to Coop of France, at least 75% of the farmers are adherents of an agricultural cooperative. The cooperatives are the grassroots structure, they actively support local farmers to develop their farm in a sustainable manner. In addition, they have recently took over a number of important companies from the non- cooperative sector and created many subsidiaries, especially in processing and marketing industries. The farmers, through their companies' subsidiaries, control more than a half of the processing industry of agricultural products.

It allows the members to have control over their production strategies, their marketing strategies and payments, taking into account the production consumer requirements, both quantitative and qualitative, thus it is possible to improve their position on the market to gain productivity.

Spain

The functioning of the cooperatives and producer organisations has been regulated since 1999 by a National Law as a general framework for cooperation in all sectors, including agriculture, and also by regional laws which provide the legal aspects on the regions concerned. The cooperatives are listed in the Registry of Cooperative Societies, recognized and constituted as legal entities. There is also specific legislation on the agricultural producer groups (APAs).

Since 1990 the consolidation of second-degree cooperatives and consortia has started. A large number of second degree cooperatives have been created, especially in regions with a long tradition in cooperation. Furthermore, constituted as autonomous regions, in some of them a regional legal framework was created in order to establish specific characteristics and rules of their regional models. The cooperatives are approved and recognized by the regional structure.

Agricultural cooperatives are characterized by a series of specific aspects which have a special significance from the social and economic point of view and have a positive effect on their production or consumption.

Cooperatives have a strategic position in the agri-food chain, providing a strong link between production, industry and trade, enabling the farmers to participate in the processing and marketing stages. Cooperatives can concentrate the supply and improve the producers' position in relation with customers and/or suppliers, they are present in all the different stages in the agri-food chain and the producers can bring the added value generated by the chain, to the total production of the sector.

The main goal of the Spanish agriculture cooperation is to have market power and high turnover. Nowadays, Spanish farm cooperatives play a key role, not only in social aspects but also from the economic perspective, and experienced a significant business development.

As in other European countries, this evolution has taken place in the context of a progressive reduction in the number of cooperatives through a continuous process of consolidation and integration.

As regards the way these are managed, the multi-purpose cooperative model is widespread among Spanish farm cooperative. The cooperatives are divided into different sections or business departments that manage their own supplies, processing and marketing.

In most cases the cooperatives also have specific sections that provide farming related supplies and services to its members. In some Autonomous Regions the cooperatives also have a loan section which uses the money deposited in the members' savings accounts to provide funding to the members and to the cooperative itself.

In Extremadura, Valencia, Catalonia and Andalusia there is specific regional legislation on financial aspects.

Italy

The Italian legislation is the one that stresses the social functions of cooperation. Cooperatives are recognized in conformity with Article 45 of the Italian Constitution. Concerning the agricultural sector, the first legal act was issued in 1971, amended several times for a continuous improvement of the legal provisions. The latest is the Law no. 59/1992 which allowed the cooperatives to have members who only share capitals and issue special privilege share.

Also, there is a special legislation on producer organisations, Legislative Decree no. 228/2001 and on this legal basis the POs could be recognized by the Ministry of Agriculture.

As a part of history, the Italian cooperative movement has a long tradition, started in the second half of the 19th century. The Italian cooperative movement had its first expansion at the beginning of the 20th century, during the so-called golden age of cooperation. In that period, it strengthened those features which characterized the movement in the long run, in particular the ideological segmentation, of different umbrella organisations and competition of each another. The largest umbrella organisations were at that time Legacoop and Confcooperative. The fascism regime tried first to destroy cooperatives but then decided to place the movement under the fascist control. With the restoration of democracy, the ideological partition of the Italian cooperative movement was brought back and Legacoop and Confcooperative were immediately reorganised.

Germany

The cooperatives are regulated by the Cooperative Association Law which provides the extension of the supporting purposes (Förderzweck) to cultural and social activities, the reinforcement of autonomy statute, the admission of the members as well as minimum capital requirements, and simplification of the formal accounting and facilitation of the cooperative setting-up.

In Germany there has been a tradition of cooperatives for almost 150 years. In 1862 Friedrich Wilhelm Raiffeisen founded credit cooperatives as loan bank associations and in short time began trading in commodities. Nowadays, each farmer is involved in a form of cooperative in different sectors (agriculture, banking, housing or retail).

They were founded with the aim of utilising economies of scale, enabling the development of market power and, by working together, of accelerating innovation and implementation of technical advances in processing and marketing. Today, these considerations continue to be the purposes of Raiffeisen cooperatives.

The newest cooperatives law is from 2006 ''Cooperative Association Law'' and provides the extension of the supporting purpose (Förderzweck) to cultural and social activities; reinforcement of statute autonomy; admission of members as well as the minimum capital requirements; simplification of the formal accounting and the facilitation of setting up cooperatives.

The principal aim of a cooperative is to provide a better access to the market and services for their members. The joint business operations are based on the principles of self-help, self-administration and self-responsibility and on the values of democracy, equity and solidarity. (Art.1, Sec.1 Cooperative Association Law).

Finland

The Cooperatives Act no. 1488/2001- OSUUSKUNTALAKI, is a general framework which applies to all economic sectors.

Finland has the strongest cooperative memberships and 59% of the population are members of a cooperative.

Finland has always been a country of small and free farmers, owning their farms, with a tradition of working together and cooperating. Cooperation spread rapidly after the formation of the Pellervo Society (now Finn Coop Pellervo) in 1899.

Finland differs from most other countries because a top-down approach was applied and at the beginning a central organisation was created before cooperatives. All cooperatives are based on the following values: responsibility, democracy, equality, equity, solidarity, honesty, openness, social responsibility and caring for others. The cooperative principles are guidelines that translate the values into practice: voluntary and open membership, democratic member control, member economic participation, autonomy and independence, education, training and information, co-operation among cooperatives and concern for community.

Basically, all cooperatives values were implemented in the Cooperatives Act.

Due to the fact that Finland has a long tradition in cooperatives sector they do not apply a special legislation for agricultural sector, the general framework in cooperatives applies also to all sectors.

The cooperatives Act (1488/2001; OSUUSKUNTALAKI) is applied to all types of cooperatives. The cooperative is defined as an organisation whose membership and share capital were not determined in advance. The purpose of a co-operative shall be to promote the economic and business interests of its members by way of the pursuit of economic activity where the members make use of the services provided by the cooperative or services that the cooperative arranges through a subsidiary.

The Netherlands

A cooperative is a legal entity under the Dutch Company Law. Co-operatives are associations (in Dutch: “verenigingen”) and have to be legally constituted on the basis of a notarial document.

The origin of the cooperative movement in the Netherlands was identified from last quarter of the 19th century. From the moment of setting-up the first cooperatives, farmers were frequently supported in their projects by the existing agricultural organizations (trade unions, agricultural societies, and others), encouraged by educational public programs and promoted by the principles of solidarity and mutual help. This cultural context, together with the Dutch agricultural social structure, characterized by its internal cohesion and its low level of polarization, are two possible factors to be considered for the rapid evolution and prosperity of the first cooperative.

The economic situation of the country, favourable for the agricultural sector was of a great importance for the organization of agricultural credit. The first agricultural cooperative is from 1877, when farmers from Zeeland joined to create a cooperative for buying chemical fertilizers at good quality and low price. The cooperative was called "WelbegrepenEigenbelang" (well-understood self-interest).

In 1886 the first milk cooperative was set-up, in 1887 the first cooperative for vegetables, and in 1896 the first agricultural credit bank.

Most of these organizations were single purpose cooperatives operating in a particular sector for a product or a set of products, accomplished a well-defined task in the market and pursued exclusively economic goals.

Until 1934, there was no representative organization of the cooperative movement.

The Cooperative National Council (NCR: NationaleCoöperatieveRaad) was established after 1934 by the central organizations of cooperatives of agricultural supply and consumption and the credit banks.

After the Second World War, an increasing concentration of cooperatives developed through mergers which, though necessary to maintain their balancing role till then, could generate a weakening of the links between the affiliated members and the cooperative.

This evolution ran parallel to the increasing role played by the central organizations and the NCR.

Cooperatives were set up as instruments of countervailing power and came into being only when farmers'position on the market demanded corrective action. For that reason cooperative are always highly specialized and single – purpose. The position of Dutch government toward cooperatives has always been a neutral one.

The Netherlands is governed by a special regulation but placed within the framework of other general regulation.

A Dutch Cooperative ("the Coop") is a special form of association with a separate legal personality, which is governed by certain specific rules and the general rules applicable to Dutch associations.

This mandatory statutory framework does not include the same rigid capital protection provisions. Therefore, there is quite some flexibility in tailoring the articles of association of a Coop to individual needs. Further, there is no statutory requirement for a Coop to maintain a minimum amount of capital.

A co-operative is an entity which has legal personality under the Dutch company law. Co-operatives are associations (in Dutch: "verenigingen") and have to be incorporated on the basis of a notarial deed.

Coops are normally owned and controlled by their members, for example agricultural cooperatives and banking cooperatives. Because of the tax benefits and the flexible legal framework, Dutch law requires at least two persons act as incorporators of the Coop. The Coop has legal personality and may hold assets in its own name.

Romania

The need for association in agriculture emerged immediately after the Revolution of 1989, after which Romania established several legal acts concerning the cooperation sector.

The Law no. 36/1991 was the first act issued in this field after the communist regime in which the activity of cooperatives was regulated, using a similar structure like the cooperatives but without the use of the term "cooperative". The law referred to the agricultural companies, but they were not considered to have a commercial character, being constituted as cooperatives on the principle of a one man one vote (Article 62). The law also repealed the legislation under which the former agricultural cooperatives operated (Decree of the State Council of the RSR No. 346/1977 on the approval of the statutes of the cooperative organizations in agriculture).

The analysis of the law text shows that there are not many activities related to trade, economic activities in general.

The Law on Credit Cooperatives (Law No. 109/1996 on the organization and functioning of consumer co-operation and credit co-operation) covers agricultural activities to a certain extent, but not sufficiently.

Later, after 2000, the cooperative sector was regulated by two laws: the Law 566/2004 - the Agricultural Cooperative Law, which regulates only the sector of agricultural cooperation and the Law 1/2005 - the Law on cooperation, which also provides for the possibility of establishing cooperatives companies for marketing (associations of natural persons constituted for marketing their own products or purchased products through direct distribution or through direct processing and distribution) and agricultural cooperative societies (associations of natural persons constituted for the purpose of joint exploitation agricultural land owned by cooperative members, jointly carrying out land consolidation, sharing machines and installations, and marketing agricultural products).

However, the new existing legal framework does not solve the major problems faced by cooperatives. These problems are related to access to European funding and, at the same time, the possibility to invest, the land fund problem, as well as the existence of fiscal facilities, or rather the removal of tax barriers that stand in the way of cooperatives.

From the analysis on the cooperative history from 2005 to 2015, the number of cooperatives constituted was relatively constant from year to year, so that in 2015 there were 741 cooperatives, a small number compared to the one of small and medium size holdings that could be organized as market access cooperatives.

However it is found that the legislation for producer groups entered into force in 2005 (the Government Ordinance no. 37/2005) and having in mind the financial incentive which is granted for the setting-up and the functioning, a future development of this type of farmers' association will be possible. The Ministry of Agriculture can recognize the following legal forms as producer groups or producer organizations: companies, according to Law no. 31/1990, republished, as amended; agricultural societies and other forms of association in agriculture, according to Law no. 36/1991; associations and foundations, according to Government Ordinance no. 26/2000 on associations and foundations, approved with amendments by Law no. 246/2005; agricultural cooperatives, agricultural cooperatives according to Law no. 566/2004; any other legal form of association, as required by law.

Poland

The Polish Act issued by the Ministry of Agriculture includes also provisions concerning the producers groups and their associations. Producer groups are recognised on/by each territorial administrative entity (voievodship) and are listed into a Regional Registry.

Until 1990, the cooperatives system, in fact collectivisation of the communist regime, was responsible for the marketing of almost the entire agricultural production of individual farms from Poland. Like in other EU countries, the cooperatives were present also in sectors like processing or providing various services for farmers. A major difference between Poland and other MS is the fact that cooperatives were not under the full control of the farmers. The state interfered strongly in cooperative businesses.

Such a policy which persisted for almost half a century conducted to the lack of farmers' confidence in their own cooperatives. After 1989 a large number of cooperatives encountered economic problems because of government policies and the farmers' lack of interest so many of the cooperatives encountered financial problems and were forced to enter into liquidation.

Nevertheless, recently, the farmers' attitude is to come back to their cooperatives. Another interesting phenomenon over the last years was the positive trend in set-up of producer groups, mainly in vegetables–fruit sectors, but also in cereals or meat sectors. They are registered as companies, associations, self-employed individuals, in some cases also as cooperatives. The Parliament issued a special Law on producer groups and they founded their own union organized along the cooperative pattern.

Laying down the foundations of a modern agricultural market is crucial for Poland. A good opportunity to restoring the market is farmer cooperatives and producer groups.

The statistics provided by the Polish National Council of Cooperatives showed that in 2010, there were 91 production cooperatives, 72 in dairy sector, 11 in fruits and vegetable sector and 7 in cereals. In addition, currently the revival of agricultural producers' groups is taking place in horticulture and also in cereals and meat sectors. There are about 350 such groups in Poland, registered as companies, associations, individual economic entities, as well as cooperatives.

In 15 September 2000 the Polish Act regarding the producers groups and their associations was issued. The producers group is defined as any individual or organizational unit without legal personality or legal persons carrying on a farm within the meaning of tax agriculture, involved in agricultural activities in the special branches of production of CAP, in order to adapt agricultural production to market conditions, improve management effectiveness, production planning with particular emphasis on the quality and quantity, concentration supply and organizing sales of agricultural products, as well as protection environment. The producers group may establish a special fund.

The National Council Chambers for Agriculture and agricultural organizations with national coverage determine, by regulation, a list of products and the minimum annual volume of commodity production and the minimum number of members of the group taking into account: regional disparities; concentration of supply, the situation on the market for agricultural products; the current state of organization of producers in the group.

Producer groups are recognised by each territorial/administrative entity – voievodships and listed into a regional registry.

Groups of agricultural producers may organize themselves into groups-associations of agricultural producers to organize and coordinate the supply of means of production, the disposal of products or the storage and processing of products. They can also have activities in promotion, market analyses or representation at national level.

The risk and cost of cooperation could be: losing the independence, higher uncertainty (enterprise and partners' behaviour) and necessity of investments. A one member vote decision making rules with the exception of cooperative that use a voting system relating to capital shares.

Ireland

In Ireland the agricultural cooperation has been regulated, since 1893, by an ordinary company law - Industrial and Provident Societies Act.

Horace Plunkett set up the first cooperative creamery in 1889. Plunkett considered that if farmers are willing to prosper they needed to combine together in business organisations and started to promote the setting up of cooperative - creameries and cooperative - agricultural societies.

By 1894, 33 cooperatives were established and Plunkett recognised the need to formalize these structures. On April 18th, 1894, the inaugural meeting of the Irish Agricultural Organisation Society- known as IAOS was held. IAOS (later on ICOS) was designed to help farmers in organising each branch of their industry cooperative and represents them on large questions through a central body.

Arising from Plunkett's efforts to encourage the government to provide the necessary educational and technical assistance, a Recess Committee was convened that recommended the setting up of a Department of Agriculture and Technical Instruction.

The numbers of co-operatives in Ireland grew to 1,114 by 1920. The creameries and agriculture societies were firstly established, but there were closely followed by the Agricultural Credit Cooperatives.

Starting to 1950 IAOS was involved in the setting up of the cooperative livestock markets around the country. In total there are over 150 coops, which are affiliated to ICOS. These co-operatives have more than 150,000 individual members.

In Ireland was not issued a specific regulation for any type of agricultural association. This sector has been regulated, since 1893, by an ordinary company law - Industrial and provident societies act.

In conformity with the Industrial and provident societies act an:

agricultural co-operative society means a society whose business is wholly or substantially agricultural and whose majority of members are mainly engaged in farming and derive the principal part of their livelihood from farming, and in which the acceptance of deposits and the making of loans constitute an insubstantial part of the business of or are incidental to, or are intended to assist the carrying on or the development of, the society's principal business.

A cooperative is a business that operates for the benefit of its user members. Like everywhere the cooperatives are based on the values of fairness, equity, democracy and mutual support. They offer an alternative to the traditional business model and are owned and controlled democratically by their members. By pooling their resources together, members are able to increase their purchasing power, boost their marketing efforts and reduce the risks that can often be associated with new businesses. Co-operatives operate in a variety of fields but especially they have success in agricultural sector, with major involvement in dairy processing, farm supplies, livestock and other services. In general, cooperatives in Ireland are taxed in the same way as other corporate businesses, although there are some minor differences in some areas. Currently the dairy sector is nearly 100% dominated by cooperatives while cooperatives control two-thirds of the livestock sector.

Denmark

The Danish Public Companies Act is the legal bases for setting-up agricultural cooperatives. In fact, there is no cooperative law in Denmark. In Denmark, the organisational model which governs the meat production, processing and marketing is the cooperatives model, existing from 120 years ago.

The cooperatives were created to provide market access to any small farmer who could not penetrate the market on his own power. Consequently, the upswing of cooperatives contributed to the development of small holdings.

Cooperatives are governed in a democratic way, usually each member has one vote.

Currently in Denmark it is easy to start a cooperative: they have to be registered as an enterprise and for benefits from special co-operative company tax legislation and it is important to have at least 10 members.

Cooperative organisations continue to be an important feature of the Danish agriculture, not only in the large monopolistic agroindustrial enterprises, but also as a tool of co-operation for the new groups of specialised farmers, and interest groups involved in developing new market opportunities.

The cooperatives organised in its various forms within the agricultural sector, at the present time, have changed to become a rational and useful economic tool for particular interest groups.

The accent is on agro-business concentration and could be important to underline the example of first meat cooperative from top ten Europe - Danish Crown which become a company with a monopolistic position on the domestic pig market that has undergone technological changes in primary production and increased the total volume.

This development has followed a series of mergers and amalgamations, aimed to ensure that the industry remained highly competitive by international standards.

Vertical integration of cooperatives with primary agriculture

A general feature for Danish agro-industrial field is its high degree of vertical integration with primary agriculture through the farmers' cooperative movement and its related institutions and organisations. It is an agro-industrial and an agro-political complex developed by the farmer cooperative movement (Ingemann 2001: p. 6).

The cooperative financial and democratic principles have been used for more than a century as the basis for the most enterprises within processing, distribution and marketing of products. The cooperative farmers' movement had and continues to have a very big influence in Danish agro-industry.

Analysing the legal base of cooperation in agriculture, 4 different types of legal framework are identified at national level:

- European member states with an extensive legal framework such as France, Spain and Italy.
- A second model is adopted by The Netherlands, Germany or Finland, countries where the cooperatives are also governed by special regulations but placed within the framework of other national general laws.
- A third group is formed by countries that have no specific regulations for cooperative societies, or agricultural producer organisations and are governed by ordinary company law. This is the case of Denmark and Ireland.
- In Poland and Romania a specific legislation was developed on producers groups which was issued by the responsible institution in charge with the agricultural sector.
- A last group comprised by countries without any specific regulations for cooperative societies/agricultural producer organisations or any type of association in agriculture, and their activity are governed by ordinary company laws. The case of Ireland and Denmark.

The models described have some common characteristics and at the same time many differences based on the history background and previous developing in terms of cooperation.

In most countries the structure described consists of and functions according to the cooperative principle and the obligation of operating exclusively or primarily with and for their members, in any cases national legislations differ based on whether they assign a social purpose to cooperation.

2. Organisation of the European Agricultural Cooperative

In order to identify a model, the question which can be addressed is related to the identification of an organisational model of the cooperatives which could be applied as a successful model among Member States.

- Can an organisational model of the cooperatives be identified?

After the examination of the context from the Member States already named, the conclusion was that in terms of economic development, there are major differences.

The differences raised also, for instance, in terms of management and organisation and the MS have adopted different models:

- **The Northern Europe Countries:** A high degree of integration through mergers, a small number of producers organizations, highly specialized and very large;
- **Southern Europe Countries:** The atomization has increased, an integration model based especially on 2nd degree, smaller dimension of trading structures;
- **Eastern Europe Countries, New Member States:** They seek to develop models.

Nevertheless, perhaps the most important factor has been the recognition of cooperatives as the business structure that best fits the demands of farm policy. The core objective of this policy is to make farmers prime agents in their own future. This has given them a greater institutional role in many aspects of policy implementation.

Southern Europe countries are characterized by an increasing of atomization, integration models based especially on organizations of small dimensions, and is the case of Italy or Spain.

In order to improve this situation, several business association formulas have been developed. For instance, the number of second-degree structures increased. Most farm cooperatives belong to an organization of this type which could be considered a key element in the Spanish farm cooperative movement.

Some of those second-degree organizations are the driving force of a number of mergers and integration processes that were initiated in different agricultural sub-sectors. Also, cooperatives company groups were set up led by second-degree cooperatives in partnership with some first-degree cooperatives or other corporate partners. Those groups founded a number of trading companies that have enabled them to increase their presence on the markets and help them to position them among the leaders.

The Spanish cooperatives maintain the main democratic principle: one member/one vote, independently of each farmer contribution to the total volume of productive activity with emphasis on the principle of mutual help and solidarity.

Eastern Europe country models are still in development process; this is the case of Romania and Poland.

In the last decade producer organisations and producer groups in Poland have grown up and for fruit and vegetables sector they hold a market share of 10-12% and 72% of the dairy market.

In Romania the socio-economic analysis of rural areas reveals the structural problems of the Romanian agriculture, problems which were addressed also by the National Development Programme 2014-2020: an excessive fragmentation of properties and the large number of small-scale agricultural holdings with a low technical endowment, low agricultural performance and a rural economy predominantly based on agriculture, bipolar structure of agricultural farms- an unfavourable element of the Romanian agriculture, with significant discrepancies between the productivity of the agricultural sector and the production potential, the Romanian agricultural productivity registering a level below the EU27 average.

As a main important tool for helping the Romanian agriculture to address the problems identified – the National Programme for Rural Development identified in its strategy a real need - consisting in measures for stimulating the association of the small and medium farms.

Also, the SWOT analyses of the programme – identified as a weakness - the poor cooperation between farmers, reflected in a low degree of association with implications for the viability of the farms.

Therefore, specific measures to address this need – and to support the setting up of associative structures were established in NRDP - were dedicated measures to support the farmers' association like the setting up of producer's groups in agriculture and fruits sector and Support for Cooperation- horizontal and vertical cooperation among supply chain actors – projects for development of local markets and local products- projects proposed by partnerships in which cooperatives can also be involved .

From the structural point of view, the Member States adopted different organisational models:

High degree of integration through mergers, a small number of producers organizations, highly specialized and of very large size. Countries with the higher average size of cooperatives are identified such as Denmark, Finland, Ireland, France, or Germany and the Netherlands with large market share in the sector:

- In Denmark there are 2 big cooperatives involved in pig meat industry which controlled almost 86% of the total meat market and on dairy sector the cooperatives controlled 96% of the total;
- In Ireland, livestock marketing cooperatives - market cooperatives controlled 70% of animals sales and the dairy sector is nearly 100% dominated by cooperatives;
- In Finland, cooperatives have a strong position in food chain from the production stage to food delivery; two largest chains have a combined common market share of more than 80%;
- In France the multipurpose cooperatives with large size, having different involvements not just in the meat food chain, controlled also other type of agricultural businesses; in many sectors (much) more than 50% of the products come from cooperatives and 94% market share in pig meat;
- In Germany the involvement of the cooperatives in the dairy market is about 65% and for cereals 50%.

There are also territorial models developed in countries where the regions have considerable political power: France, Spain or Germany or Poland.

The largest producer organisation in France is a group of cooperatives. The meat channel is dominated by the Breton cooperatives.

Spain makes economic development possible in areas where it is difficult to establish activities from outside the primary sector.

From the economic development perspective, the importance of cooperatives and producer organisations in each country depends on their economic development and in general, and as consideration of the author, a high economic development is associated with developed structures of a considerable size. Also, it is strongly linked with the main economical features of the organisations.

SOLUTION AND RECOMMENDATION

In the Member States which were subject of the present study, cooperation has a rich or more recent history.

In Eastern European states, in the context of the structural problems of agricultural sector (small size and fragmentation of farms, large number of small holdings, etc.), climate change effects and a low level of market requirements, the association and cooperation between farmers are essential in strengthening the negotiation power, procuring equipment and agricultural technologies, increasing access to credit, introducing innovations and new management ideas.

Thus, the essential need to support the cooperative structure and to encourage association is an important measure to strengthen their role in relation to other economic actors for an efficient production which can help to increase the quality of production and better inputs supply.

Farmers need to cooperate in organized structures to ensure their connection to the retail market in order to improve and adapt production to both market requirements (including those arising from the quantitative, qualitative and uninterrupted supply and food safety) and the consumer preferences. These structures have as main objective the commercialization of members' products, thus relieving the producer from the burden of selling the production for which there is neither time nor knowledge necessary to carry out in accordance with demand. As a result, the cooperative can be an effective solution to the challenges mentioned above.

It is also clear that in some societies the cooperative is not only an important element in the life of the farming and rural community, but also an important social instrument for information, learning and networking. In the New Member States the cooperative is now, in this respect, business oriented and totally different from the situation under socialist rule.

In order to improve the cooperation sector it is recommended to implement national facilities (such as tax reductions) to encourage the setting-up of cooperatives and producer groups. Another measure that will facilitate is improving the communication/information at the local level. All the existing support measures provided need dissemination, publicity and communication.

The interest of the associative structure in common investments is very low and has to be much more studied as a social phenomen and compared to other Member States in order to identify some points to find measures for stimulating the development of producers associations.

Farmers interested in starting an associative structure have to take care of the leadership and the human capital. It requires social, economic and organisational skills and resources, and sufficient time and capacity for building the organisation. The legal institutions, the cooperative legislation and the cooperative traditions can greatly support this process of the setting-up the associative forms in agriculture in different ways.

CONCLUSION

The strongest models of cooperation are met in the Northern European countries, due to their long history and involvement in cooperation system. The producer organisations do not reveal the same level of development in all European countries in most of the case study they are in a developing process.

The paper did not identify a general, integrative model.

As it was shown the Northern countries with a long history in cooperatives are much better developed and involved on the market.

In the French model presented as well as in Spain, very well organized structures with common characteristics could be identified, especially concerning the power of producer groups, mostly linked with the economic development of the region concerned.

Spanish and Italian models present the same characteristics at the beginning, a larger number of producer organisations which started to decrease and became powerfully due to the merges and construction of two or even three tiers organisations.

In case of Dutch and German models, there are structures which tried to rise upon the first meaning of the cooperatives. They started to be transformed into businesses and allow to be involved also, outside investors, giving them power but losing the cooperatives principles.

In case of Romania and Poland it shows models trying to be strengthened and developed. The involvement of the producers is found just at the first step, joining for marketing purposes, without any involvement in processing activities.

The explanation for the growth and current market power of European producer organisations consists in their evolution as businesses. The key to this has been their progressive consolidation, whether through the creation of cooperative groups, through second-tier cooperatives, through mergers or by establishing business structures based on working together with other types of business organizations. All this has undoubtedly made them much more competitive.

Nevertheless, perhaps the most important factor has been the recognition of cooperatives as the business structure that best fits the demands of farm policy. The core objective of this policy is to make farmers prime agents in their own future. This has given them a greater institutional role in many aspects of policy implementation.

From the analysis carried-out, some conclusions resulted that lead to the necessity to identify the most important factor that the recognition of cooperatives as business structures best fits the demands of farm policy. The core objective of this policy is to consider farmers prime agents in their own future. This could lead to identify farmers as the most important actors on the agricultural market with the major role in the agricultural policy implementation.

Anyhow from historical point of view there are Member States with a long history in cooperation which are most developed in terms of cooperation and the author could name: Finland, Denmark, Germany and France. Also it cannot ignore the importance of the social perspective and political context which identifies discrepancies between the countries being under the communist regime in the last century and those which had a free development from all the perspectives, especially socially and economically.

From the legal point of view it was identified the fact that it is not important to have a specific legislation with all the aspects concerned very clearly regulated in the case of a mature society aware of the cooperation benefits, but in the case of new Member States a clear legislation on establishing and functioning associative structures is a must.

The author considers that is not possible to identify the best model which can be successfully applied among all European countries because of many (social and economic) differences among the Member States.

Anyway the author considers that the most successful model for cooperation is the development of such structures through vertical integration in the food chain. Another important aspect identified is the size of organisation and functioning and a successful model could be merging structures constituted at the beginning from small organisation which have acquired market power. This is the ideal type of organisation which can be enriched by a better awareness of the producers and their willingness to join together for making profit and increasing their income.

In any case there is a real need of a high degree of awareness and education of farmers on benefits and need of cooperation.

The poor performance of cooperatives in some countries (particularly new Member States) has an important social background. Low levels of self-organisation and networking have far more consequences for these countries than simply constraining cooperative development.

In the new Member States, coming from former communist regime, coping with fundamental collective action problems, often in poor regions with vulnerable rural societies, reflect pioneer activities that resemble early stages of the cooperative movement in other parts of Europe. The issues in the former socialist Member States are diverse and contrasting due to differences in historical backgrounds, pre-collectivisation land reforms, post-collectivisation transformation laws, cooperative traditions and its collective memory, policy streams and the social and cultural context. However, all cases have in common that the impact of the communist legacy persists, as lacking trust was identified as a major obstacle to cooperative development.

The models described clearly demonstrate that cooperatives could add a substantial contribution to the achievement of the economic dimensions of the future sustainable development.

Sustainable development needs a well-governed State, a strong and responsible private sector, an influential and inclusive civil society, a supportive global partnership, and a vibrant social economy, including cooperatives.

Cooperatives should focus their energy on those goals and targets for which they are best suited. In doing so they should bring to the table their biggest comparative advantage, their duality as associations and businesses, and the complementarity of their triple role: the economic role, the social role, and the societal role.

REFERENCES

R193 - Promotion of Cooperatives Recommendation. (2002). *Recommendation concerning Promotion of Cooperatives*. Geneva, 90th ILC session (No. 193).

Bijman, J., & van Dijk, G. (2009). *Corporate Governance in agricultural cooperatives: a Dutch perspective*, Wageningen University.

Cofacerating. (2008). *Coop de France.* Report Bétail et Viande.

COGECA- European agri-cooperatives. (2014). *Development of agricultural cooperatives in the EU*, Brussels, PUB (14) 112:3.

Coop de France. (2009). *Repère Economique.* Author.

Coop de France. (n.d.). Retrieved from www.coopdefrance.coop

Cooperation. (n.d.). *Responsible finish ownership*. Retrieved from www.pellervo.fi

Copa Cogeca. (n.d.). Retrieved from www.copa-cogeca.be

Duch meat association. (n.d.). Retrieved from www.cov.nl

Guitierrez, M. S. (2005). A cooperative Synthesis of 20th Century Agricultural Cooperative Movements in Europe. *Journal of Rural Cooperation.*

ICOS Cooperative Statistics. (2008). *114th Annual Report and accounts.* Irish Cooperative Organisation Society Ltd.

Igual, J., Francisco, J., & Meliá Martí, E. (2008). Social Economy and the Cooperative Movement in Europe: Contributions to a New Vision of Agriculture and Rural Development in the Europe of The 27 CIRIEC-España. *C.I.R.I.E.C. España*, (62), 147–172.

Institutional Characteristics of the German Agro-Food-ChainmW. (n.d.). Grosskopfm University of Hohenheim, Stuttgart.

Marcuello, C., & Lopez, A. (2003). *Agricultural Cooperatives: a wealth, employment, and social involvement generator*. Academic Press.

Menzani, T., & Zamgni, V. (2009). Cooperative networks in the Italian economy. *Enterprise and Society*, 4.

Ministerul Agriculturii si Dezvoltarii Rurale- Direcţia Generală de Dezvoltare Rurală. (2015, May 26). *Programul Naţional de Dezvoltare Rurală 2014-2020 (PNDR).*

Ministry of Agricultural, Food and Forestry Policies (Italy). (n.d.). Retrieved from www.politicheagricole.it

Ministry of Agriculture and Rural Development of the Republic of Poland. (n.d.). Retrieved from www.minrol.gov.pl

Ministry of Agriculture and Rural Development Policies (Romania). (n.d.). Retrieved from www.madr.ro

Popescu, G. (2014). *Cooperaţia în agricultură de la argumentul istoric la transferul de cunoaştere.* Bucureşti.

PVE product boards for livestock; meat and eggs. (n.d.). Retrieved from www.pve.nl

The German Cooperatives in Europe DGRV – DeutscherGenossenschafts- und Raiffeisenverbande V. (n.d.). *German Cooperative and Raiffeisen Confederation*. Retrieved from www.dgrv.de

Chapter 11
Public-Private Partnership Model for Supporting Traditional Producers

Ioana Claudia Todirica
Bucharest University of Economic Studies, Romania

Anca Marina Dinu
Bucharest University of Economic Studies, Romania

Henriette Cristiana Calin
Bucharest University of Economic Studies, Romania

ABSTRACT

Traditional products market in Romania is starting to become competitive as a result of the adoption of new legislative framework (Order 724/2013), which requires new conditions regarding traditional foods definition and limitation of produced quantity. At the end of 2016, there were 556 registered traditional products nationwide as follows: 456 limited liability company, 64 individual, 14 family business, 17 individual, and 15 individual enterprise. Given that there is a limitation in terms of traditional foods production, we must consider the effects of land taxation over the entrepreneur's income in this field. The aim of this study is to identify an effective model of public-private partnership that could support the traditional producers, while taking into consideration the role of financial instruments in order to remove the barriers between the farmer and the government. One of these forms is the association or creating groups of manufacturers.

DOI: 10.4018/978-1-5225-5739-5.ch011

Copyright © 2019, IGI Global. Copying or distributing in print or electronic forms without written permission of IGI Global is prohibited.

INTRODUCTION

The existence of agricultural associations contributed to the reduction of agriculture disparities between the countries less developed and those developed, especially after European Union integration. Thus, for a better functioning of the agricultural sector, associations have a very important role in increasing the market access of producers, balancing the relations between supply and demand. These also influence economic policies, which are defined by relations between the manufacturer and the state.

The concept of public-private partnership generally expresses a way of cooperation between public authorities and the private sector (non-governmental organizations or companies), with the aim of carrying out projects aimed at local development from all points of view. The sustainable development of rural areas is one of the main objectives of European Union member states. It is necessary to promote a general harmonious development, contributing to the reducing of disparities between levels of development of the different regions.

In the current context of global economic and financial crisis, in which all European Union member states are in competition to attract available funding on the market is very important to stimulate the process of attracting private investments in large public interest projects and very important at local level.

Public-private partnership should be seen not as a miracle of this period, but as an option viable, useful, among other existing traditional models. Over time, practice has developed a great deal diversity of public-private partnerships created by the dynamism of development, the multitude of peculiarities contractual arrangements and circumstantial circumstances.

Fragmented agriculture cannot yield the globalized market of today and a solution to resolve this structural problem is the development of associativity in the countryside. For historical reasons and beyond, the cooperation in agriculture has a bad reputation in Romania. However, in order to capitalize fully on the development, potential offered by the Common Agricultural Policy in the financial years 2014-2020, the state must stimulate the association of agricultural producers.

In order to benefit these measures, farmers have to go into the formalized/taxed economy. But doing this, they are unfairly competing with producers who do not register their production and with importers who obviate the payment of taxes. Thus, by tolerating tax evasion and encouraging the informal economy (for example through producer certificates), the Romanian state diminishes the incentive power of the association measures, which it also finances.

BACKGROUND

Traditional foods are a valuable part of Romanian agriculture sector and also an important part of the national culture, identity and heritage. The aim of the study is to document on traditional foods at global, European and national scale in order to identify a public-private partnership model that can support Romanian producers and their value-added products.

The study focused on the European and national regulations that certifies the traditional foods and public- private partnership national regulations.

As stated in order studies as well, in order to understand and predict the success in selling traditional foods we need to know the meaning of word "traditional" in consumer's belief:

Traditional foods - food produced in national territory and for which local raw materials are used; which does not contain food additives; which presents a traditional recipe; a production and / or processing mode and a process/technologically distinct and distinct from other similar products belonging to the same categories (724/1082/360/2013 Traditional Foods Order).

European Food Information Resource (EuroFIR) study "Traditional Foods in Europe" defines traditional foods as below:

Traditional food is a food with a specific feature or features, which distinguish it clearly from other similar products of the same category in terms of the use of 'traditional ingredients' (raw materials of primary products) or 'traditional composition' or 'traditional type of production and/or processing method'.

MAIN FOCUS OF THE CHAPTER

As the rural area occupies a fairly large share of the total area of any country, it is necessary to stimulate the economy by creating more diversified activities. For example, organic, traditional and mountain foods can create added value and also promote the area they come from, thus encouraging both gastronomic and rural tourism. It should be taken into consideration that the future of agriculture at EU level is guaranteed by the use of both traditional and industrial production technologies (Szczepaniak, & Tereszczuk, 2016, p. 72).

Issues, Controversies, Problems

Nowadays, rural areas are facing several limitations: lack of a solid infrastructure system that can direct to agriculture development, lack of an agro-tourism development program, agriculture artificial industrialization, all of the above are the reasons behind the rural demographic decline, rural aging and lack of young families (European Union, 2017).

In the word-wide rural areas, most of the peasants are mainly focused on agricultural activities: subsistence and semi-subsistence farms. In the developed villages, located in the area of expansion of major cities, there ca be found several non-agriculture activities such as: industry, factories and tourism. Due to the lack of investors, most of the villages fail to overcome the underdevelopment stage they are in. In order to attract the investors, the government should facilitate and support them by creating specific financial instruments and public-private partnership models.

TRADITIONAL PRODUCTS: FROM NATIONAL IDENTITY TO GLOBALIZATION

In order to support the traditional producers, at a global level, traditional foods are certified by different regulations agreed in each country. The certification purpose is to gain the consumers trust regarding foods high quality, valued added, traditional recipe and manufacturing processes and safety. It is considered that the certification might increase the selling of the products on the market.

Considering the fact that European Union (EU) regulations are very strict, in order for extra community products to be imported and released on the internal market they need to comply EU regulations.

Countries like China, Colombia, Thailand, Dominican Republic and Morocco managed to comply with EU regulations and now they are selling their products (spices and meat in most of the cases) on European market (Eurostat Database, 2017). In total, there are 19 food products outside EU registered and other six are in process of certification. It is well known the fact that in the last few years European Union emphasized the importance of this certification and classified them in three categories based on some specific criteria:

1. **Protected Designation of Origin (PDO):** The highest quality is offered by the above certification, the product comes from a narrow region with a certain tradition in the preparation of the delicacy, the manufactures and the ingredients are also local, strictly controlled.
2. **Protected Geographical Indication (PGI):** The second level of certification is PGI, a protected PGI (Protected Geographical Indication), a more permissive certification, which guarantees only the traditional recipe used for preparation, the raw material and the manual manufacturing processes may come from outside the region of origin of that product.
3. **Traditional Specialty Guaranteed (TSG):** At the opposite end, TSG (Traditional Specialty Guaranteed) certifies one thing: the product was made according to a traditional recipe of one specific country. Both work and ingredients can come from a much larger area.

At EU level (Table 1) there are 1372 traditional products, of which 618 PDOs, 702 PGIs and 50 TSCs (Table 1). Most are animal origin products: meat and cheese products. Fruit and vegetable products are the second most important product of their pastry products. The number of traditional products is steadily increasing. At the EU level, Romania has managed to register three traditional foods: two PGI: magiun de Topoloveni (plum jam of Topoloveni) and Salam de Sibiu (Sibiu Salami) and one PDO: Telemea de Ibăneşti (Telemea of Ibăneşti). Four other products are still in the process of registration: PGI certification - Protected Designation of Origin: Cârnati de Pleşcoi (Sausages Plescoi), an assortment of plum jam Topoloveni, Novac afumat din Ţara Bârsei (Novac smoked from Barsa) and Scrumbie de Dunăre afumată (smoked Danube shad).

At a national level, in Romania, traditional foods are certified under 724/2013 Order meeting some specific criteria:

In terms of manufacturing process: food manufactured on national territory, using local raw materials, following a traditional recipe, a production and / or processing method and traditional operation technology that distinguishes from other similar products from the same category.

In terms of quantity: `production capacity achieved - reported for a year or 365 days, which may not exceed an average of 150 kg or liters per day total traditional product certified and no more than 400 kg / liters per day total certified traditional products, except for production of traditional bread and bakery products - which may not exceed the average quantity of 300 kg per day total certified traditional product and not more than 800 kg per day total traditional certified products (724/2013 Order).

There are 577 traditional foods registered under 724/2013 Order in Romania between 2014 and 2017(Table 2). Most of them are registered in 2014 (307 products), the trend being descendant until 2017 when only 21 products were registered. This type of certificate confirms the below benefits:

- A traditional product preserves the cultural heritage of rural areas and local traditions and customs: local celebrations and festivals where ports, customs and products are being promoted.

Table 1. Traditional certified foods at EU level between 2014 and 2017

Product Category	Number
Class 1.1. Fresh meat (and offal)	163
Class 1.2. Meat products (cooked, salted, smoked, etc.)	178
Class 1.3. Cheeses	236
Class 1.4. Other products of animal origin	47
Class 1.5. Oils and fats (butter, margarine, oil, etc.)	133
Class 1.5. Oils and fats (butter, margarine, oils, etc.)	1
Class 1.6. Fruit, vegetables and cereals fresh or processed	370
Class 1.6. Fruit, vegetables and cereals, fresh or processed	1
Class 1.7. Fresh fish, molluscs and crustaceans and products derived therefrom	3
Class 1.7. Fresh fish, molluscs, and crustaceans and products derived therefrom	44
Class 1.8. Other products of Annex I of the Treaty	4
Class 1.8. Other products of Annex I of the Treaty (spices etc.)	55
Class 2.1. Beer	6
Class 2.1. Beers	20
Class 2.3. Beverages made from plant extracts	0
Class 2.3. Confectionery, bread, pastry, cakes, biscuits and other baker's wares	14
Class 2.4. Bread, pastry, cakes, confectionery, biscuits and other baker's wares	73
Class 2.4. Pasta, whether or not cooked or stuffed	1
Class 2.5. Natural gums and resins	2
Class 2.6. Mustard paste	2
Class 2.7. Pasta	9
Class 3.1. Hay	1
Class 3.2. Essential oils	4
Class 3.4. Cochineal (raw product of animal origin)	1
Class 3.5. Flowers and ornamental plants	3
Class 3.6. Wool	1

Source: Eurostat Database, 2017

- Traditional foods support local economy due to the fact that the income obtained returns to farmers and by doing this it sustains their activity in the future. For many farmers, these type of activities are the main source of income.
- Products obtained through traditional practices are of better quality because the raw materials are natural.

The main challenge for Romanian producers is the fact that the law has a constrain regarding the quantity produced and manual processes are required in order to comply with the regulations.

Table 2. Traditional certified foods in Romania under Order 724/2013 between 2014 and 2017

Crt.nr.	Category/ Year	2014	2015	2016	2017	Total
1	Beverages	5	15	0	1	21
2	Meat and meat products	127	79	14	4	224
3	Milk and dairy products	78	46	1	6	131
4	Fruit- Vegetable	44	13	20	7	84
5	Breads, bakery and pastries	38	41	6	3	88
6	Fish	15	7	5	0	27
7	Others	0	2	0	0	2
	TOTAL	307	203	46	21	577

Source: Own calculations using data from National Register of Traditional Foods, 2017

Considering the above we must take into consideration the impact that globalization phenomena can have on this type of products, especially with the industrialization of the production at a global level (Szczepaniak, & Tereszczuk, 2016).

Globalization has an influence in all branches; we can see the impacts of globalization on agriculture having both positive and negative aspects. Stating with the development, the technology exchange has become increasingly useful in the field of agriculture. New technology and knowledge help people to produce large seeds and fertilizers. With the introduction of new working machines, practicing agriculture has become much easier. Globalization has helped to increase production and production across the border and across the continent also helps us to make profits (McMichael, & Ritzer, 2007).

At the same time, globalization phenomenon, creates a negative impact on agriculture: due to the development and introduction of new machinery, the unemployment rate is rising; with the time passing by, fertilizer destroyed soil fertility. This phenomenon also determines prices in the growth of food crops by promoting commercial crops. With the removal of governmental restrictions by liberalization, it has become increasingly useful for developed countries to gain growing profits.

In a global agricultural economy, large farms will continue to replace the small farm in the world market. Many small farms depend on sales of internationally marketed goods to provide cash in development, as well as less developed countries.

The most important aspect of the functioning of agriculture can be its non-cash contributions to the quality of their lives. In less developed countries, the major non-cash contribution of farms can be food, clothing, and while in other countries it can be a healthy environment, privacy and security, as well as an independent lifestyle.

However, in both cases, the economic viability of the farm may depend on the cash income from the sale of international commodities. In terms of globalization and corporate colonialism, small independent family farms simply will not have access to internationally traded goods markets. In essence, all these products will be produced under full contracts offered by corporations linked to one of the global food groups. Just higher growth will be able to secure such contracts, as well as in many countries, such operations may be held collectively.

In a global agricultural economy, small farms will be replaced by large farms, which in turn will be controlled by giant multinational corporations. Small farmers simply will not be able to compete in a free market, in a global economy. But even more important than that are the ecological and cultural limits that are essential for the long-term sustainability of agriculture. Industrialization strengthens the process of globalization, with industrial products at a much lower price than those produced by traditional methods. Despite the globalization process of traditional products in the world, the tendency of consumers to return to their roots and tradition is steadily growing.

Traditional products are an important part of European culture, identity and heritage (European Commission, 2007). In order to maintain and develop the traditional products market, improvements in consumer safety are mandatory. At the same time, in this action, it is necessary to keep in mind the traditions of the manufacturing process.

From a globalization standpoint, much emphasis is placed on returning to the sense of national identity, promoting tradition more and more. Traditional products target a narrow niche because their added value is reflected in the sales price (European Union, 2015).

Considering the above, in order to support this type of producers and their production, a model has to be created. This model should be implemented at national level, from the local authorities to the small farms.

Barriers to the Development of Traditional Foods Selling System

From an economic standpoint it is important to understand the key obstacle regarding traditional foods access to the market and how policy measures can overcome this type of obstacles and how they can support the producers in order to strengthen the rural economy.

1. **Quantity Limitation and Distribution Limitation:** As most of the traditional foods producers are using raw materials from their own farms it can be difficult to meet higher demand. It is difficult for small producers to focus on raw materials production since they send most of their time outside the farm, selling their products on the market.

Usually, traditional foods are being sold on the market by the producer itself. The lack of a distribution channel can be found as a limitation and as an impediment in increasing the supply and demand for this type of products.

2. **Lack of Accounting and Marketing Skills:** Lack of knowledge and lack of experience in accounting and marketing area are among the most important reasons why the traditional producers are facing issues in extending their business and increasing their income. To overcome this, traditional producers should be instructed and maybe supported and integrated in a producer's cooperative or association.
3. **Foods Traceability System:** If the law does not provide the specific measures, traditional producers should be included in a food traceability system in order to gain the customer trust. This type of project can be cost- prohibitive for the small producers this is why the government should intervene and create some measures that can support the producers.

4. **Taxation:** According to farmers, taxation is one of the main subjects when speaking about barriers. It is quite hard to develop in Romania and is different from nation to nation having its own policy in all developed or developing countries.

The agriculture taxation influences the farmer's decision about their crops, its future and their commercial facts. Because there are different ways to tax agriculture, it can be developed the ones those affect the most. For example, income tax affects production or investment decisions, but is too hard to verify if the farmers give the correct self-declared incomes. But the total income of a household is more likely to be close to the reality than the income made from farming alone. It is demonstrated that the farming income is easier to approximate than the normal income of the people.

When treating taxation in agriculture for the traditional products, the authors can also speak about the funds that farmers can obtain. But to have access to funds, the farmers must achieve two conditions: in the process of production must be used only local raw material in accordance with the requirements of European Union legislation and the second condition is to have been certified as a traditional product. Every year, Ministry of Agriculture publish a renew list of traditional products. According to some experts in agriculture, the process for the registration is absolutely necessary for product name protection, but also to help the consumers to receive information on the character of the product. Moreover, traditional products are also tools to promote the region they belong, keeping the identity and increasing interest to discover the value of the place. Another advantage of the record is the possibility of obtaining funds for supporting the processing of traditional products, for a better market representation, even if, some of the products have a limited marketing share.

This next part will be about Value Added Tax (VAT), which is a complementary tax policy. The efficiency of collection is generally higher than in case of corporation tax because the tax is automatically charged in price at each product. The tax is paid by the final consumer, while the corporation tax is paid by the producer. Romania introduced the Value Added Tax system in January 1993, replacing the tax on the movement of goods. The next table will give a short history of rates applied in Romania.

Table 3. Romanian Value Added Tax

Period	Normal VAT	Reduced VAT
1st July 1993	18%	0
1993-1998	18%	9%*, 0%**
1998-2000	22%	11%
2000-2004	19%	0
2004-2010	19%	9%*
Till 2008	19%	5%***
2010-2016	24%	9%, 5%
2016-2017	20%	9%, 5%
1st January 2017	19%	9%, 5%

Source: Ministry of Public Finance

*Reduced VAT applies for entrance tickets to historical monuments, books, magazines, medicines

**VAT exemption applies to activities provided by law, to the export of goods, intra-community supplies of goods in Romania, international passenger transport, deliveries of goods under suspensive customs procedures

***Buildings for some homes, for social purposes or that have an area less than 120 m^2

Economic theory teaches us that a low VAT leads to an increased consumption and thus to economic growth. This effect depends on the extent to which the Value Added Tax is passed to the final consumer as well as on his consumption preferences.

The effect of lowering VAT depends on consumer sensitivity at the prices on the shelf. European data shows that demand for food is generally at least sensitive to price fluctuations (Copenhagen Economics, p. 43). Households will most likely adjust their consumption to other product categories to offset the decline or increase in disposable income. However, several European countries have a low value added tax rate for food: 0% in the UK, 4% in Italy, 5% in Poland, 8.5% in Slovenia, 6% in Belgium and 7% in Germany. In 2011, a Romanian consumer spends on average 41% of the total income on food consumption. The value is among the highest in the EU, after Bulgaria (49.6%), but well above the weight recorded in Hungary (29.4%) or in Germany (18.5%) (Institute of National Statistics & Darby, 2010; Figures for European countries refer to 2010).

From this perspective, while reducing the pressure on the state budget, the efficiency of the VAT increase from 19% to 24% without making a distinction between food and non-food goods can be questioned. Considering the high polarization of revenues in the Romanian economy, it seems plausible that the effect of the VAT increase is felt more strongly among the low-purchasing households, which represent a significant proportion in Romania. Increasing VAT generally results in a rise in prices.

Under an unchanged level of consumption for commodities (food), disposable income for other expenditures will decrease proportionally to the share of these products in total expenditure. Given that low-income households will allocate a higher proportion of commodity revenue than the rest of their population, the purchasing power will have more to suffer.

Having in discussion the agricultural income tax the authors found that because of the small number of the traditional products there are small incomes coming from taxes (from the agricultural income statistics and studies, National Institute of Statistics 2005-2017). This raises an important question concerning how taxes and tax policy could be usage to encourage traditional products to solve or to reach their objectives. For this moment traditional products have the same tax policy as the normal agricultural products, with the exception of bakery products (all kinds of bread, as well as the croissants, buns, bars, rolls) and wheat flour (white wheat flour, semi-white wheat flour).

Of the total of 579 traditional products, almost 14% are the bakery ones. Due to the VAT reduction on flour and bread, the production raised, improving the profitability, increasing the number of active companies and the number of employees, on the same level of consumption. (Ministry of Public Finance, "The Evaluation of Macroeconomics Effects of the VAT Reduction for Flour and Bread", 2014).

Table 4. Annual average prices of the main agricultural products, lei/kg, lei/product (eggs)

Food Product/ Year	2005	2006	2007	2008	2009	2010	2011	2012	2013	2014	2015	2016	2017
Wheat	0.36	0.34	0.61	0.66	0.47	0.59	0.88	0.91	0.85	0.76	0.74	0.63	0.64
Corn	0.37	0.38	0.77	0.97	0.67	0.71	1	1.04	1	0.76	0.76	0.74	0.67
Pork meat	4.69	4.19	3.55	4.63	5.1	4.93	5.21	6.07	6.16	5.81	5.23	5.34	5.67
Eggs	0.33	0.32	0.36	0.41	0.45	0.46	0.46	0.59	0.48	0.48	0.52	0.51	0.56
Cottage cheese	7.44	7.95	8.46	10.47	11.93	11.63	12.94	13.66	14.78	15.04	14.55	14.44	15.07

Source: National Institute of Statistics

Having all this, it can be appreciated that not only VAT was reached, but also the income tax and health tax have an important increase. In addition, Ministry of Public Finance estimates that tax evasion on bread and bakery products have registered a reduction of about 300 million lei. (Thematic Publication Number 38 – Traditional Foods, Ministry of Agriculture and Rural Development, http://madr.ro/docs/dezvoltare-rurala/rndr/buletine-tematice/PT38.pdf).

Therefore, is this the act that all traditional products must implement in order to gain their objectives and also the state objectives? It seems that the product which does not pay any taxes is not a priority for its nationality. (Ministry of Public Finance, "The Evaluation of Macroeconomics Effects of the VAT Reduction for Flour and Bread", 2014).

To calculate the income tax of a general agricultural product it should be used two values, as the total area cultivated in hectares and the income rule applicable to the type of crop and the county. In Romania, there are some limits used to calculate the tax, usually for the first hectares cultivated there is no need to pay tax, but from the next hectares. The limits are different for all crops, as in the example below: Cereals: 2 ha, Oil plants: 2 ha, Potato: 2 ha, Sugar beet: 2 ha, Tobacco: 1 ha, Vegetables in the field: 0.5 ha, Vegetables in protected areas: 0.2 ha, Grain legumes: 1.5 ha, Fruit trees: 1.5 ha, Vineyard: 0.5 ha, Flowers and ornamental plants: 0.5 ha.

The agriculture income tax for livestock and bee breeders is similar to the crops, with some different limits, as: Cows and buffaloes: 2, Sheep and goats: 10, Pigs for fattening: 6, Bees: 50 families, Poultry: 100. The final result of the income tax for both crops and animals is calculated by applying 16 percent not to the annual revenue, but to the income rules which are mentioned above, in the Table 5, which are different from county to county, but also from year to year. The income rules were introduced in Romania in 2013, as a national income rule and not different as is now. This income shall be established by 15th of February of the year for which they are due. Therefore, farmers are not obligated to keep accounting. The tax will be calculated on the basis of income rules regardless of whether the actual income is lower or higher.

The income tax is paid by the farmer who effectively use the land and not by the owner in the case of lease.

The last part will show how taxes are calculated for the both, vegetal and animal holdings, having the values above for each county.

Example 1, a vegetable farm of 100 hectares of cereals in Vaslui: the first 2 hectares are not taxed, as was mentioned above, but the rest 98 hectares. The annual revenue is: 98 ha x 182 lei/ha (the value for the county took as example) = 17.836 lei and the tax is 16% x 17.836 lei = 2.853,76 lei.

Example 2, an animal farm with 102 cows in Suceava: the first 2 heads are not taxed, but the rest 100 of them. The annual revenue is: 100 heads x 175 lei = 17.500 lei and the tax is 16% x 17.500 lei = 2.800 lei.

ASSOCIATION IN AGRICULTURE: THE SOLUTION FOR TRADITIONAL PRODUCT MANUFACTURERS

The relationships between association and cooperation are diversified and represent a variety of links established between agriculture and other branches of economy.

In Romania, under current legislation, agriculture has the following forms of association:

Table 5. Income rules in Romania, 2015

County	Income Rules lei/hectares
Alba	200 lei
Arad	150 lei
Argeş	170 lei
Bacău	191 lei
Bihor	225 lei
Bistriţa-Năsăud	203 lei
Botoşani	190 lei
Braşov	165 lei
Brăila	360 lei
Buzău	164 lei
Caraş-Severin	182 lei
Călăraşi	200 lei
Cluj	264 lei
Constanţa	205 lei
Covasna	226 lei
Dâmboviţa	100 lei
Dolj	210 lei
Galaţi	173 lei
Giurgiu	187 lei
Gorj	201 lei
Harghita	152 lei
Hunedoara	180 lei
Ialomiţa	205 lei
Iaşi	190 lei
Maramureş	100 lei
Mehedinţi	250 lei
Mureş	162 lei
Neamţ	168 lei
Olt	200 lei
Prahova	190 lei
Satu Mare	206 lei
Sălaj	95 lei
Sibiu	265 lei
Suceava	175 lei
Teleorman	212 lei
Timiş	250 lei
Tulcea	242 lei
Vaslui	182 lei
Vâlcea	208 lei
Vrancea	158 lei

Source: National Agency for Fiscal Administration, 2017

- Agricultural companies and other forms of association in agriculture (Law No. 36/1991 on agricultural companies and other forms of association in agriculture, as subsequently amended and supplemented);
- Associations (Government Ordinance No. 26/2000 on associations and foundations, approved by Law 246/2005, as subsequently amended and supplemented);
- Agricultural cooperatives (Law 566/2004 on agricultural cooperation, as subsequently amended and supplemented);
- Producer groups (Government Ordinance no. 37/2005 of producer groups and organizations recognition and function for the agricultural and forestry products marketing).

The Agricultural Society, known as the most important form of association in agriculture, is a privately-owned, variable-capital company, with an unlimited number of associates. Its main object is the agricultural exploitation of land, animals and other means brought to society, as well as the investment of agricultural interest. The obligations of the agricultural society are guaranteed by the social patrimony, associations being only responsible with their own social parts.

The association is made up of three or more people who do community activities for a non-patrimonial personal interest.

The Agricultural Cooperative is an association of individuals and legal entities and operates with a minimum of five persons and performs an economic, technical and social activity intended to supply goods, services and jobs mainly to its members.

There are two types of agricultural cooperatives:

1. First grade cooperatives, made up of individuals associations;
2. Second-class cooperatives consisting of legal entities constituted from the first degree cooperatives, or individuals and legal entities.

Upon entry into the agricultural cooperative, movable and immovable property constituted as a contribution to the share capital, shall be valued in order to determine the value of the subscribed shares in kind by the cooperating members.

In the case of cooperatives of the agricultural land management, goods brought into use and lands of any kind remain the property of cooperative members, the agricultural cooperative having the right of its using.

The activities of agricultural cooperatives are mainly focused on the provision of services, purchases and sales, processing of agricultural products, manufactures and small industries in agriculture, exploitation and management of agricultural, forestry, fish and livestock, financing, mutual assistance and insurance agricultural.

Producer groups are associations of farmers who market their products in common. These groups ensure production planned and adapted to market demand, according to quality and quantity conditions. They promote the supply and marketing of products obtained by their members. The Group has the task of optimizing production costs, setting producer prices and promoting the use of cultivation practices, production techniques and environmentally sound waste management practices.

Advantages of producer groups:

- Reduction of production costs (by purchasing equipment that is shared by all members);

- Use of modern and competitive technologies for waste production and management in order to comply with the environmental and biodiversity standards that we must call ourselves as EU members;
- Ability to plan and change production based on quasi-quantitative demand on the market;
- Facilitating access to European funds and contracting of bank loans;
- Facilitating communication between farmers;
- Enhancement of bargaining capacity to obtain better prices;
- More efficient production promotion on the domestic and external markets.

The Common Agricultural Policy first accepted producer groups in the 1970s as an alternative to some proposals by Sicco Mansholt (European Commissioner for Agriculture) on optimal sizing of agricultural holdings. At present, the philosophy work of producer groups is very little interested in the size of holdings and their ability to fit and comply with certain criteria and conditions of response to market requirements.

In Romania, producer groups were first recognized in 2005 as part of the legislative and institutional preparations for integration into the European Union. Producer groups share the associative spirit and enable the manufacturer to choose the most appropriate legal form.

It has become a truism that fragmented agriculture cannot yield the globalized market of today and a solution to resolve this structural problem is the development of associativity in the countryside. For historical reasons, the cooperation in agriculture has a bad reputation in Romania. However, in order to capitalize fully on the development potential offered by the Common Agricultural Policy in the new financial year 2014-2020, the state must stimulate the association of agricultural producers.

By association, manufacturers benefit from raw material supply, being helped with product processing and an efficient equipment for utilization. Funding can be accessed that is not available to individuals. Being associated, producers can improve marketing or sales and the realization and capitalization of high-value natural products (HNV), which are cultivated in areas called high nature value agricultural landscapes.

High nature value agricultural landscapes are areas in rural areas where traditional farming is the main economic activity and a key factor in nature conservation (http://www.gazetadeagricultura.info). These lands are characterized by the presence of natural and semi-natural vegetation (grazing), generally very rich in species, and are integrated, in some cases, into an extensive and uninterrupted mosaic landscape that includes natural elements (such as snow grooves, hedges vineyards or stone, wooded or shrubs, streams) and groves of cultivated land and orchards. Traditional agricultural practices have helped to preserve many areas of this type in Romania (but also in Europe), delivering a range of public benefits (goods and services), such as valuable cultural landscapes, high quality water and food, quality of life, opportunities for recreation, carbon storage or flood control. These landscapes also deserve support for their economic and agricultural productivity that ensures the lives of many farmers' communities in Romania.

Examples of Good Practice

The Ministry of Agriculture and Rural Development in partnership with the mayors organizes fairs to promote traditional products, encouraging small producers and help them sell their products. Many manufacturers claim that they can't sell their products without these fairs.

Among the most important annual fairs in Romania, where traditional product manufacturers can expose and sell their merchandise, are the following:

- **INDAGRA:** Is the most important agricultural fair. At its 22th edition, INDAGRA is a complex manifestation, with the purpose of promoting the Romanian agriculture, domestic zoo technical potential and traditional products,

The most recent edition of the *Indagra Fair* has been attended by more than 500 exhibitors from all sectors of agriculture, with an area of exposure of about 41.000 square meters.

Indagra is organized by *Romexpo* in partnership with the Chambers of Commerce and Industry of Romania.

- **AGRARIA:** Another important event dedicated to the agricultural field takes place in Cluj County. The largest agricultural fair in Transylvania, now in the 23th edition, aims every year to be an important meeting place for farmers, specialists, managers and authorities in the field of agriculture.

Each year, at *Agraria*, over 290 exhibitors expose their main products and services in the agricultural field, with an area of more than 20.000 square meters.

- **AGROMALIM:** The most important agricultural event in western Romania, at its 25th edition of the International Agriculture, Food and Packaging Event, over 250 exhibitors from about ten countries expose their stands annually. The *Agromalim* program includes conferences, seminars, business meetings, innovative product, service presentations and traditional products.

International exhibitions are a great addition to traditional producers too. They become known on the international market and can sign contracts with foreign distributors. Associations, based on proposals submitted to the Ministry of Agriculture and Rural Development, may be selected for participation. However, a major problem faced by manufacturers is the cost. In some areas of the country (e.g. Baia Mare), there are associations that bear entire costs to the producers.

The "Produs de Cluj/Cluj Product" Association, an organization under the authority of the Cluj County Council, participated in the 80th edition of the Grüne Woche (Green Week) International Exhibition in Berlin. Organized annually, Grüne Woche is a unique manifestation with its tradition, size and duration, being considered as the largest platform for the world food industry. More than 1.600 exhibitors from 69 countries participated in this year's event, and the trade fair was attended by 80 ministers. The event has registered more than 500.000 visitors, attracted by the quality, but also by the novelty of the products.

"We Grow Romania Together" Association, supported by Patria Bank launched the first virtual "rural mall" in Romania. Loyal customers to traditional products have the opportunity to supply 100% natural products. The "Rural Mall" is an innovative concept where villagers can easily sell their products. Producers group in partnerships on small geographic areas and open a product "stand" on an online platform where they will receive the orders. Manufacturers present their weekly offer in personalized and individual stands, receiving the necessary help. Partnership can also access European funds to acquire the necessary equipment for the collection, sorting and distribution of traded products.

Obstacles in Association and Solutions to Overcome Them

The minimum existence of the forms of association in Romania is due to the reluctance and low interest of the agricultural producers towards the associative forms due to the low awareness and the lack of information on the advantages resulting from the association of the different degree of training of the participants. The difficulties encountered by small agricultural producers and the justification of the need for association exist because of the difficult cooperation with the service units, especially if the area of the holding is reduced, as well as the poor development of the marketing structures.

The insufficient development of associative forms in agriculture is due to low social capital, insufficient promotion of functional initiatives of this kind and lack of the vision of the support that associations would need (consultancy, organizational development, business planning).

Association in agriculture is a growing phenomenon, but there are many barriers to be. Although, association is based on interpersonal relationships, the shared capital provided by cooperative members is low. Inappropriate communication of farmers in the same community, lack of information on the benefits of association and lack of support from local authorities are real issues that slow down the association process. These initiatives should support the development of the capacities of associative and cooperative forms through training and consultancy actions, include rural associations in social economy support programs, set up specialized departments within agricultural chambers, finance warehouses and proximity markets, support short production chains and provide direct tax incentives.

Table 6. Obstacles and solutions in agricultural associations

Obstacles	Solutions
Mentality: Lack of democratic practice, distrust, tax aversion (including payment of contributions), difficulty seeing beyond self-interest, lack of vision.	Systems of revision, control of transparency, very clear tasks for members, good practice models, study visits, small-age education, more information.
Lack of knowledge about association.	Support structures to provide information and facilitation for association in agriculture;
Lack of knowledge about business management.	Support structures to provide business management consultancy; Motivating payment for cooperative employees.
Insufficient funds for training and financing of cooperatives.	Accessing funding and granting tax incentives.
Double taxation (first is the transaction between cooperative and cooperative member and after that they sale to third parties).	Redefining the relationship between the cooperating member and the cooperative by changing the legislation.
Lack of encouragement from the authorities.	Establishment of other supporting structures of associative forms
Reduced affiliation to national and international organizations (affiliation fees to international organizations are very high).	Subsidies by the state of subscriptions for affiliation to European networks
Difficult to compete with imported products (they come at very low prices).	Common marketing with other cooperatives; Creation of short distribution chains; Financial support (grants).
Distortions caused by European funds	More information on European funding and others.
Lack of leaders in agriculture.	Measures to make cooperatives and other associative forms, that could be true leadership nurseries; Leadership training sessions on business management and other topics of interest.

Source: www.agrointel.com

Even if association is based on interpersonal relationships, the shared capital provided by cooperative members is low. Poor communication among farmers in the same community, lack of information on the benefits of association and lack of support from local authorities, are just three of the issues that slow down the association process and, as suggested by the Romanian Center for European Policies, they should take priority solving through public policies in agriculture. These initiatives should support the development of the capacities of associative and cooperative forms through training and consultancy actions, include rural associations in social economy support programs, set up specialized departments within agricultural chambers, finance warehouses and proximity markets, support short production chains and provide direct tax incentives.

FUTURE RESEARCH DIRECTIONS

Due to the rural areas particularities there are several issues that should be taken into consideration while conducting a future research. It can be observed that in the rural areas, where traditional producers are based, the population growing trend is negative. The main reason is the migration: from rural to urban or to foreign countries. Considering this, the rural population is aging making the process of transmitting the tradition to the young difficult. Without instruments that can facilitate local businesses, rural areas can't revive. Future research directions can be focus on the rural population development strategies in order to gain more knowledge and in order to advise instruments that can be applied by the government.

CONCLUSION

Large number of agricultural holdings and the overwhelming share of those without legal personality with areas up to 2 hectares are considered to be the major structural and financial disadvantages of Romanian agriculture. In spite of European and national programs designed to correct these disadvantages, changes over time are insignificant. This suggests that, beyond the failure of some development programs, we have to rely on realities and explore our own development paths based on the traditional way of production and on the special relationship that the Romanian has with the territory. The much-desired land mix, which in principle ensures increased productivity, has its own limits, and large outputs are detrimental to product quality and biodiversity.

In the register of trade in Romania are registered 579 traditional products. The practices of traditional agriculture do not put in danger the plant's life and animal living in the area and also they do not have a negative effect on the environment.

Traditional products help the national economy, because the farmers are receiving back the money and help sustaining future activities of them, but also help the patrimony of the national areas and preserve local traditions as local holidays which promote traditions, products and customs. Traditional products have also a better quality and healthy.

The great challenge of rural development programs is the fiscal part of the rural economy and, in particular, of agricultural production. It is questionable whether the taxing of agricultural production would mean more than the costs induced by the effects of these policies, the irreversible transformation of the country into a rather economic logic, ignoring its complex functions.

Analyzing the financial part that contributes to the final product, the authors found that the decrease in Value added tax leads to a bigger consumption of the population. The history of Romania's VAT, can show some different value from 1993 till present with benefits in some cases.

A potential benefit of lowering VAT would be to reduce tax evasion. The high share of agriculture in Gross Domestic Product is associated with a higher level of tax evasion.

Romania is among the European countries with the lowest income from taxes and social contributions, only 28% of GDP in 2010 compared to 45% in France and 44% in Austria. The difference stems from higher taxes in these countries, but also from the high level of tax evasion in Romania.

The development of rural communities is not targeted excluding decision-makers at central level, regional or national. To ensure development of rural areas is necessary mobilizing all stakeholders (local authorities, organizations and representatives of civil society) and, implicitly, the formation of a solid partnership between them. Rural development requires an approach to the rural issue beyond what would seem to be on superficial things or focusing strictly on the agricultural sector. Rural development is focuses on the whole system with productive potential: from the use of water resources, including waste water management, to irrigation and land improvement. Agriculture, forestry and forestry animal husbandry is considered not isolated, but parts of a system, integrating them with the development of human capital, infrastructure and social development.

The role of public- private partnership is related to exposure the main challenges at the local level, setting priorities, identifying solutions development and implementation of measures and strategies integrated. Strategies are based on links between participating factors with multiple effects on local development and general programs at regional, national and community level, designed for recovery strengths or social "trumps" environmental.

Traditional products market development, especially in the peasant households, is one of the strategic objectives that could serve as a stimulus for the development of the family farm and, implicitly, of the rural households. Due to the legislation, the infrastructure, and the lack of financial incentives, Romania has quite a few traditional products managed to pass all the homologation stages at national and European Union level. Thus, at the level of the national agricultural policy, it is necessary to create a coherent and responsible legislation in order to recognize all the traditional products made in the peasant farms. They also need to be promoted, monitored and marketed in modern formulas.

Association is based on interpersonal relationship, but the shared capital provided by cooperative members is low. Poor communication among farmers in the same community, lack of information on the benefits of association and lack of support from local authorities, are just three of the issues that slow down the association process. These initiatives should support the development of the capacities of associative and cooperative forms through training and consultancy actions, include rural associations in social economy support programs, set up specialized departments within agricultural chambers, finance warehouses and proximity markets, support short production chains and provide direct tax incentives.

Scheme 9.1 - *Establishment of Producer Groups in the Agricultural Sector* from the actual National Rural Development Program encourages farmers in the agricultural and forestry sector to associate, receive recognition of the producer group, and then receive the grant. The main purpose of the association is to sell the products jointly, thus reducing the costs of selling and obtaining a better price from the buyer. Evan though it stimulates the entrepreneurship spirit, this scheme is the least accessed due to the fact that the farmers associate this with the communism.

The main role of producer groups is to jointly market the products of individual farmers. Producer groups fulfilling certain conditions can obtain subsidies from the European Union. Producer groups are considered to be a chance for farmers to strengthen their economic power by negotiating product prices, lowering transport costs, obtaining the profit typically earned by intermediaries.

Producers entering a group of producers should consider the following: loss of independence, increased uncertainty, need for investment.

REFERENCES

Copenhagen Economics. (2007). *Study on Reduced VAT applied to Goods and Services in the Member States of the European Union*. Academic Press.

European Commission. (n.d.). Retrieved from http://ec.europa.eu/agriculture/quality/door/list.html

Eurostat Database. (n.d.). Retrieved from ec.europa.eu

Farm Accountancy Data Network Database. (n.d.). Retrieved from http://ec.europa.eu/agriculture/rica/database/database_en.cfm

Giescke & Nhi. (2010). *A General Framework for Measuring VAT Compliance Rates*. General Paper No. G-206 August 2010 Centre of Policy Studies and the Impact Project.

Harold, V. D. M. (2007). *Exploring agricultural taxation in Europe*. Wageningen University.

Indira, R. (2004). Taxing agriculture in a developing country: a possible approach. Academic Press.

Jitea, I-M. (2010). *Personal Agricultural Income Taxation. A Comparative Analysis between Several New Member States*. Academic Press.

Ministry of Agriculture and Rural Development. (n.d.). Retrieved from http://www.madr.ro

Ministry of Public Finance. (2014). *The evaluation of macroeconomics effects of the VAT reduction for flour and bread*. Author.

National Agency for Fiscal Administration. (n.d.). Retrieved from www.anaf.ro

Order 724/1082/360/2013 –traditional food attestation

Ordinance no. 37/2005 on the recognition and functioning of producer groups for the marketing of

Paturot, D., Mellbye, K., & Brys, B. (2013). *Average Personal Income Tax Rate and Wedge Progression in OECD Countries*. Academic Press.

Philip, Mc., & George, R. (2007). *The Blackwell Companion to Globalization*. New York: Wiley-Blackwell.

Popescu, G. (2007). *Cooperation in agriculture, from the land market to the transfer of knowledge*. Iasi: Terra Nostra.

Popescu, G. (2013). *Old issues, new relationships in agriculture*. Bucharest, Romania: Romanian Academy Publishing.

Rankin, M., & Nogales, E. G. (2016). *Public-private partnerships for agribusiness development, A review of international experiences*. Rome: Food and Agriculture Organization of the United Nations.

Stanciu, M. (2013). Study concerning the capitalization on local level of traditional products obtained from sheep. Academic Press.

Szczepaniak, I., & Tereszczuk, M. (2016). *Regional and Traditional Products as a Way to Improve agricultural and forestry products*. Competitiveness of Polish Food Producers in Foreign Markets, Thematic publication number 38 – Traditional foods, Ministry of Agriculture and Rural Development. Retrieved from http://madr.ro/docs/dezvoltare-rurala/rndr/buletine-tematice/PT38.pdf

KEY TERMS AND DEFINITIONS

Cooperatives: Autonomous society of natural and/or juridical persons, as the case may be, a private legal person constituted on the basis of the free consent expressed by the parties in order to promote the interests of the cooperative members in accordance with the cooperative principles organized and functioning according to the provisions of this law.

Development Strategy: All the major objectives of the organization on a long-term basis, the main ways of realization, along with the allocated resources, in order to obtain the competitive advantage according to the mission of the organization.

Farm Income: The return (both monetary and non-monetary) to farm operators for their labor, after all production expenses have been paid.

Food Traceability System: The ability to track the route movement of a food through the specific stages of production, processing, and distribution.

Producer Groups: Is formed and operates on the free initiative of the producers, based on the unit of interest and action of the group under the following conditions: they are composed of at least 5 members; prove through the bookkeeping a minimum value of the marketed production, for the product group for which recognition is requested, of at least 10 thousand euro, equivalent in lei; complies with the provisions of art. 6 par. (1) of the Government Ordinance no. 37/2005, as amended and supplemented; have a centralized system of accounting, billing, registration and quantitative, qualitative and value tracking of members' production.

RNPT: Romania National Register of Traditional Foods, a record where all registered traditional foods can be found.

Rural Limitation: Lack of a solid infrastructure system that can direct to agriculture development, lack of an agro-tourism development program, agriculture artificial industrialization, all of the above are the reasons behind the rural demographic decline, rural aging, and lack of young families.

Section 3

Solutions for the Future

Chapter 12
European Agriculture in 2020: Solutions for a Sustainable Development

Silvia Elena Cristache
The Bucharest University of Economic Studies, Romania

Florica Georgeta Rotaru
The Bucharest University of Economic Studies, Romania

Elena Cristina Rotaru
Ovidius University, Romania

ABSTRACT

The chapter is structured in two parts: the first part presents the agricultural policy of the European Union in accordance with the Treaty of Lisbon and the Europe 2020 Strategy. At the same time, the National Rural Development Program of Romania is presented. In the second part of the chapter an econometric analysis of the agricultural sector is made. The linear multifactorial regression model presented in the second part of the chapter was based on data provided by Eurostat: surface cultivated with pesticides, GDP/capita, net investment in agriculture. The model proposed in this chapter explains and highlights the existence of a positive, linear, and significant relationship between the variables included in the model. The multiple linear regression model indicates that, with the increase of one million euros per GDP, the value of the surface cultivated with pesticides in Romania will grow by an average of 132.08 hectares.

INTRODUCTION

Launched in 1962, the EU's Common Agricultural Policy (CAP) is a partnership between agriculture and society, between Europe and its farmers.

Its main objectives are: to increase agricultural productivity, to guarantee the security of food supply at reasonable prices for EU citizens and to ensure a certain standard of living for EU farmers. In the period 2014-2020, the measures taken under this policy would ensure better internet services and quality infrastructure for about 18 million citizens in the rural area, i.e. 6.4% of the total rural population. About half of the lands in the EU are used in agriculture (Agriculture, A Partnership between Europe and Farmers, 2017).

DOI: 10.4018/978-1-5225-5739-5.ch012

Copyright © 2019, IGI Global. Copying or distributing in print or electronic forms without written permission of IGI Global is prohibited.

That is why agriculture and food production are essential to our economy and society. In the 28-member states, there are about 11 million agricultural farms and approximately 22 million people who work in this segment. In the EU countries, the agricultural and food industry segments provide about 44 million jobs. This common European policy is a must because, in Romania, natural resources are in good condition, but risk management systems and services which the potential climate and sanitary risks can be reduced, are insufficient. Romania also faces difficulties in reaching her socio-economic potential of the agri-food and rural areas.

As far as the objectives of the National Sustainable Development Plan are concerned, they are in line with the EU's Common Agricultural Policy priorities and the Europe 2020 Strategy, with an emphasis on increasing competitiveness and diversifying jobs in rural areas. The main objective of the 2014-2020 Strategy for Romania is to increase competitiveness in agriculture, forestry and the food segment, which is essential to the economic development and jobs' diversity in the rural areas, being the central element of this strategy. The EU strategy for food safety is based on: rules on the safety of food for humans and animals, independent and publicly available scientific advice, control measures, the consumer's right to choose having complete information. Last, but not least, through measures of rural promotion and development, the European Commission and EU countries support farmers who are involved in projects meant to increase food quality. Regarding the structure of EU exports by industries in 2015 we noticed that agriculture ranked 4th, ahead of important segments, such as construction industry, transport, etc. (see Figure 1)

The Treaty of Lisbon does not confer upon the Union additional exclusive prerogatives. However, the new text modifies the way in which the Union exercises its current prerogatives, as well as adopting new ones, encouraging the participation and protection of citizens, creating a new institutional order that modifies the decision-making processes with the purpose of increasing their efficiency and transparency. In this way, it also ensures a higher level of parliamentary control and democratic accountability.

Under the Treaty of Lisbon, the European Union has significantly intervened in the field of legislation, food security, environment protection, etc. The European Union's Common Agricultural Policy has been adopted by all member states (European Commission, 2016).

Figure 1. Structure of EU countries exports in 2015
Source: European Commission, 2015 (EU-28)

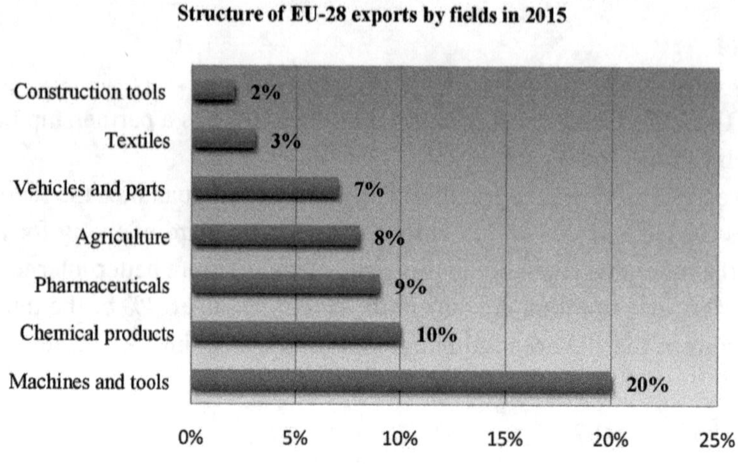

The main goal of this common policy is to introduce several measures and guidelines needed for farmers to fulfill their role in society, namely the production of food people need for a decent living. Therefore, food products must be affordable, of good quality, safe, varied and in sufficient quantities. Also, due to the implementation of the EU's Common Agricultural Policy (CAP), Europeans can enjoy the benefits of food security.

Also, in all EU member states, farmers maintain the vitality of rural areas and preserve their specific lifestyle (specific food, leisure services, traditional families, a certain lifestyle and attitude towards work, etc.). Furthermore, it's worth mentioning the fact that, after Romania joined the European Union, it was necessary to increase the efficiency of its service, education and agricultural segments, i.e. those segments most used by the population and having significant effects on Romania's sustainable development strategy. In this context, a key way to achieve this goal was to ensure a proper balance between economic, environmental and social needs, which define and characterize sustainability in any country (European Commission, 2017).

The main objective of the European Union's Common Agricultural Policy is to increase agricultural output to ensure the security of food supply, at reasonable prices for consumers and a reasonable standard of living for EU farmers. Market measures and income support are funded exclusively from the EU budget, while rural development measures are based on multi-annual programming and are co-financed by member states. In conclusion, European citizens are the main beneficiaries of the EU's Common Agricultural Policy, because there is always enough food in shops and supermarkets, at prices that are generally affordable and environmentally oriented. At the same time, the rural development measures and directives provided by the European Union's Common Agricultural Policy represent a significant factor of change and progress for farmers in agricultural holdings, enabling them to raise livestock, cultivate the earth and ensure a higher level of food security (Douglas, 2010).

BACKGROUND

The title of this chapter (i.e. "Agriculture 2020 - Sustainable Development Solutions") is a paradigm that authorities will soon face, to make sure there is no discrepancy between farming in the European Union and the rudimentary way of farming still widely practiced in Romania.

At a microeconomic level, there remains some uncertainty with respect to the fact that conventional agriculture would be less profitable than organic farming. If farmers were convinced that this new type of farming would bring significant profits in a relatively short time and with acceptable costs, then Romania could quickly align with the European Union's desideratum, namely sustainable growth

In this context, unlike the EU cooperatives, farms in Romania are working more in production and less in processing or marketing. Romania is on the penultimate place in the Union, as far as the capitalization of agricultural holdings is concerned – this modest capitalization results in a low degree of technology use in agricultural holdings in Romania (PwC, 2017).

The estimation of the main economic indicators shaping the profitability of organic agriculture as compared to the conventional one was developed and presented in time by numerous researches such as: (Girardin, Bockstaller, & Werf, 1999; Rigby, Woodhouse,Young, & Burton, 2001; Van derWerf & Petit, 2002; Halberg, Verscuur, & Googlass, 2005; Halberg, 2012).

Regarding the econometric analysis introduced in the second part of this chapter, it represents a macroeconomic approach. In this macroeconomic analysis, the key issue discussed was the fact that intensive agriculture is meeting to a significant extent the current global needs of agriculture but with a negative environmental and ecosystem impact. Therefore, the variables included in the regression model were macroeconomic data, which mainly target conventional agriculture or supporting activities.

Conventional (traditional) agriculture also represents the model of intensive or productive agriculture, pursued in the last 40 years, with the following main features and peculiarities (Morganand & Murdoch, 2000): economic and strategic thinking, political and administrative governance, technology invention to increase production and productivity. What defines conventional agriculture is its efficiency due to high yields resulting from the significant allocation of the following factors: machines, pesticides, fertilizers, selected seeds, etc.

At the opposite end, organic farming is considered a remarkable phenomenon because it emerged without any form of governmental support, scientific institutions, enlargement services or special legislation. Organic farming involves many ecological, social and economic advantages (Cobb, Feber, Hopkins, Stockdale, O'Riordan, Clements, Firbank-Goulding, & Macdonald, 1999) for food safety and health (Dyson, 1996). This is the reason why the title of the chapter emphasizes the need to include this sector in the strategy for sustainable development by 2020.

Although this phenomenon reflects the fact that Romania's economic structure is getting closer to that of other member states, the share of the agricultural segment is still more than three times larger than the average in the EU. In accordance with the National Strategy for Agriculture on the Medium and Long-Term (2014), Romania is distinguished by the variety of her pedoclimatic conditions, which are proper for obtaining diversified, high quality and quantity agricultural productions that can cover an important segment of domestic demand for agri-food products and not only.

The main objectives of the National Rural Development Program of Romania in the period 2014 - 2020 entail:

- Improving the economic performance of all farms and facilitating the restructuration and modernization of farms, to increase market participation and market orientation, as well as to diversify agricultural activities.
- Facilitating the entry in the agricultural segment of farmers who are better qualified for this work, e.g. the new generation.
- Improving the competitiveness of primary producers through a better integration in the agri-food chain, due to quality planning, an increased added value of agricultural products, as well as the promotion of producer groups and inter-professional organizations on the local markets and on short supply routes.

The successful implementation of the National Rural Development Program of Romania (NRDP) for the period 2014-2020 will not only depend on the existence of financial support or the implementation of well-structured measures, but also on the dissemination of information, success models as well as innovative methods and best practices for projects that promote the development of rural businesses and communities in a coherent approach. The National Rural Development Program of Romania (NRDP) for the period 2014 – 2020 continues its efforts for rural development, having the following strategic objectives: to restructure and increase the viability of agricultural holdings; to develop a sustainable management of natural resources and to fight climate change; to diversify economic activities, to create jobs and to improve the infrastructure, services and the quality of life in rural areas.

These objectives are in line with the National Strategy for Medium and Long-Term Development of the Agri-Food Segment 2020-2030, aiming at: accelerating the structural transition towards an economically viable agriculture with environmentally friendly agricultural practices; increasing the coverage of food consumption from domestic production to regaining the status of net agro-food exporter; limiting the carbon footprint of agriculture and promoting a climate-change resilient agriculture; improving living standards in rural areas; developing education/counseling partnerships, ICT and improving the performance of the agricultural administration.

The objectives of the NRDP are in line with the priorities of the European Union's Common Agricultural Policy (CAP) and the Europe 2020 Strategy, with a focus on increasing competitiveness and diversifying jobs in rural areas. Agriculture is considered an industry in developed EU countries, after it was heavily supported by public resources to reach a high level of performance and stability. However, there are concerns about the variation of agricultural production, due to climate change, the volatility of prices in agriculture – under the pressure of searching alternative energy resources (Von Braun, 2010) and speculative actions. Also between the European Union member states, the synchronization of the business cycles is a prerequisite for the effective implementation of common policies. Nevertheless, the significant differences between the new and old member states, also given by the specific features of agriculture in each country, reduce the level of synchronization. (Da-Rocha & Restuccia, 2006)

Tangible investments in Romanian agriculture have increased in recent years, despite the economic crisis. Thus, the largest producers of machines, seeds and seedlings, chemical fertilizers, etc. have invested millions of euros in agriculture. These investors count both on the growth and development of the local agriculture and on Romania's export potential.

Since the beginning of 2012, the Ministry of Agriculture and Rural Development (MADR) has spent 502 million euros from European funds and subsidies for farmers provided by Brussels. (Ministry of Agriculture, 2012)

National accounts are a coherent and consistent set of macroeconomic indicators that provide an overview of the economic situation. These are widely used in economic analyses and prognoses, as well as in policy-making. Eurostat publishes data on annual and quarterly national accounts, data on annual and quarterly accounts by segment, as well as data on the supply/use, entry/exit tables, each of them being presented with associated metadata.

If we consider the Gross Domestic Product as a standard of a nation's situation, there are many aspects, as far as the development of a society is concerned, that can be approached by potential policy makers, in any country, to establish a set of measures needed for the sustainable development and growth of the national economic potential.

The Gross Domestic Product is often considered the best statistical data for measuring the performance of an economy. The share of agriculture in Romania's Gross Domestic Product has declined steadily over the last 20 years. Thus, from 22.6 in 1993 to below 5% of GDP in 2015. Although the contribution of agriculture to GDP is decreasing, Romania still has the largest share of the agricultural sector in the GDP structure of all EU countries, three and a half times higher than the European average. In this context, the authors of this chapter chose GDP/inhabitant as the independent variable of the multiple linear regression model.

The main crop cultivated area is the land that was sown or planted in the field, in the reference agricultural year or in previous years (biennial, triennial or perennial crops), and occupying the land for the longest period of the year, according to the National Institute of Statistics in Romania (National Institute of Statistics, 2015).

Romania was ranked first among European Union member states in terms of the area cultivated with maize, but, as far as the output is concerned, the country ranks second, after France. This conclusion was reached after analyzing the tentative data on the output of main crops in 2014, published by the National Institute of Statistics. According to Eurostat, approximately 60% (8.2 million ha) of the total agricultural land is cultivated arable land; about 66% (5.4 million ha) of this area is cultivated with cereals, mainly wheat and maize; other uses of the agricultural area include livestock breeding, livestock pastures, hayfields, etc. (PwC, 2017).

The main obstacle to the technological endowment of the Romanian farmers is the reduced area occupied by agricultural holdings, which does not allow for the efficient use of modern technological means and does not justify the investment therein due to the small amounts of harvest obtained. Another problem is the lack of access to the necessary infrastructure, such as liquid fuel. In Romania, in 2016 the surface treated with herbicides had the largest share in the surface treated with plant protection products in solid form (69.1%), followed by the one treated with fungicides (15.9%) and the one with insecticides (14.1%) (National Institute of Statistics, 2016).

That's why in 2017 Romania's financial institutions specializing in lending to farmers (Agricover Credit, Payment and Intervention Agency for Agriculture) have financially supported the use of new products (particularly pesticides) by farmers. These new pesticides included: a fungicide based on strobilurin for cereals, a rape and sunflower fungicide, an herbicide for cereals, and a corn herbicide. The use of these pesticides has as main objectives the reduction of costs and the increase of agricultural output in Romania (Zimmermann, 2015).

Therefore, the authors of this chapter introduced in the multiple linear regression model the surface cultivated with pesticides as the dependent variable.

The integration of the Romanian economy into the European market has led to an increase in the national dependence on foreign goods and capital, on large European industrial and financial centers, on connections to production means that use foreign capital, while increasing the size of the internal and foreign material flows.

According to the information released by Ion Tita in the Cuget free newspaper in September 2017, the amount of payments made by the Agency for Rural Investment Financing (AFIR) in August 2017 to the beneficiaries of the investment projects financed under the NRDP 2020 amounts to EUR 147.97 million, i.e. the absorption target for that month (Cuget Free Newspaper, 2017).

Much of the investment in agriculture in Romania comes from European funds, but they are also insufficiently used due to the lack of transparency, information, the lack of writing capabilities and the guarantee of investment credits. According to the Annual Report issued by the Council of Taxation of Romania since 2016 tax evasion in the agricultural sector has the second highest share of total tax evasion in Romania, i.e. 9% or about 1.7 billion euros.

Last, but not least, the analysis of investment developments in the agricultural segment is another significant criterion, depending on which sustainable economic growth can be achieved and existing technologies can be improved, so that Romania will be in line with the requirements of the EU's Common Agricultural Policy and the Europe 2020 Strategy.

ECONOMETRIC ANALYSIS OF DATA SERIES

Trends in the evolution of the agricultural sector in Romania are complemented by an analysis of the connection and interdependence between the indicators that describe the agricultural sector in Romania: net investments in agriculture, forestry and fisheries, GDP/capita, surface cultivated with pesticides, the agricultural output in the period 2001 -2016.

This econometric analysis, presented in the second part of this chapter, was based on the study of the correlations that can be established following the application of the regression and parametric correlation methods. The authors used the database management and analysis softwares Excel and EViews for this analysis (Andrei & Bourbonnais, 2008).

In the elaboration of the multiple linear regression model (Anova) the authors started from the following hypothesis: a variable Y is dependent on a series of factorial independent variables X_1, X_2, ... X_i,... X_n.

The general model of multiple linear regression used in this chapter to analyze the dependency between the dependent variable (surface cultivated with pesticides), Y and k-1 independent variables x_2, x_3... x_k (net investments in agriculture, forestry and fisheries, the GDP/capita) and a random variable ε_i is described by the following general equation:

$$y_i = \beta_0 + \beta_1 x_{i1} + \beta_2 x_{i2} + ... + \beta_k x_{ik} + \varepsilon_i, where\, i = \overline{1, N} \tag{1}$$

For each statistical "i"-unit of the total collection, the dependent valid y is represented by a separate equation, resulting in a system of "N" linear equations:

$$\begin{aligned} y_1 &= \beta_0 + \beta_1 x_{11} + \beta_2 x_{12} + ... + \beta_k x_{1k} + \varepsilon_1 \\ y_2 &= \beta_0 + \beta_1 x_{21} + \beta_2 x_{22} + ... + \beta_k x_{2k} + \varepsilon_2 \\ &\ldots\ldots\ldots\ldots \\ y_N &= \beta_0 + \beta_1 x_{N1} + \beta_2 x_{N2} + ... + \beta_k x_{Nk} + \varepsilon_N \end{aligned} \tag{2}$$

where:

Y = (y_1... y_N) – the column vector of the endogenous variable having the size (*N*, 1);
β= (β_1... β_K) - is a column vector of parameters (β_j) of the regression model having the size (k+1.1)
ε= (ε_1... ε_N) – is a column vector of the random variable, having the size (*N*, 1)

The disadvantage of the regression method is that it does not consider the connections between the variables that are considered independent. Multiple linear regression model parameters at the level of the general collectivity were determined within the chapter as estimators based on sample data according to the following general mathematical relation:

$$y_i = b_0 + b_1 x_{i1} + b_2 x_{i2} + ... + b_k x_{ik} + \varepsilon_i, wherei = \overline{1, n} \tag{3}$$

Also for the econometric analysis of the above-mentioned indicators and for the estimation of the parameters of the econometric model within the chapter the OLS method was used. By applying statistical tests known to verify the significance of model parameters, model significance and OLS assumptions, the results obtained by the authors were statistically significant for a significance level of 5% for all the independent variables included in the model.

From the correlation chart, generated using the EViews software package, it resulted that the points in the chart network (see Figure 2) are evenly distributed, so we can conclude that the connection between the independent variables that describe the Romanian agricultural segment and the cultivated area where pesticides were used in Romania, as a dependent variable, is linear and direct.

At the same time, in the correlation graph, we can notice that the cultivated area where pesticides were used in Romania, in the period 2001-2016, was also influenced by factors other than those included in the linear multifactorial regression model. The influence of these factors, which are considered as random, can be eliminated by adjustments, i.e. by establishing the theoretical regression line. It should be noted that the variables included in the linear multiple regression model were deflated by the average annual euro rate. Thus, the influence of inflation was removed. In this context, the model was not considered as stationary by the authors of the chapter, the authors considering that the deflated data led to the obtaining a valid and statistically correct linear multiple regression model (Cristache, 2003).

This fact was subsequently confirmed by the results obtained.

Issues, Controversies, Problems

Today, more than 50 years after the launch of the European Union's Common Agricultural Policy, the EU is faced with new challenges that can be classified as follows: ensuring a sustainable food production that will contribute to feeding the world's population, as the number of people is expected to grow significantly in the future; fighting climate change and developing a sustainable management of natural resources; supporting rural areas across the EU and maintaining the vitality of the rural economy.

Figure 2. Actual, Fitted, Residual Graph
Source: Own calculations using Views (see Appendix 1)

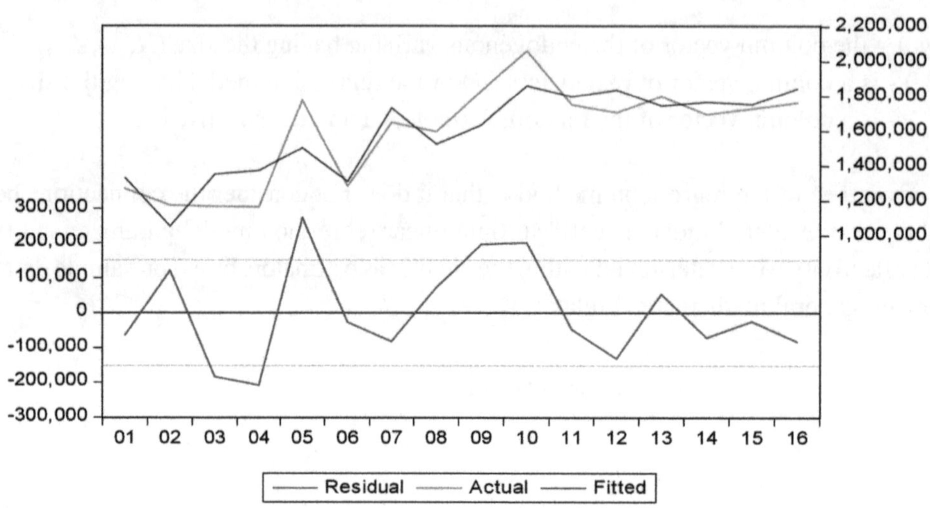

As far as the negative side of European Union's Common Agricultural Policy is concerned, one should note that the budget of the Common Agricultural Policy has declined drastically over the last 30 years, if we refer to its share in the EU budget - from almost 75% to less than 40%. Furthermore, during this period, 18 other countries have become members of the European Union (the number of farmers has increased by at least 50%) (European Commission, 2010).

A survey of the Presidential Commission for Public Development and Agricultural Policies shows that, in terms of capital endowment, Romanian agriculture is at a similar stage to that of the Inner Six countries in theyears1965–1970.

The econometric analysis introduced in the second part of this chapter was based on the macroeconomic approach, that intensive agriculture satisfies to a significant extent the current global needs but with negative effects on the environment. Therefore, the variables included in the regression model were macroeconomic data, which mainly targeted conventional agriculture or supporting activities.

Although Romania has a large agricultural area and a reasonable utilization rate, production yield is the main limiting factor in the agricultural sector. Romania is ranked 6th in the EU in terms of available agricultural area and the land usage is high compared to the EU average, which indicates that the main limiting factor is the agricultural yield. As a paradox, although the share of agriculture in GDP is decreasing, the added gross value to GDP from agriculture in Romania is one of the highest in the EU. That is why the authors of this chapter considered the inclusion of GDP in the linear multiple regression models as an independent variable.

On average, 4.4% of the EU's total employed population work in agriculture, with higher percentages in the Central and Eastern European countries, while in the Western European countries the share of the population working in agriculture is below the European average. Approximately 84% of the Romanian agricultural workers are classified as non-salaried (employers, self-employed, unpaid family workers and members of cooperative associations); by contrast, the EU-average share of non-salaried workers in the agricultural sector is 72%, and in some countries even much lowers (Spain - 50%, France - 63.1%, Germany - 55.8%). The evolution of the Distribution of Gross Added Value per sectors of activity reveals a continuous decrease in the share of agriculture (9.52% of the Added Gross Value in 2005, 6.0% in 2012) in favor of the secondary (35.90% in 2005, 42.3% in 2012) and tertiary sectors of the economy (54.58% in 2005, 51.6% in 2012) (European Commission, 2010).

Another characteristic of the Romanian agricultural sector from the perspective of human resources is the fact that there is no clearly defined professional status of the farmer. Agriculture is also strongly affected by the "black market" phenomenon, but this is harder to quantify, due to subsistence agriculture and unpaid family work (PwC, 2017).

Pesticides are therefore one of the factors that make it possible to produce an adequate amount of food from both vegetable and animal origin, for the growing population over the limitation of the arable area. In line with the above, the authors of this chapter considered it necessary to use the surface cultivated with pesticides as the dependent variable within the multifactorial linear regression model. *The main objective of the chapter* was to prove that for the Romanian agriculture, the process of transforming important crop areas from conventional farming to organic farming can be a favorable decision to increase the income of rural farmers. They encounter several restrictions due to the practice of subsistence agriculture as well as due to the lack of capital investments and territorial fragmentation still existing in Romania. The potential of Romanian agriculture to produce agricultural products under organic farming conditions accounts for at least 15% of the agricultural area of the country, according to specialists' estimates.

Despite the growing evidence of the serious problems posed by chemical pesticides, substantial policy change measures have not yet been taken to reduce their adverse environmental effects. This situation can be considered a failure on a European level. Thus, the transition from an agricultural system based on the intensive use of pesticides to an organic farming model requires significant political and financial support (Zimmermann, 2015).

Currently, most farmers are involved in a system that promotes the process of industrialization and specialization of agricultural holdings, often without considering the serious economic and environmental impact thereof. In conclusion, the transition from an agricultural system based on the intensive use of pesticides to an organic farming model requires significant political and financial support. Most farmers will only be able to adopt organic farming practices by systematically implementing effective support mechanisms. At the same time, the ecological and economic problems currently faced by agriculture in Romania can be solved only by reducing the use of pesticides and, ultimately, by converting the current agricultural systems to a system based on organic farming practices.

1. Econometric Analysis of Data Series

a. Parameter Estimation of the Model and Interpretation of Parameters

In this chapter we personalize the linear multiple regression model, by taking into account the two independent variables included in the model (GDP / capita, and net investments in agriculture) and the dependent variable (surface cultivated with pesticides), the following system of equations is obtained:

$$\begin{aligned} &\sum_{i=1}^{n} y_i = n\widehat{b_0} + \widehat{b_1}\sum_{i=1}^{n} x_1 + \widehat{b_2}\sum_{i=1}^{n} x_2 \\ &\sum_{i=1}^{n} x_1 y_i = \widehat{b_0}\sum_{i=1}^{n} x_i + \widehat{b_1}\sum_{i=1}^{n} x_1^2 + \widehat{b_2}\sum_{i=1}^{n} x_1 * x_2 \\ &\sum_{i=1}^{n} x_2 y_i = \widehat{b_0}\sum_{i=1}^{n} x_2 + \widehat{b_1}\sum_{i=1}^{n} x_1 * x_2 + \widehat{b_2}\sum_{i=1}^{n} x_2^2 \end{aligned} \tag{4}$$

To estimate the parameters of the linear multifactorial regression model equation, we will use *the least squares method*, to reduce the noticed errors.

$$\min \sum_{i=1}^{n} e_i^2 = \min \sum_{i=1}^{n} \left(yi - \left(\widehat{b_0} + \widehat{b_1}x1 + \widehat{b_2}x2\right)\right)^2 \tag{5}$$

Thus, by applying a multiple linear regression model and by using the EViews software package, we obtained the following results on the theoretical linear adjustment equation, summarized in Table 1.

The parameter $\widehat{b_0}$ is the ordinate estimator at origin. This is the value of the surface cultivated with pesticides (dependent variable) when the GDP / capita (independent variable) and the net investments in agriculture are null. Such a situation is impossible in the real economy and the surface cultivated with pesticides were used is also influenced by other factors – not only the GDP per capita and net investments in agriculture, in Romania, between 2001 and 2016.

Table 1. Correlation between GDP/capita, net investments in agriculture as independent variables and the surface cultivated with pesticides, as a dependent variable in Romania, during 2001-2016

Dependent Variable: SURFACE_PESTICIDES				
Method: Least Squares Date: 09/27/17 Time: 1:31 p.m. Sample: 2001 2016 Included observations: 16				
SURFACE_PESTICIDES = C (1) + C (2) * GDP_CAPITA + C (3) *				
NET_INVESTMENTS_AGR				
	Coefficient	**Std. Error**	**t-Statistic**	**Prob.**
C (1)	136686.	161492.8	8.463905	0.0000
C (2)	132.0809	21.20287	6.229389	0.0000
C (3)	-599.5218	224.6230	-2.669014	0.0193
R-squared	0.750241	Mean dependent var.		1588740.
Adjusted R-squared	0.711816	S.D. dependent var.		281458.2
S.E. of regression	151094.6	Akaike info criterion		26.85656
F-statistic	19.52504	Durbin-Watson stat		2.074237
Prob(F-statistic)	0.000121			

Source: Own calculations using EViews
→ SURFACE_PESTICIDES=1366860 + 132.08 * GDP_CAPITA-599.52 * NET_INVESTMENTS_AGR

But, if it were influenced only by the two independent variables included in the linear multifactor regression model, when the independent variables were null the value of the cultivated area where pesticides were used would be 1366860 hectares.

This can also be described as an average given that $t_{\hat{b}_0}$ =8.46 and the significance threshold *P-value* is 0.00<0.05, which means that this coefficient is significant.

Parameter $\widehat{b}_1$ is the estimator of the regression line slope. It shows that, given an increase of one unit of the GDP per capita (i.e. one million euros), the value of the surface cultivated with pesticides, in Romania, will grow on average by 132.08 hectares. Since $t_{\hat{b}_1} = 6.22$ and the significance threshold *P-value* is 0.00<0.05, it means that this coefficient is significant. Having a positive value, it also indicates that the connection between the two considered variables is direct.

Parameter $\widehat{b}_2$ is the estimator of the regression line slope. It shows that, given an increase of one unit of the net investments in agriculture (i.e. one million euros), the value of the surface cultivated with pesticides, in Romania, will decrease on average by approximately 600 hectares.

Having a negative value, it also indicates that the connection between the two considered variables is indirect. Since $t_{\hat{b}_2}$ = -2. 66 and the significance threshold *P-value* is 0,019 < 0.05 it means that this coefficient is significant. Thus, the obtained linear multifactorial regression function is:

$$\rightarrow \hat{y} = 1366860 + 132.08x_1 - 599.52x_2 \quad (6)$$

b. Parameter Estimation in the Confidence Interval of the Model

Parameter estimation by confidence intervals is made by using the Excel software package, which confirmed the results obtained by using the EViews software package, in Table 1 (see Table 2).

The confidence interval for parameter α is determined according to the relation:

$$\hat{b}_0 - t_{\frac{\alpha}{2},n-k-1} \cdot s_{\hat{b}_0} \leq \alpha \leq \hat{b}_0 + t_{\frac{\alpha}{2},n-k-1} \cdot s_{\hat{b}_0} \tag{7}$$

For applying this calculation formula, the following stages are to be completed:

Stage 1: Needs to determine the critical value of the Student test, namely:

$$t_{\left(\frac{\alpha}{2};n-k-1\right)} = t\left(0,025;13\right) = 2,65$$

Stage 2: Needs to determine the square average deviation of the estimator **α** based on the sample data. This stage is completed by using the Excel or EViews software packages (see Table 1 and Table 2)$\Rightarrow S_{\widehat{b1}} = 161{,}492.80$

Stage 3: Calculation formula 6 is applied, with the following results: α∈ (1,017,975.73; 1,715,743.70)

Stage 4: Conclusions can be drawn: given a 95% probability, there is enough statistical evidence to state that estimator $\widehat{b_0}$ originates from a population where α∈ (1,017,975.73; 1,715,743.70). (see Table 2)

The confidence interval for parameter β1 is determined according to the following relation:

$$\hat{b}_1 - t_{\frac{\alpha}{2},n-k-1} \cdot s_{\hat{b}_1} \leq \beta_1 \leq \hat{b}_1 + t_{\frac{\alpha}{2},n-k-1} \cdot s_{\hat{b}_1} \tag{8}$$

For applying this calculation formula, the following stages must be completed:

Stage 1: Needs to determine the critical value of the Student test, namely:

$$t_{\frac{\alpha}{2},n-k-1} = t\left(0,025;13\right) = 2.65$$

Stage 2: Needs to determine the square average deviation of the estimator **α** based on the sample data. This stage is completed by using the Excel or EViews software packages (see Table 1 and Table 2) $\Rightarrow S_{\widehat{b_1}} = 21.20$

Stage 3: Calculation formula 7 is applied, with the following results: β1 ∈ (86.27; 177.89)

Stage 4: Conclusions can be drawn: given a 95% probability, there is enough statistical evidence to state that estimator $\widehat{b_1}$ originates from a population where β1 ∈ (86.27; 177.89).

The confidence interval for parameter β2 is determined according to the following relation:

$$\hat{b}_2 - t_{\frac{\alpha}{2},n-k-1} \cdot s_{\hat{b}_2} \leq \beta_2 \leq \hat{b}_2 + t_{\frac{\alpha}{2},n-k-1} \cdot s_{\hat{b}_2} \tag{9}$$

For applying this calculation formula, the following stages are to be completed:

Stage 1: Needs to determine the critical value of the Student test, namely:

$$t_{\frac{\alpha}{2},n-k-1} = t\left(0,025;13\right) = 2.65$$

Stage 2: Needs to determine the square average deviation of the estimator **α** based on the sample data. This stage is completed by using the Excel or EViews software packages (see Table 1 and Table 2) $\Rightarrow S_{\widehat{b_2}} = 224.62$

Stage 3: Calculation formula 8 is applied, with the following results: β1∈ (-1084.79; -114.25)

Stage 4: Conclusions can be drawn: given a 95% probability, there is enough statistical evidence to state that estimator $\widehat{b_2}$ originates from a population where β2 ∈ (-1084.79; -114.25). An additional argument for the conclusion that parameters $\beta_{0,}$ β_1 and β_2 are statistically significant is that the confidence interval does not include the zero value. (see Table 2)

2. Testing the Significance of the Correlation and Parameters of the Multiple Linear Regression Model

In order to measure the concomitant influence of the independent variables (GDP / capita and net investments in agriculture) on the dependent variable (surface cultivated with pesticides) we shall use the multiple correlation ratio or the multiple correlation coefficient, in the case of linear connections. The multiple correlation ratio measures the connection and how intense it is, according to multifactorial correlation models.

a. Testing the Significance of Correlation

The intensity of correlation (Jaba, 2003) is analyzed, in this paper, by using a synthetic indicator called *correlation ratio*. It is calculated by formula 10:

Table 2. Parameter estimation by confidence interval

	Coefficients	Standard Error	t Stat	Lower 95%	Upper 95%
Intercept	$\widehat{b}_0 = 1366859{,}71$	$S_{\widehat{b}_0} = 161492{,}80$	$t_{calc} = \frac{\widehat{b}_0}{S_{\widehat{b}_0}}$ 8,4	$\hat{b}_0 - t_{\frac{\alpha}{2},n-k-1} * S_{\hat{b}_0}$ 1017975,73	$\hat{b}_0 + t_{\frac{\alpha}{2},n-k-1} * S_{\hat{b}_0}$ 1715743,70
GDP / capita (million euro)	$\widehat{b}_1 = 132{,}08$	$S_{\widehat{b}_1} = 21{,}20$	$t_{calc} = \frac{\widehat{b}_1}{S_{\widehat{b}_1}}$ 6,22	$\hat{b}_1 - t_{\frac{\alpha}{2},n-k-1} * S_{\hat{b}_1}$ 86,27	$\hat{b}_1 + t_{\frac{\alpha}{2},n-k-1} * S_{\hat{b}_1}$ 177,89
Net investment in agriculture, forestry and fishing (million Euro)	$\widehat{b}_2 = -599{,}52$	$S_{\widehat{b}_2} = 224{,}62$	$t_{calc} = \frac{\widehat{b}_2}{S_{\widehat{b}_2}}$ -2,66	$\hat{b}_2 - t_{\frac{\alpha}{2},n-k-1} * S_{\hat{b}_2}$ -1084,79	$\hat{b}_2 + t_{\frac{\alpha}{2},n-k-1} * S_{\hat{b}_2}$ -114,25

Source: own calculations

$$R_{y/x_1,x_2\ldots x_k} = \sqrt{\frac{\sum_{i=1}^{n}\left(\hat{Y}-\bar{y}\right)^2}{\sum_{i=1}^{n}\left(y_i-\bar{y}\right)^2}} = \sqrt{1 - \frac{\sum_{i=1}^{n}\left(y_i-\hat{Y}\right)^2}{\sum_{i=1}^{n}\left(y_i-\bar{y}\right)^2}} \tag{10}$$

The correlation ratio is a quantitative value that describes the relation between two or more variables. It ranges between (-1, 1), where extreme values assume a perfect relation between the variables, while value 0 represents a complete lack of linear relation. A more proper interpretation, of the values we obtained, is made by comparing the obtained result with certain pre-established values, listed in correlation tables, based on the number of subjects, the type of connection and the wanted significance threshold. This can be determined within the chapter by using the EViews software packages (see Table 1) and Excel (see Table 3) with the following results:

The connection between the variables included in the linear multifactorial correlation model in Romania, in 2001-2016, measured by the multiple correlation ratios, equals 0.87. The authors of this chapter consider that there is a direct and significant linear between the correlated variables. Independent variables included in the linear multiple correlation model (GDP /capita, net investments in agriculture) explain 75% of the variance in the surface cultivated with pesticides. The difference of 25% is influenced by other factors (agricultural output, net average earning in agriculture, etc.). In order to test the correlation ratio, we shall use the Fisher test.

Knowing that the significance threshold is $\alpha = 0,05$ and $k = 2$ (the model has two influencing factors - GDP / capita and net investments) we establish that:

- The Fisher-Snedeckor distribution table, with a significance level of 5% and depending on the number of freedom degrees $v_1 = k = 2$ and $v_2 = n - k - 1 = 13$: determines the critical value (in the table): $F_{\alpha;k;n-k-1} = F_{0,05;2;13} = 3.80$

Table 3. Testing the significance of correlation

Regression Statistics		Indicator Formulas
Multiple R	0.87	$R_{y,x_1,x_2} = \sqrt{\dfrac{\sum_{i=1}^{n}(\hat{y}_i - \bar{y})^2}{\sum_{i=1}^{n}(y_i - \bar{y})^2}} = \sqrt{1 - \dfrac{\sum_{i=1}^{n}(y_i - \hat{y}_i)^2}{\sum_{i=1}^{n}(y_i - \bar{y})^2}}$
R Square	0.75	$R^2_{y,x_1,x_2} = \dfrac{\Delta^2_{y/x}}{\Delta^2_y} = \dfrac{\sum_{i=1}^{n}(\hat{y}_i - \bar{y})^2}{\sum_{i=1}^{n}(y_i - \bar{y})^2}$
Adjusted R Square	0.71	$\bar{R}^2_{y,x_1,x_2} = 1 - \dfrac{\Delta^2_e / n-k-1}{\Delta^2_y / n-1}$
Standard Error	151094.58	$s_e = \sqrt{\dfrac{\Delta^2_e}{n-k-1}} = \sqrt{\dfrac{\sum_{i=1}^{n}(y_i - \hat{y}_i)^2}{n-k-1}}$
Observations	16.00	n

Source: own calculations

- If the calculated value of the Fisher test is greater than the critical value ($F_c > F_{\alpha;k;n-k-1}$) then the multiple correlation ratio of the collection from which the 16 unit sample was drawn is significantly different from zero, meaning that it is statistically significant.

The determination of test statistics (F_c) is based on relation 11, with the following result:

$$F_c = \frac{R^2}{1-R^2} \cdot \frac{n-k-1}{k} = \frac{0,75}{1-0,75} \cdot \frac{13}{2} = 19.5 \tag{11}$$

Since $F_c\ (19.5) > F_{0,05;\,2;\,13}\ (3.80)$, we can estimate that the multiple correlation ratio of the collection from which the 16-unit sample was drawn is significantly different from zero, meaning that it is statistically significant. We also consider that the independent variables included in the linear multifactorial regression model have a significant influence on the variation in the dependent variable (surface cultivated with pesticides), thus contributing to its average annual growth.

b. Testing the Significance of the Multiple Regression Model Parameters

The estimation of the parameter significance is made by using the Excel software package, which confirmed the previous results in Table 2, when the EViews program package was used (see Table 1).

- **Testing the Significance of Parameter α:** Given that $t_{\left(\frac{\alpha}{2};n-k-1\right)}$, and that the critical value of the Student test is lower than its calculated value ($t_{calc} > t_{\left(\frac{\alpha}{2};n-k-1\right)} \Rightarrow 8.46 > 2.65$) we conclude that it is likely that estimator $\widehat{b_0}$ originates from a population where α≠0 with a probability of 95%, which means α is significantly different from zero, being statistically significant.
- **Testing the Significance of Parameter β1:** Given that $t_{\left(\frac{\alpha}{2};n-k-1\right)}$ and that the critical value of the Student test is lower than its calculated value ($t_{calc} > t_{\left(\frac{\alpha}{2};n-k-1\right)} \Rightarrow 6.22 > 2.65$) we conclude that it is likely that estimator $\widehat{b_1}$ originates from a population where β1≠0 with a probability of 95%, which means β1 is significantly different from zero, being statistically significant.
- **Testing the Significance of Parameter β2:** Given that $t_{\left(\frac{\alpha}{2};n-k-1\right)}$, and that the critical value of the Student test is lower than its calculated value ($t_{calc} > t_{\left(\frac{\alpha}{2};n-k-1\right)} \Rightarrow -2.66 > -2.65$) we conclude that it is likely that estimator $\widehat{b_2}$ originates from a population where β2≠0, with a probability of 95%, which means β2 is significantly different from zero, being statistically significant.

3. Implementation of the ANOVA Analysis and Interpretation of Results

The linear multiple significance econometric model (the quality of adjustment) is checked by means of dispersion analysis method or variance (ANOVA) and of the Fisher-Snedecor test (*F*). This method entails the breakdown of the total variance (the sum of squares of total deviations - SST) into the residual, non-explicit variance (the sum of squares of the errors - SSE) and explicit change (the sum of squares of deviations resulted from regression - SSR).

$$\sum_{i=1}^{n}\left(y_i-\overline{y}\right)^2=\sum_{i=1}^{n}\left(y_i-\hat{y}_i\right)^2+\sum_{i=1}^{n}\left(\hat{y}_i-\overline{y}\right)^2 \Rightarrow SST=SSE+SSR \text{(See Appendix 2)} \quad (12)$$

where predictable component

$$:\hat{y}_i=\hat{b}_0+\hat{b}_1x_i \quad (13)$$

Dispersion analysis is based on the separation of the total dispersion of a data series into component dispersions that can be attributed to the various involved factors. The variance that can be attributed to a factor is separated into the estimated variance of the sample, this being a normal dispersion, resulting from the effect of sampling variations upon the sampled data (Andrei & Bourbonnais, 2008).

By using the probability distributions F, it can be determined whether the calculated value exceeds the corresponding value at the significance level $\alpha = 0.05$ of F, for each size of freedom degrees (Malhotra, & Peterson, 2006).

Also, to test the validity of linear multifactorial regression model on the agricultural segment in Romania during 2001-2016, we have two hypotheses:

Null Hypothesis (H_0): The model is not statistically valid (the distribution of theoretical values due to the exogenous variable *does not differ significantly* from the distribution of values caused by random factors)

$$S^2_{y/x} = S^2_e \tag{14}$$

Alternative Hypothesis (H_1): The model is statistically valid (the distribution of theoretical values due to the exogenous variable *does differ significantly* from the distribution of values caused by random factors)

$$S^2_{y/x} \neq S^2_e \tag{15}$$

If the calculated value of the Fisher test is higher than the critical value (tabular), then the authors of this chapter can assert that the model is valid, statistically identified correctly, and the independent variables significantly affect the variance of the dependent variable, i.e. the pesticides were used.

By using the Excel software package, the following results were obtained (see Table 4).

In conclusion, the Fisher-Snedecor test indicates that the obtained results are significant for a significance threshold of 5%. Since $F_{calc} = 19.50 > F_{0,05,2,13} = 3.80$ and *Significance F (the significance threshold)* is 0.0001 (the value being less than 0.05), authors this chapter can say that the linear multifactor regression model is valid and correctly identified from the statistical perspective, depicting a real connection between the correlated variables. The model can also be used to analyze the dependency between the variables included in the model.

4. Testing Classical Statistical Hypothesis on the Multifactorial Regression Model

a. Testing the Linearity of the Proposed Model

In order to test the linearity of the multiple linear regression model, we shall determine the coefficient of correlation and coefficient of determination. The *coefficient of correlation* ($r_{y/x1, x2, x3}$) is a quantitative value that describes the relation between two or more variables. It varies between (-1; 1), where extreme values assume a perfect relation between variables (Kenkel, 1995). A more proper interpretation, of the values we obtained, is made by comparing the obtained result with certain pre-established values, listed in correlation tables, based on the number of subjects, the type of connection and the wanted significance

Table 4. Validity of the multiple linear regression model using the Fisher criterion

ANOVA	df	SS	MS	F_{calc}	Significance F
Regression	df_1=k=2	$SSR = \sum_{i=1}^{n}\left(\hat{y}_i - \overline{y}\right)^2$ 891496607124,96	$s_{y/x}^2 = \frac{SSR}{k}$ 445748303562,48	$F_{calc} = s_{y/x}^2 / s_e^2$ =19.53	0,0001 $< \alpha = 0,05$ (the H_0 hypothesis is rejected; the model is valid)
Residual	df_2=n-k-1=16-2-1=13	$SSR = \sum_{i=1}^{n}\left(y_i - \hat{y}_i\right)^2$ 296784423558,48	$s_e^2 = \frac{SSE}{n-k-1}$ 22829571042,96		
Total	df=df_1+df_2=15	$SST = \sum_{i=1}^{n}\left(y_i - \overline{y}\right)^2$ 1188281030683,44			

Source: Own calculation

where:

1. The number of freedom degrees (df).

df_1 = the number of freedom degrees due to factorial variance, k being the number of independent variables that were considered (for linear multiple regression with two independent variables included in the model, the value of k will be 2)

df_2 = the number of freedom degrees due to the residual variance

df = the number of freedom degrees due to the total variance

n = number of observations included in the linear multifactorial regression model = 16 (see Appendix 1)

2. $SSR = \Delta_{y/x}^2 = \sum\left(\hat{y}_i - \overline{y}\right)^2$ = (explained) factorial variance, the sum of the deviations parameters caused by regression.

$SSE = \Delta_e^2 = \sum\left(y_i - \hat{y}_i\right)^2$ = (unexplained) residual variance, the sum of squares of errors

$SST = \Delta_y^2 = \sum\left(y_i - \overline{y}\right)^2$ = total variance, the sum of squares of total deviations

$S_{y/x}^2$ = estimator of the corrected factorial dispersion

S_e^2 = estimator of the corrected residual dispersion

threshold. The value of the coefficient of correlation depends on the shape of the regression line, for which reason this indicator is significant for linear type correlations and less significant for non-linear type correlations. In this case, authors have used the correlation ratio ($R_{y/x1, x2, x3}$).

If $r_{y/x1, x2, x3} = R_{y/x1, x2, x3}$ the authors consider that the correlation linkage described by the multiple linear correlation model is linear, using either the coefficient or the correlation ratio to measure the intensity of the correlation bond.

From the graphic representation (see Figures 3a and 3b) of the connection between the GDP per capita and the surface cultivated with pesticides were used in agriculture, in Romania, as well as of the connection between the net investments in agriculture and the surface cultivated with pesticides, the authors of this chapter can assume that there is a direct connection between the variables included in the multifactorial regression model, because the points placed in graphs 3a and 3b are ordered on the first bisector. Thus, the placed points are distributed relatively uniformly along the regression line, especially in figure 3 b. In conclusion the diagrams of dispersal 3.a and 3.b indicate a linear and direct link between the correlated variables. Also, one can notice in Table 3 that the coefficient of correlation (r) and the coefficient of determination (R^2) are quite close in value ($r_{y/x1,x2,x3} = 0.87$ and $R^2_{y,x1,x2,x3} = 0.76$) which means that the connection between the variables is linear and of average intensity.

Figure 3. The dependency between the GDP / capita, net investment, surface cultivated with pesticides from Romania, during 2001-2016
Source: The National Institute of Statistics – The Statistical Yearbook of Romania, Tempo/line 2012-2016

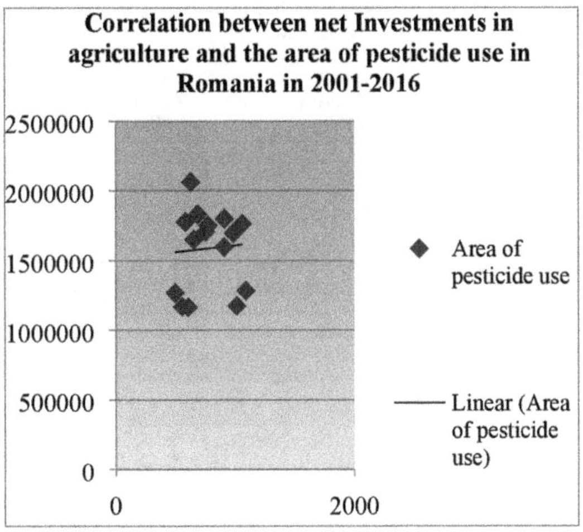

(a)

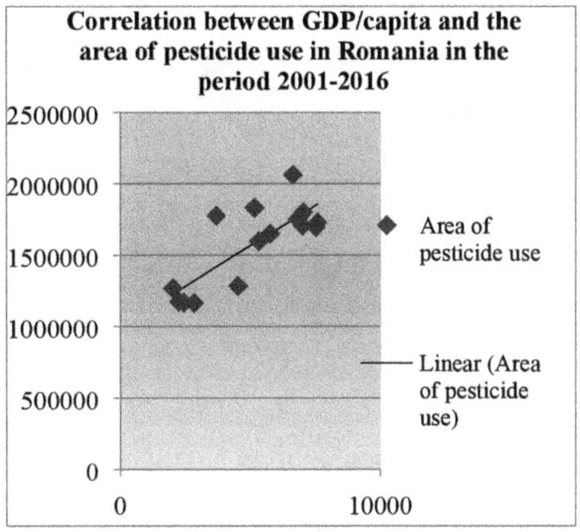

(b)

b. Testing the Error Normality

In order to check the error normality hypothesis, one can use the Jarque-Berra test, which is an asymptotic test (valid for a large volume sample), having a hi-square distribution.

If the probability *p* (*JB*) corresponding to the calculated value of the test is sufficiently low, then the error normality hypothesis is rejected, whereas otherwise, for a sufficiently high probability level, the hypothesis of error normality is accepted, or if $JB > \chi^2_{\alpha;k}$, then the hypothesis of error normality is rejected. By using the EViews software package to calculate the Jarque-Berra test (see Figure 4) the following results were obtained:

By using the EViews software package to calculate the Jarque-Berra test (see Figure 4) it results that $JB = 0.97 < \chi^2_{0,05;3} = 7.81$ and that *p* (JB) = 0.61. Since the calculated value of the J-B test is less than the table value of $\chi^2_{\alpha;3}$ and the probability that the J-B test will not exceed the table value is sufficiently high, the hypothesis of error normality can be accepted.

c. Testing the Homoscedasticity Hypothesis

The error homoscedasticity hypothesis, for this multiple linear regression model, will be checked by using the White test. The use of the White test entails the following stages:

- Parameter estimation of the initial model and calculation of the estimated values of the residual variable e_i,

Figure 4. Distribution of error normality
Source: own calculations using EViews

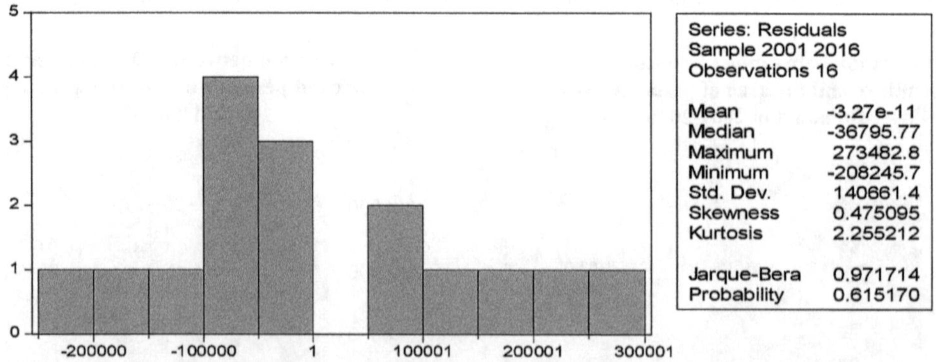

- Construction of an auxiliary regression, based on the supposition that there is a relation of dependency between the square of the error values, the exogenous variable included in the initial model and the square of its values:

$$\hat{e}_i^2 = \alpha_0 + \alpha_1 x_i + \alpha_2 x_i^2 + \omega_i \quad (16)$$

And calculating the coefficient of determination, R^2, corresponding to this auxiliary regression;

- Checking the significance of the newly constructed model parameters and, if one of them is no significant, the hypothesis of error heteroscedasticity is accepted. There are two versions in which the White test can be used:

    - The use of the Fisher-Snedecor test, based on the parameters nullity hypothesis, namely:

The null hypothesis (H_0): $\alpha_0 = \alpha_1 = \alpha_2 = 0$ and the alternative hypothesis. (H_1): $\alpha_k \neq 0$ (17)

If the null hypothesis, according to which the estimation results are no significant ($F_c < F_{\alpha;v_1;v_2}$) is accepted, then the hypothesis of homoscedasticity checks out and, in the opposite case, it signifies the presence of error heteroscedasticity.

- b. The use of LM test, calculated as the product between the number of observations corresponding to the model (n) and the coefficient of determination (R^2) that corresponds to this auxiliary regression. In general, the LM test is asymptotic, distributed as .

By using the EViews software package to determine the White test (see Table 5) the following results were obtained.

Table 5. Testing the homoscedasticity hypothesis with the White test

Heteroskedasticity Test: White		
F-statistic	1.299585	Prob. F (5,10)
Obs * R-squared	6.301811	Prob. Chi-square (5)
Scaled SS Explained	2.610953	Prob. Chi-square (5)

Source: Own calculations using EViews

After analyzing the results by using the database management software analysis (Eviews), the authors of this chapter observed that $F_{calc} = 1.29 > F_{0.05,2,13} = 3.80$, so that the homoscedaticity hypothesis checks out. The use of LM test, calculated as the product between the number of observations corresponding to the model, *n*, and the coefficient of determination, $R^2_{y, x1, x2}$:

$$LM = n \cdot R^2 \sim \chi^2_{\alpha,k} \tag{18}$$

→ LM = 6.3, so LM < $\chi^2_{0.05, 3}$ (12 < 7.81) meaning that errors are homoscedastic (error variation is constant). In conclusion, the hypothesis that the error dispersion is constant, because the obtained coefficients are not affected by the error, checks out.

d. Testing the Error Autocorrelation Hypothesis

To detect autocorrelation of errors, empirical methods such as the Breusch-Godfrey test were used in this chapter. This test will analyze the existence of an autocorrelation of order k, k ≠1. It is assumed that the regression model errors are given by the equation:

$$\varepsilon_t = \rho_1\varepsilon_{t-1} + \rho_2\varepsilon_{t-2} + \ldots + \rho_k\varepsilon_{t-k} + v_t\text{, for t = k,..., n, and } v_t \sim N(0, \sigma^2_v) \tag{19}$$

To statistically evaluate the presence of an autocorrelation of the k order, the following statistical assumptions will be used:

$$H_0 : \rho_1 = \rho_2 = \ldots = \rho_k = 0\text{; errors are not autocorrelated} \tag{20}$$

$$H_1 : \rho_1 \neq 0 \text{ or } \rho_2 \neq 0 \text{ or } \ldots \rho_s \neq 0\text{, autocorrelated errors} \tag{21}$$

The authors can conclusions, by using statistical software (EViews), that statistical probability of F is 0.33 (small) the model has an order 2 autocorrelation. (See Table 6)

The authors of this chapter can check the error independence hypothesis, in this model, by using the Durbin-Watson test, which calculates the empirical test, which calculates the empirical term and then compares this quantity “d” with two theoretical values d_1 and d_2, taken from the Durbin-Watson table,

Table 6. Testing the error autocorrelation hypothesis (Breusch-Godfrey Serial Correlation LM Test)

Calculated test values			
F-statistic	1.224885	Prob. F (2,11)	.3309
Obs. * R-squared	2.914273	Prob. Chi-Square (2)	.2329

in connection with the randomly selected threshold of significance α, with the number of exogenous variables (k) and with the observed values $(n, n \geq 15)$. The acceptance or rejection of the error independence hypothesis is based on a certain rule, which is:

- If $0 < d < d_1 \Rightarrow$ positive autocorrelation;
- If $d_1 \leq d \leq d_2 \Rightarrow$ indecisive, and the acceptance of positive autocorrelation;
- If $d_2 < d < 4 - d_2 \Rightarrow$ errors are independent;
- If $4 - d_2 \leq d \leq 4 - d_1 \Rightarrow$ indecisive, and the acceptance of negative autocorrelation
- If $4 - d_1 < d < 4 \Rightarrow$ negative autocorrelation.

Statistically, DW (see Table1) has a calculated value DW = 2.07, which we compared to the critical statistical values for α = 0.05, p = 2 and n = 16; d1 = 1.10 and d2 = 1.37, and which suggests accepting the hypothesis of independence of the residual variable values.

In this context, bilateral correlations have been established to further highlight the causal relationship between each variable as a factor of influence and the outcomes targeted. The results from the correlation matrix allowed them to be included in the linking process only when the coefficients have values greater than 0.4 (See Table 7).

In the end, the authors of this chapter conclude that the future agricultural policy of the European Union is intended to maintain the main elements of the concept of rural development, based on multi-annual plans, on co-funding by member states or their regions, as well as on food security issues and the continued implementation of the concept of sustainability.

Table 7. Correlation Matrix for model

Indicators	Surface Cultivated With Pesticides	Net Investment in Agriculture, Forestry and Fisheries (Million Euro)	GDP per Capita (Million Euro)
Surface cultivated with pesticides	1		
Net investment in agriculture, forestry and fisheries (million Euro)	0,42	1	
GDP per capita (million euro)	0,80	0,62	1

Source: Own calculation using Excel

FUTURE RESEARCH DIRECTIONS

In the future, this chapter should be expanded with other econometric models to quantify the links between the macroeconomic indicators that characterize the conventional and organic agricultural sectors. These comparative models can then be compared based on the Akaike test, so that the authors may choose the most suitable for the characterization of the agricultural sector. The comparative criterion of econometric models – Akaike (AIC) – is a major estimator of the relative quality of models for a given set of data. Given the use of data models, AIC estimates the quality of each model in relation to all other models. Thus, AIC provides a means of selecting and classifying the model. Concerning the concepts and main aspects to be considered in the future, the development of the Romanian agricultural sector can be classified according to the following criteria: promoting sustainable rural development, optimizing the size of agricultural holdings, increasing farmer training, implementing tax cuts. Therefore, in the future, public authorities need to use and implement a wide range of information channels to attract young people into the agricultural sector, such as web solutions and online socializing solutions. In conclusion, the problems faced by the Romanian agriculture limit the development of the agricultural sector in terms of the requirements of a sustainable / ecological economy. At the same time, a regression panel model at European level would be needed to deepen the analysis of the agricultural sector. This analysis would complement and develop the results obtained in Romania regarding the econometric characterization of the intensive agricultural sector that is still practiced on large surfaces.

CONCLUSION

Although focused on satisfying domestic demand, the Romanian agriculture benefits from an important and diverse natural potential. Nevertheless, the agricultural sector is still at the beginning of a long and difficult process of modernization and restructuring aimed at capitalizing the region's important agricultural potential more efficiently. Europe is well renowned for its wide range of agri-food products. That is why the EU has developed a whole range of instruments, to promote the quality and safety of food and agricultural products. These tools address marketing standards, product quality identification systems, certification systems and hygiene standards. If the number of manufacturers, the main beneficiaries of these policies, and the areas converted to arable crops will continue to decline in the future then the purpose of agricultural support will be questioned. Although the introduction of the EU's Common Agricultural Policy is not yet directly perceived by Romanian farmers the expectations regarding the stabilization of policies and the financing of investments and other transitional measures have created a positive attitude of farmers and the public towards the Agricultural Policy of the EU. It is difficult for Romania to ensure that current mass production practices in agriculture are aligned to the standards of a green economy, as this requires additional costs or risks, especially due to the low use of modern technologies and techniques. (Aceleanu, 2016).

The two indicators (agricultural employment and added gross value) show that the Romanian agriculture still has a low efficiency. This is explained by the usage of outdated technologies, fragmentation of land and underdeveloped workers (compared to those in other industries) with changes in the labor market can be partially explained by the migration of these workers to other sectors. Organic farming is, in these cases, a successful solution.

REFERENCES

Aceleanu, M. I. (2016). Sustainability and Competitiveness of Romanian Farms through Organic Agriculture. *Sustainability Journal*, *8*(245), 4–6.

Andrei, T., & Bourbonnais, R. (2008). *Econometrics*. Bucharest: Economica Publishing House.

Andrei, T., Stancu, S., Iacob, A. I., & Tusa, E. (2008). *Introduction to econometrics using Eviews*. Bucharest: Economica Publishing House.

Cobb, D., Feber, R., Hopkins, A., Stockdale, L., O'Riordan, T., Clements, B., ... Macdonald, D. (1999). Integrating theenvironmental and economic consequences of converting to organic agri-culture: Evidence from a case study. *Land Use Policy*, *16*(4), 207–221. doi:10.1016/S0264-8377(99)00023-X

Cristache, S. E. (2003). *Statistical Methods of Calculating and Analyzing Economic Efficiency in Commerce* (Doctoral Thesis). ASE Bucharest Publishing House.

Da-Rocha, J. M., & Restuccia, D. (2006). The Role of Agriculture in Aggregate Business Cycle. *Review of Economic Dynamics*, *9*(3), 455–482. doi:10.1016/j.red.2005.12.002

Douglas, K. (2010). *Romania and the Common Agricultural Policy*. Denver University. Retrieved August 23, 2017, from http://sar.org.ro/wp-content/uploads/2013/01/Romania-and-the-Common-Agricultural-Policy.pdf

Dyson, T. (1996). *Population and Food: Global Trends and Future Prospects*. London, UK: Rout-ledge.

European Commission. (2010). *Europe 2020 - a European Strategy for Smart Sustainable and Inclusive Growth*. Retrieved August 23, 2017, from http://ec.europa.eu/eu2020/pdf

European Commission. (2013). *A European Strategy for Smart, Green and Inclusive Growth* (Communication from the European Commission, 2020). Retrieved August 23, 2017 http://eurlex.europa.eu/LexUriServ/LexUriServ.do?uri=

European Commission. (2016). *Let us understand the Policies of the European Union: Agriculture* (the Directorate General for Citizen Communication and Information). Retrieved September 10, 2017, from https://europa.eu/european-union/topics_en

European Commission. (2017). *Agriculture. A Partnership between Europe and Farmers (CAP)*. Retrieved August 23, 2017, from http://europa.eu/pol/pdf/flipbook/en/agriculture_en.pdf

EViews. (2010). QMS Quantitative Micro Software (Version 7.0) [software]. Available from http://www.eviews.com/EViews7/EViews7Student/evstud7.html

Girardin, P., Bockstaller, C., Werf, H.V.D. (1999). Indicators: tools to evaluate theenvironmental impacts of farming systems. *Journal Sustain. Agric.*, *13*(4), 5–21. 0310.1300/J064v13n04

Halberg, N. (2012). Assessment of the environmental sustainability of organic farm-ing: Definitions, indicators and the major challenges. *Canadian Journal of Plant Science*, *92*(6), 981–996. RetrievedFebruary102017. doi:10.4141/cjps2012-035

Halberg, N., Verscuur, G., & Googlass, G. (2005). Farm level environmentalindicators; are they useful? An overview of green accounting sys-tems for European farms. *Agriculture, Ecosystems & Environment, 105*(1–2), 195–212. doi:10.1016/j.agee.2004.04.003

Ion, T. C. (in press). 36,000 investment projects in agriculture worth 2.16 billion euros. *Cuget liber newspaper.* Retrieved September 2017, from https: www.cugetliber.ro/stiri-economie, on-line.

Jaba, E. (2003). *Statistics* (3rd ed.). Bucharest: Economic Publishing House.

Kenkel, J. L. (1995). *Introductory Statistics for Management and Economics* (4th ed.). South-Western Publishing Group.

Kotler, Ph., Jain, D., & Maesincee, S. (2009). *Marketing Moves: A New Approach to Profits, Growth, and Renewal.* Bucharest: Meteor Press Publishing House.

Malhotra, K. N., & Peterson, M. (2006). *Basic Marketing Research.* Englewood Cliffs, NJ: Prentice Hall.

Ministry of Agriculture. (2012). *National Rural Development Plan 2014-2020.* Retrieved September 10, 2017, from http://www.madr.ro/pndr-2014-2020/instrument-financiar-de-creditare.html

Ministry of Agriculture and Rural Development. (2015). *Sector development strategy medium and long-term agri-food horizon 2020-2030.* Retrieved September 10, 2017, from http://www.madr.ro/docs/agricultura/strategiaagroalimentara-2020-2030.pdf

Morgan, K., & Murdoch, J. (2000). Organic vs. conventional agriculture: Knowl-edge, power and innovation in the food chain. *Geoforum Elsevier, 31*(2), 159–173. doi:10.1016/S0016-7185(99)00029-9

National Bank of Romania. (n.d.). Retrieved August 23, 2017, from http://www.bnr.ro/Cursul-de-schimb-3544.aspx

National Institute of Statistics. (2015). *Tempo/line.* Retrieved August 23, 2017 from http://statistici.insse.ro

National Institute of Statistics. (2017). *Tempo/line.* Retrieved August 23, 2017 from http://statistici.insse.ro

PricewaterhouseCoopers (PwC). (2017*). Potential of agricultural sector development in Romania.* PricewaterhouseCoopers Management Consultants. Retrieved September 10, 2017, from https://www.juridice.ro/wp-content/uploads/2017/03/Raport_PwC-agricultura.pdf

Rigby, D., Woodhouse, P., Young, T., & Burton, M. (2001). Constructing a farm levelindicator of sustainable agricultural practice. *Ecologic. Econ. Elsevier, 39*(3), 463–478. doi:10.1016/S0921-8009(01)00245-2

Van der Werf, H.M., Petit, J.(2002). Evaluation of the environmental impact of agriculture at the farm level: a comparison and analysis of 12 indicator-basedmethods. *Agric. Ecosyst. Environ., 93*(1), 131–145. (01)00354-1.10.1016/S0167-8809

Von Braun, J. (2010). *Globalization of Food and Agriculture and the Poor.* Oxford University Press.

Zimmermann, D. (2015). *Europe's pesticide addiction.* Greenpeace report. Retrieved February 10, 2017, from https://www.greenpeace.org/archive-international/en/press/releases/2015/Europes-pesticide-addiction-threatens-our-environment---Greenpeace-report/

KEY TERMS AND DEFINITIONS

Confidence Level: Refers to the possibility of a parameter that lies within a specified range of values, which is denoted as c. Moreover, the confidence level is connected with the level of significance.

Correlation: Expresses the power of the link between two variables (the intensity of the relationship).

Econometric Analysis: The statistical and mathematical analysis of economic relationships, often serving as a basis for economic forecasting.

Error Autocorrelation: Presence of a correlation between values residual variable.

Homoscedasticity: Describes a situation in which the error term (that is, the "noise" or random disturbance in the relationship between the independent variables and the dependent variable) is the same across all values of the independent variables.

Least Squares Method: Is a statistical method used to determine a line of best fit by minimizing the sum of squares created by a mathematical function. A "square" is determined by squaring the distance between a data point and the regression line.

Null Hypothesis: Is a general statement or default position that there is no relationship between to measured phenomena, or no association among groups.

Regression: Is the mathematical expression that allows the estimation of a variable depending on at least one other variable.

APPENDIX 1

Table 8.

Year	Net Investments in Agriculture, Forestry and Fishing (Million)	Surface Cultivated With Pesticides	GDP / Capita (Million)
2001	498.62	1268694	2002
2002	1013.57	1174950	2232
2003	557.5	1168538	2419
2004	608.93	1165487	2816
2005	585.72	1776305	3688
2006	1090.54	1284362	4530
2007	656.88	1649500	5743
2008	908.44	1597382	5318
2009	689	1830777	5153
2010	631.8	2061336	6602
2011	775.17	1750803	6778
2012	754.58	1706104	6980
2013	910.39	1799582	7000
2014	985.25	1695816	7500
2015	1022.64	1728672	7565
2016	1060.1	1761527	8421

Dața Source: National Institute Of Statistics From Romania, Tempo_Line

APPENDIX 2

Using the Excel software was determined in the second part of the chapter, the values adjusted on the basis of the regression equation of the linear multifactor regression model and the residual component or error values. These are presented in Table 9.

Table 9.

RESIDUAL OUTPUT		
Observation n	***Predicted area – pesticides*** $\hat{y}_i$	***Residuals*** e_i
1	1332352,132	-63658,13212
2	1054006,977	120943,0232
3	1352130,028	-183592,0284
4	1373732,744	-208245,7439
5	1502822,203	273482,7966
6	1311383,724	-27021,72413
7	1731586,51	-82086,51029
8	1524636,41	72745,58984
9	1634402,129	196374,8707
10	1860080,024	201255,9763
11	1797372,82	-46569,82012
12	1836397,319	-130293,3194
13	1745627,442	53954,55847
14	1766787,695	-70971,69489
15	1752956,833	-24284,83318
16	1843560,009	-82033,00866

Data Source: Excel Software

Chapter 13
Organic Agriculture: Opportunities and Trends

Mirela Stoian
Bucharest University of Economic Studies, Romania

Diana Caprita
Bucharest University of Economic Studies, Romania

ABSTRACT

Promoting sustainability, including the production and consumption of food, is badly needed nowadays, given the fact that consumers are increasingly concerned about protecting their health, through a thorough verification of food quality. From this perspective, organic food may represent a viable solution for a healthier future. Currently, we are witnessing a substantial increase in the number of countries, organizations, and companies encouraging organic farming, an economic activity that involves environmentally friendly agricultural practices. The main objective of this chapter is to reveal the growing importance of organic farming to the food markets. This research will also focus on presenting a very detailed analysis of the defining elements of organic agriculture, such as the evolution of certified organic surfaces, both contributory and disfavoring factors of the developing organic agriculture, and last but not least, overall outlook for global consumption of certified organic products.

INTRODUCTION

21st century consumer is concerned about both food quality and environmental protection. Considering this new perspective, producers have adequately responded to the stimulated demand. New market opportunities have developed as part of a business strategy to address consumer concerns, particularly in the European Union and the United States.

Major food companies see the processing, handling, stocking and promoting of organic foods as part of a positive public image, with consistent benefits in the demand section. Thus, preponderant concern for healthy ingredients increase domestic sales and become a serious component of marketing campaigns. Retailers of all sizes are using this strategy, aggressively promoting organic foods and sustaining organic markets, with major food retailing chains now accounting for an important share of the retail markets, for both fresh and processed foods.

DOI: 10.4018/978-1-5225-5739-5.ch013

Copyright © 2019, IGI Global. Copying or distributing in print or electronic forms without written permission of IGI Global is prohibited.

Consumers are becoming more and more skeptical about the safety of conventional foods and the credibility of industrial agriculture is seriously questioned. The habit of using growth regulators (such as Alar in the United States) stimulated consumer's interest in organic food. The crisis over dioxin-contaminated food and livestock diseases (such as Bovine Spongiform Encephalopathy (BSE) and foot-and-mouth in Europe) further increased demand for organic food. Consumer surveys organized in almost every country show a market segment that demands an alternative to genetically modified foods. Governments have responded to these concerns by setting targets for the expansion of organic production. Although the progress made in this sector is consistent, there are still efficient methods which can be used in order to improve the current unsatisfactory results. Regarding the cultivated areas which are respecting the principles of organic agriculture, these are still insufficient to meet the growing demand for organic products. According to Research Institute of Organic Agriculture (FiBL) and IFOAM – Organics International, in 2015 the total area of ecological agricultural land amounted 50.9 million hectares, unevenly distributed, as follows: Oceania 45%, Europe 25%, Latin America 13%, Asia 8%, North America 6% and Africa 3%. Compared with the last report, regarding the result from 2014, it should be pointed out the fact that the overall area which represents ecological agricultural land has increased with 6.5 million ha, the largest growth recorded until now. The ranking of the top three states with remarkable results in this field is maintained for several years in the same formula: Australia (22.7 million ha with 97% used for grazing), Argentina (3.1 million ha), US (2 million ha) and the top is completed by Spain, China, Italy, France, Uruguay, India and Germany. Eleven countries have reached the point in which the organic agricultural land represents more than 10% of their total farmland. In this respect, the countries with the largest share are: Liechtenstein (30.2%), Austria (21.3%), Sweden (16.9%) and the rest of the list is completed by: Estonia (16.5%), Sao Tome and Principe (13.8%), Switzerland (13.1%), Latvia (12.8%), Falkland Islands (12.5%), Italy (11.7%), Czech Republic (11.3%), Finland (10%). Beside organic agricultural land there are vast, non-agricultural territories which are also considered organic (forests, aquaculture, wild collection, grazing) totalizing 39.7 million hectares (data collected in 2015). The countries in which we find most part of this surface are: Finland (12.2 million ha), Zambia (6.6 million ha) and India (3.7 million ha).

Also confirming the increased interest in organic agriculture, FiBL officially informs that no less than 2.4 million organic farmers from 172 countries are responsible with providing the necessary amount of organic products, both food and drinks, and most important, ¾ of them are from developing countries. Most of the producers are concentrated in countries like India (585.200), Ethiopia (203.602) and Mexico (200.039). According to FiBL - IFOAM study, the value of global organic market in 2015 was estimated at 81.6 billion US dollars. If we take into consideration the fact that in 2000, the value of this market was estimated at 17.9 billion US dollars, we can easily conclude that the global market of organic products is considerably growing and the obvious and undisputed leaders for setting this trend are US, with 35.9 billion euros, followed by Germany (8.6 billion euros), France (5.5 billion euros) and China (4.7 billion euros). As we can observe by analyzing the trends in the most important markets, double digit growth rates are met in most of the major markets. According to a FiBL survey from 2017, based on national data sources and data from certifiers Global market, in Europe organic markets continue on a rising trend, with an estimated value of 30 billion euros (reached in 2015), of which an overwhelming percentage is due to European Union members (90%). The result of the survey also emphasized the leaders of the European market, namely Germany (8.6 billion euros in 2015, with a significant growth of 11% compared with the value registered in 2014), France (5.5 billion euros, with an increase of 15% compared with 2014) and Italy (2.3 billion euros, a 15% growth compared with 2014).

The appetite of the final consumer for organic products is on the rise, and they are increasingly willing to spend on such products at the expense of those obtained through the use of chemical additives. As the date collected on the FiBL survey show, the countries with the highest per capita consumption of organic products are: Switzerland (262 euros), Denmark (191 euros) and Sweden (177 euros).

Undoubtedly, the collective concern about the quality and safety of food has become the major driving force in the development of organic agriculture, mostly in developed countries. While some may question the validity of consumer concerns, there is no doubt that these have a significant contribution to the growth of the organic sector.

BACKGROUND

Given the spectacular evolution during the recent years, organic farming has often become the subject of careful researches, conducted from both individuals (researchers) and organizations. Therefore, seeking sources and development resources, researcher's base their work on statistical data collected and centralized by international organizations with reputation in this field, such as FiBL or IFOAM. Surveys and statistics published annually by the two organizations, coupled with the information each country collects and reports annually, are important in every analysis. Also, reports by world-renowned consultancy companies analyzing the agricultural market, such as PwC, Ernst and Young, KPMG or Deloitte provide conclusive data that can serve as the starting basis for any research. Also, the information provided by the European Commission on legislation gives us the full picture of legislative and procedural issues, so necessary to understand the entire complex of factors that govern the opportunities and trends of organic farming. Starting from EU regulation No 834/2007 each member state has the responsibility to transpose the provisions of European law into national law. Organic agriculture plays an important role in globally achieving sustainability objectives. Organic market, like any other food market is mainly driven by demand fluctuations. Intensive market penetration and also implementing marketing strategies are utmost important since without supermarket chains being part of the market outlet, organic food can't be easily reached by the final consumer, thus the market development is threatened (Fischler, 2012). The incredible advances made by organic farming can also be tracked through the evolution of legislation in place. If 15 years ago only 2 international benchmark for organic agriculture were functioning, in 2012 almost 350 in 150 countries were involved in inspecting and regulating organic products (Crucefix, 2012). For some time, the development of the sector has been jeopardized by overregulation and international cooperation in terms of accreditation documents. Moreover, the lack of cooperation, harmonization and mutual recognition has affected the activities of private companies in the organic farming sector, especially in developing countries. A company that was planning to export to more markets, needed more than one certificate to be able to work, the bodies that supervised them, imposing different requirements, which, in turn, were supervised by several accreditation bodies or government authorities. The imposition of differentiated organic standards for each export market and the obligation for manufacturers to apply not just one, but several, each adding considerable cost to the final price, was the main argument that acted as a driving force towards cooperative and unitary regulation.

ORGANIC AGRICULTURE: A WORLDWIDE SOLUTION FOR SUSTAINABLE DEVELOPMENT

Benefits Provided by Organic Farming

As defined by the Codex Alimentarius Commission, organic agriculture represents,

a holistic production management system that avoids use of synthetic fertilizers, pesticides and genetically-modified organisms, minimizes pollution of air, soil and water, and optimizes the health and productivity of interdependent communities of plants, animals and people. (EL-HAGE Scialabba, 2013)

Given that currently 10 million ha of agricultural land is wasted annually as a result of unsustainable agricultural practices, by implementing organic management measures positive impacts on environment are generated, for both soil and water. The benefits are undoubtedly spectacular when analyzing the statistics. Besides increasing soil fertility or reducing water pollution, less energy is needed in organic production (10-70% in Europe and 29-37% in USA) while the irrigation requirements are less with 20-60% in organic farms.

Although theirs is few scientific evidence regarding the qualitative superiority of the organic food, there is a strong belief on consumers that fundaments the increasing demand on this type of products. The perception that organic farming generates nutritionally superior food has exponentially spread, especially over the past 5-6 years. Moreover, only a small number of research studies have scientifically compared the nutritional compositions of organically and conventionally produced crops, with even a smaller number of studies comparing products of animal origin (meat, milk and dairy products) resulting from the two agricultural systems. Very few differences in composition have been observed, although there were reported significant differences in the higher nitrate content and lower vitamin C content of traditional products, especially for leaf vegetables (Williams, 2002).

Thus, more and more consumers are convinced that another advantage associated with the consumption of organic products is the contribution to the general health of the consumer due to high nutrient content of organic products and low intake of contaminants. Since organic products have higher levels of health-promoting phytonutrients, vitamins and minerals, and at the same time a lower level of insecticide residuals, consumers believe in the health benefits provided an organic menu (Crinnion, 2010). Also, during the last 15 years, due to food crisis such as mad cow disease, foot-and-mouth epidemic or Belgian dioxin scandal, European consumers have lost their trust in the food derived from conventional methods of production, being more and more interested in trying organic products (Maquet, Siderer, & Anklam, 2005). On the contrary, organic products have never generated any type of health problems to any consumer, and for this reason organic farming is gaining each day new adepts.

Consuming organic food is an excellent way to reduce exposure to contaminants commonly found in foods that have been grown using conventional farming practices. These contaminants may include not only pesticides (considered to be potential cancer agents) but also heavy metals (lead, mercury), solvents (toluene and benzene). It can be objectively deduced that minimizing exposure to these potential toxins brings important health benefits to any person. Although there are still on-going studies trying to scientifically compare organic and non-organic products, there are few results that are being recognized in the US, by FDA and US Department of Agriculture, regarding organic milk and organic tomatoes. Nutritionally, the organic version of the two products is superior than the non-organic version (Williams,

2002). According to a 10 year study, conducted by University of California, organic tomatoes contain more antioxidants (79% higher in quercetin and 97% higher in kaempferol), with effects in reducing heart disease and also reducing the chances of developing cancer.

Other generally accepted benefits in consuming organic products are: lower levels of insecticide, reduced pesticide exposure levels, better taste of the products, ensuring environmental protection by not using harmful chemicals during the production process, respecting animal welfare. A study conducted on preschoolers has shown that organic diet significantly reduces the total amount of organophosphate pesticide (OP) residues found in their urine, and also other chemical compounds: malathion and chlorpyrifos (Crinnion, 2010). This mineral compound was linked as a cause for several serious diseases as autism and ADHD. In this context, we can clearly see that the benefits of choosing organic products are way beyond financial arguments. It is the quality of the products, the nutritional composition and even the taste, which make the consumer's choice easier.

Issues, Controversies, Problems

Starting from the growing concern of people for a healthy lifestyle, organic farming has seen a remarkable development in recent years. Food security has become one of the country's priorities as a result of people's hunger for healthy agro-food products, for foods whose traceability is known, verified and certified as such.

The growth of demand has naturally been reflected in a reunion of the supply of organic products at competitive prices established precisely as a result of the demand/supply interaction. With an average price higher than non-organic products, organic products provide benefits that go beyond the financial boundary, and which may be considered as a health investment over time. Thus, organic consumption has two great advantages in terms of ensuring consumer's health, namely reduced intake of contaminants and high value of nutrients, in relation to those traditionally grown.

Manufacturers quickly understood the people's need for food security and promptly responded. Thus, according to data provided by FIBL, centralized at the level of 2015, both the number of producers and the total area dedicated to organic crops have increased significantly, so that organic farming currently accounts for 1.1% of world agriculture. Organic farming may also represent a growth solution, especially for developing countries. Currently 2.4 million producers, of which ¾ are active in developing countries, cultivate a total of 50.9 million hectares of organic products, with 6.5 million hectares more than in the previous year. This increase of 14.6% is felt globally, respecting the distribution weights in the continents.

As we can analyze in Table 1, the repartition of farmland in unevenly. Oceania concentrates an important area of agricultural land dedicated to organic farming, with 44.79%, followed by Europe with 22.19%. When we analyze each country's results in terms of organic farming, best results are found in Australia (22.69 million hectares) followed by Argentina (3.07 mil. ha.) and United States of America (2.03 mil. ha). When considering the available agricultural area, only 11 states allocate more than 10% to organic farming: Liechtenstein (30.2%), Austria (21.3%), Sweden (16.9%), Estonia (16.5%) Sao Tome and Principe (13.8%), Switzerland (13.1%), Latvia (12.8%), Malvinas (12.5%), Italy (11.7%), Czech Republic (11.3%), Finland (10%).

Over time, the development of organic farming has been gradual, with constant growth, so that in the period 2010-2015 the total ecologically exploited area increased from 35.7 million ha to 50.9 million ha (an increase of 45.2%). This demonstrates a significant increase in global demand and a trend that will set new consumer behavior in the years to come, basically the shift towards sustainable consumption.

Table 1. Repartition of organic farmland worldwide, in 2015

Continent	Organic Farmland (mil. ha)
Europe	12.7
Asia	4
Africa	1.7
North America	3
Latin America	6.7
Oceania	22.8

Source: FiBL, 2017

After Oceania, where the extremely rapid development of organic farming can be noted (between 2006 and 2015, the area thus cultivated increased from 12.1 million ha to 22.8 million ha), Europe is remarkable for its important contribution in increasing the share of organic farming. Thus, during the same period, 2007-2015, the organic cultivated area grew from 7.8 million ha to 12.7 million ha (an increase of 62.8%). In Table 2 we find the annual evolution of the total agricultural areas cultivated ecologically on the old continent.

In the period 2014-2015, the most speculative increases in the implementation of organic farming were recorded in Australia (+ 4350 thousand ha), The United States of America (+475 thousand ha) and India (+ 460 thousand ha). In Europe, Member States registered a significant increase in Spain (+ 258 thousand ha), France (+ 256 thousand ha) and Italy (+105 thousand ha). Also noteworthy is the position of the Russian Federation in the development of organic farming, with an increase in the ecologically cultivated area of 139 thousand ha in 2014-2015 (FiBL, 2017).

Regarding the distribution of these increases according to the typology of agricultural crops, we can see preponderance for permanent grassland, followed by arable crops and permanent crops.

Another criterion that we can use to analyze the degree of development for organic farming during the last years is the total number of farmers working in organic farming. Yearly this number has grown globally, reaching 2.4 million farmers in 2015, but there are countries with outstanding results. Thus, the ranking according to the number of farmers is led by India (585.200), Ethiopia (203.602) and Mexico (200.039). In Europe, the ranking is headed by Italy, with 52.609 farmers, ranked 10th globally. Compared to Italy, in 2015 in Romania there were 4 times fewer farmers, respectively 12,231 organic ecological operators. In fact, with respect to the number of operators, we can see an interesting global distribution

Table 2. Organic farmland in Europe between 2007 and 2015

Year	Organic Farmland
2007	7.8 mil ha
2009	9.2 mil ha
2011	10.5 mil. ha
2013	11.4 mil. ha
2015	12.7 mil. ha

Source: FiBL, 2017

Table 3. The evolution of organic farming according to the type of crops

Year	Permanent Grassland (mil. ha)	Arable Crops (mil. ha)	Permanent Crops (mil. ha)
2010	23.1	6.5	2.6
2011	22.6	7.6	2.9
2012	22.6	7.9	3.2
2013	27	8.5	3.3
2014	28.3	8.8	3.4
2015	33.1	10	4

Source: FiBL, 2017

at continental level. Thus, Asia dominates, with 35% of the total number of operators followed by Africa (30%) Latin America (19%) Europe (14%), while Oceania and North America have only 1% each (FiBL, 2017). It must be noted that, by adding up the number of operators in Asia, Africa and Latin America, we get a result representing the overwhelming majority of operators in organic farming (84%).

In Figure 1 we can see the distribution of the ecologically cultivated agricultural lands according to the typology of the crop, the data collected being related to the year 2015.

The Road to Organic Farming: Case Study - Romania

In a ranking of the EU Member States, based on their agricultural land, Romania ranks the 6th place. After being a benchmark of agricultural production for years, both in terms of quantity and quality of its products, with leading positions in the grain exports rankings, Romania has always been recognized as a state with huge potential in developing economic growth by practicing agriculture. Of the total surface area of 23.8 million hectares, representing the territory of Romania, the utilized agriculture area (UAA) represents 55.9%, amounting to 13.3 million ha.

Out of this total, 8.3 million ha are arable land. In other words, arable land accounts for 62.5% of the total agricultural area. In fact, if we report the total area of arable land to the number of inhabitants

Figure 1. Organic land use types in 2015
Source: FiBL, 2017

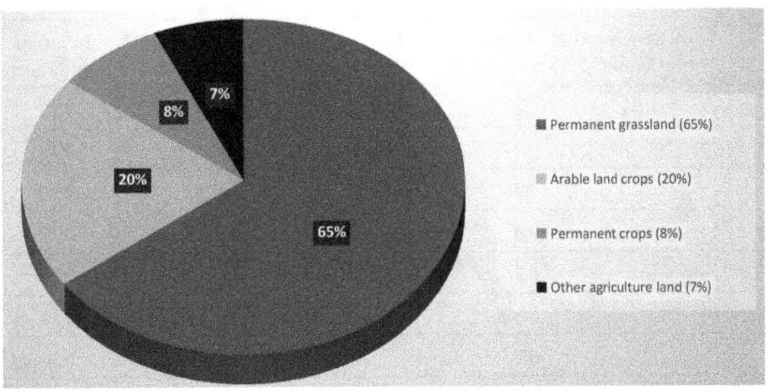

(according to INS, on January 1st 2017, the Romanian resident population was estimated at 19.310.216), we find that each Romanian has 0.41 ha of arable land, way above the European average of 0.212 ha / inhabitant (CCIR, 2014). This comparative advantage is not necessarily capitalized, for various reasons, especially due to the lack of funding, the poorly qualified workforce, the legislative and bureaucratic deficiencies that hinder reform processes that have begun but are poorly implemented. In Romania, agriculture has always been a significant contributor to GDP growth. The transition from communism to capitalism implied structural changes of the economy, a natural evolution from an industrial-agrarian one, towards a service-oriented one. This is the reason why the share of agriculture in GDP has been declining from year to year. If we compare the results from 1993 to the ones in 2015, we can see that the percentage has shrunk 4.5 times. As can be observed from the statistics presented in Table 4, over the last 5 years, between 2010 and 2015, the contribution of agriculture in the production of Gross Domestic Product has values between 4.8% and 6.4%. Although in the past few years the percentage has continuously declined, agriculture's contribution on Romania's GDP is still quite high, practically 3.5 times higher than the European average of 1.7%.

In recent years, the demand for organic products has been constantly growing globally, and Romania is not making a discordant note on this topic. As a result of national awareness campaigns that promote healthy lifestyles, more and more consumers have turned their attention towards organic products, willing to make a significant improvement in their living conditions. However, the road to organic agriculture as a viable solution for economic development, based on a national strategy with effective measures to support this sector, is still a tedious one. At this point, in Romania, organic farming is in full development process, with a fluctuating path in 2010-2016. Since it is still considered a niche field, organic farming remains a controversial choice for the average Romanian farmer, being viewed mostly with skepticism. Scared by the high costs associated with the certification process and legal uncertainty, the average farmer avoids taking the risks of transition from traditional farming to organic agriculture. Fearing the unknown, many of the Romanian farmers are still loyal to traditionalist farming, although this strategy does not bring most of them the expected gain. So, the share of organic farming in Romania is extremely low, with a variable percentage from year to year but never exceeding 2%. Thus, according to the data provided by the Ministry of Agriculture and Rural Development in 2016, from the total 13.3 million ha of agricultural land, only 226.309 ha were ecologically cultivated, which means that organic land only represents 1.7%. The evolution of this share over the last 10 years has been fluctuating, as can be seen in Table 5.

Table 4. The share of agriculture in the GDP formation in Romania during 2010-2015

Year	The Evolution of the Gross Added Value of Agriculture in GDP	Share of Employed Population in Agriculture in the Total Employed Population in Romania
1993	22,6	35.2
2010	6.4	29.1
2011	7.5	29.2
2012	5,3	29.3
2013	6.1	27.9
2014	5.3	27.3
2015	4,8	*no centralized data

Source: INS, PwC, 2017

Table 5. Evolution of Organic Agriculture in Romania, 2010-2016

Year	Number of Certified Operators in Organic Farming	Organic Land (Ha)	Total Utilised Agricultural Area (ha)	Weight in Total Utilised Agricultural area (%)
2010	3.155	182.706	14.156.480	1,29
2011	9.703	229.946	13.981.620	1,64
2012	15.544	288.261	13.733.140	2,09
2013	15.194	301.148	13.904.640	2,16
2014	14.470	289.251,79	13.830.420	2,09
2015	12.231	245.923,9	13.858.420	1,77
2016	10.562	226.309	13.520.850	1,67

Source: MADR, 2017

As shown by the statistical analysis, the evolution of organic farming in Romania has experienced an interesting fluctuation. Thus, during 2010-2012, both the number of operators and the total area allocated to organic farming (organic land) have increased considerably. Specifically, the number of operators increased 5 times, while the surface increased by 57.78%. Starting in 2012, both the number of operators and the surface have decreased, with each passing year. Thus, in 2016, the level of the two indicators reached comparative values with those registered in 2011. Practically, the area allocated to organic farming in 2016 is 3637 ha less than in 2011. It is interesting to analyze what caused such a regression, considering that the pace of evolution in the previous period (2010-2012) was positive, with very high growth rates. This shifting can be explained by the many factors that have influenced the market. For example, the economic crisis (which in Romania was felt with a time lag compared to the euro area) as well as the reluctance of farmers who opt for traditional farming because of the difficulty in selling organic products, the costs associated with certification - considered as Being high - as well as volatile legislation or legislative vacuum - are obstacles that the farmer encounters in his current work. However, returning to 2017, organic farming is considered to be a primary objective, a key element in the medium-term economic development strategy. In the governance program set for the period 2016-2020, we find concretely the strategic objective of increasing the organic land up to 500,000 ha by 2020. But this desideratum, which basically requires doubling the total organic land, can't be achieved in the actual conditions. The average Romanian farmer has already experienced deep disappointment generated by the lack of financial support from the government. Thus, an explanation for the turning point in 2012 could be precisely the fact that farmers have realized that the amount of subsidies granted for each eco-cultivated ha is insignificant related to the needs. In Romania, agricultural subsidies have been granted since 2004 and during 2005-2007 they functioned within the national SAPARD program. Since 2007, many other European grants have been accessed, but until 2011 no financial bonus for conversion to organic farming has been granted. The spectacular evolution of organic farming and the granting of these conversion funds have led to a significant number of farmers deciding to do it but, in the absence of consistent financial support, this conversion has not been a long-term one. Thus, many farmers decided to give up organic farming when they realized that the support was lower than expected.

PROBLEMS IN IMPLEMENTING ORGANIC AGRICULTURE

In spite of a growing interest of the end-users for organic agricultural products, both at European level and in Romania, the development of this segment faces obstacles rather difficult to overcome. Considering the vast agricultural area and the high rate of utilization, we can unequivocally say that the main factor limiting the development of Romanian agriculture is the yield of production. Particularly, organic farming is embracing a whole set of impediments during the development process. These obstacles are either of a financial nature, or of a professional nature. On entering the market, organic farmers must overcome first the financial obstacles: namely the difficulty of attracting the necessary funds for modernization investments and adequate equipment, or the cost generated by the certification of organic products. At the same time, difficulties regarding lack of qualified and specialized work force, or technical obstacles such as fragmentation of organic land, appear and have to be resolved in order to stay competitive. Nevertheless, organic farmers have to deal with bureaucratic problems, when the process of environmental certification is difficult and involving a complicated and tedious procedure. Also the problems in developing organic agriculture might be cultural ones, generated by the customary consumption of traditional uncertified products and also by the customary and non-differentiated social perception of both organic and non-certified products, as a result of a lack of information. Given that the cost of the certification process - costs varying from state to state, affecting the final price of the organic products. Organic farming therefore implies additional costs that represent real barriers for farmers at the onset of this new type of production. At the same time, the structure of the market on which agriculture is active is still such as to discourage new enthusiasts. By having an oligopoly structure, the present market does not offer too many alternatives to organic farmers. All these problems are obstacles that hinder the development of organic farming and the growth of the organic land in Romania and also in other developing countries.

Under these circumstances, the European Union's financial support for farmers complying with the CAP principles, protecting the environment and ensuring food security for over 2222 million consumers can provide real support for organic farmers. In a survey conducted by European Commission, in accordance with Eurobarometer methods, 27.822 EU citizens, from different social and demographic categories, have declared their support regarding the financial aid for organic farmers. Thus, the majority would like to see EU support farmers who carry out agricultural practices which are beneficial to both environment and food quality (Eurobarometer, 2016).

Production Yield

Annual agricultural production is a function defined by three variables, namely: the available agricultural area, the utilization rate of the available agricultural area and, last but not least, the yield of agricultural production. Considering that the first two criteria place Romania at the forefront of the Member States (of the 14.4 million ha available, 96.3% were used and the remaining 3.7% unused), the conclusion is that the yield of production is unsatisfactory, seriously affecting the results obtained in agriculture. Although the contribution of the agricultural sector to GDP has fallen significantly over the past 25 years (from 22.6% in 1993 to 4.8% in 2015), the number of employees in the agricultural sector remained similar, which means a perpetuation of the lack of systematic efficiency. Thus, in 2014, the labor force in agriculture represents 27.3% of the active population of Romania, 6 times more than the European average of 4.4%. The low degree of professionalism represents a major obstacle in the transition to organic farming, and at this moment in Romania, the situation of well-trained workers can represent a real problem. At the same time, undeclared work doubled by the lack of adequate wages significantly contributes to an unfavorable context of natural evolution towards organic farming. According to INS data, 85% of the total labor force in agriculture is unbalanced, working on its own land, thus practicing subsistence agriculture. If we take into account that the European average is 72%, while in countries with a higher productivity the percentage of the unskilled labor force is around 50%, we can observe the dangerous gap. If in 2014 the population occupied in agriculture is comparable to that of the early 1990s, the contribution of agriculture in GDP has fallen about 4 times. The causes of this decline in Romanian agricultural output are found in the lack of professionalism, in the absence of major investments in equipment, in the absence of adaptation and flexibility on demand (including the more serious orientation towards organic products, increasingly sought by final consumers), as well as the low ability to access European funds as well as the difficulty of access to credit banking sources. The transition to organic farming can't be done without identifying efficient solutions for each of these barriers against modernization.

Thus, through major investments in this sector, as well as through a modern approach, focused on market demands and environmental orientation, it is possible that in the coming years there will be an increase in efficiency, namely an absolutely necessary increase of the gross added value per cultivated hectare. At this moment, Romania produces 600 euro/ hectare cultivated, with over 40% less than the added gross value/ ha recorded in Western European countries, where over 1,000 euros are gained with each cultivated hectare. When it comes to people employed in agriculture, we find the same huge decay in relation to the European average. In particular, the gross added value/ employee in Romania is statistically estimated at 2.893 euros, while the European average is 6 times higher, respectively 17.165 euros (PwC, 2017). If we consider the results recorded by states with outstanding agricultural results, we find that France has a gross added value/employee 17 times higher, while Spain exceeds 11 times in this ranking.

Fragmentation of Agricultural Land

The prospects for the development of organic farming in Romania are deeply affected by the high degree of agricultural land fragmentation. Despite the extremely generous agricultural area available, Romania is one of the worst fragmented member states. Thus, Romania is the third country in the European Union, depending on the degree of fragmentation of agricultural land, after Malta and Cyprus, which do not have a totally low agricultural area.

Thus, three quarters of the total number of farms exploit agricultural land with an area of less than 2 hectares. Moreover, 98% of all Romanian farms are under 10 ha. On the opposite side, we find large farms, which exploits over 100 ha of land. Unfortunately, this category of farms is statistically insignificant, representing only 0.5% of the total number of farms.

However, the agricultural area that this minority is processing represents half of the total agricultural area at national level (49%).

Given the fact that the vast majority of farms are small, they can't benefit from the advantages of a business that has large areas of land, such as: scale economy, easy access to financing, workforce with adequate training, technology and modernization. Scale economy would allow farmers to increase their profits and gain more money for future investments. The economic principle is a very well-known one: the bigger the fragmentation, the less chances to produce efficiently due to scale economy and the inverse relationship between the total quantity produced and the cost per unit (fixed costs per unit). Both traditional production and organic production are affected by the big number of small farms in Romania.

Although Romanian farmers exploit the 6th agricultural area in Europe, their productivity is still low. According to a PwC studio, of the 13.9 million hectares of agricultural land used, 60% (8.2 million ha) is represented by agricultural land, of which 2/3 is used for cereal crops. However, cereal crops generate only 25% of the total value of agricultural production (14.2 billion in 2015 - the latest centralized data). These weak results are due to a cumulus of factors, of which fragmentation is the most important. In the absence of land concentration in large farms, skilled and trained managers, economies of scale can't be achieved and the economic performance of agricultural activity will remain at the weaker level recorded today. The high degree of fragmentation faced by Romania has extremely damaging consequences on yield per hectare. That is why it is worth mentioning that 68% of the farms report annual production below 2000 EUR. 29.9% record a standard annual production of between 2000 and 14999 euros, while 0.9% admit that they get between 14,999 and 24,999 euros (PwC, 2017).

Difficult Access to Funding Sources

Another important difficulty in developing organic farming is heavy access to funding sources. Whether we are talking about the difficulty of accessing bank loans as a result of the high degree of fragmentation which translates into a limited ability to quickly repay short time loans (generated by a very low annual production) or to the failure of Romanian farmers to access European non-reimbursable funds or the low subsidies, the financial availability of the farmers is extremely limited, and this can easily be seen in the degree of modernization and reduced technology, in the innovativeness of the equipment, in the total investments made. The high share of small farmers (98% of farms manage less than 10 hectares of arable land) and the standard production produced annually are important premises for granting a credit to farmers. Their ability to repay the credits was statistically analyzed by an INS study which says that for 2014 only 12% of indebted companies managed to pay their debts earlier than 90 days, while about 40% were consented to delays in short-term loans.

Thus, looking at the deadlines for repayment of loans, we notice that a small farmer returns 5 times later a credit received. Specifically, banks have reported that the average time to recover outstanding debts from small businesses was 669 days, according to the data collected in 2014. For large companies, this indicator did not exceed 142 days. Regarding the time required to pay current debts, it differs significantly between small firms (1,725 days), medium ones with cultivated areas ranging from 500-2000 ha (416 days) and large firms, which cultivate over 2000 ha (303 days). Analyzing this statistical data,

Figure 2. Distribution of farms according to annual production
Source: PwC, INS, 2017

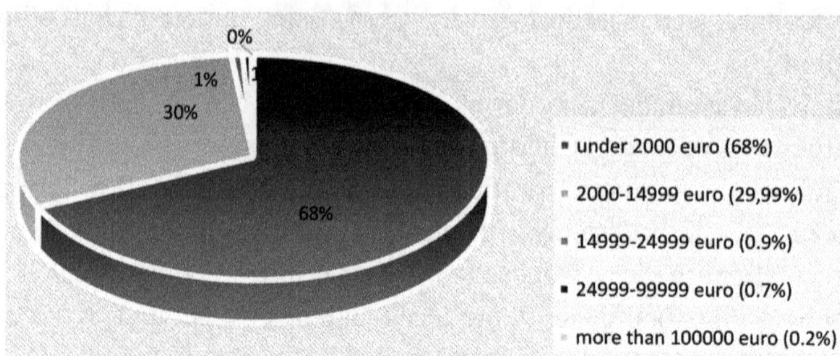

banks are granting agricultural loans with difficulty, and when they do so, the amounts are well below European averages. Precisely, the agricultural credits granted to the Romanian farms have an average value of 110 euro / ha, while in the EU the average value of credits to agricultural holdings is around 2000 euro / ha (PwC, 2017).

As regards subsidies received from the state, the Romanian farmer is also in the queue of the European rankings. In 2013, state budget spending on agriculture was estimated at 93 euros / ha, which puts us the penultimate place in the EU, above Hungary, with 89 euros / ha. In comparison, in the Czech Republic the state granted 465 euro / ha and Italy 397 / ha.

Aged Farmers and Poorly Qualified Workforce

European farmers have to accomplish a huge responsibility: producing sufficient, high quality food, at affordable prices, for almost 500 million European inhabitants. According to a briefing from European Parliament, the EU farming population is estimated at 22.2 million regular employees (around 8.7 million full time equivalent employees). The problem starts to appear as soon as we start to analyze the structure and the distribution of the farming population. Thus, the biggest concern is the fact that the farming population is worryingly ageing, and the younger workers, beside insufficient, are also poorly educated. The development of organic farming can't be done in the absence of a skilled labor force, employed in legal forms, and last but not least, properly remunerated. Currently, the whole Europe still has a high percentage of workers in the field working without legal forms, non-taxed, with a low degree of training, active in agriculture based only on the low knowledge gained after employment.

According to the European Parliament briefing, Eurostat defines three levels of agricultural education and training: practical experience, basic training and full agricultural training (at least two years of full time training). Eurostat data collected in 2013 (latest official rapport regarding this issue) shows that almost 70% of the farmers (69.8% to be precise) have gained all their knowledge from practical experience alone, whilst the rest are divided between farmers who have attended and completed at least one course in agriculture (21.7%) and farmers who received full education (8.5%). Regarding the distribution of the farmers by education, it is important to mention that there is a correlation between farmer's education and the size of the farm. Thus, the percentage of fully trained farm managers increases proportionally with the size of the farm, so that only 4% of fully trained farmers worked in small size farms (with an output <14.999 euros) whilst 34% of them were employed in large holdings (output >250.000 euros). Also, an interesting remark regarding farmer's education and training is that women are less educated in his sector, and the differences are significant, as we can see in Table 6.

Table 6. Share of farm managers according to education and gender (%)

Education Level	Women	Men
Practical experience only	79	68
Basic Training	16	22
Full agricultural training	5	10

Source: European Parliament Briefing, 2017

Besides poor training, another pressing issue that could negatively influence the development of agriculture in general, including the ecological one in particular, is the high average age of employees in the sector. Thus, at European level, in an overwhelming proportion, people who work in agriculture are over 50 years old (43.3%). The rest of the farmers have ages between 25 and 49 years old (49.6%), while the youngsters, with ages under 25 represent only 7% (PwC, 2017). Using Eurostat data from the European Parliament briefing we can see a detailed distribution of farmers according to age in Figure 3:

Analyzing the age structure of European agrarians we can understand why their evolution towards new forms of agriculture, respectively their adaptation to new methods of production, is reluctant. It is well known that, along with age, the resistance to change and modernization increases. Statistically speaking, it is much more likely that young agrarians will embrace new forms of production, risk-oriented production, towards the development of new consumer markets. Given that organic farming serves a segment of consumption that can still be classified as "niche", switching from traditional farming to organic farming implies taking a risk. In Romania, farmers' distribution, both in terms of professionalization and age categories, strongly highlights the structural problems faced by the agricultural sector, problems that can't be solved without effective reforming measures, applied with surgical rigor. Aging and lack of training are clearly highlighted in statistical data: 96.4% of farmers say they have no training beyond practical experience, while 40.7% of farmers are over 50 years old. Starting from these premises, to which are added the financing difficulties, the reduced support from the state, the cumbersome legislation, the bureaucratization and the high fragmentation, it is hard to believe that organic agriculture can be a serious direction of development.

SOLUTIONS AND RECOMMENDATIONS

In the context of sustainable global environmental policies, organic farming can be the right solution to ensure both economic growth and respect for the environment, as well as people's need for food security. Considered an investment in health, organic farming is growing more and more, both as a number of producers and as a surface, thus covering the high demand for such products. Trade barriers and higher prices compared to products of the same range without environmental certification are impediments that organic farming has to overcome in order to consolidate the market. It is highly important to realize that,

Figure 3. Distribution of European farmers according to age
Source: Eurostat, European Parliament, 2017

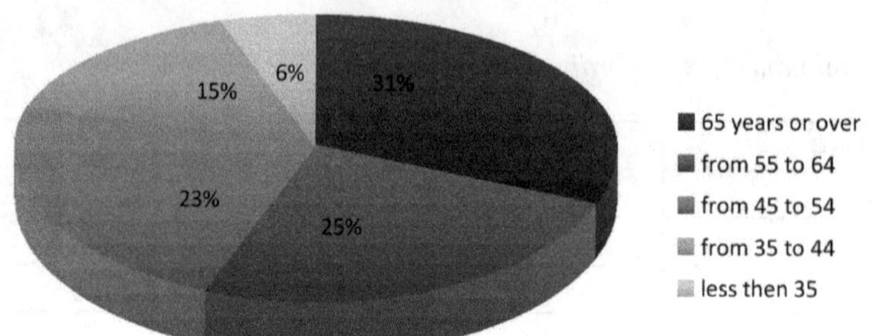

in 10 years from now, the main challenges for the organic sector could significantly change. Besides harmonization and equivalence efforts, another type of problems from within the sector could generate more difficulties. For instance, new CSR concepts could emerge, transforming organic products from niche to mainstream (Rundgren, 2012). Also, developing the IT tools used in monitoring the activity, increasing regulation process and growing fatigue caused by it could influence the organic market in the years to come.

Awareness campaigns and access to information can be the means by which organic farming can gain more and more followers in the years to come. Also, reducing the costs associated with the certification process could be a solution to facilitate the development of organic farming, with beneficial effects on quality of life and longevity. Public policies and awareness campaigns, educating children from the early ages on the importance of knowing food quality can definitely change the evolution of future generations. Undoubtedly, the planet is at a crossroads, and today's choices will be reflected in the future, with high impact for the future generations. Considering the importance of encouraging organic farming through integrated global strategies, we ensure the balance between the economy and the environment, as well as the economic growth, mostly the developing ones. In the short, medium and long term, effective measures can be implemented so that the economic and fiscal context in which the farmer operates will become a good one for his development. Among the objectives of any European country in charge, we find food safety elements. Ensuring food needs, both quantitatively and qualitatively, has become a priority for any EU Member State. In Romania, the current Government has assumed the objective of ensuring healthy food, both inside and outside the country, by increasing agricultural exports. At the same time, it is desirable to increase the number of young farmers, which would allow both a change of mentality and an acceleration of the initiated reform processes through the implementation of new management techniques, namely bringing a higher professional degree in agriculture.

For instance, tax incentives can be provided in order to encourage youth employment. For example, when hiring 3 young people (aged under 35) it is proposed to eliminate the tax payer for those employees during their employment, but not more than 5 years. This measure would have a double effect, beyond the rejuvenation and professionalizing of the workforce, it would also reduce both undeclared work phenomenon and tax evasion. It is estimated that only by implementing this measure, 8.000 jobs would have been generated, with an estimated budgetary impact of 25 million Euros (Governance Program, 2016). Without any doubt, all proposals inherently support the systematic growth of organic farming. Beyond these general directions, the medium and long-term objectives concern mainly respecting the payment terms on agricultural subsidies, not mention the need to increase the amounts of subsidies, since the present values of the allocated funds are both insufficient and incomparable to the support given to agriculture in other Member States. Also, farmers are complaining the even those insufficient funds are allocated with large delays, affecting the entire process of production, by generating additional costs. At the same time, by eliminating any charge on utilized agricultural area would generate a very well deserved increase on earnings for farmers, so much needed especially for organic farmers. Another measure which needs to be implemented as soon as possible is the opening of a national laboratory for the certification of organic and traditional products. This laboratory would mean creating the necessary framework to significantly increase the number of protected products. It is estimated that an opening would determine a generous increase of the protected products, from 3 to 14, by 2020.

FUTURE RESEARCH DIRECTIONS

Taking into account both the favoring circumstances for globally developing organic farming, in the medium and long term, as well as the factors that constrain this development, implementing a slower growth rhythm, the evolution of organic farming in the next period is mostly expected to pursue a positive trend. In this concern, a global percentage increase in the share of organic farming will be achieved, mainly due to the growing concerns of society for a healthy lifestyle, a concern that translates into a significant increase in demand for organic products. Intersecting the demand, the offer will also increase, generating a price that is often considered prohibitive for certain market segment. Statistically, the prices of organic products are superior to those of similar non-certified products, which considerably affects a potentially rapid increase in the demand of this products, in the years to come. Future research could be focused precisely on analyzing the impact of the higher price of organic products on the rhythm of demand growth for such products. By using econometrics, we can determine the degree of influence between the two variables - the price and the rhythm of the demand growth. By computing the correlation index between the two variables, it will be possible to measure the extent to which they mutually influence each other. In this way, one could be dethroned or confirmed, namely the assumption that, most often, costumer- although he knows the superior net value of an organic product - prefers a similar product from a non-certified range, for strictly financial advantages.

Also, it is important for researchers to supervise the entire evolution of organic markets on medium and long term. Most probably, ten years from now, the main questions regarding organic products would be different from the ones that are today on the table. New health or Corporate Social Responsibility concepts are about to be introduced, which could eventually drive organic food intro exclusivity or even mainstream it (Rundgren, 2012). It would be most interesting to rigorously and comparatively analyze the evolution of these concepts, by reference to the demand for organic food, in the years to come.

CONCLUSION

In recent years, organic farming has been recognized worldwide, as a result of a global development of this segment, both in terms of the number of operators and organic land. Thus, centralized data for 2015 shows that there is an organic cultivated area of 50.9 million hectares globally, with 6.5 million ha more than in 2014 and 3.6 times higher than it was reported 16 years ago, in 1999. Although an important share of about 75% of this agricultural area is concentrated on the territory of the top 10 actors on the market, headed by Australia, there is an important growth potential in every state where we currently find dedicated farmers decided to implement organic farming. In fact, regarding the total number of organic farmers, a fabulous growth was recorded in 2015, when globally were registered no less than 2.4 million operators, about 160,000 more than in 2014, which means an increase of 21% . We could talk about a supply boom if we take into consideration the fact that, compared to 1999, the number of global operators increased by 1000% (FiBL, 2017). This increased interest in organic farming is a natural response to consumer needs of quality food, all over the world. The whole benefit complex, as we presented in this paper, is generating an increased interest for organic products, and the statistics show increased values in both cultivated area and demand. The contributing factors of the generated growth, both in demand and supply, are rather referring to a social transformation of the collective mind, as a result of a globally public debate on health topics and also following a very complex process of correct and complete information of the customer, regarding the composition and traceability of agro-food products, through awareness-raising campaigns conducted at national or international level by NGOs concerned with healthy eating. All of these elements have generated significant mutations in demand, and at the same time an instant feedback on the organic market, by rising the supply of organic certified products. Organic farming has managed to offer a decent response to sustainable consumption trend, as shown by the statistical data presented in the previous sections. Regarding the development of organic farming, we have seen that this is taking place at a sustained pace in the recent years, with different frequencies depending on the continent and the country. The evolution in each state depends both on the economic and fiscal context and also on the facilities granted to farmers, the agricultural areas used and the available workforce. At European level, where the demand for organic products is highest, the pace of organic farming is slower than the growth rate of demand for such products. Besides, Europe is the continent where the consumption ratio organic products/capita is the highest. Precisely, this exact demand structure will lead to an increase in ecologically exploited agricultural land, especially in developing countries. In fact, these countries can use organic farming as an engine of economic development, being statistically proven that ¾ of the total number of organic farmers are operating in these states.

The constraints that farmers face when entering the organic market are very likely to affect an exponential development of this sector. Whether we are talking about financial constraints (the difficulty of attracting investments, limited capacity to obtain agricultural credits, the additional expenditures related to the certification process, the lack of government financial support, the level of subsidies granted per hectare cultivated, the inability to access European funds), technical constraints (obsolete fleet, severe fragmentation of agricultural areas), administrative constraints (excessive bureaucratization, legislative vacuum, lack of a clearly defined professional status of the farmer), constraints related to professional training of employees (low training, aging) or social constraints (adequate information so that the final consumers really know the benefits of organic products in order to be able to efficiently compare certified and uncertified products of the same range), each of these are conclusive obstacles to the development of organic farming.

The additional costs of the organic farming production process propagate along the entire distribution chain, inevitably being found on the shelf. This is what generates the so-called customer dilemma, and statistically it can be seen that the large majority of them prefer the regular product of the same range, against the certified, more expensive one. Therefore, in emerging economies, the development of organic farming is fluctuating. This is also the case of Romania, which, although it had experienced a significant increase in this segment between 2010 and 2012, changed the switch sharply, reaching in 2016 similar figures to those corresponding to 2011, both in terms of the number of operators and the area cultivated.

However, the medium to long-term developing trend will be rising, as organic farming can be a viable solution for economic growth in developing countries. In this respect, it should be noted that among the 2.4 million producers (which cultivate about 51 million hectares of organic products, 1.1% of world agriculture), three quarters are found in the developing countries. Their achievements depend globally on the food security needs of the most demanding consumers, those who prefer healthy, organic products, willing to pay the correct price to get all these intrinsic benefits. In turn, these manufacturers are confronted with emerging economy issues, ranging from volatile legislation to political instability that translates into economic instability, exchange rate fluctuations, or economic crises.

ACKNOWLEDGMENT

This research received no specific grant from any funding agency in the public, commercial, or not-for-profit sectors.

REFERENCES

Advisory Consulting. (2017). *Potentialul dezvoltarii sectorului agricol din Romania*. Retrieved August 28, 2017, from https://www.juridice.ro/wp-content/uploads/2017/03/Raport_PwC-agricultura.pdf

Camera de Comert si Industrie a Romaniei. (2014). *Statistica – Agricultura*. Retrieved August 21, 2017, from https://ccir.ro/wp-content/uploads/2014/11/statistica-agricultura.pdf

Crinnion, J. W. (2010). Organic food contain higher level of certain nutrients, lower level of pesticides and may provide health benefits for the consumer. In *Environmental Medicine* (pp. 4-12). Academic Press. Retrieved on 20 August 2017 from http://www.altmedrev.com/publications/15/1/4.pdf

Crucefix, D. (2012). 10 years back and 10 years forward. In Let the good market flow- Global Organic Market Access in 2012 and beyond (pp 10-13). UNCTAD, FAO & IFOAM.

El-Hage Sclalabba, N. (2013). Organic Agriculture's Contribution to Sustainability. *Crop Management*, *12*. doi:10.1094/CM-2013-0429-09-PS

Eurobarometer conducted by European Commission. (2016). *Europeans, agriculture and the Common Agricultural Policy*. Retrieved August 25, 2017 from https://ec.europa.eu/agriculture/survey_ro

European Parliament Briefing. (2017). *Agricultural education and lifelong training in the EU*. Retrieved December 29, 2017, from http://www.europarl.europa.eu/RegData/etudes/BRIE/2017/608788/EPRS_BRI(2017)608788_EN.pdf

Fibl &IFOAM- Organics International. (2017). *The world of organic agriculture 2017*. Retrieved August 25, 2017, from www.organic-world.net

Fischler, F. (2012). Keynote speech: The role of organic farming in our societies. In Let the good market flow- Global Organic Market Access in 2012 and beyond. (pp. 9-10). UNCTAD, FAO & IFOAM.

Maquet, A., Siderer, Y., & Anklam, E. (2005). Need to research to support consumer confidence in the growing organic food market. *Trends in Food Science & Technology*, 332–343.

Ministerul Agriculturi si Dezvoltarii Rurale. (2016). *Programul de Guvernare 2016-2020. Sprijinirea fermierilor romani*. Retrieved 20 August, 2017, from http://www.madr.ro/docs/agricultura/Programul_de_Guvernare_pentru_agricultura.pdf

Ministerul Agriculturii si Dezvoltarii Rurale. (2015). *Date generale despre agricultura Romaniei*. Retrieved August 21, 2017, from http://www.madr.ro/docs/agricultura/agricultura-romaniei-2015.pdf

Ministry of Agriculture and Rural Development. (2017). *Organic farming*. Retrieved 20 August, 2017, from www.madr.ro/ro/agricultura-ecologica.html

PwC. (2017). *Potentialul dezvoltarii sectorului agricol din Romania.* Retrieved 20 august 2017, from https://www.juridice.ro/wp-content/uploads/2017/03/Raport_PwC-agricultura.pdf

Rundgren, G. (2012). Future: what will happen, what should happen. In Let the good market flow- Global Organic Market Access in 2012 and beyond (pp. 9-10). UNCTAD, FAO & IFOAM.

Williams, C. (2002). Nutritional quality of organic food: Shades of grey or shades of green? *The Proceedings of the Nutrition Society*, *61*(1), 19–24. doi:10.1079/PNS2001126 PMID:12002790

KEY TERMS AND DEFINITIONS

Agro-Food Market: All relations related to stock-purchase acts, including the demand and supply of an agri-food product, as well as the place of meeting between sellers and buyers of the product.

Annual Agricultural Production: Is defined as a function of three variables: available agricultural area, utilized agricultural area (UAA), and last but not least, economic yield of agricultural production.

Circular Economy: The circular economy is a new economic paradigm in which the entire production process is organized in such a way that no waste is generated. Thus, waste is reintegrated into the economic flow as secondary raw material, generating both economic growth and environmental protection.

Food Safety: Involves ensuring the legal framework and the necessary rules for the proper checking of the quality of the agri-food products as well as the careful control of their origin, respectively the traceability transparency.

Food Security: Represents the guarantee for all social categories access to a satisfying diet, compatible with a healthy and active life. The factors that influence food security are the socio-political context, the results obtained in the agri-food sector, the social protection measures, and the measure in which health and hygiene are insured.

Organic Agriculture: Represents the type of agriculture that aims to produce healthy food, according to European quality standards, in full correlation with the protection of the environment, respecting the principles of sustainable development.

Organic Product: Is the result of the production process, based on the raw materials obtained in organic agriculture.

SDG (Sustainable Development Goals): A set of 17 global goals, established by UN in 2015, in order to end poverty, ensure economic growth and protect de environment. Each goal has a specific target to meet in the next 15 years, as part of the 2030 Agenda.

Sustainable Consumption: Sustainable consumption means changing consumers' habits by orienting their preferences to products that correspond to a sustainable lifestyle, ecological products that have been obtained through economic processes based on sustainable development principles. Sustainable consumption implies a fundamental transformation of consumers' values and lifestyles by prioritizing environmental protection and accountability for both quantity and quality of products consumed.

Sustainable Development: Is the economic development which maintains a fair balance between the increasing needs of this generation and the need to preserve the natural resources and to protect the environment on which our future generations depend.

Chapter 14
The Role of Higher Education in Supporting Sustainable Rural Entrepreneurship: A Case Study

Maria Claudia Diaconeasa
The Bucharest University of Economic Studies, Romania

Florentina Constantin
The Bucharest University of Economic Studies, Romania

ABSTRACT

This chapter refers to an empirical research that aims to analyze how the knowledge and skills acquired during undergraduate studies have been perceived by the graduate students and also how the higher education contributes to the development of entrepreneurship in rural areas. The research method identified as being more appropriate was a quantitative research based on an online survey established in accordance with the objectives set up. The participants in the research were the graduate students of the Faculty of Agri-food and Environmental Economy within The Bucharest University of Economic Studies. The findings are based on participants' open statements that could be important arguments to be taken into consideration for improving the decision making in higher education which would lead to the professional development of the graduates.

INTRODUCTION

In Romania the rural area represents approximately half of the total surface and population (NSI, 2016). This area tends to be described by a low level of education, poor infrastructure and a low level of income generated especially by agriculture (NRDP, 2014).

It is reasonable to expect that such a large part of Romania – as a member of the European Union (EU) - should be engaged in activities related to the knowledge market. Since 2000, the EU has promoted the knowledge economy as the new era of global economy and has required equal participation from

DOI: 10.4018/978-1-5225-5739-5.ch014

Copyright © 2019, IGI Global. Copying or distributing in print or electronic forms without written permission of IGI Global is prohibited.

all its members (Lisbon Agenda, 2000). The population could get involved in using new agricultural technologies, rural tourism as branches of entrepreneurship in the rural area or engage in taking decisions for the local community.

Bitterly, such expectations are challenged by the real situation of the rural area. It is acknowledged both by authorities and researchers that rural people are not sufficiently prepared to be an active part of the knowledge economy and its market. As a result, a set of solutions must be found and implemented by the authorities in order to bridge the gap between the current rural area and the desirable one.

Considering that entrepreneurship is acknowledged as a process that enhances *wealth through innovation and exploitation of opportunities* (Nasution, Mavondo, Matanda & Ndubisi, 2011) and that it requires characteristics like risk taking, proactiveness and autonomy of a person, is it possible to speak of it in rural Romania? Even if the current situation is not a desirable one, researchers (Fromhold-Eisebith & Werker, 2013; Bercovitz & Feldman, 2006) imply that the higher education institutions have an important role in enhancing the entrepreneurial development of a region.

In this case, the Faculty of Agri-food and Environmental Economy (FAEE), as part of The Bucharest University of Economic Studies, has the advantage of being the only one with such a profile in Romania. Arguably, the Faculty must fulfil their role in the field of agriculture and rural development and thus, to contribute to ensuring long term food security by educating the rural youth and by triggering its entrepreneurial potential.

The purpose of this paper is to compare the perceptions of FAEE's graduates on the role of this institution with the Faculty's goals and potential of providing a connection between the rural areas' challenges and the knowledge economy by educating new generations of entrepreneurs. One can justifiably assume that the most graduates who choose to work in rural areas build a career in agriculture and the knowledge acquired during their studies can contribute to the development of the rural area.

The graduates' perceptions and the correlations they make between their choice of school, the knowledge and skills needed in the field of work and their possibilities of developing a business have been gathered in a piece of quantitative research based on an online survey. Generating the findings will involve:

- Calculating the proportion of respondents that work in the field of agricultural, food or environmental economy and the proportion of respondents that are rural entrepreneurs;
- Gauging the extent to which the knowledge and skills acquired during undergraduate education has helped the respondents in their current situation;
- Analyzing the respondents' beliefs and attitudes on the usefulness of their education in relation to their career.

BACKGROUND

Entrepreneurship is defined as *any attempt to create a new business enterprise or to expand an existing business by an individual, a team of individuals, or an established business* (Zacharis, Bygrave, & Shepherd, 2000). The concept of new business implies also new information, new relations and new products, which creates a direct connection between entrepreneurship and the knowledge-based economy that is currently replacing the industrial economy. Even more, Kayne (1999) considers entrepreneurs to be *individuals who blend innovation with sound business practices to commercialize new products and services that result in high-growth firms*.

Powell and Snellman (2004) define the knowledge economy as *production and services based on knowledge intensive activities that contribute to an accelerated pace of technical and scientific advance, as well as rapid obsolescence*, the most important aspect of this new economy that people face is a reliance on intellectual capabilities greater than on physical activities and natural resources. Information and the human possibility of using it represent strategical resources for developing and survival of any economic activity, the progress that we all witness is the proof of the increasing of the knowledge's role in the global economy (Popescu, 2014).

Knowledge assets are represented by stocks of knowledge, applied experience, organizational strategy, etc. from which services are expected to flow for a period, hard to specify in advance, but could, in theory, last forever (Boisot, 2002; Vreja & Talpău, 2012). It is obvious that technology has a huge impact on the development of the knowledge economy, but the key actor in it is the human and its ability to master the technology to create profits.

Knowledge economy is seen as one main reason for the growing wage inequality and the educational wage differentials (Levy & Murnane, 1992; Morris & Western, 1999). The explanation for that is simple, the gains in productivity given by better educated workers and better technology lead to a higher demand for educated workers. In this case, the low skilled jobs have the tendency to disappear and to be replaced by technology, so a substantial decrease of less educated people occurs (Powell & Snellman, 2004).

While the OECD reports in 2000 that *firms who introduce new practices such as employee involvement, flatter management structures and team work tend to enjoy higher productivity gains than other firms*, the EU takes things further and builds long term strategies like Lisbon 2000, Europe 2010 or Horizon 2020 that include objectives of smart grow, sustainable grow or inclusive grow.

In the EU's acceptance, smart growth refers to developing a knowledge, research and innovation-based economy, sustainable growth represents promoting an economy that is efficient in lowering the resource consumption and inclusive grow refers to an economy with higher employment rate, able to generate social cohesion (EU Commission, 2010). The objectives set for this smart growth or development are not very tight and they regard in general education, innovation, information and communication technology. As a counter position, Naldi et al. (2015) came with the concept of smart specialization that refers to focusing on the most promising economic branches of a region and fully developing them, the other branches should follow the lead to growth, this concept has a strong connection to knowledge transfer.

Researchers tackle the problem of measuring the results of investing in higher education. The Economist asks if the 2.7% of GPD American investment in higher education is paying off or is a waste of money. To them it would seem that the employers are more interested in the name of the university an employee graduated rather than his/her skills and knowledge and that encourages the cost restrictions of the most desired universities and not an increase of the highly skilled workers (The Economist, 2016). On the other hand, in the UK (Dearden, Fitzsimons & Wyness, 2014) demonstrate that the subsidies for poorer students do contribute to the increase of student's participation in higher education.

In this case, the authors ask themselves: is higher education an important factor in developing rural entrepreneurship? The answer is simple, yes. Because being either an entrepreneur or a highly paid employee in the knowledge economy requires a solid educational base. Implicitly, finishing a tertiary education institution gives a person a superior qualification and it should ensure a starting point for him/her to work and provide a higher profit for any business.

Authors (Altbach & Salmi, 2011) have conclude that universities have earned their place as central institutions in the knowledge economies of today's world, an elite role they had lost for a long time.

In Romania, the National Education Law 1/2011 states that the mission of higher education is to generate and transfer knowledge to the society by initial and continuous formation at university level with the specific purpose of personal development and professional insertion of a person and satisfying the competence needed by the socio-economic environment. Yet, in the 2016 European Monitor for education and training, Romania's summary for higher education is dry: *tertiary educational attainment has risen so that it is now very close to the national target, but it is still one of the lowest in the EU and ensuring labor market relevance of higher education is a challenge*. Also, the EU mentions that the public expenditure on education is very low in Romania, so there is no need for a sophisticated analysis to understand that one can't expect to have highly skilled entrepreneurs and workers in a developing knowledge economy with a low national investment in the educational system.

By looking at the Skills Panorama database (Skills Panorama, 2017) one can see that in 2014 Romania was the third in the rank of highest level of skills obsolesce with a percentage of 62.9%. The government implication is essential in changing this kind of situation, as Salmi (2009) observes.

The concern of what future skills will be needed has not remained at institutional level, but many private actors have started researching into this problem and websites have dedicated it more and more space. For example, The World Economic Forum website has given a lot of attention to the future skills topic. They emphasize the necessity of skills such as: active learning; oral expression; reading comprehension; written expression and ICT literacy as content skills and active listening; critical thinking and monitoring self and others as process skills, the two categories form the basic skills. There are also cross-functional skills formed of social skills, system skills; complex problem-solving skills; resource management skills and technical skills.

In correlation to the business environment and developing the entrepreneurial spirit, other authors (Letena, Landonic & Van Looy, 2014) state that the proximity of a university is beneficial for firms. Even more, Fromhold-Eisebith (1992) considers that when the firms profit of a university's proximity through keeping contact with former professors or colleagues or by using testing and measuring equipment they create information flows. This results in effective spin-off entrepreneurship. Certainly, not all universities generate the same amount of entrepreneurial spirit. Authors like Di Gregorio and Shane (2003) consider that a series of factors like discipline focus, orientation towards private funded research and development projects and the support the university offers in raising students' awareness of entrepreneurship opportunities define its importance in enhancing the entrepreneurial spirit of a region.

Altbach and Salmi (2011) consider that *universities play a key societal role by serving as cultural institutions, center for social commentary and criticism, and intellectual hubs* and for the developing countries, finding the effective mechanisms *to participate in the global knowledge network on equal basis with the top academic institutions in the world* is a major challenge. The fact that Romanian universities have started an evaluation process and that some of them are visible at an international level is a good sign in the direction of taking this challenge.

The evolution of the number of students is important for all the EU members since they have a 2020 target to accomplish. While there is a general target for the EU, of 40% of the population between 30 and 34 years old to have graduated a form of higher education (Europe2020 Indicators, 2017), there are individual targets for each member. Thus, in Romania, the target is set at 26.7% and it has been almost reached, since the percentage was 25.6% in 2015. Romania has assumed one of the lowest targets in the EU only Italy has a lower target.

The preferred study fields for the Romanian students are economics, public administration, law and engineering. These fields gather more than 45% of the students in 2016 (NSI, 2016). Agricultural studies are placed at the end of the preference list, along with educational studies. The idea of a rural or agricultural business is not as desirable as an urban one.

Also, the fact that students prefer traditional universities to increase their security in getting a job (Lenton, 2013; Drydakis, 2016) is true for Romanian students as well. They prefer historical university centers with traditional universities. Bucharest, the capital of Romania, is the most desirable university center because of the many fields of interest available, the opportunity of getting a job during studies, number of dormitories or leisure opportunities. The statistical evidences show that approximately a quarter of the students prefer it. In 2015 more than 171 000 persons studied in Bucharest. Next on the preferred university centers were Cluj-Napoca with more than 65 000 students in 2015 and Iaşi with more than 55 000 students. Other known centers are Timişoara, Constanţa, Braşov, Sibiu and Oradea (NSI, 2016). The agricultural domain as a specialization has been of interest since the early period of higher education in Romania and it still is, ten traditional universities include this profile in their educational offer. Also, some of the present profiles have been adapted to include both economic and environmental knowledge.

Each one of the traditional university centers of Romania offers the basics of one multidisciplinary qualification for the farmer that should be able to be a business man, a manager, a sociologist and an engineer all in one, while Bucharest is the only one with two offers, and it has the highest number of students.

JOB EVOLUTION IN 2015-2025

The trend forecasted by the European Centre for the Development of Vocational Training (CEDEFOP) regarding the future employment trends and the employment needs of the future years shows that the difference between job opportunities and employment trends stands in what the labor market demands in terms of job openings or replacements and the number of employed people in different economic sectors. The evolution of employment in terms of qualification of employees between 2015 and 2025 will register a small increase for the low qualified personnel (6.2%), a dramatic drop (-21.3%) for the medium qualified personnel and a major increase (40%) for the highly qualified one (CEDEFOP, 2017).

In terms of employment evolution, the agricultural field should register a drop of 10% in worker numbers. Because, in the present, Romania has the highest percentage of agricultural workers, 30% (CAP Indicators, 2017), the drop is understandable, the need for qualified agricultural workers can be felt for a good period now. Also, a major drop in employee numbers should be registered in mining and quarrying industries, while the administrative, IT or real estate fields should register significant increases of more than 10%.

IS IT POSSIBLE TO CHANGE EDUCATION AND TRAINING FOR THE AGRICULTURAL SECTOR?

Considering the prognosis of the United Nations, according to which the world population will reach 8.5 billion by 2030 (MDGs, 2017) and the known food production processes it is obvious that something must be done in order to reinvent the agriculture so it can feed the growing population. But is this possible?

In the given situation and considering these projections, one could easily assume that Romanian farmers would take advantage of the opportunity and prepare themselves or their children to properly manage their farms as business and be important parts of the future supply of worldwide agricultural products. The current percentage of farmers with full agricultural training is 0.5% in Romania, while the EU average is of 13.9%, according to CAP context indicators (2017).

Correlating the fact that there is an increase in demand for qualitative agricultural products and the high agricultural area that Romania has, this could be an opportunity to develop in order to properly respond to that demand. The lack of understanding this as an opportunity and missing the start for having skilled agricultural workers soon may cause Romania further lagging.

THE ROLE OF FACULTY OF AGRI-FOOD AND ENVIRONMENTAL ECONOMY IN DEVELOPING ENTREPRENEURIAL SPIRIT IN THE RURAL AREA

By highly skilled agricultural workers one shouldn't understand that a person will need special qualifications for planting the seeds, but that person will have to know the consumption trends, to be connected to the markets, to have good skills in marketing, management, accounting and logistics, to be aware of climate changes, sustainability options and scientific research, to have legislative and financial knowledge and negotiation skills apart from knowing the best agricultural practices and the newest technologies available. Moreover, a farmer should care for the community and the future of it. Because farming isn't just an occupation, nowadays it has become a business that must make profits while respecting high environmental and social standards that are not meant only for the present, but for the future.

For a population almost equally spread between rural and urban, a very large area of agricultural land and potential for animal breeding and a high demand for highly qualified agricultural workers, one could assume that the educational offer would be more focused and willing to contribute to the of a competitive and comprehensive generation of farmers. Yet, only one higher education school felt the opportunity to mix tradition with the future, the Faculty of Agri-food and Environmental Economics (FAEE), an institution that modified its curricula to include the basic skills for the wanted farmer as a business man.

The further study will try to find the opinions of graduates, regarding their satisfaction with the curricula, the lectures content and way of delivering and if they believe that this faculty has been helpful in developing skills and knowledge requested in the present labor market as part of the knowledge market.

Needed skills encountered by both entrepreneurs and employees can be split into cognitive skills, such as: compliance with professional ethics; objective/results presentation; achieving new knowledge from daily experience; Integrating new technologies in regular work processes; adapting to change/innovation; setting, following and reviewing a strategy; taking decisions in a short time and evaluating the long term impact of those decisions; correct delegation of tasks and involvement in designing the future of the company; and practical skills, such as: fast problem identification; problem solving; contribution to team problem solving; prioritizing individual tasks; setting and meeting deadlines; report writing; efficient communication with the team and colleagues; working well under stress; balance between professional and personal life and accepting failure as a possibility. There cannot be an exact delimitation between cognitive and practical skills, they intertwine, and one can't just have the cognitive without the practical ones. If a practical skill can be thought or intensified through knowledge, a cognitive one is developed or may be developed by practical experience.

ANALYSIS OF GRADUATES' PERCEPTION OF THE KNOWLEDGE AND SKILLS ACQUIRED DURING UNDERGRADUATE STUDIES

In order to draw an accurate image of the FAEE's response to the labor market requests regarding the graduates' skills and knowledge in the field of agriculture, food industry and environmental economics, the authors considered that eliciting the opinions of those directly involved in the process is most relevant. More specifically, the opinions of the FAEE's graduates who are currently working were collected. As these respondents went through all the stages from student to employees or entrepreneurs, their opinion regarding the relevance of undergraduate training for getting a job is a powerful information resource for this study.

Designing the Questionnaire and Methodology

The fact that the present society builds its life on the basis of real time knowledge transfer and makes decisions based on the truthfulness of information regarding a certain topic has generated the present study: how does the university environment respond to the request for knowledge, skills and practical training that comes from the labor market of the present knowledge economy, with an accent on the Faculty of Agri-food and Environmental Economy? The labor market is mostly formed of private companies which must keep up with international standards and innovation to survive. Implicitly, they ask the same from their employees.

Romania, as a member state of the EU for more than ten years, must align to the concept of "knowledge society" as a whole, not just on certain groups or areas, and considering that academia is one that should be first aligned with European standards, the authors want to see what is the opinion of those who went through the system and had to find a job or start a business based on received training in the field of agri-food or environmental economics.

The selected method for data collection in the present study is direct research, a method that asks for the participation of the group, not just observation. The data collected by the authors is mostly of quantitative nature, the perceptions and opinions of the respondents on the selected topic were processed by the authors and resumed also to a quantitative analysis. The authors chose to consider only the graduates of FAEE undergraduate program for gathering diverse answers which led to relevant conclusions, since the direct research method involves collecting data directly from their carriers.

The direct research method chosen is a survey, applied through a self-administrated structured questionnaire as a tool. The questionnaire was distributed online, and it allows obtaining relevant results.

The questionnaire was introduced in the Google forms platform and posted online via Facebook groups created by the graduates of several generations. Since answering the questionnaire was at the free will of the graduates' group members the study falls under the category of pilot research and is empirically based. The main group selected has members from many generations of FAEE graduates, and there were also selected three smaller groups, of the 2008, 2012 and 2016 generations of graduates. The Facebook groups where the questionnaire was distributed have a frequent activity and were chosen to get a fair opinion of the respondents. Even if the respondents' answers cannot be extrapolated to the entire community of FAEE graduates, they provide a powerful image of the perception on the quality of lectures and seminars offered by this school.

Research Hypothesis and Research Questions

The initial hypothesis of this study is that the FAEE has a real contribution to the training of the future entrepreneurs and employees in the fields of agri-food and environmental economics through offering a relevant curriculum and developing important skills for the students. The main objective of this research is confirming or rejecting the hypothesis.

The research objectives of this study are:

- Establishing how many of the respondents could find a job or start a business in the field of agri-food and environmental economics;
- Gauging the extent to which the knowledge and skills acquired during undergraduate education has helped the respondents in their current situation;
- Analyzing the respondents' beliefs and attitudes on the usefulness of their education in relation to their career.

The research questions of this study are the following:

1. What is the motivation of choosing a carrier in agri-food or environmental economics?
2. Is the FAEE's curricula adapted to the knowledge and labor market demands in the field of agri-food or environmental economics?
3. Does the business environment collaborate with the academic one to standardize specific skills and knowledge concepts for the future employees or entrepreneurs in the field?

The necessary data for this study were gathered through a structured questionnaire named *The response of the Faculty of Agri-food and Environmental Economy to the Labor Market Demand in the Field of Agri-Food and Environmental Economics*. The questionnaire combines various types of questions and invites respondents to provide several extensive answers on their personal learning experience and evaluation of learning outcomes. It also contains questions that require simple direct answers and complex structured questions that invite respondents to express their opinion by rating several statements. The questionnaire is presented in Appendix.

To achieve the above-mentioned objectives, the authors designed a questionnaire of 21 questions, structured so as that the logical strand of the questions is derived from the previous answer possibilities. For obtaining an answer to all the questions there were used both answering options and open answers. The questions are structured into four sets as it follows: the first set, has the role of finding if the respondents are fit for the study and to set up their general profile. The second set aims at finding the motivations of choice for undergraduate program, working place and the satisfaction or dissatisfaction of the choice made and what improvements would they make. The third set of questions aims to find to what extent the respondents are aware of the skills required of an entrepreneur/employee and what importance they give to each of them. Also, it is set out to find what skills do the respondents associate with something they learned during undergraduate program. The final set is aimed at seeing if the respondents consider that the involvement of the business environment in the academic environment as useful and what possible partnerships they propose.

The first question "Are you a FAEE graduate?" aims to select the right respondents for the study: if the answer is no, then the respondent has finished the questionnaire. The question "When did you graduate from FAEE?" has the role of allowing a categorization of the respondents. The next question "Which is your residence area?" aims at finding if the FAEE is an attractive school for rural youth more than for the urban. The fourth question "What is your age category?" has the role of finding if there are important differences of opinion based on the respondents' age. The fifth question aims at finding if the graduates of FAEE are restricted to a certain income.

The first question of the second set "What motivated your choice of FAEE against other possibilities?" has the purpose of registering the possible thought mechanisms of young people when they had to decide their future and seeing to what extent was it a true choice for them or if they were influenced in a certain way. The next question "At the moment, do you consider you made the right choice or would you like to change it? Please motivate your answer." aims to make the respondent acknowledge all the progress made since student days and to evaluate the present situation in order to see it's satisfaction with the choice made. The following question "Are you occupied in a field related to your specialization? If not, why did you chose to change the field?" aims to see how many of the respondents are currently occupied in a field related to their training and if not, what triggered the choice of change. The question "Are you an employee or an entrepreneur?" has the purpose to see how many of the respondents have developed an entrepreneurial spirit since the rural area has many opportunities in this respect. The following question "What time was necessary for finding a job after graduation?" has the purpose of revealing an average time needed by FAEE graduates in finding a job and it is complemented by the next one "What obstacles did you meet in finding a satisfactory job?" - intended to find the most common obstacles the FAEE graduates meet in their path of work. The next question "Do you believe the courses you followed during undergraduate program have helped you in finding a job? In what way?" aims at finding if there are any courses that the graduates find useful in a practical manner and it is complemented by the next question "What specific courses do you find useful? In what situations?" to see if they associate something they learned during a course with a certain life situation which they solved with that knowledge. The next question "Do you consider that this faculties' profile is able to bring competitive advantages to its graduates? If not, what changes would you make in order to make FAEE competitive?" aims at finding the respondents opinion regarding the possible advantages the FAEE may offer to its students and what improvements would they suggest to increase those advantages. The question "If you ever considered opening a business related to this field, what sources of information would you use?" aims at finding to what extent the research skills have been triggered during the undergraduate program and if the respondents would call upon their professors for information.

The third set of questions starts with "What importance do you believe the following skills have in finding and keeping a desirable job?", a question that aims to see what skills do the respondents find more important for their jobs, and the second "Which of those skills and to what extent were triggered or developed during undergraduate period?" aims to find their opinion on the possibility an undergraduate program has in developing such skills. The final question of the set "What possibilities of developing these skills during undergraduate program do you see" aims at finding the respondents opinions on how to improve the skills development during undergraduate programs.

The final set of questions, turns to the practical aspect of an undergraduate program and the possibility of a real connection between theory and practice. The first question of the set "Do you consider that the profile companies are satisfied with the level of training of the new graduates? Please motivate your answer" aims to elicit the perception of the respondents regarding the level of training of new graduates. The final question "Do you consider the agri-food and environmental economics related companies are involved in the students training? In what way?" has the purpose of finding if the employers who generally have high demands of the young graduates get involved in the education and training of students in their field of interest, knowing that universities have limited ways of ensuring practical experience on their own.

Questionnaire Results

From the 51 respondents only 38 were in the researched category and continued responding to the questionnaire. From the first set of questions the authors extracted the respondents' profile. Most of the respondents are young, three quarters of them are 25 to 28 years old. The most present generation in this study is the 2012 graduates' generation, with 36.8% of the answers. A quarter of the respondents come from the rural area, a high percentage considering the low numbers of rural students, which can be translated in the high importance this school has in developing the rural entrepreneurial spirit. 7.8% of the respondents are rural entrepreneurs, while only 5.2% are urban entrepreneurs.

Regarding the incomes, 27% of the respondents are below the medium wage in Romania, 55% have a medium wage and only 18% are above a 4000 RON level.

Regarding the motivation of choice for the field of study of the respondents: studying the agri-food sector was the motivation for four of the respondents, half of them from the rural area. This means that the need for highly qualified people in this sector is equally felt in both areas. A pragmatic motivation was considered by nine out of ten rural area graduates and for 21 out of 28 urban area graduates. This means that this faculty has a high importance for the potential students and its pragmatic approach on current economic issues is a plus in their vision. Considering the respondents satisfaction with the choice they made, we may say that most of the respondents, 60.5%, consider they made a good choice because currently they have a job, or they own a business in the field. 15.7% have some doubts about their choice, and they would give some serious thought if they had the chance of a new start, while 23.6% would definitely make another choice.

47.3% of the respondents currently work in the field of agri-food or environmental economics, in state institutions, consulting firms and private business or in their own farms. 18.4% have some connection to the graduated field, they work in the financial field or they have some kind of managerial position. The other 34.2% of the respondents have switched fields due to the lack of job opportunities, better payment in other fields or since they discovered another passion which they turned into a job.

68% of the respondents are currently employees, 13% are entrepreneurs, 11% are both employees and entrepreneurs, and only 8% didn't have a job at the moment of filling the questionnaire.

There are more entrepreneurs in the rural area (3) than in the urban area (2), while both type of activities are held by 3 graduates from the urban area. Also, the 3 respondents that didn't have a job live in the urban area. 7 respondents from the rural area and 20 from the urban area are currently employees.

The form of entrepreneurship is not well developed, even if the school has more than one course that should be inspiring for the students, after graduation they prefer the safety of a job to the risk of entrepreneurship.

Next question regards the period passed since graduation until finding a good job, in their perception. Most of the respondents (31.5%) had a job during the undergraduate program and decided to keep the job after graduating. 18.4% have spent 1 to 3 months after graduating for finding a job, a period considered small and beneficial for them in order to relax from the stress of the graduation exam. 23.6% of the graduates have spent 4 to 12 months finding a job and 26.3% spent more than one year for this matter. While those who spent little time in finding a job have a stronger connection to the FAEE field (7), only 3 of those who spent more than one year in finding a good job have connections to the field of study.

The most important obstacle in finding a job mentioned by the respondents is the difference between the lectures, which have a more theoretical and historical base rather than a pragmatic one, and real market requests. The following obstacle turns to the additional training modules needed to rise to the market requests, an activity that takes both money and time from the graduates. Other obstacles mentioned regard the differences between the job offer and the employers' requests at the job or the feeling of not being prepared enough. 31.6% of the respondents considered that they had no important obstacles in getting a good job.

The following set of questions aims to find if the lectures seemed useful to the graduates, especially in correlation to their current jobs. Half of the respondents consider that they had something to learn from the lectures, even if it wasn't a specific concept, they remained with a logical way of solving problems, of organizing, presenting or reporting an issue. 15% of the respondents consider that they have a partial use for the lectures, while 5% have little use for the lectures, most of them mention the general knowledge they now use. 10% mention that a more practical approach would have made the lectures truly useful, while 18.4% reveal that they had no use for any of the lectures during undergraduate program. The lectures with a high relevance in the graduates' opinion are those which offer a general base of knowledge, concepts that can be applied in many circumstances, rather than specific ones.

Regarding the advantages this faculty might bring for its students in a competitive economy, 16 of the respondents considered that this faculty does bring a competitive advantage. Their answers rely mostly on the uniqueness of the program and the provided development of the environmental sector. 12 of the respondents consider that this faculty has the potential of offering a competitive advantage for its graduates under some conditions, like: "If more emphasis was placed on European funds because it is a topic of current and future interest" (O41); "If they would opt for a career in agriculture, the profile would have many advantages because of the information they have gained" (O39) or "Yes, if we had a developed agro-food industry, with solid production units, to capitalize on the agricultural potential of this country" (O19).

Basically, the respondents consider that a more practical approach, the development of agriculture and environmental sectors and a specific interest in this field are the recipe for competitive graduates. The rest of the respondents 26%, consider that the faculty doesn't bring anything special for its graduates.

Considering the changes fit for increasing the quality of the faculty, the main idea of the respondents stands for introducing more practice opportunities, through partnerships with the business environment, real study cases or internships. Even more, one participant suggests international exchanges with western professors, for the students to have a wider understanding of concepts and approaches. Other ideas include emphasis on niche specializations, updated courses, while two participants recall the necessity for a strong teacher training and basic skills that each student should learn and use.

The pragmatism of the graduates can be observed from their answers to the next question, even if they had multiple answers, most of the respondents would turn to guides and legislation (78%) first when opening a business and second, to consultants (73%). The internet is the third option (63%). Appreciably, the internet is not the first option, the quality of the information available being difficult to establish, so the graduates consider finding more reliable sources of information. Courses and books are an option for gathering information for 47% of the respondents, while only 34% would turn to their professors for advice.

The next two questions have the purpose of evaluating the importance of soft-skills in the respondents' opinion and seeing how much they believe these soft-skills have been triggered or developed during their undergraduate program at FAEE. Both cognitive and practical skills have been mentioned to the respondents. The respondents had to give a mark from one – the least – and five – the most – according to the importance they give to the mentioned soft-skills.

The most important skill, probably because of the difficulty of achieving it, is the balance between the personal and professional life. Young people tend to forget about their personal life in the struggle for success, but the fact that the respondents are the millennials generation (Banon, Ford & Meltzer, 2011) helps them in realizing the importance of personal life. Next, in the order of importance come finding key components in problem solving and setting and meeting deadlines – skills that often include more than one person, so the responsibility for solving problems and setting feasible deadlines is extended to other people as well. A high importance is involvement in designing the future of the company. Objective or results presentation was graded with three by most of the respondents. That may mean a higher emphasis is put on achieving the results than on presenting them. Yet, a good presentation of objectives gives a team more understanding of their tasks and a good presentation of the results gives the team motivation. Report writing has an equal spread of opinions between two and four, this skill less important than others in the respondents' opinion.

The next question has the aim of finding if any of those skills have been triggered or developed by the activities in undergraduate program. There is an obvious difference between the use and need of the skills in the workplace and the undergraduates' programs' possibility of developing them, most of the skills are evaluated by the respondents as half developed. Still, there is no skill entirely undeveloped. The skill with the highest development is communication with the team, an indicator that all the team projects have a good and practical result, apart from grading students.

The least developed skills are balance between personal and professional life and involvement in the future of the company. The first can be more developed if the students imagine their studies as a real job, but even so, the real struggle of balancing job success and personal life can't be understood until being put in that position. The second can only be included in some well thought case studies. A practical module couldn't work in this case, no company would give such an important role to an intern or student.

When asked what chances of a better development of these skills do they see fit, the respondents considered that practice is the key. Through longer practice modules, paid practice, workshops, a close relation with the business environment, 47% of the respondents consider that skills would be better developed. Other opinions regard designing case studies in accordance to the reality of the economy, innovation in teaching methods and team projects with reality inspired objectives. There is one last opinion, coming from 4 of the respondents that is directed to the professors, they consider that a better teacher training is needed, to update the skills of the professors first and then develop the students ones as a result.

Through the following question the authors aimed at finding the respondents opinion on how satisfied the companies in this economical field with the graduates' level of training and knowledge are, since there are entrepreneurs and people in managerial positions who can offer an answer to this question. For those able to answer the question, the general belief is that the companies are not really satisfied with the graduates. The prevalence of theory against practice makes the companies take the role of learning organizations and there aren't many which are happy or willing to take that role. Yet, companies have a reduced role in training the undergraduates. The most mentioned way of business involvement in students training was being a practice partner for the faculty, 47% of the respondents have heard of this option, other 34% know companies which offer internships for interested students, 21% have heard of companies who offer scholarships and 11% know that there are companies who participate in lectures or give presentations to students. 26% of the respondents don't know companies involved in any of the mentioned ways.

A limitation of the study is that respondents did not have the option of specifying if the same company is involved in more than one way and the quality of the practice modules offered wasn't taken into consideration.

Questionnaire Conclusions

The initial hypothesis of the study is not entirely confirmed, the contribution of FAEE to the training of future employees and entrepreneurs in the agri-food and environmental economics is recognized, but the majority of the respondents consider that the lectures should be updated, connected to the dynamics of the economy and accompanied by case studies and practical applications designed by well trained teachers, in a relevant way for the students, in order to develop the soft-skills more than they do at the moment. Also, a great emphasis is put on practice. Ensuring a relevant practice module for the students represents, in the respondents' opinion, the kick start of their career. The faculty has a high importance for the potential students due to its pragmatic approach on current economic issues.

Most of the respondents admit that general theoretical concepts ensured by basic economics lectures can be used for any domain and help them in their current jobs, while more specific lectures are highly used by those who work in state agencies, consultancy in this field and use EU funding for their business. Regarding the skills developed by the school, they are at half of their potential, according to the respondents, but it is important that all the considered skills are in some way triggered.

Regarding the business environments' cooperation with the academic one in order to standardize some specific skills and knowledge concepts for the future employees or entrepreneurs in the field, there is little interest. The main activity is offering practice places for students. Both state agencies and private companies allow students to come and observe their activity. There are fewer internship places, since they imply giving the intern specific tasks, someone to teach him/her, and sometimes payment. The easy ways of cooperating with the academic environment are preferred by the business one, such an attitude leads to the companies becoming learning organizations each time they hire new graduates.

FUTURE RESEARCH DIRECTIONS

The results of this pilot study give the authors motivation to reproduce this empirical research on a large scale, with statistical representativeness. Ideally, this questionnaire could be followed by a series of in-depth interviews in a genuine mixed methods approach. Such an approach may be a useful continuation of the research reported here.

CONCLUSION AND IMPLICATIONS OF THE STUDY

Adjusting to the knowledge economy poses significant difficulties for each person since humans' capacity of understanding innovation is used to its maximum. It should be easier for young people since they have the opportunity to be born into this world ruled by information. Yet, the real challenge comes for their teachers, to be able to keep up with the technology dynamics and still to pass on the bases and history of all the novelty.

The role of knowledge in the global economy and the ability of using it in new ways has become more important than it ever was before. The key actor in developing the knowledge economy is the human and its ability to master the technology to create profits. The human capital of a firm must be relevant in this context, it consists of the abilities, knowledge, skills and qualifications the people working in a firm have obtained through education and continuous development.

The emphasis put on higher education comes from the superior qualification given by completing such a program, which should ensure a starting point for a working person in the knowledge economy. Yet, the evaluation for Romania given by the EU Monitor (2016) shows that we stand at the bottom of the ranking. In this context, the authors considered if the program addressing rural entrepreneurship is prepared enough, regarding the curricula and the teaching methods, for being an active participant in this string of changes. Asking the graduates about their opinion on the program and suggestions for designing a better one is a first step in keeping a real connection between the market demands and the school offer.

Making universities central actors for designing new economic and social strategies is the key chosen by many countries for developing soft-skills and basic concepts for graduates that should be relevant for the labor market. A partnership between businesses, public and academic sectors should be able to ensure well prepared graduates.

Agriculture represents an important part of the economy for Romania, a part that employs almost 30% of the population, has low productivity, low use of technology and employees that aren't trained, but they work based on practical experience. The current trend asks for farmers who can be highly skilled marketers, negotiators, managers and environmentalists. Moreover, they should care for the community and the future of it. Yet, to get formal training in this specialization there are limited options available in Romania. The Faculty of Agri-food and Environmental Economy within The Bucharest University of Economic Studies managed to include both economic and environmental knowledge in the curricula. Even if the profile sounds more interesting for rural youth, it doesn't pose any restrictions for the urban youth.

Considering the perceptions of FAEE graduates regarding the schools 'possibilities of developing the soft-skills the agricultural and environmental sectors need from the future employees or entrepreneurs, the main results are: the potential of the school is amazing because it is a niche, but some important adjustments should be made to reach that potential.

Some improvements and conclusions drawn from the respondents' answers are:

- The curricula should be updated to keep track of the changes in the national and international economic landscape;
- More emphasis should be put on knowledge with general use, like general economics or statistics;
- Specialized lectures have an important role for the students, but they would understand more through practical applications than through theory;
- More emphasis should be put on stages of practice, the graduates suggest that the key for having some experience and increasing your possibilities for getting a job in the field is a good practice module;
- Having well trained professors, able to integrate novelty in practical applications and case studies is a plus, so constant modules of training for them would benefit both the students and the school;
- Partnerships with the business environment for developing workshops, applications and case studies based on real situations would mean a better connection for the students with the market and skills development.

A quarter of the respondents of this study are located in the rural area, so the role of this school in developing the agricultural and environmental specialists in the rural area is acknowledged. Yet, the promotion of the program targeted to high schools in the rural area should increase the number of rural students and further, have a positive impact in developing the rural area by delivering skilled business managers and entrepreneurs.

The rural youth attracting strategy should emphasize the opportunities of each sector, the importance of developing the rural area – and so, showing the rural youth its high stakes role. Also, considering the low incomes of the rural population, the strategy should rely on the number of budget places, the number of dorm rooms with low costs and not least on the scholarships that recently have increased to almost a minimum wage value.

Attracting former students that now have a high managerial position in a private or public institution or that are now entrepreneurs into having a role in the schools' dynamics means increasing the opportunities of good practice places and a real connection with the market. This idea is already taking place through some of the professors, but more emphasis should be put on it by all faculty.

Developing a set of short time courses, designed especially for niche specializations, differentiated into postgraduate studies or introduction to basic concepts courses, is an idea that might help both the farm managers for developing their business and the recognition and promotion of the school.

Overall, the importance of the school in developing the entrepreneurial spirit in the rural area is acknowledged. Still, it could always could be improved. A good communication with the students and graduates ensures the school a down to earth approach and real time opportunities for upgrading its lectures, applications and answering both students and market needs.

Considering that a school curriculum may be changed periodically, such a study may be of interest when adding or deleting some of the lectures. Also, the request for longer practice modules should be taken into consideration by the school board. It is true that for emphasizing practice the private business environment should have a greater openness to partnerships with the academic environment. In this regard, the business environment might be interested of this study.

REFERENCES

Altbach, P. G., & Salmi, J. (Eds.). (2011). *The road to academic excellence: The making of world-class research universities*. World Bank Publications. doi:10.1596/978-0-8213-8805-1

Bannon, S., Ford, K., & Meltzer, L. (2011). Understanding Millennials in the workplace. *The CPA Journal*, *81*(11), 61.

Bercovitz, J., & Feldman, M. (2006). Entrepreneurial universities and technology transfer: A conceptual framework for understanding knowledge-based economic development. *The Journal of Technology Transfer*, *31*(1), 175–188. doi:10.100710961-005-5029-z

Boisot, M. (2002). The creation and sharing of knowledge. *The strategic management of intellectual capital and organizational knowledge*, 65-77.

CAP Context indicators. (2017). *CAP Context indicators*. Retrieved June 23, 2017, from https://ec.europa.eu/agriculture/sites/agriculture/files/cap-indicators/context/2016/indicator-table_en.pdf

CEDEFOP. (2017). *Employment trends*. Retrieved August 17, 2017, from http://www.cedefop.europa.eu/en/publications-and-resources/data-visualisations/employment-trends

Dearden, L., Fitzsimons, E., & Wyness, G. (2014). Money for nothing: Estimating the impact of student aid on participation in higher education. *Economics of Education Review*, *43*, 66–78. doi:10.1016/j.econedurev.2014.09.005

Di Gregorio, D., & Shane, S. (2003). Why do some universities generate more start-ups than others? *Research Policy*, *32*(2), 209–227. doi:10.1016/S0048-7333(02)00097-5

Drydakis, N. (2016). The effect of university attended on graduates' labor market prospects: A field study of Great Britain. *Economics of Education Review*, *52*, 192–208. doi:10.1016/j.econedurev.2016.03.001

EU Commision. (2010). *Europa Agenda*. Retrieved November 14, 2015, from https://ec.europa.eu/research/innovation-union/pdf/innovation-union-communication_en.pdf

EU Monitor. (2016). *Education statistics*. Retrieved March 23, 2017, from https://ec.europa.eu/education/sites/education/files/monitor2016_en.pdf

European Union. (2010). *Horizon 2020*. Retrieved September 5, 2017, from http://ec.europa.eu/eurostat/web/europe-2020-indicators/europe-2020-strategy/main-tables

Fromhold-Eisebith, M. (1992). Regionalwirtschaftliche Effekte des Wissens-und Technologietransfers der Rheinisch-Westfälischen Technischen Hochschule Aachen. *Geographische Zeitschrift*, 230–244.

Fromhold-Eisebith, M., & Werker, C. (2013). Universities' functions in knowledge transfer: A geographical perspective. *The Annals of Regional Science*, *51*(3), 621–643. doi:10.100700168-013-0559-z

Kayne, J. (1999). *State entrepreneurship policies and programs*. Academic Press.

Law 1/2011. (n.d.). Retrieved from http://www.fen.ro/legea_educatiei_final.pdf

Lenton, P. (2015). Determining student satisfaction: An economic analysis of the National Student Survey. *Economics of Education Review*, *47*, 118–127. doi:10.1016/j.econedurev.2015.05.001

Leten, B., Landoni, P., & Van Looy, B. (2014). Science or graduates: How do firms benefit from the proximity of universities? *Research Policy*, *43*(8), 1398–1412. doi:10.1016/j.respol.2014.03.005

Levy, F., & Murnane, R. J. (1992). US earnings levels and earnings inequality: A review of recent trends and proposed explanations. *Journal of Economic Literature*, *30*(3), 1333–1381.

Morgenstern, M. (in press). Artificial intelligence: The impact on jobs. Automation and anxiety. *The economist*.

Morris, M., & Western, B. (1999). Inequality in Earnings at the Close of the Twentieth Century. *Annual Review of Sociology*, *25*(1), 623–657. doi:10.1146/annurev.soc.25.1.623

Naldi, L., Nilsson, P., Westlund, H., & Wixe, S. (2015). What is smart rural development? *Journal of Rural Studies*, *40*, 90–101. doi:10.1016/j.jrurstud.2015.06.006

Nasution, H. N., Mavondo, F. T., Matanda, M. J., & Ndubisi, N. O. (2011). Entrepreneurship: Its relationship with market orientation and learning orientation and as antecedents to innovation and customer value. *Industrial Marketing Management*, *40*(3), 336–345. doi:10.1016/j.indmarman.2010.08.002

National Rural Development Programme. (2014). *National Rural Development Programme*. Retrieved December 10, 2014, from http://madr.ro/docs/dezvoltare-rurala/programare-2014-2020/PNDR_2014_-_2020_01.07.2014.pdf

National Statistics Institute. (2016), Tempo online, Educație. *Studenți și cursanți înscriși în învățământul superior, pe grupe de specializări*. Retrieved July 12, 2017, from http://statistici.insse.ro/shop/index.jsp?page=tempo3&lang=ro&ind=SCL103L

OECD. (2017). *Skills beyond school*. Retrieved July 12, 2017, from http://www.oecd.org/education/skills-beyond-school/

Popescu, G. (2014). *Cooperation in agriculture from the historic argument to the knowledge transfer*. Bucuresti: Editura Academiei Romane.

Powell, W. W., & Snellman, K. (2004). The knowledge economy. *Annual Review of Sociology*, *30*(1), 199–220. doi:10.1146/annurev.soc.29.010202.100037

Salmi, J. (2009). *The challenge of establishing world-class universities*. World Bank Publications. doi:10.1596/978-0-8213-7865-6

Skills Panorama. (2017). *Skills Panorama*. Retrieved July 18, 2017, from http://skillspanorama.cedefop.europa.eu/en/

Treaty, L. (2007, December 17). Treaty of Lisbon amending the Treaty on European Union and the Treaty establishing the European Community. *Official Journal of the European Union*.

United Nations. (2017). *Milenium Development Goals*. Retrieved July 25, 2017, from http://www.un.org/millenniumgoals/2015_MDG_Report/pdf/MDG%202015%20rev%20(July%201).pdf<

Vreja, L. O., & Talpau, M. (2012). Intangible assets as strategic resources of the knowledge based Organization. *Metalurgia International, 17*(1), 175–180.

World Economic Forum. (2017). *Future of Jobs*. Retrieved August 20, 2017, from http://www3.weforum.org/docs/WEF_Future_of_Jobs.pdf

Zacharis, A. L., Bygrave, W. D., & Shepherd, D. A. (2000). *Global Entrepreneurship Monitor: National Entrepreneurship Assessment*. United States, 2000 Executive Report. Kauffman Center for Entrepreneurial Leadership at the Ewing Marion Kauffman Foundation, 3.

KEY TERMS AND DEFINITIONS

Business Involvement: Partnering activities that the private environment could have with the academic environment in order to develop the knowledge and soft skills of the future entrepreneurs and employees.

Knowledge Economy: The new era of economy that relies on information as a resource and on peoples' abilities of transforming information into new products and services.

Practice Modules: Short periods when students go to an assigned company and learn the practical appliance of the studied theory.

Rural Entrepreneurship: A new business located in the rural area that positively influences the development of the area.

Soft-Skills: Human abilities of understanding, using and transmitting information in order to work more easy and effective.

APPENDIX: QUESTIONNAIRE

1. Are you a FAEE graduate?
   a. Yes
   b. No
2. When did you graduated from FAEE?

____

3. Which is your residence area?
   a. Urban
   b. Rural
4. What is your age category?
   a. 22 – 24 y.o.
   b. 25 – 28 y.o.
   c. 29 – 32 y.o.
   d. + 32 y.o.
5. What is your income category?
   a. Under 1250 ron
   b. 1250 ron – 2000 ron
   c. 2000 – 4000 ron
   d. + 4000 ron
6. What motivated your choice of FAEE against other possibilities?

____

7. At the moment, do you consider you made the right choice or would you like to change it? Please motivate your answer.

____

8. Are you occupied in a field related to your specialization? If not, why did you choose to change the field?

____

9. At the moment, you are an:
   a. Employee
   b. Entrepreneur
   c. Both
   d. None of the above

10. What time was necessary for finding a job after graduation?
    a. I was hired before graduation and I chose to keep the job
    b. 1 – 3 months
    c. 4 – 12 months
    d. + 1 year
11. What obstacles did you meet in finding a satisfactory job?
    a. I wasn't prepared enough
    b. The employer's requests were different than the job description
    c. The theory had no relevance so I had to take other training modules
    d. There is a big difference between classes and real market requests
    e. I had no important obstacles
    f. Other answers
12. Do you believe the courses you followed during undergraduate program have helped you in finding a job? In what way?

____

13. What specific courses did you find useful? In what situations?

____

14. Do you consider that this faculty's profile is able to bring competitive advantages to its graduates?

____

15. If not, what changes would you make in order to make FAEE competitive?

____

16. If you ever considered opening a business related to this field, what sources of information would you use?
    a. The internet
    b. Courses and books
    c. Guides from related authorities – like the National Rural Development Program
    d. Professors from FAEE
    e. Consultants
    f. Other sources
17. What importance (on a scale from 1 – mostly unimportant to 5 – very important) do you believe the skills shown in Table 1 have in finding and keeping a desirable job?
18. Which of those skills (Table 2) and to what extent were triggered or developed during undergraduate period? (from 1 – it wasn't developed at all to 5 – it was well developed)

Table 1. Importance of soft-skills

Skill	1	2	3	4	5
Fast problem identification					
Finding key components in problem solving					
Contribution to team problem solving					
Prioritizing individual tasks					
Setting and meeting deadlines					
Report writing					
Compliance with professional ethics					
Efficient communication with the team and colleagues					
Objective/results presentation					
Working good under stress					
Balance between professional and personal life					
Achieving new knowledge from daily experience					
Integrating new technologies in regular work processes					
Adapting to change/innovation					
Setting and following a strategy					
Strategy revision based on new information					
Taking decisions in a short time					
Evaluating the long term impact of the decisions					
Correct delegation of tasks					
Involvement in designing the future of the company					
Accepting failure as a possibility					

19. What possibilities of developing these skills during undergraduate program do you see?

____

20. Do you consider that the profile companies are satisfied with the level of training of the new graduates? Please motivate your answer.

____

21. Do you consider the agri-food and environmental economics related companies are involved in the students training? In what way?
    a. They are practice partners for the interested institutions
    b. They offer internships
    c. They offer scholarships
    d. They participate in lectures or presentations for students
    e. None of the above
    f. Other

Table 2. Level of development of the soft skills during undergraduate program

Skill	1	2	3	4	5
Fast problem identification					
Finding key components in problem solving					
Contribution to team problem solving					
Prioritizing individual tasks					
Setting and meeting deadlines					
Report writing					
Compliance with professional ethics					
Efficient communication with the team and colleagues					
Objective/results presentation					
Working good under stress					
Balance between professional and personal life					
Achieving new knowledge from daily experience					
Integrating new technologies in regular work processes					
Adapting to change/innovation					
Setting and following a strategy					
strategy revision based on new information					
Taking decisions in a short time					
Evaluating the long term impact of the decisions					
Correct delegation of tasks					
Involvement in designing the future of the company					
Accepting failure as a possibility					

Chapter 15
The Effects of the Transition Period on the Knowledge Transfer Market in Romanian Agriculture

Gabriel Popescu
Bucharest University of Economic Studies, Romania

Simona Bara
Bucharest University of Economic Studies, Romania

Elena Preda
Bucharest University of Economic Studies, Romania

ABSTRACT

As the process of globalization intensifies, the demand for agri-food products is growing and much technical progress is needed. In the process of manufacturing agri-food products, a special role is played by the knowledge transfer market. At the same time, its functionality is given by the size, quality, and structure of the main components: supply and demand of information and knowledge, linkages between the two components. Linkage vectors between the two components coordinate the process of knowledge and information transferfrom the producers (research entities, universities, and the business sector) to the beneficiaries (the agricultural producers). The chapter captures the evolution of these actors involved in the knowledge transfer market of the Romanian agriculture.

INTRODUCTION

In 2016 the agriculture contributed with over 15 billion euros to the socio-economic development of Romania. This sector pushed our country to the top of the largest agricultural producers in the European Union. It is also worth noting that over the past ten years Romania has recorded high yields in field crops. For 2017, the specialists estimate higher yields, except for the southwest part of the country

DOI: 10.4018/978-1-5225-5739-5.ch015

Copyright © 2019, IGI Global. Copying or distributing in print or electronic forms without written permission of IGI Global is prohibited.

(Banat Region), where drought has compromised sunflower and maize crops. According with the World Economic Forum, at the global level the sectoral results are marked by the moment of economic sector recovery, because technological and geopolitical forces continuously shape and redefine economic, social and political order. The Global Competitiveness Report 2017-2018 emphasizes the poor values of indicators related to research sector in Romania, that reveals the existence of unused skilled resources.

The increase of quality and quantity of agricultural production as well as the activities efficiency have a significant contribution to the sector development, ensuring also food security. This is related to implementation of scientific results; the absorption capacity of the newest information and knowledge in agricultural activities; functionality of the relationship between information and knowledge producers, on the one hand, and theirs consumers, on the other hand.

THE SCIENTIFIC KNOWLEDGE MARKET FOR AGRICULTURE

The market of scientific knowledge for Romanian agriculture is an important component of the general market. The market of scientific knowledge is a part of the classic configuration, placed it on the face, on the one hand, two established actors (research, as a producer of information and knowledges, and agricultural producers, as recipients/ beneficiaries of the results from the research activities) and, on the other hand, the vectors for linking the both actor mentioned, in a way the mean for communication between them, as follows:

- Information and knowledge producers, which have a unanimous and quasi-determinant role in the current stage, these being represented by research institutes and research centers as well as agronomic and veterinary universities, the business environment and various NGOs which have as their main object of interest the scientific research that have as result producing of information and knowledges.

Consumers or beneficiaries of information and knowledge are represented by landowners, agricultural holdings - indifferent of their age, school education, organization and / or farm profile of the farmer, etc.

The vectors of the links between information and knowledge producers and consumers of them and / or beneficiaries of information and knowledges which are the "channels" through which information and knowledges circulate from producers to consumers, and them are represented by: educational environment, consultancy, extension, mass -media, etc.

In the knowledge transfer market of agriculture, forestry and food industry the demand for information and knowledgea *is determined by a number of factors*, among which a significant role is played by the following:

- The general level of economic and social development;
- Status and functionality of specific legislative and institutional structures reffered to the domain;
- The general degree of absorption - within each field of activity of the sector - of the technical and scientific progress elements;
- The personal characteristics of the actors involved, their economic status and their profile. This category of factors includes: Information and Knowledge Producers - R & D institutes and special-

Figure 1. The scheme of the vectors which assure of the link between supply and demand for information and knowledge
Source: (Popescu & Bara, 2017)

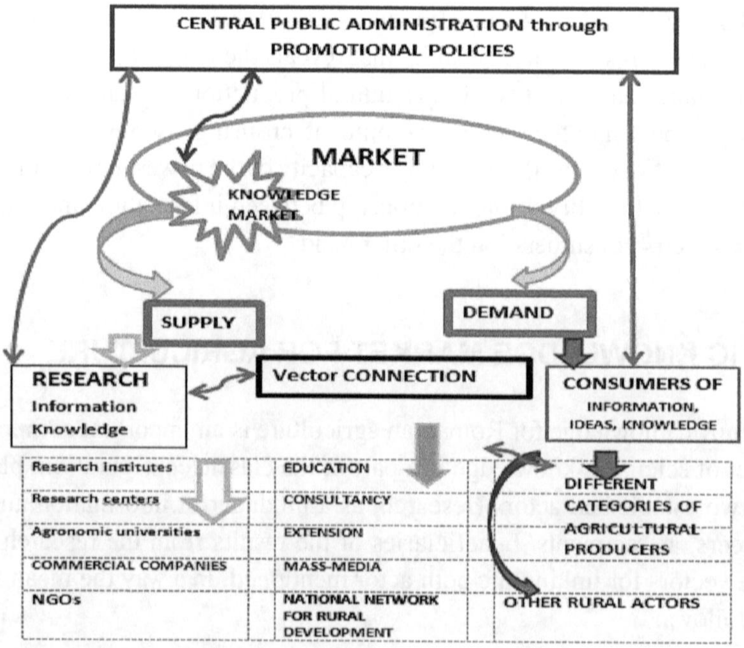

ized universities; agricultural producers represented by subsistence and semi-subsistence farms, commercial companies, cooperatives, associations, etc.);

- Institutional structures of professional, associative and / or cooperatist type - existing in rural areas, as well as the ability to accede to them or not to farmers and other stakeholders interested in acquiring information and knowledges from the knowledge transfer market;
- The ability of agricultural producers to access / purchase the specific products of the R & D activity available on the knowledge market.

1.The Producers of Information and Knowledge

Human Resources in Agricultural and Forestry Research

The main responsibility, but not the only one, regarding the nebulosity that has marked almost all the fields of socio-human research, including that of agriculture, forestry and agrarian research, since 1989 and up to now, is - unambiguously - chronic under-financing of the sector, a disastrous impact phenomenon, first of all, on the quantity and quality of the human factor, as well as on the research activities results.

The post-communist society which is in a crisis of identity and in the background of deep economic and financial imbalances has hardly support the burden of the non-productive sectors (including research, education and health); as one effect on it has been reduced untill humiliation the earnings of researchers, teachers, doctors, engineers, etc. It is only in the middle of 2017 the Government takes some attempts to make few salary adjustments for these categories.

Any analysis, with little respect for the truth, shows us that in the 27 years of transition from a socio-economic system to another (from the system of totalitarian economy to the capitalist one), the evolution of public organizational structures from the research agricultural and forestry, including the human research factor, has been conducted in a simple scenario, but unproductive one, and with an inefficient resource consumption (as many as on was!).*The analysis of the evolution of the total number of employees from the research activity* (number of persons in full-time equivalent) registered increases (by 19.7% in 2015 compared to 2010), while the number of scientific researchers decreased (by -11.7% over the same period).

In terms of *performance sectors*, the number of employees in full-time equivalent in the research activity was as follows: *within the government sector* (public one), the number of employees envolved in the research activity increased (by 38.8% in 2015 compared to 2010) as well as the number of researchers (by 19.1% in the same period); in the *business sector* the total number of employees envolved in the research activity increased (by 22.5% in 2015 compared to 2010) and the number of researchers decreased (by -27.7%); in the *higher education sector* - on the whole - the total number of employees in the research activity recorded a slight decrease (-0.5% in 2015 as compared to 2010), while the number of researchers was seriously reduced (by -21.4%); in the *non-profit private sector* both the total number of employees in the research activity and the number of scientific researchers decreased (by -19.0% in 2015 compared to 2010 and the number of scientific researchers by -6.5% in the same refference period).

The number of scientific researchers, in full-time equivalent, at the end of 2015, was 27,253. On scientific domains, the structure of the scientific researchers (at the end of 2015) was the following: engineering and technological sciences 44.2% of the total number; 19.3% natural and exact science; 12.8% social sciences; 9.7% medical science; *8.8% agricultural science*; 5.1% of humanities.

In 2015, compared to 2010, this indicator decreased by 11.2%. On scientific fields, the dynamics of the number of researchers, at the end of the year, was different, as follows: *increased the number of researchers at the end of the year in agricultural science (by 11.2%)* and in the field of engineering and technological sciences (2,9%); the number of scientific scientists at the end of the year was down in the humanities (-50.5%), social sciences (-34.9%) and medical sciences (-23.9%).

The indicator for the total number of scientific researchers per 10,000 occupied persons, in 2015, was 32.7 - below compared with one from 2010 (36,7).

For the agricultural sector, the number of scientific researchers from agriculture per 10,000 people employed in the sector in 2015 was 11.0 (scientific researchers from agriculture per 10,000 people employed in the sector). It should be noted that in 2010 the level of this indicator was of 6.2 scientific researchers in agriculture per 10,000 persons employed within the branch. So there on can note an increase of 1.8 times. This increase is also explained by the continuous reduction of the civil population employed in agriculture.

The same phenomenon is found in *the ratio between the number of scientific researchers in the agricultural field per 10,000 persons employed in agriculture and the total number of scientific researchers per 10,000 people employed in the economy* (which increased of 2 times).

Table 1. Total number of employees and number of researchers envolved in the R & D activity (including researchers; technicians and assimilated ones; other categories of employees) on performance sectors and scientific fields

	2010	2011	2012	2013	2014	2015	Dynamics2015/10, %	Structure, %	
								2010	2015
Employees - total. *Number in full-time equivalents*	26,171	29,749	31,135	32,507	31,391	31,331	119.7	100	100
Researchers	19,780	16,080	18,016	18,576	18,109	17,459	88.3	100	100
Business environment	8,271	10,002	10,887	10,514	10,437	10,128	122.5	31.6	32.3
Researchers	*5,853*	*3,518*	*4,956*	*5,333*	*5,244*	*4,234*	*72.3*	*29.6*	*24.3*
Governamental environment	8,704	10,675	11,381	12,336	11,866	12,080	138.8	33.3	38.6
Researchers	*5,590*	*5,846*	*6,372*	*6,583*	*6,409*	*6,659*	*119.1*	*28.3*	*38.1*
Tertiary environment	9,054	8,879	8,710	9,523	8,966	9,008	99.5	34.6	28.8
Researchers	*8,245*	*6,563*	*6,591*	*6,578*	*6,378*	*6,480*	*78.6*	*41.7*	*37.1*
Private non-profit environment	142	193	157	134	122	115	81.0	0.5	0.4
Researchers	*92*	*153*	*97*	*82*	*78*	*86*	*93.5*	*0.5*	*0.5*
Number of Researchers by Scientific Domain, at the End of the Year									
Total	30.707	25.489	27.838	27.600	27.535	27.253	88,8	100	100
Natural sciences and exact ones	5,163	5,448	4,789	4,986	4,032	5,259	101.9	16.8	19.3
Engineering Sciences and technology	11,718	10,122	13,063	13,157	12,904	12,053	102.9	38.2	44.2
Medical Sciences	3,491	3,010	2,572	2,621	2,736	2,656	76.1	11.4	9.7
Agricultural sciences	*2,154*	*1,293*	*1,252*	*2,398*	*2,525*	*2,396*	*111.2*	*7.0*	*8.8*
Social Sciences	5,376	4,112	4,428	2,484	4,204	3,500	65.1	17.5	12.8
Humanities	2,805	1,504	1,734	1,954	1,134	1,389	49.5	9.1	5.1

Source: Processed by "Romania's Statistical Yearbook", NIS, 2017a

Changes Within Human Resources From Agricultural and Forestry Research Under the Impact of the Transition Process

The main human resource from agricultural and forestry research and development domain - represented by the scientific researchers - in the last 27 years of transition process has been the subject of different processes and phenomena such as:

1. *There was a category of scientific researchers*, not a few, who were lost either naturally - when due to their age they retired or were drawn by the transformations from the economy or by the mirage of a better living within the big multinational companies either one of them choosed to go to work abroad or others were "thrown overboard" by the "bodyguards" to those who have become masters

Table 2. The Number of Researchers per 10,000 Occupied Persons and Number of Researchers in Agricultural Sciences per 10,000 People Employed in Agriculture

	2010	2011	2012	2013	2014	2015	Dynamics 2015/10, %
Researchers per 10,000 civilian employees	36.7	30.5	32.5	32.4	32.7	32.7	89.1
Share of researchers in agricultural science in the total	7.0	5.1	4.5	8.7	9.2	8.8	1.8
Researchers in agricultural sciences per 10,000 persons occupied in agriculture	6.2	4.0	4.2	8.9	10.3	11.0	176.9
The ratio between researchers in agricultural sciences per 10,000 personne occupied in agriculture and the total researchers per 10,000 personnes employeed	0.17	0.12	0.13	0.27	0.31	0.34	200.0

Source: Processed by "Romania's Statistical Yearbook", NIS, 2017a

of property owned by the research institutes or research centers which have been sold through the privatization process and the people emploeed become unemployed.

Obviously, researchers included in this category can be considered as "stranded losses" for the R & D sector, for the economy - in general.

With them departures from the research system and by redundancies, especially in the early years after 1989, there has also been a phenomenon of social "demonetization" of scientific research profession and, by extension, this phenomenon has affected even some people with extremely creative minds.

2. *A second category of human resources involved in agricultural and forestry research, but less than the first, includes those scientists who have proven themselves capable of reversion (through relational deviations or behavioral abilities or political capital and not only) into a businessmen or in a political or in a trade union leader and in other similar categories.* It is remarkable that *some of this people assumed the "role of locomotive" in the new society that emerged after 1989.* It should be noted that the majority of them were in the previous period lieders - there are only few exceptions - they were involved within the units of research management, or within ministerial or within production entities or even militant of the Romanian Communist Party with significant / consistent leadership functions.

These findings on the human resource from agricultural and forestry research do not seek to criticize the accumulations or scientific skills of the individuals concerned or their current social position or good intentions. Not! Our aim is to underline that these people, from a rational point of view, in the new socio-economic context could not provide a realistic theoretical foundation that could support the transition to a capitalist market economy arise after 1989. The fact that these people have been educated and trained, have worked - and what is very important for many - have also believed in the superiority of the socialist socio-economic model and then, after 1989, suddenly they were forced to abandon everything for them had value, could not lead to a smooth transition process, easy and without personal, professional and institutional consequences.

In fact, if the results of Romanian scientific research before 1989 were performing, then according to the law "mutatis mutandis" and the economy was performing. However, the socio-economic realities that emerged since 1989 in relation to the previous period have been in opposition to this logical construction, which makes any other interpretation purely speculative.

3. A third category of human resources from the agricultural and forestry research and from development activity is made up by the professionals who have honestly continued their activity in the existing institutional structures, as much as they have remained after the transition of the reforming wave from the economy.

The fact that these people continued to work in spite of the minimum income received, of the decredibility of the system in which they operate, of the diminishing public support, and others, has numerous interpretations that can be motivated by various arguments,

that relate more to an internal motivation, or a personal onr, or a cultural reason and less due a general phenomenology of a social, political or even economic nature. The stubbornness of these scientific researchers to continue working in the system, although in many ways the results of their work bear critical interpretations, represents a perticular value. Through them, the flame of the creative power of this nation could be maintained, which gives us the hope that the relaunch of agricultural and forestry research has real chances of accomplishment.

Undoubtedly, *scientific research, like the whole society, needs elite*, heads to professionalize the public decision-making, including research, also, elites as social models based on a real scientific, moral and professional values, elites to increase the credibility of Romania in front of the states of the world. In this context, scientific research must be the main pool of recruitment and formation of the Romanian elites, which are so diluted and atypical today. *And all the scientific research are the ones that have a decisive role in the orientation towards the knowledge society, as a strategic target of the European Union towards 2020.*

In the general context outlined above, the agricultural research does not make a dissonant note. However, the natural, economic and social particularittis of Romanian agriculture, as well as the post-1989 reformist process, have also impacted on the results of the research sector from the branch.

Current Expenditures in R&D Activity

Current expenditures on R & D activity, by performance sectors and types of research, amounted to ROL 2,762.1 million in 2015, with 35.1% more than in 2010. As a whole, within Romanian research activity, in terms of research type structure, were predominany the applicative scientific research (49.4% of the total R & D expenditure in 2015) over the fundamental research (31.5%). Types of research: fundamental research; applied research; experimental development.

The analyse of the structure of current expenditures from R & D activity by performance sectors and by types of research, in 2015, reveals the following issues: *in the business sector*, the applicative research (63.6% of total R & D expenditure) is dominant, followed by experimental development (25.8%); *in the governmental sector*, the highest share of expenditures for R & D is allocated to fundamental research (48.1%) and on the second place to applied research (39.0%); *in the tertiary environment* the applicative research is dominant (60.0%), followed by applicative scientific research (27.4%) and to a considerable distance by the experimental one (12.6%); *in the non-profit private sector* dominant are fundamental research (51.6%), followed by applicative research (37.6%).

By performance sectors, in 2015, *the current expenditure structure for R & D activity* was as follows: 48.4% of the total expenditures are ones of the business sector; 38.6% are within the governmental sector; 12.7% are within the higher education sector; 0.3% are within the nonprofit private sector.

Comparing the evolution of the total expenditures level, in 2015 as compared to 2010, on can see: an increase of the current expenditures from the R & D activity from the business sector by 11.4%; the decrease of the current expenditures for the research-development activities within the higher education sector by 11.3%.

Total expenditures for the R & D activity allocated to agriculture by the business sector represent 1.2% of the total, respectively ROL 17.9 million in 2015 (compared to ROL 10.95 million in 2013). The growth trend of these expenditures, in 2015 compared to 2013, was significant (1.64 times, in 2015 compared to 2013).

In 2015, *the analysis of the structure regarding total current and capital expenditures for R & D activity in the business sector* reflects the following issues: on a total, the current expenditures for the R & D activity, in the business sector, had the highest share - them represented 87.4%, while the capital expenditures for the R & D activity held 12.6%.

In 2015 compared to 2013, current total R & D expenditures, in the business sector, fell by 1.5 percentage points (pp), while the capital expenditures increased by the same level (+ 1.5 pp). These developments in total R & D spending, in the business environment, can be characterized as only for the "maintenance" at a subsistence level of the research activity and no to support the development.

The same trends were recorded in expenditures for the research and development, in the business sector from agriculture, forestry and fisheries in 2015 compared to 2013; Thus, in the sector current expenses increased by 2.6 pp, while capital expenditures declined (-2.6pp).

Information and Knowledge Production

Scientific knowledge - as a direct product of research activity - is master of a double representation in the economy: first of all, it is an intangible asset; on second, it is a production factor.

As an intangible asset the scientific knowledge has some essential characteristics, such as: it is a part of the national patrimony with a greatest dynamics; on has a high degree of volatility, because when

Table 3. Total current expenditure for research and development activity by type of research (in Thousand RON on current prices)

	2010	2011	2012	2013	2014	2015	Dynamics, 2015/10, %	Structure, %	
								2010	2015
Total	2,045,132	2,251,480	2,355,492	2,064,485	2,228,378	2,762,085	135.1	100	100
Fundamental research	913,033	943,269	935,201	793,011	805,365	869,361	95.2	44.6	31.5
Applicative research	994,602	944,573	1,021,278	865,623	992,438	1,364,596	137.2	48.7	49.4
Experimental development	137,497	363,638	399,013	405,851	430,575	528,128	384.1	6.7	19.1
Business Environment									
Total	757,366	824,913	926,145	671,971	910,495	1,336,832	176.5	100	100
Fundamental research	123,423	104,786	52,234	28,945	155,062	142,156	115.2	16.3	10.6
Applicative research	546,888	522,875	612,244	422,129	510,095	849,604	155.4	72.2	63.6
Experimental development	87,055	197,252	261,667	220,897	245,338	345,072	396.4	11.5	25.8
Governamental Environment									
Total	**787,571**	**924,465**	**1,000,607**	**1,004,495**	**974,154**	**1,064,936**	**135.2**	**100**	**100**
Fundamental research	453,391	521,446	610,223	526,332	449,543	511,675	112.9	57.5	48.1
Applicative research	297,500	285,068	291,760	331,258	383,014	415,433	139.6	37.8	39.0
Experimental development	36,680	117,951	98,624	146,905	141,597	137,828	375.8	4.7	12.9
Tertiary Environment									
Total	**490,740**	**493,722**	**418,105**	**379,073**	**334,566**	**351,527**	**71.6**	**100**	**100**
Fundamental research	334,701	310,887	267,711	234,035	197,463	210,994	63.0	68.2	60.0
Applicative research	146,871	134,606	111,775	107,414	93,702	96,255	65.5	29.9	27.4
Experimental development	9,168	48,229	38,619	37,624	43,401	44,278	483.0	1.9	12.6
Non-Profit Environment									
Total	9,455	8,380	10,635	8,946	9,163	8,790	93.0	100	100
Fundamental research	1,518	6,150	5,033	3,699	3,297	4,536	298.8	16.1	51.6
Applicative research	3,343	2,024	5,499	4,822	5,627	3,304	98.8	35.4	37.6
Experimental development	4,594	206	103	425	239	950	20.7	48.5	10.8

Source: Processed by "Romania's Statistical Yearbook", NIS, 2017a

Table 4. The total current and capital expenditure for R & D and for R & D for agriculture, forestry and fisheries from the business sector (in Thousands RON on current prices)

	2013	2014	2015	2015/10, %	Structure, %	
					2013	2015
Total Expenditures						
Total	755,710	1,059,424	1,529,866	202.4	100	100
Out of which: Agriculture, forestry and fishing	10,950	12,998	17,913	163.6	1.45	1.17
Out of which: The share of Agriculture, forestry and fishing in the total	1.45	1.23	1.17	-0.28*)	x	x
Current Expenditures						
Total	671,971	910,495	1,336,832	198.9	100	100
Out of which: Agriculture, forestry and fishing	10,171	12,915	17,105	168.2	1.51	1.28
Out of which: The share of Agriculture, forestry and fishing in the total	1.51	1.42	1.28	-0.23*)	x	x
Capital Expenditures						
Total	83,739	148,929	193,034	230.5	100	100
Out of which: Agriculture, forestry and fishing	779	783	808	103.7	0.93	0.42
Out of which: The share of Agriculture, forestry and fishing in the total (%)	0.93	0.06	0.42	-0.51	x	x

Source: Processed by "Romania's Statistical Yearbook", NIS, 2017a

Figure 2. The structure of total current expenditures and capital expenditures in the business sector for the total research and development activity, as well as for agriculture, forestry and fishing.
Source: Processed by "Romania's Statistical Yearbook", NIS, 2017a

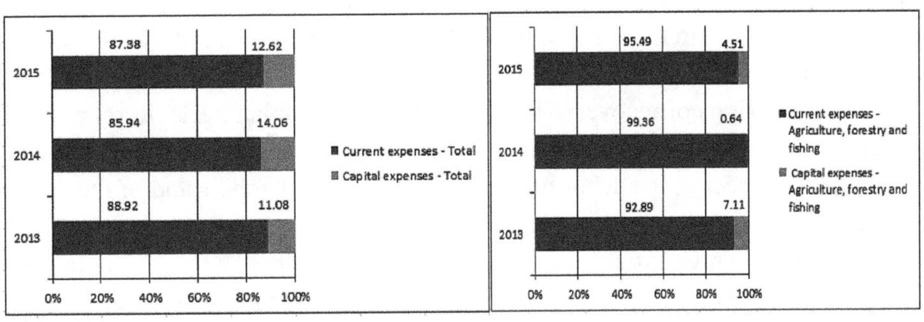

it receives the quality of public good, it quits of control of the national authority; on has a high speed of movement under the influence of the current technical-scientific regime. The direct consequence of these features is that, as the globalization process intensifies and go on, scientific knowledge migrates from the weak economies to the developed countries. In other words, scientific knowledge on a natural way is polarized inside of the rich world - because only it can facilitates the "escape" of brains and ideas from the origin countries, frequently in such formulas that exceed the limits of morals and / or of their value equivalent.

Scientific knowledge as a production factor will be certainly - and not over much time - on be out of the scope of the abstract world and it will acquire concrete valences when it will be considered as an expense component within the cost of all products. Under such circumstances, the center of interest in sustaining and the funding of the research activities will move on from the public budget to the private one, and the responsibility concerning the research sector will come, in particular, from the economic agents.

1. In the category of information and knowledges producers for agriculture, forestry, fishing and agri-food industries, the Academy of Agricultural and Forestry Sciences (AAFS/ ASAS) is the most representative institution, but it is not a unique one, in the field of agricultural research.

According to the lawno. 45/2009, AAFS/ ASAS is an autonomous institution of public law with with legal personality, is the scientific and research consecration forum within the fundamental and applied domanis for agriculture and its branches. To this aim, AAFS/ ASAS collaborates with the Ministry of Agriculture, Forests and Rural Development, the Ministry of Education, Research and Innovation, the Ministry of Environment, the Romanian Academy, the Consortium of Agricultural Sciences Universities, as well as with other institutions and organizations for achieving the objectives of its domains of activity.

Organizationally, AAFS/ ASAS co-ordinates or subordinates several research institutes and research centers, which are the target to cover the entire Romanian territory, for which their profile has been established according to the productive specificity of the areas where they were located.

Romanian agricultural research, in the modern scientific formulas, has its beginnings, quite shy, in the early years of the First World War. Then it was a remarkable evolution during the interwar period, especially in the third decade of the last century.On note that in the current institutional architecture, the Romanian agricultural research has had broadly maintained its old configuration from the communist period.

After a century of existence the institution represented by the AAFS/ ASAS has recorded considerable accumulations in the fielf of science and in the material domain, also. The steadily rising dynamics of Romanian agricultural research has come into a declining slope immediately after the abandonment of the command economy; this trend has had remained throughout the transition period and at present, it does not yet show any signs of recovery. The crisis of the Romanian agricultural research was not a single fact. This was, as a phenomenon appeared in all Romanian research and was in connection with the underfinancing of the sector, the departure of the scientific researchers and of the decreasing of the research results.

By the *Law no. 45/2009 regarding the organization and functioning of the Academy of Agricultural and Forestry Sciences Gheorghe Ionescu-Sisesti and of the research and development system in the fields of agriculture, forestry and food industry* on was decided that the reorganization of all entities to be done within 60 days from the date of the entry into force of the Law, by seting-up of a Governmental Decisions (GDs/ HG) for each unit placed in the coordination or subordination of the AAFS/ ASAS. It should be noted that only two GDs/ HGs have been developed till the middle of 2017; for the remaining units (46) the GDs/ HG were being developed in the summer of 2017, and currently them are in the process of being approved by the competent public institutions. Also, these delays explain the difficulties encountered by the R & D system from agriculture, forestry and from the food industry.

Box 1. The structure of the R & D system from Romania, with an accent on agricultural research

The structure of the research system in Romania includes:
1. National Institute for Research and Development (NIRD/ INCD) in coordination
a. NIRD/ INCD in coordination with the Ministry of Research and Innovation;
b. NIRDs/ INCD with agricultural and forestry profile are:
i. National Institute for Agricultural Development Research Fundulea - NIRD/ INCDA Fundulea
ii. Institute for Research and Development for Plant Protection - IRDPP/ ICDPP
iii. Institute for Research and Development for Forest - IRDF/ ICAS
iv. The Institute for Research and Development for the Pastures Brasov - IRDP Brasov/ ICDP Brasov
v. National Research and Development Institute for Pedology, Agrochemistry and Environmental Protection - IRDPAEP/ ICPA
vi. Research and Development Unit for Fruit Growing from Constanta
vii. Research Institute for Vegetable and Floriculture from Vidra
viii. Institute of Food Bioresources from Bucharest -
c. NIRD/ INCD within coordination of ministries
i. Ministry of Agriculture and Rural Development (MADR) for sectoral research in the field of agriculture.
2. RDI institutions subordinated to the Ministry of Research and Innovation - which include:
a. The Romanian Space Agency (ROSA)
b. Institute of Atomic Physics (IAF/ IFA)
3. RDI institutions under the authority of the Ministry of Research and Innovation, which has partner institutions under authority, which include:
a. The “INCERTRANS Transport Research Institute” - Ltd./ S.A.
b. Institute for Research and Design of Bearings and Assemblage Organs - ICPROA-S.A. Braşov
4. Institute, centers and foundations which are subordinated to the Romanian Academy.
These entities have been organized since 1995 and them were initially classified as commercial companies operating as self-governing regies. At present, the national R & D institutes are organized according to Law no. 324/2003 which endorsement the Government Ordinance no. 57/2002 regarding scientific research and technological development.

National research and development institutes have as their main object of operation the research and development activities and function on the basis of economic management and financial autonomy, them compute amortization and manage their accounting in an economic regime; they have their own patrimony and administer the public and private patrimony of the state in order to ensure the performance of them activities; on operates in the coordination of a specialized body of the central public administration. The Institutes participate in the elaboration of development strategies in their specific domain, carries out research and development activities in order to achieve the objectives included in the National Strategy for the Research, on represent a reference basis for scientific and technological competence, expertise, human resources improvement and scientific and technical documentation. The national institutes may carry out secondary commercial and production activities according to its own regulations.

Source: Processing of Law no. 324/2003 for the approval of the Government Ordinance no. 57/2002 concerning the scientific research and technological development http://www.research.gov.ro/ro/articol/4514/sistemul-de-cercetare-institutii-international-de cercetare-dezvoltare-in-coordonare # Presentation (MRI, 2017)

In Romania, in 2015, the number of researchers from agriculture, forestry and fishing was 2.396 and they represented 8.8% of the total number of researchers. This evolution of the number of researchers in agriculture, forestry and pisciculture has been a winding one. Thus, in 2015 as compared to 2007, the number of researchers in agriculture, forestry and fishing was with 25.6% higher, although the hole total socio - professional category of researchers recorded a decrease (-11.3%). On also mention that in the same reference period there was also an increase in the share of researchers in agriculture, forestry and fish farming in the total number of researchers (with +2.6 pp). Regarding the latter indicator, in 2015 as compared to 2000 the share of the researchers in agriculture, forestry and fish farming in the total number of researchers increased by 3.4 pp and in 2015 as compared to 1995 the share of the researchers in agriculture, forestry and pisciculture in the total number diminished with 1.5pp.

In order to stop the negative evolutions of the number of scientific researchers, in the autumn of 2017, some public measures were taken in view to increase the salaries of Romanian specialists involved in the research, development and innovation (RDI) projects accomplished on the basis of contracts with public funding;

Figure 3. The evolution of the number of researchers from agriculture, forestry and pisciculture in Romania, between 1995-2015
Source: NIS, Tempo Online, Accessed on October 10, 2017; "Statistical Yearbook of Romania. Time series ", (NIS, 2017c)

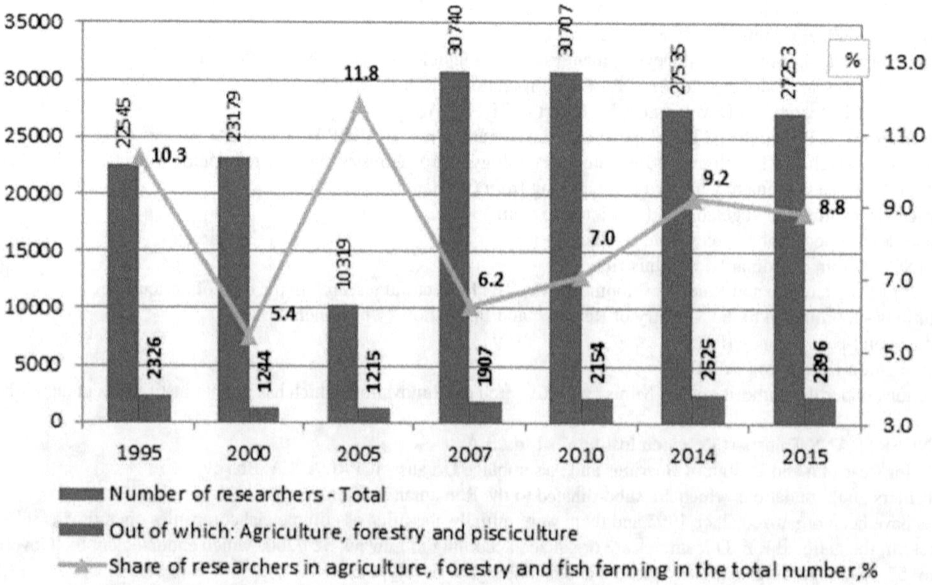

This decision are expected to lead to changes in financial capacity and to sustain stabilization of researchers in the country. Thus, in the next period, the direct monthly salary costs of researchers (for public funded research contracts) will be updated, with out exceed the level of 1,500 euro per month for the first category of activities (activities involving a high level of creativity and / or experience and skills management / management), euro 900 per month for the second category of activities (activities which requiring in-depth knowledge of the methods of analysis and synthesis as well as skills for their use) and euro 500 per month for the last category of activities (support activities, for example for professional positions of technician steps I, II, III and trainees - TI, TII, TIII, TS, student, etc. for which on can have only medium education is admitted).

The above-mentioned measures addressed to the research staff, although "do not involve additional budgetary allocations - because the payments will be made within the research budgets for both ongoing contracts and those to be launched" - are nonetheless to encourage the implementation of the ambitious objectives of the "National Strategy for Research, Development and Innovation During 2014-2020" and them nor can contribute to the diminishing of the "brain migration" phenomenon to countries with a better salary for scientific researchers.

Another phenomenon that affected the undesirable developments in the field of scientific research was,

the influence of the privatization of agriculture in the domain of land relations, because, on the one hand, by give up to the agricultural cooperatives and to the state property rights for the lands confiscated by the Communists in 1945 -1949 and, on the other hand, on put in practice the system of land concessionaires for the state-owned institutes or for the research centers.

As a result of these actions, much part of the agricultural area allocated to research in agriculture, forestry and in the food industry went into in the commercial circuit. For example, in 1990 at the disposal of the agricultural research were about 160,000 hectares, and now there are about 30,000 hectares, it means a reduction of more than 80% of the areas for research in agriculture, forestry and the food industry.

Another problem facing of all the Romanian research is its poor involvement at European level. For example, in this moment in Romania there are only three entities in the list of units recognized by the European Commission - Eurostat as research entities which can delivery micro - data reffered to the Romanian research activities.

This reality can be considered either lack of information or - why not - indifference.

2. The *agronomic and veterinary universities from Bucharest, Cluj, Iași, Craiova and Timișoara can be, also, included in the category of producers of scientific information and knowledges from Romania.* Certainly, in the years after 1989 these universities did not remark - through their scientific production - at the level of the AAFS/ ASAS institutes and centers, because their activities were focused on didactic, educational and editorial issues rather than on their proper scientific research.

The five universities of agronomic sciences and veterinary medicine from Bucharest, Cluj, Iași, Craiova and Timișoara are the result of an academic, scientific and administrative progress supported by

*Table 5. The list of entities registered and recognized by Eurostat for Romania in view to delivery microdata about the research activity *)*

Country	City	Research Entity – The Nume in English	Research Entity – The Official Nume
RO	Bucharest	National Scientific Research Institute for Labour and Social Protection	Institutul National de Cercetare Stiintifica in domeniul Muncii si Protectiei Sociale
	Bucharest	The Romanian Academic Society	Societatea Academică din România
	Bucharest	University of Bucharest	Universitatea din Bucuresti

Source: http://ec.europa.eu/eurostat/documents/203647/771732/Recognised-research-entities.pdf (EC, 2017a)
*) Extracts from "List of research entities recognized by Eurostat as entities for delivery of micro-data on research activity".

Box 2. Procedure for the Recognition by Eurostat of the Entities they can Delivery Micro-Data on the Research Activity - Few Line

In view to request admission of entities from a country in the "List of entities registered and recognized by Eurostat for the transmission of micro-data on research", it is necessary to consult the document "How to apply to access to micro-data?".
In this context, stakeholders may present their own indicators and analysis proposals. This information sent to Eurostat is evaluated (by Eurostat and national statistical authorities) and then they can become publicly available.
Eurostat appreciates that by involving as many as possible research units from a country in the process of delivering of micro-data on their activity, it is possible to improve the support and credibly, internationally, and to provide a more realistic reflection of the national research situation.

Source: Processing by http://ec.europa.eu/eurostat/web/microdata/overview (EC, 2017b)

the great names of national education and agronomic science. At present, the agronomic and veterinary medicine universities are public ones.

In Romania, the first agronomic university was AVU/ USAMV Bucharest, which debuted as the first agronomic school (Pantelimon School) founded 164 years ago.

The number of graduates from all agronomic and veterinary universities who have studied (2013) is approximate 150,000 students excluding Agronomy University Craiova, accounting for about 5.0% of the total students.

The existent universities develop their educational supply and manage the study programs, which can be carried out in several fundamental domain.

At the same time, the agronomic and veterinary universities organize and carry out the admissions competitions for the undergraduate and master's degree studies and manage the entire activity of the students during the study cycles. Within the research centers and laboratories of the faculties there is also the scientific activity of doctoral students enrolled in the doctoral schools of AVUs/ USAMV, as well as that of post-doctoral students.

Upon graduation, AVUs/ USAMV graduates have various professional perspectives, such as: agricultural companies and associations; agricultural chambers; paying agencies and intervention ones for agriculture; phytosanitary customs police; research institutes; higher education or pre-university education; agencies and rural development offices; town halls - the direction of sustainable development programs; certification and assessment of agricultural land and products; agricultural advisory offices; food control offices; ecology and environment protection; topography and cartography; breeding and selection units for animals; sanitary veterinary control of food; rural veterinary medicine; on the skilled labor market from the European Union, etc.

Table 6. The number of graduates from agronomic and veterinary universities (2011/2012) who have studied under subsidized on the state budget and tuition fee, per study cycle (ref. 1 Jan2013)

	Total Graduates (GL;MG;PhDG*)			**Graduates License (GL)**	**Master Graduates (MG)**	**PhD Graduates (PhDG)**
	Total	**State Budget**	**Tuition fee**			
Agronomic and Veterinary University (AVU/ USAMV) Bucuresti	3.482	1.284	2.198	2.662	705	115
AVU/ USAMV Cluj	1.531	1.077	454	903	519	109
AVU/ USAMV Iasi	930	711	219	608	258	64
AVU/ USAMV Timisoara	1.532	833	699	1.052	464	16
Agronomy University, Craiova**)	:	:	:	:	:	:
Total Graduates at National Level	**148.904**	**80.052**	**68.852**	**87.808**	**5.5967**	**5.129**
Total AVUs/ USAMV-uri exclusive the Agronomy University Craiova	**7.475**	**3.905**	**3.570**	**5.225**	**1.946**	**304**
The share of graduates from AVUs, %	5,0	4,9	5,2	6,0	3,5	5,9

Source: Data Processing from NCHEF/ CNFIS - National Council for Higher Education Financing, http://www.cnfis.ro/wp-content/uploads/2012/08/TotalAbsolventi_CicluriLMD_1ian20131.pdf

*) GL-License, MG-master, PhDG - doctorate. **) For Agronomy University, of Craiova thre is no data within National Council for Higher Education Financing (NCHEF/ CNFIS).

Figure 4. The structure of the number of graduates from agronomic and veterinary universities (2011/2012) (in %)
Source: Data Processed from NCHEF/ CNFIS, http://www.cnfis.ro/wp-content/uploads/2012/08/TotalAbsolventi_CicluriLMD_1ian20131.pdf

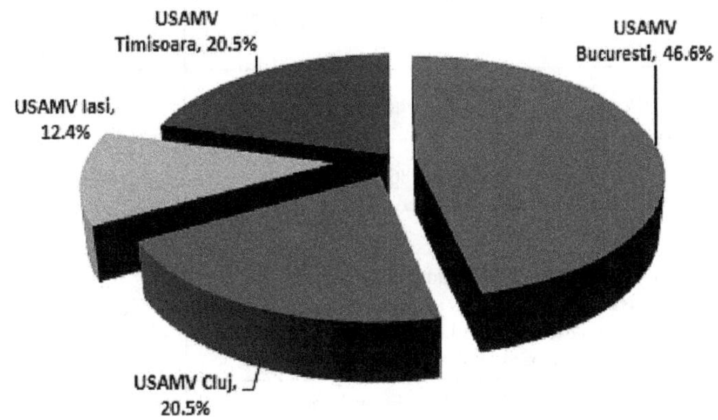

Young graduates from AVUs/ USAMV, however, to find a job in accord with professional training is a hard problem, which is why they either turn to other areas or go to other countries.

The university management - which occurred after 1990 - have also been cases where it maked some mistakes or errors and them came under the criminal law. In this context, for example, it is notorious the selling of over 200 hectares of land belonging to the Băneasa Didactic Farm within AVU/USAMV Bucharest, which was under the public law. We do not want to replace the Court's Decision, we do not judge the legality of this action! We can stress, however, that the responsibility of the university management in training process of young generations, the decision to renounce/to sell the Băneasa Didactic Farm at AVU/ USAMV Bucharest was an antinational act, which has compromising the future quality of agricultural specialist for many generations of agronomic engineers, veterinarians, or agrarian economists.

3. It should be mentioned that in Romania there are also a large international specialized companies in the field of agriculture. They participate direct or indirect within the local market of information and knowledge producers not only with their certified products which are the result of a long-standing practice at European and international level, but also with a strong marketing experience which permit them to enter on the information and knowledge market much easier than the national producers mentioned above.

For example, in 2015, the top 11 international companies operating in the Romanian agricultural sector together had a turnover of nearly 12 billion lei (the equivalent of 2.5 billion euros, At the course of 2015, 4.5211 lei per euro, National Bank of Romania). We appreciate that this is also the result of at least two major directions:

- Normative public acts issued in 2002 and 2003 (Government Emergency Ordinance /OUG/ no.157/2002, Law no.350 / 2003 for completing the Law no.16 / 1994 on leasing), which subsidizes the production achieved by firms on very large land areas; they also had the role of stimulating the land lease process;

Table 7. The "Top 500" companies with foreign capital, with the largest business turnover, in 2015

Crt. No.	Name of Company	Affiliation	Turnover,		%	Range
			RONMill.	EuroMill.		
1.	ADM România Trading SRL, Bucureşti	Part of the parent company founded in 1902 in the USA	2380	524	20,5	44
2.	Ameropa Grains SA România, Constanţa	Part of the Ameropa Switzerland Group	2376	526	20,5	45
3.	Cargill Agriculture, Bucureşti	US company founded in 1865	2092	463	18,0	51
4.	Nidera Bucureşti	Chinese-Dutch Company	1560	345	13,3	78
5.	Bunge România, Buzău	The company was founded in 1818 in the Netherlands. Today is an International Company headquartered in the USA. On the local market on was imposed oil brands Floriol and Unisol	294	65	11,5	86
6.	Agricover – Ilfov	Company founded in 1990 by an Iranian established in Romania. The company provides complete solutions for agriculture (seeds, pesticides, fertilizers - about 55% of the business) and trade in cereals and livestock (about 45%)	1330	263	3,0	98
7.	Monsanto România, Ilfov	USA - producer and supplier of agricultural seeds and technologies	346	77	3,0	355
8.	Agrana Buzău	Austria, the sugar industry	344	76	2,7	357
9.	Pioneer HI-Bred România, Ilfov	USA - producer and supplier of agricultural seeds and technologies	313	69	2,6	409
10.	Zahărul Oradea, Bihor	German shareholder, sugar industry	300	66	2,4	430
11.	Agrisol Internaţional, Prahova	Lebanese shareholder - chicken growth and slaughtering	276	61	2,5	471
	Total	x	**11611**	**2535**	**100**	x

- Contemporary agrarian policies aimed to stimulate the sale of the obtained production and the formation of farms on large areas held by an owner or a company. Thus, the owner / business becomes strongly linked by global production networks, contributing to the emergence of areas specializing in productivist agriculture - what are functioning as real "development engines".

In this context, international companies which are acting in the agricultural sector from Romania bring with them the technical progress, also, and at the same time they produce the diminishing of the traditional rural household role which are operating yet under the self-consumption regime.

The Main Results of Information and Knowledge Production

1. *One of the direct effects of the workforce and of the level of expensis from R & D activity is found out in the total number of research projects carried out in 2015*, which was 7,872 project, returning an average expenditure for research and development of 246,021 lei per project.

Out of the total number of scientific research projects developed in Romania, 10.1% were from agricultural research (794 projects, includes only the projects within the government sector and with the higher education sector*),* with an average of expenditure of 218.406 lei per project (with 11, 2% below the total expenses average).

2. *Another effect of the state in which the research and development activity function is found in the patent applications submitted by the research and development institutes.* In 2015, were submitted 1,053 patent applications, a number decreasing by 25.7% compared to 2010. By type of applicants, most of the patent applications (980 patent applications, respectively 93.1% of the total) were submitted by the Romanians in the following structure: 50.0% of the total number of patent applications submitted with State Office for Inventions and Trademarks (SOIT/ OSIM) belonged to individuals; 18.6% of enterprises; 16.0% of research institutes; 8.5% of educational institutions.

Table 8. Number of patent applications submitted between 2010-2015

	2010	2011	2012	2013	2014	2015	2015/10, %	Structure, %	
								2010	2015
Total									
Patent applications submitted in Romania- total	**1418**	**1462**	**1077**	**1046**	**1036**	**1053**	**74.3**	**100**	**100**
By Categories of Applicants:									
-Romanian applicants:	1382	1424	1022	995	952	980	70.9	97.5	93.1
Individuals	567	640	431	496	523	526	92.8	40.0	50.0
Enterprises	135	139	158	157	132	197	145.9	9.5	18.6
Research Institute	334	357	208	136	156	168	50.3	23.6	16.0
Education Institute	346	288	225	206	141	89	25.7	24.4	8.5
- Foreign applicants	36	38	55	51	84	73	202.8	2.5	6.9
Out of which in the technical field "Life needs"	308	322	264	248	242	274	89.0	x	x
Share of Patent Claims in the "Life Needs" domain in Total Claims, %	21.7	22.0	24.5	23.7	23.4	26.0	4.3%	x	x
Patent applications submitted by RO to the European Patent Office	35	60	72	85	102	110	318.6	x	x
Average number of patent applications submitted with the EPO by an EU Member State (the average)	2027	2052	2028	2027	2027	2024	100	x	x
Share of patent applications submitted to EPO by RO in the total Patent applications, %	2.4	4.1	6.6	8.1	9.8	10.4	8.0*)	x	x

Source: Eurostat, http://ec.europa.eu/eurostat/tgm/table.do?tab=table&init=1&language=en&pcode=tsc00033&plugin=1. Source: Processed on "The Romania's Statistical Yearbook" (NIS, 2017a)

Romania's patent applications sumitted by Romania to the European Patent Office (EPO) was 110 applications in 2007, respective by 3.2 times more than in 2010, but very little compared to the EU average (5.4% patent applications submitted to the EPO by an EU Membre State).

The evolution of this indicator, in 2015 as compared to 2010, expresses: a decrease in the total number of patent applications filed with SOIT/ OSIM by the solicitants from Romania (by -4.4pp), and by categories of applicants the progress of the research institutes are significant (-7.6pp) and of education institutions (-15.9); there was an increase in the total number of patent applications submitted to SOIT/ OSIM by individuals (by + 10.0pp), by enterprises (by + 9.1pp), as well as by foreign applicants (+ 4.1pp).

Patent applications filed by R & D institutions from agriculture, forestry and fishing, classified as the domain "Life needs", were 274 patent applications in 2015 and accounted 26.0% of the total. In 2015 as compared to 2013, regarding this indicator on mention: the dynamics of indicator was over total (89.0% versus 74.3%); there was, also, a tendency to increase the specific weight in the total number of patent applications submitted by R & D institutions from agriculture, forestry and fishing (from 21.7% to 26.0%).

3. Each agronomic and veterinary university (USAMV) edits and publishes yearly the volumes with the scientific papers of its faculties (Agriculture, Horticulture, Animal Husbandry, Veterinary Medicine, Land Improvement and Environmental Engineering, Biotechnology, Management and Economic Engineering in Agriculture and Rural Development) as well as the scientific journals of this entities.

In recent years, most journals of agronomic and veterinary universities have been indexed in international databases such as Copernicus, Cabi, Era Ulrich's Periodicals Directory, Google Scholar, National Council of Scientific Research in Higher Education CNCSIS B +, ProQuest, DOAJ, EBSCO and online platforms.

However, research results from agronomic and veterinary medicine universities could have been more representative and genuinely useful for practice if institutional management were much more involved to capitalize - with more careful accountability and efficiency - the results of people implicated in the research activities and would have been bussy with the empowerment of human resources (teachers, researchers, specialists, students), land resources (agricultural land of any kind) and all other elements of the agricultural capital that the higher education institutions benefited.

2. Consumers or Beneficiaries of Information and Knowledge Produced Through R&D Activity: The Main Features

Mainly, the consumers or beneficiaries of agricultural, forestry and fishery information and knowledges that have been produced through R & D activity within specialized institutes, research centers and agronomic and veterinary universities are represented by the civil occupied population in the mentioned domain. In Romania, the employed civil population in agriculture, hunting, forestry and fishing, in 2015, was 2,184 thousand persons and represented 25.6% of the total occupied population.

Table 9. Evolution of the civil occupied population employed in agriculture, hunting and forestry and fishing at the NACE/ CAEN Rev.1 section from Romania, between 1995-2007. Thousand persons

	1995	2000	2005	2007	Dynamics 2015/2007, %
Civil Occupied Population - Total	9,493	8,629	8,390	8,469	89.2
Out of which: Agriculture, Forestry and Fisheries	3,270	3,575	2,678	2,518	77.0
The share of population from agriculture, forestry and fishing in the total,%	34.4	41.4	31.9	29.7	-4.7*)
	2010	**2014**	**2015**	**Dynamics 2015/2000, %**	**Dynamics 2015/1995, %**
Civil Occupied Population - Total	8,371	8,431	8,535	100.8	89.9
Out of which: Agriculture, Forestry and Fisheries	2,440	2,304	2,184	86.7	66.8
The share of population from agriculture, forestry and fishing in the total,%	29.1	27.3	25.6	-4.1*)	-8.9*)

*) Percentage points.

Source: NIS, Tempo Online, Accessed on October 10, 2017; "Statistical Yearbook of Romania", time series, 2016 (NIS, 2017c)

Figure 5. Evolution of the civil occupied population employed in agriculture, hunting and forestry and fishing at the NACE/ CAEN Rev.1 section from Romania, between 1995-2007. Thousand persons
Source: NIS, Tempo Online, Accessed on October 10, 2017; "Statistical Yearbook of Romania", time series, 2016 (NIS, 2017c)

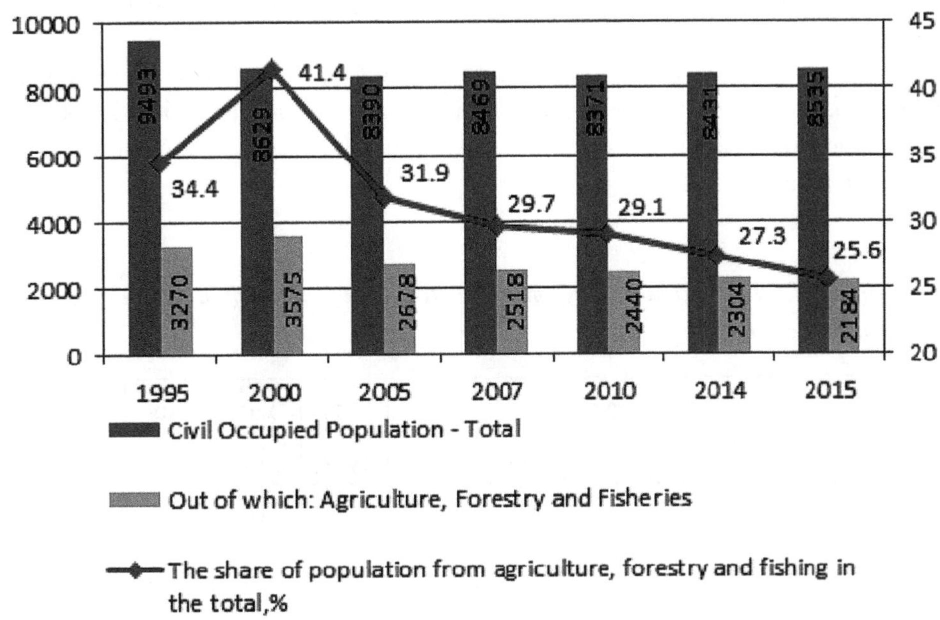

Between 1995 and 2017 there were a negative developments of both the civil occupied employment (with -11.1%) and of the civil employment in agriculture, hunting and forestry an fishing (with -33.2%).

The profile of farmers, depending on their interest in knowledge aquisition and it is specifically set up in relation to the level of training, age, income level and living standard; the nature of production responsibilities; the size of the holdings held; the degree of intensification of the production within their agricultural holding, where they activate and other issues.

The evolution of the rural population in the total school population, by education levels, which are mostly negative, explains why the labor force in the Romanian agricultural sector presents a low level of training compared with the European countries. Thus, according to European Commission data, in 2013, 96.4% of Romanian farmers said they "get the knowledge in their field strictly on the basis of practical experience", compared to 70.9% which is the average of farmers at EU level.

The evolution of the rural population by age, in 2013, according to European Commission data (Eurostat...), show that 40.7% of the inhabitants of the rural space are over 50 years old, 49.6% have between 25-50 years and 9.7% have under 24 years. This age structure of the rural population by age group is close to that of the EU (43.3% in the age group over 50 years, 49.6% in the age group with the age between 25-50 years and 7.0% in the the age group under 24 years of age).

This phenomenon of abruptly diminishing of the civil employment in agriculture, hunting, forestry and fishing, towards the total civil occupied population, is supplemented by other elements, such as: low education; reduced incomes (in 2015, on had 648.5 lei per person per month, respectively only 64.20% of the national average; in 2016, on had 632.20 lei per person per month, respectively 56.8% of the national average (NIS, 2017b).

The access of farmers, especially of the small ones, to information and expertise is also limited by the *capacity of households to cover their expenditure on their income*. Thus, in 2016, the share of agricultural households that could not cover their expenditures was 52.9% compared with only 30.4% as was the average on national level (NIS, 2017b).

Table 10. The share of school population from the rural in the total school population, by education levels (selection). Percentages

	2000	2007	2010	2014	2015	2016	2016/00 pp	2016/07pp
Total	32.69	29.31	30.28	29.94	29.69	29.23	-3.46	-0.08
Schoolboy enrolled in primary education (including special education) - classes I-IV	49.55	51.63	50.46	45.70	44.91	44.45	-5.10	-7.19
Students enrolled in gymnasium education - grades V-VIII secondary cycle 1	42.13	48.30	48.93	48.49	47.69	46.23	+4.10	-2.07
Students enrolled in high schools and theoretical colleges	5.70	4.53	4.76	4.93	4.81	4.65	-1.05	+0.12
Enrolled students - license	0.005	0.06	0.07	0.04	0.05	0.04	+0.04	-0.01

Source: TEMPO Online Processing, http://statistici.insse.ro/shop/index.jsp?page=tempo3&lang=en&ind=SCL103I (NIS, 2017c)

Figure 6. School population by residence area, during 1995-2016
Source: TEMPO Online Processing, http://statistici.insse.ro/shop/index.jsp?page=tempo3&lang=en&ind=SCL103E, (NIS, 2017c)

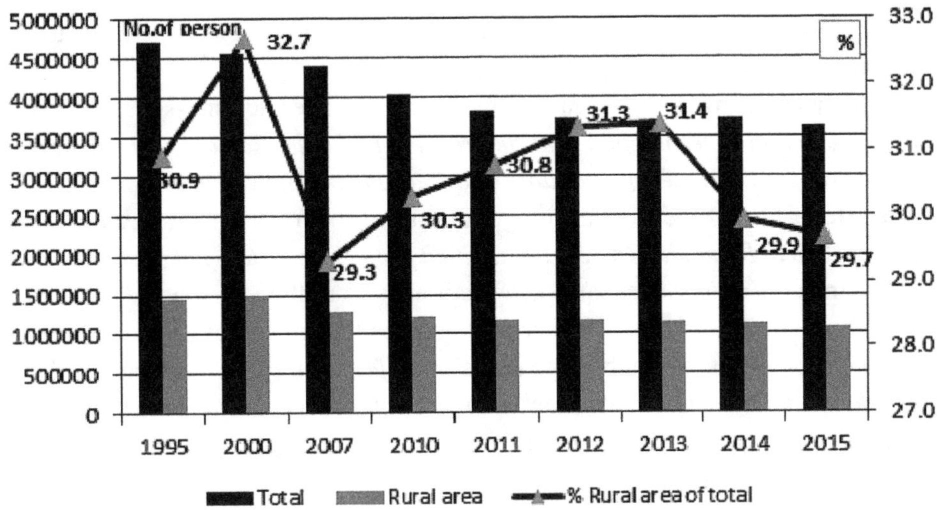

Figure 7. The households' situation according to their ability to cover the expenditure by income
Source: Processing from: "Coordinates of living standard in Romania. Population income and consumption, in 2016", (NIS, 2017b)

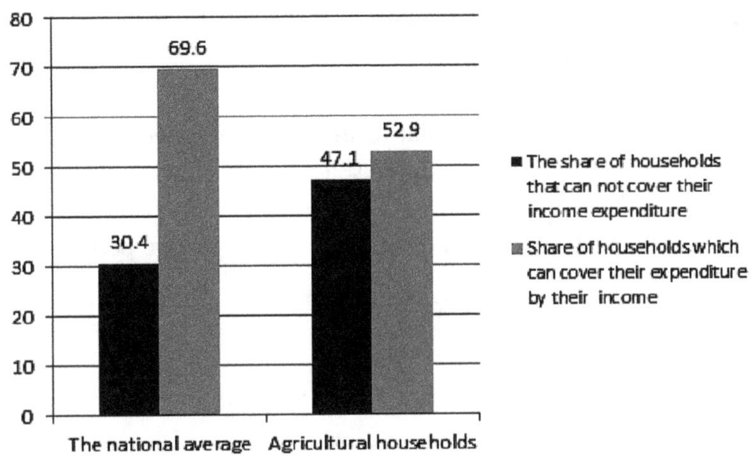

Incomes, respectively the standard of living, have a direct and significant impact on the degree of access to knowledge of the owners of the small households.

In fact, there are directly proportional interconditionality relation between currently available income and knowledge. If income increases, then, of course, prerequisites are created to increase accessibility to knowledge and vice versa.

The ratio of 1 to 0.25 between the non-agricultural households' incomes compared to those of an agricultural household, in 2010 Processing from: "Coordinates of Living Standard in Romania. Population Income and Consumption, in 2016 ",NIS, 2017, Bucharest, ISSN 1584-7454., explains why the level of school education in rural areas is well below of the urban level and therefore the low level of knowledge holded by the owners of small peasant households.

The existing production structures dominated by the subsistence and semi-subsistence farms and the low degree of *introduction of technical-scientific progress make the Romanian agriculture to function with different absorption capacities of available information and knowledge.*

Another specific feature of the Romanian agricultural sector from the perspective of the available human resources is that *in Romania is no clearly defined professional status of the farmer*, which affects - given the large number of people occupated in agriculture (2.5 million persons) - both the functionality of the fiscal system, as well as of the social security and health care system.

In addition, the sociologists avert about peasants' *manifestation of a "conservatism" behaviour as the main reason for retaining of those who work in their small households faced to the new world toward progress.* Peasant conservatism is a native fact given by: the isolation of farmers at the spatial level; their modest financial resources; the absence or low relationships with the market, because the great their majority produces mainly for self-consumption.Economically, the conservatism manifested by farmers is materialized by the inability of small subsistence and semi-subsistence peasant households to adapt to change, regardless of the form that they assume (political, economical, structural, professional, institutional, legislative, etc.).

3. Vectors for Circulation of Information and Knowledge between Producers and Consumers

Unfortunately, the contribution to the spectacular results in the plant production sector registered during the last three years does not belong to small households, but to large industrial farms. In addition, because the research results flow circulates from "up to down", for example, knowledge get off from those who produce it to those who need them and than, in a perspective period, it will be necessary to

Box 3. The profile of the farmers, depending on the size of their holdings held

The Size of the Agricultural Households	The Share of the Agricultural Households in the Total Number of Households *), %	The Share of Agricultural Holdings in the Agricultural Area Eligible for APIA Subsidies *), %
Under 1 hectare	0.27	0.02
1-5 hectares	75.79	16.28
5-30 hectares	20.00	18.30
30-50 hectares	1.44	5.23
Over 50 de hectares	2.50	60.17

*) The structure of agricultural holdings refers to all 3.8 million entities.

**) The Agency for Payments and Intervention for Agriculture - APIA - supports almost one million farms.

Source: Daia, P. "ZE Agribusiness: How Can Romanian Farmers Get Out of the Raw Exports Race?", Financial Journal Conference, October 30, 2017, (Daia, 2017)

take a new step towards opening up the farmers' research and only after that to identify the appropriate ways of comunicationg information and knowledge to consumers, which are most the small farmers.

Regarding the circulation of information and knowledge between producers and consumers of them, the following points should be made:

1. *A first way of absorbing the information and knowledge is the traditional way* on which the "new", regardless of its format, has a minimal degree of manifestation, and the own experience is the main channel of transmission/ communication and acquiring of the knowledge for the activity of production.
2. *The second way of absorbing the information and knowledge obtained by scientific research activity is through the practice of the industrial production system, which is joined by the technological system in which the scientific information acquires the quality as an important factor for the production increase and, through it, of the economic efficiency.*

In the second case, the knowledge puts the own traditional experience on the second level, this time being of interest the modern channels of information and acquisition of knowledge, namely: education; consulting; mass-media and newer social-media devices, etc.

At present, it is clear that Romanian agriculture is accelerating towards the high-performance industrial sector, which is supported, in particular, by foreign firms that have bought or leased large agricultural areas that they cultivate in a performance regime. Or the results of this sector, demonstrated by the increases in the recorded outputs, were based, above all, on the judicial use of various elements of the technical progress. From this point of view, *it can be said that on the Romanian market there is real demand for the "scientific product" as the main factor of progress in agriculture.*

Also, by *orienting the sectorial development to the organic farming segment, there is an increased need to use with great attention the diverse components of technical progress,* which will sustain the increasing of the demand for information and knowledge produced by the R & D institutes and by reserach centers, by universities, as well as by other entities active in the domain.

Generaly speaking, on is noted that in Romania, in the development strategies for agriculture and rural environment, the segments through which the information communication is provided, appeared with a minimum degree of visibility, because they are an erroneously considered to be influenced by the knowledge market, where a decisive role are the producers of information and knowledge.

In fact, consumers, and along with them, the channels of information communication among producers and consumers (of information and knowledge) require greatest efforts and involvement, especially from part of the public authorities.

The most eloquent example is the Law of Gravity. For nearly two centuries, since Isac Newton's formulation of this Law, several generation of students have endeavored to understand and to know it. On is clear that the efforts at the level of knowledge consumer are much larger, immeasurably higher than those taken on the moment of discovery. By extension, the previous example is in agriculture, but on reffer to other issues. For example, the creation of a new plant variety or a new animal race asks much less to spent than the efforts necessary for them assimilation and extension in the great agricultural production.

There are a variety of actors involved and / or likely to be involved in the processes of information circulation/ communication among producers and consumers. In the field of agriculture, forestry and fishing in the processes of circulation of information and knowledge are distinguished: the associative and cooperative system; the education system, including the university (which also has the role of producer of information and knowledge through them research activity developed); consultancy and extension; National Rural Development Network (NNRD/ RNDR); media, etc.

In the knowledge market, the vectors through which information flows circulate amongst producers and consumers have the following features:

1. Generally, *they do not belong to prodecers or consumers of information*. That is why the functionality of the vectors through which the circulation of information among producers and consumers takes place is independent of the interests of the two actors (producers, consumers).
2. *The main channels through which the circulation of information and knowledge - namely by education, consultancy, extension, a part of the specialized media, National Rural Development Network and others - are still largely under the responsibility of the public authorities; in Romania is considered that the promotion of the new within the branch must be in accordance with the interest of the agrarian and agricultural policy makers.* Unfortunately public authorities, inclusive the Ministry of Agriculture, do not give a real attention to set-up the real chanells for the circulation of information and knowledge produced by the research institutes and universities.
3. *The private sector has emerged and developed in Romania in the fields of education, consultancy and means of communication within the mass-media.* However, it should be noted that *all these entities have made limited progress in developing their specific activities concerning circulating information and knowledge gained by agricultural and forestry research to producers.*
4. In Romania, associative and cooperative organizations - components of the European social model and of the single market - are in various stages of development. Although they have a crucial role to play in implementing public policies and programs at different levels - paying a particular attention to education, training, promotion of entrepreneurship and of continuing training programs for agriculture, forestry and fishery - yet the rural population still remains reluctant faced to the associative and cooperative system. In Romania there are:
    a. 356 agricultural cooperatives (16.0% of the total number of cooperatives), with 247 employees (0.8% of the Romanian total staff), with assets of almost 172 billion lei (over 38 billion euros, respectively 15.3% of the total sector), revenues of 317 billion lei (70.4 billion euros, respectively 18.0% of the total sector) Source: INS 2009 - 2012, FDSC-IES processing, taken over and processed from the "Social Economy Atlas of Romania, 2014" by Cristina Barna, Institute of Social Economy (IES), ISBN: 978-973-0-17000 -9.

Guiding and regulating the cooperatist sector through two central public administration structures - the Ministry of Agriculture and Rural Development, on the one hand, and the Ministry of Economy, Commerce and Business Environment - the Business Environment Directorate - and the absence of a homogeneous legislative framework, and unbalanced allocations from public funds (national and European ones) maintain mistrust in the system.

In 2015, at the national level there were 41 County Agricultural Chambers, financed by public funds, but also from own incomes - by providing consultancy or by carrying out different courses or events. Also, other 14 Agricultural Chambers, were set-up by the Governamental Ordinance no.58/2013 (23Government Emergency Ordinance no.58/2013 amending and supplementing the Law no. 283/2010 on chambers for agriculture, food industry, fishing, forestry and rural development and it abrogations the Art. II of Law no. 122/2012 for amending and completing the Law no. 283/2010 on chambers for agriculture, forestry and rural development. Government Emergency Ordinance no. 58/2013, was published in the Official Monitor, Part I no. 347 of 12 June 2013) and them were subordinated to the National Agricultural Chamber.

b. The poor functionality of the Agricultural Chambers induced dissatisfaction to different representatives of farmers and it induced the perception that the Ministry of Agriculture "... wants to escapes by some expenses that on had with the former consultancy activity and intends to pass them to the agricultural producers ..." Processed by Mihai Lungu: „Camerele Agricole nu mai reprezintă interesele fermierilor", http://asociatia.bio/?p=8433
c. Local Action Groups (LAGs) are based on the private-public partnership. The founding members of LAGs are: Local Public Administrations; representatives of the economic sector; representatives of the civil society.

In fact, the LAGs are working as a bridge between the European Commission and local communities, namely the LAGs, which are as NGO bodies with legal personality (http://www.madr.ro/docs/dezvoltare-rurala/Axa_LEADER/gal-romania-update-16.07.2013.pdf) and with responsibilities in the development of the local development strategies.

5. Extension activity to agriculture or the agricultural advisory services includes the whole set of organizations that support people involved in agricultural production and them facilitates: the link with markets and other actors from the agri-food chain; providing reliable and relevant information, skills and technologies to improve production results; improving productivity by expanding technology transfer; developing human and social capital, increasing skills and knowledge on production and processing, facilitating access to markets; the organization of farmers and producer groups; the adoption of good practices, and practices of managing natural resources, etc.
6. A modern format for agricultural consultancy is represented by the information platforms. For example, the University of Agricultural Sciences and Veterinary Medicine (USAMV) from Cluj-Napoca, through the Faculty of Agriculture, has developed and Launched on March 1, 2017, in Bucharest the first online agricultural consultancy platform from Romania specialized in integrated agricultural input management called "*AGRIM Platform:* www.agrim.ro" (https://www.gazetadeagricultura.info/informatii-utile/573-consultanta/19761-prima-platforma-on-line-de-consultanta-agricola-din-romania-specializata-pe-managementul-integrat-al-inputurilor-agricole.html). The purpose of setting up the AGRIM Platform:

Box 4. Territorial Rural Development Centers, Tools for Knowledge Transferring in the Developed Countries

In the developed countries, the research market is more complex and functional. Classical information and knowledge producers, the same as us, authorize the responsibility of distributing their results to the territorial development centers.
In turn, agricultural producers who are organized into co-operative and / or associative structures that also have responsibilities in identifying, taking over and disseminating information from the territorial centers for rural development. These results in a market, a public and private functional partnership between different organisational structures wich have a distinct size, with flexible activities, where the objectives are structured so that the information transmits can be used.

a. Helps Romanian farmers in choosing high-quality inputs (seeds and biological material, fertilizers, phyto-sanitary products);
b. Initiates promotional actions (current best agricultural practices for adapting to climate conditions, inputs of input providers, etc);
c. It functions as a forum for providing on-line consultancy on the fields of competence of the professors from faculty of agriculture from the usamv cluj-napoca;
d. Is a study tool for the students from agricultural faculties in the country - through information bagging across the spectrum of agricultural inputs, as well as free access to information by a forum sector.

7. *The National Rural Development Network* (NRDN/ RNDR) was established according to the EC Regulation no. 1698/2005, art. 68. The aim of setting up the NRDN is to ensure better integration of agricultural, forestry and social-economic approaches into the territorial dynamics / contexts. NRDN/ RNDR groups the organizations and public administrations involved in the rural development and its actions has the aim: identifying and analyzing best practices and informing farmers about these; network management; organizing exchanges of experience and know-how; preparing the training programs for constitution of local action groups; providing technical assistance for actors co-operation. The funding of the NRDN/ RNDR is made from the European Agricultural Fund for Rural Development.
8. On believe that in the future by setting up at the level of research institutes / research centers a "*system of subscriptions to agricultural advisory services*", it is possible to create real premises for reviving, on new bases, the connections between agricultural and forestry research, on the one hand, and the majority of small agricultural households, on the other.

Figure 8. The scheme for subscriptions to agricultural advisory services for consumers of information and knowledge by individual households or commercial companies, authorized individuals, associations, cooperatives, other consumers

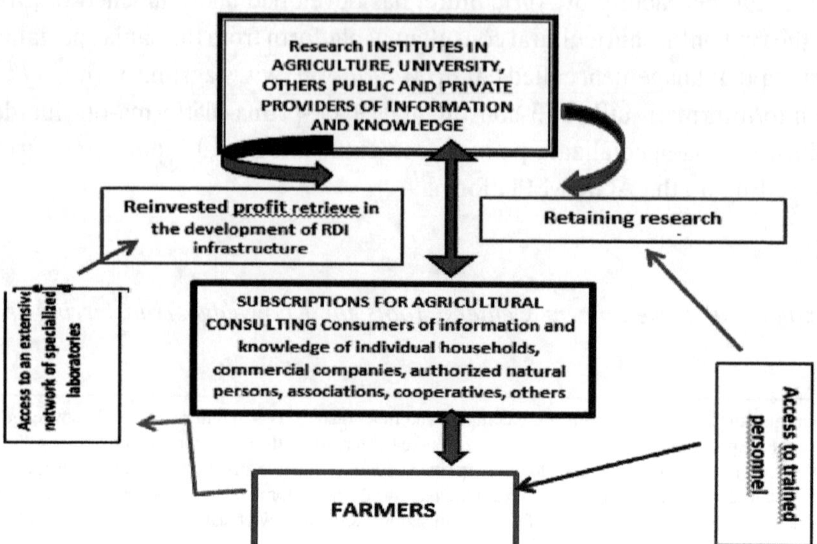

For the start, agricultural / forestry research institutes / centers, depending on the specifics of each individual actor, can develop and place on their own institutional site applications with information about the types of services they can offer to small agricultural households. A such initiative can be successfully implemented with the extension of the use of social media devices, for example, mobile phones, as well as the fact that now everything is interconnected and interdependent in the Universe, also.

On this way, our understanding of the world and how we interact with each other acquires real valences for a large number of farmers, contributing to unitary dissemination and by various forms of materialization of information and knowledge achived into the research activities until to the level of an agricultural producer (in fact, the last one are the real consumers of information and knowledge achievd by research).

As the system of subscriptions to agricultural advisory services on develops, they will be able to diversify them according to the requirements of the market, respectively of the agricultural producers.

In addition, such an activity is likely to contribute to the motivation, from new positions, of the qualified human resources existing in the agricultural / forestry research and in development institutes, but also will increase the confidence of the agricultural producers in the scientific research results.

Additionally, counseling and consultancy activities can be restarted by launching this project. In the ten years since Romania joined to the European Union, these activities (counseling and consulting activities) went slowly, slowly and them became static. Moreover, the existing capacities have been directed especially to completing grant applications and certification, which has led to the loss of technical competence and experience in the field. The proposed project seeks to disappear some of the existing deficiencies in present agricultural advisory activities. At the same time, the role of this services can be reconsidered and strengthened through partnerships with research staff for to get back to the way for which *agricultural advisory and consultancy activities were initially conceived, namely mediators between scientific research and practice.*

CONCLUSION

1. Currently, the knowledge market in Romanian agriculture *is dominated by two actors, who naturally "do not can see" because there are "big to small" relationships between them.*
2. The information producer, regardless of its institutional nature - the Academy of Agricultural and Forestry Sciences (AAFS/ ASAS), research institutes or research centers or universities - is maintained in a organizational configuration related the ones from the communist period, when the research institutions were sized to work for large agricultural production units - agricultural cooperativea or state enterprises. *In fact, the Communists organized the transfer of information and knowledge from producers of them to the consumers on "big-to-big" relationships, regardless of the field of activity - production, services, market, etc.*
3. At present, the potential consumers of information, in a decisive ratio, are the 3.8 million of peasant households whose average size, which is a result of the process of building and restoring of the land ownership rights, is about 3.5 hectares, and in these entities are practicing traditional agricultural systems.

At the same time, after the dissolution of the large agricultural units, as a result of privatization, agricultural research remained in a similar organizational structure like to the period before the 1990s, without seeking viable solutions to the requirements of the small agricultural producers which appeared as a result of the application of the land legislation regarding the establishment and reconstruction of private property. That is why "agricultural research has become like an "island" for the traditional farming and, moreover, it is constantly facing to the lack of public support for its operation.

Having in view of the above, on consider that the main "culprit" (if a guilty one is being sought) for the Romanian research - namely, whether or not consciously on maintaining the old organizational structures which were prior before the 1989 - is the managerial incapacity to reform the system and to link it by the new market relationships, inclusive to link it by informtion and knowledge markets.

4. Certainly, the current state of the agricultural research, as an agricultural policy issue, has many aspects reflecting the crisis that this sector is undergoing, as well as the need to redefine a new strategy where strengths are centered on the directions that can potentiate and support the performance of the vast majority of small agricultural producers.

The key to revival of Romanian research, with particular reference to agricultural scientific research, is *to concentrate on market requirements and less on budget allocations, which are currently known not only to be relatively limited but also difficult to assure.*

In this context, on consider that the good example of foreign companies, which are the large suppliers of a quality biological material, about their approach to the knowledge and information requirements for small farms is not without interest for us. While these companies intensify their contacts and contracts in different forms with farmers - through symposiums, seminars, round tables, working visits, etc. - the "classical" agricultural research institutes and centers from Romania are becoming more and more isolated. For this reason, the danger of a "gradual compression" of these exponent of the research, even to extinction, is not a false alarm, a metaphor, but it can become a reality.

5. In the context of *the globalization process increased*, scientific knowledge, as a direct product of the research activity, in its duality position of an intangible asset and of a production factor, not only *migrates to developed countries, but also its financing is a part of a process, more broadly, to diminish the role of the public budget in the favor of private sector and the responsibility of research, including agricultural, is increasingly taken over by large multinational companies.*
6. The Romanian agriculture, which is developing at a visible pace towards the high-performing industrial sector - so there is a real demand for the scientific product - needs a rethinking of the circulation of the information and knowledge produced by the local research system. In this context, a first step towards a new attitude lies with the Romanian researcher and only afterwards to the farmer.

The relationship between the research and agricultural production is demanded by both sides, but with initiatives caming from the supply-side. In the absence of the public research effort, the Romanian farmer: (a) has recourse to the experience, which is not only much easier, but also it is predominantly traditional and it leads to low yield, and; (b) either the farmer makes use of the foreign input market, which is the technological progress, but they are much more expensive than domestic ones and sometimes even unsafe for the most small agricultural producers, because sometimes them are not known the commercial conditions.

7. In order to meet the socio-economic, technological and cultural challenges, further information and new knowledge are needed, which necessity to be transferred - in the most diverse forms, depending on the training and economic capacity of the consumers, namely the agricultural production units. But for the large majority of agricultural producers, the economic environment is increasingly difficult, the competitiveness is harder to sustain and the environmental restrictions are what they need with accuracy is another unknown issue.
8. The way in which agricultural, forestry and food industry scientific research from Romania can contribute to the creation, exchange and implementation of new knowledge, new technologies and new products requires the development of new ways of organizing, learning and cooperating.

In Romania, although the "National Strategy for Research, Development and Innovation 2014-2020" (Government Decision no. 929/2014 on the approval of the National Strategy for Research and Development and Innovation 2014-2020, published in the Official Monitor, Part I no. 785 of 28.10.2014.) *supports professionalisation in the technology transfer segment of public scientific research organizations and other innovation oriented organizations (*Professionalisation measures on the technology transfer segment: (a) setting up specialized tools for staff involved in technology transfer; (b) developing tools to develop technology transfer capacity at university level, focusing on the training of specialized human resources; (c) establishing tools for the development of incubators and technology transfer centers at regional level; (d) the development of trading platforms for the supply and demand of intellectual property, and; (e) initiating and using specific tools for the temporary hiring of highly qualified people to encourage the development of human resources flows from the public sector (institutes, universities) to firms with RDI activity.). However, their translation to agriculture, forestry, anf food industry occures with a low rithm relative with other sectors from the economy.

In this context, on considered that it is a particular interest to focus on creating a Knowledge Market Knowledge System in agriculture, which could have the following objectives: (a) to improve the knowledge base from the level of knowledge producers and to disseminate them in a transparent manner; (b) developing the capacity to provide detailed knowledge for basic commodity, and; (c) facilitating the links amongst producer of information - the transmission vector for information - consumer of information, in parallel with the public authorities involvement in sustaining the coordination of a strong development information and knowledge market.

9. In the perspective of *the development of subscription-based dissemination activities of some products resulting from the R&D activity, the repositioning of the agricultural and forestry research is carried out and, on the other hand, on can prioritize the production of information and knowledge and, also, of the future activities of research entities.* In the medium and long term, implementation of this measure can influence: research institute partnerships; the choice of future research directions; providing additional financial resources at the level of each agricultural and forestry research units.

REFERENCES

Bara, S., Zaharia, A., Teodor, C., & Preda, E. (2017). *Possible ways to stimulate the human resource within romanian agricultural and forestry research*. Paper presented at "The International Conference Competitiveness of Agro-Food and Environmental Economy", CAFEE 2017, Bucharest, Romania.

Daia, P. (2017). ZE Agribusiness: How Can Romanian Farmers Get Out of the Raw Exports Race? Presentation at the *Financial Journal Conference*.

European Commission. (2017b). *Access to microdata. Overview*. EUROSTAT. Retrieved from http://ec.europa.eu/eurostat/web/ microdata/overview

European Commission (EC). (2017a). *Recognised-research-entities*. EUROSTAT. Retrieved from http://ec.europa.eu/eurostat/documents/203647/ 771732/Recognised-research-entities.pdf

Government of Romania. (2014). *Government Decision no.929/2014 approving the National Strategy for Research, Development and Innovation 2014-2020*. Official Monitor, Part I no. 785 of 28.10.2014. Retrieved from http://www.cdi2020.ro/wpcontent/uploads/2014/10/ HOT%C4%82R%C3%82RE-nr-929-din-21-octombrie-2014-privind-aprobarea-Strategiei-na%C5%A3ionale-de-cercetare-dezvoltare-%C5%9Fi-inovare-2014-2020.pdf, accessed on 9 October 2017

Ministry of Research and Innovation (MRI). (2017). *Institute Naţionale de Cercetare Dezvoltare (INCD) în coordonare*. Retrieved from http://www.research.gov.ro/ro/articol/4514/sistemul-de-cercetare-institute-nationale-de-cercetare-dezvoltare-in-coordonare#Prezentare

National Institute of Statistics. (2017b). *Coordinates of living standard in Romania. Population income and consumption, in 2016*. Bucharest: Author.

National Institute of Statistics. (2017c). *Tempo online*. Author.

National Institute of Statistics (NIS). (2017a). *Statistical Yearbook of Romania*. NIS.

Popescu, G., & Bara, S. (2017). *The Knowledge Transfer Economy (Unpublished Lecture Notes)*. Bucharest: Bucharest University of Economic Studies.

Romanian Law no. 45/2009 regarding the organization and functioning of the Academy of Agricultural and Forestry Sciences Gheorghe Ionescu-Sisesti and of the research and development system in the fields of agriculture, forestry and food industry. (2009). Published in Official Montor of Romania.

World Economic Forum. (2017). *Global Competitiveness Report 2017-2018, in the framework of the System Initiative on Shaping the Future of Economic Progress*. Retrieved from http://www3.weforum.org/docs/GCR2017-2018/05FullReport/TheGlobalCompetitivenessReport2017%E2%80%932018.pdf

Chapter 16
Technology Scouting and Inventions Patenting With Impact on the Agrifood Future:
INACO – Institutional Innovation for Competitiveness in Romania

Andreea Paul
Bucharest University of Economic Studies, Romania & Initiative for Competitiveness in Romania (INACO), Romania

ABSTRACT

This chapter is the sketch of a possible pattern of the future world in which any kind of business will be developed in a completely new human, technological, agricultural, and commercial context, heavily and quickly changed from the one we live in now. The first objective of this chapter is to scout for the mega-technology trends that will reshape completely the future business and jobs, focusing on the agrifood industry. The second objective is to tackle the main challenges to patent inventions in terms of costs and timing in Romania, relative to other countries, and raise pragmatic recommendations. The third objective is to describe the institutional innovation called INACO (the Initiative for Competitiveness), a think-tank dedicated to tackle the challenges and opportunities of the future economy and how can a country such as Romania stay competitive in a more and more competitive world.

INTRODUCTION

The tomorrow's hyper-technological world has emerged and tempts us with its various scientific discoveries and economic strategies from today. In Romania, there are already signs of this very near future - in today's leading industries - such as IT - but also in traditional areas like agriculture. Few know that, in fact, strategically, agriculture is the area where innovative technologies apply for the first time - as proof of the fundamental importance of securing the food resources of the world's countries.

DOI: 10.4018/978-1-5225-5739-5.ch016

Copyright © 2019, IGI Global. Copying or distributing in print or electronic forms without written permission of IGI Global is prohibited.

For any economy, any public policy or even for every business to be profitable and competitive tomorrow, all need to know the global trends that today's world is facing. Information is in itself a certain value - more valuable than natural or financial resources - especially in the perspective of technological, social, agrifood or climatic evolutions etc. Exceptionally fast. That is why the developed nations - China, the United States, Japan, and not only[1]- use supercomputers to pinpoint the potential of the near future economy as accurately as possible.

This chapter is the sketch of a possible pattern of the world over just a few years in which any kind of business will be conducted in a completely new human, technological, agricultural, commercial context. Completely changed from the one we live in now. The first objective of this chapter is to scout for the mega-technology trends that will reshape the future business and work places, focusing on the agrifood industry, and for the first time in the scientific literature focusing on Romania from this perspective. The second objective of this chapter is to tackle the main challenges to patent inventions as costs and timing in Romania, in relative terms to other countries, and raise pragmatic recommendations. The third and last objective of this chapter is to discuss a Romanian institutional innovation called INACO – the Initiative for Competitiveness, a think-tank dedicated to tackle the challenges and opportunities of the future economy and how a country such as Romania or a business or an individual can stay competitive in a more and more competitive world. We end up with solutions and recommendations.

MEGA-TECHNOLOGY TRENDS TO RESHAPE THE FUTURE ECONOMY

The future economy raises intricate issues related to the speed and goals of sustainable development, controversies in accepting the mega-technology trends in our lives, moral and ethical debates related to robots and humankind; thus, relevant problems for the digital and technological illiterate people, many solutions and opportunities in the meantime for people, businesses, regions and countries.

In the 1970s, researches have been conducted using the concept of technology forecasting, broadening it afterwards towards technology foresight. In the 1990s, concepts like technology intelligence, sharing economy, circular economy, green or bio economy, optimal climate policy and economic development, flexible working or gig economy have emerged and fuelled the scientific literature in the last two decades, shaping the future economy, public and private policy-making.

Our purpose is to focus on the technology scouting, researched very little by now, and defined as relevant core technology developments and trends, with disruptive potential and with high relevance on the competitiveness of public and private decision-makers and practioners, by detecting advances in technology in their early stage and patents transferred into the real economy. The costs for patenting and administrative procedures are relevant issues from this perspective, that is why we will tackle it pragmatically and comparatively, as well as an institutional systemic approach that we describe in this chapter. Scouting globally and locally for technology early applied into the real economy enhances the motivation and facilitates the technological competitive advantage of private and public research, development, innovation, investments and macro- and micro-management strategy to cope with the future, but also of consumers as early buyers and changes in the markets, nevertheless legislation, attitudes and socio-cultural shifts. Our approach is thus larger than that proposed by Rohrbeck (2007) who defines technology scouting, *as a systematic approach by companies whereby they assign part of their staff or employ external consultants to gather information in the field of science and technology (S&T) and to help facilitate or execute Technology Sourcing. Technology Scouting is either directed at a specific technological area or undirected in order to identify relevant developments in technological white spaces. Technology Scouting relies on formal and informal information sources including networks of experts.*

Starting from the latest and most important events, inventions and innovations, global information and reports - such as the World Economic Forum's Center for the Fourth Industrial Revolution and New Vision for Agriculture recent reports (2017), the PricewaterhouseCoopers report issued at the end of February 2017, the Singaporean Committee on the Future Economy issued in the same month this year, Europe on the Move report published in May 2017 (EC, 2017), the high-level Global Commission on the Future of Work launched in august 2017 in Geneva, even the earlier US Future Economy Initiative (Sheeran, Ackerman, Enelow, Goodstein, Hahnel, Power, & Schor, 2014) or the created Ministry of the Future in Sweden in 2014 (Ma-Dupont, 2016) – this chapter aims to suggest some current directions that will reshape the future economy, business and labour force skills needed, generally valid, also in Romania, due to technology dynamics and globalization.

We focus on the impact on the future economy of robotics, additive manufacturing/3D printing, biotechnology, artificial intelligence, blockchain system, internet of things, environmental evolution, future resources, transports and labour force of the future on the short and long run, with relevant impact on the agrifood industry in the world and, particularly, in Romania.

Projects on smart agrifood systems have been heavily financed by the European Commission in the last financial framework (EC DG for Agriculture and Rural Developments - FP 7), especially for robotics, data, micro-nano-bio systems, GNSS, Fiware accelerators, thematic networks and other projects, and keeps on going in the Horizon 2020 research-development-innovation financing.

1. Robotics

The first industrial robot appeared in 1961 and the first industrial robot to move completely in six directions in 1971. Evolution has since been huge, especially in the third millennium, from industrial robots that played the role of a worker who was doing a technological operation, a production band, towards today robots that perform extremely complex operations at their workstation. In different industries, robots of different categories have taken over almost all of the production process from humans – e.g. in the world's largest industrial air-conditioning plant Midea in China, 800 robots have replaced 24,000 workers (see Midea Robot Factory video, 2016). The largest Ikea packaging plant in Sweden is fully automated and has a single operator (see Motoman robots packing IKEA video, 2012). Intelligent military robots are already on battlefields - the United States, China and Israel being world leaders in their use on the ground (see AiirSource Military video, 2013). The robot cop in traffic will soon replace people (see Kiwi in Munich video, 2016, for *Real-life Robocop: Dubai to get android police force by 2017*). Taxi drones becomes operational this year with one passenger (see Tech Medium, 2017, see First Real Flying Car in Dubai Ehang 184 video) and in few years with two or more passengers (see Engadget, 2017, video) etc.

In agriculture, autonomous tractors/unmanned combines were introduced, supervised by a single operator who co-ordinates from a computer dozens of farm machinery on the field (see the CNH Industrial Autonomous Tractor video, 2016). Such tractors are already used in Romania for agricultural work in regions such as Banat and in the Big Island of Brăila (Pambuccian, 2017).

A Romanian-made robot called "Banat" was in the top ten agricultural robots in Europe at the Field Robot Event 2014, which took place in Bernburg – Strenzfeld, Germany. The little robot was presented for the first time at the first edition of the AgriPlanta Vest exhibition by the project coordinator Sorin Bungescu, Assoc. Prof. Eng. USAMVB Timisoara (Agroinfo, 2014).

Autonomous agri-robots can oversee crop development, identify the needs of each plant on the field they are supervising 24 hours a day, 7 days per week and 365 days a year. Robots can constantly monitor the various animal livestock, while powering with solar panels (see LadyBird Robot and not only at New Scientist video, 2016). The drones in agriculture can supervise but also herbicide, spraying with fertilizers etc. huge areas of land (see Farmbot harvest robot video, 2016). Orders given by a computer to a different farm dredge are based on the data collected by the drones themselves that analyzed the land to calculate how much fertilizer or herbicides each square meter of land is needed (see DJI MG-1S - Agricultural Wonder Drone video, 2017).

In the near future, replacing labour force in actions requiring repetitiveness, physical effort, risk, and precision with robots will be almost complete. In addition to handling objects, wrapping and smart storage, multifunctional robot networks appear. Even cafeteria services are already be replaced by robots (TechCrunch, 2017).

The future in using industrial robots means the potential disappearance of even the production plants. Industrial robots can be rented, can be moved anywhere on demand, to produce/assemble, as appropriate, the desired "on-site" products.

The automotive industry currently uses nearly 40% of the global industrial robots, being by far the most robotised industry, followed by electronics, metal industry, chemicals and plastics, and afterwards by the food industry - all five sectors in one quantifying for about 80% of the total industrial robots used in the world in 2015, according to the International Federation of Robotics (The Economist, 2017).

2. Additive Manufacturing / 3D Printing

Making the most diverse objects through the 3D printing process will have a global spread due to the speed and potential of "customizing" each product as opposed to industrial manufacturing. The prospect is that every man can produce objects or food from various materials in a single copy for his own needs, thus through personalisation.

Human organs are sampled from stem cells that can be accepted by the human body and personalized for each individual patient. 3D bones, ears, bladders were produced and a kidney was attempted (Descoperă.ro, 2016). The future means that every man can have his own organs as personalized "spare parts".

In the field of nutrition - and therefore a threat to traditional farming - we are already printing 3D different types of foods with the quantities of personalized nutrients for each client.

Today, 3D is being "printed" in very fast-moving machine parts or airplanes with complex joints (see Aeronews, 2015, how the General Electric prints a 3D mini jet engine), in housing or office buildings record construction with forms that could not be achieved by other processes (New China, 2014). 3D dresses are printed in art and fashion.

In the field of nutrition - and therefore a threat to traditional farming - we are already printing 3D different types of foods with the quantities of personalized nutrients for each client (ABC Rural, 2016). Intelligent food or digital beverages, smart packaging or virtual lemonade with electrodes to mimic the

flavor and LED to mimic the color are not just successful experiments, but real start-ups (VentureRadar, 2015; Reuters, 2015; PackagingDigest, 2017; Ranasinghe, 2017).

3. Biotechnology

Genetic discoveries have produced outstanding results in many areas, beyond biotechnology applications in agriculture, nutrition or pharmaceuticals. Research on the human genome is very advanced and the methods of applying the advances of computer science and other contemporary technologies are constantly improving. Last year, synthetic meat was produced - that is, meat that does not belong to any animal but was produced by growing muscle cells created in the laboratory - in an attempt to solve the problems of food resources and the slaughter of animals (Plant Based News, 2016).

In 2014, phyllorescence plants were created by genetic bioengineering (Wall Street Journal, 2014). Research continues to try to produce luminescent trees with which "natural" public lighting is desired in the future.

The first synthetic organisms - viruses, microbes, bacteria with DNA artificially made by scientists - appeared. Procedures for extracting DNA fragments with problems with healthy fragments have paved the way for a potential prolongation of human life for decades or, in a far farther future, with centuries. Bionic eye or bionic ear are already realities. In 2007, the first interface between the neurons cells and the machinery through cortical implant is carried out. With the help of a connection made with the brain, with a special helmet, researchers have recently managed to mentally divert devices such as drones (Associated Press, 2016). The neural connections made in 2016 made it possible for paralyzed patients who have never moved to be able to command only the devices that helped them go for the first time in their lives (World Science Festival, 2016). Current research is moving towards telepathic communication between two human brains.

4. Artificial Intelligence

Contemporary technological developments have created computers with artificial intelligence (AI) that go beyond any predictions. Fundamental issues related to the morality and ethics of the use of AI are already being addressed. The automated driver of cars without a driver in 2020 will appear on all roads in the world (Mercedes and BMW have billions of dollar projects) will choose to sacrifice passengers if it is inevitable to have more casualties than car crashes. The AI is being debated at the UN to ban autonomous military drones and robots. It is forbidden to assemble the robots around the globe by intelligent robots, assembling being done only by humans, in order to avoid the robots to self-produce themselves.

Face recognition is used at airports, to fight terrorism, at street supervision, in hypermarkets and customer shoppers (NECIT Solutions, 2012). The cities of the future will be Smart Cities, where the actions of all residents will be pursued through AI to ensure security, serving utilities and products, etc (Amsterdam Smart City, 2013). The use of AI has innumerable potentials, which have already begun to be used in various business areas.

The use of AI, along with automation, will have a huge impact on the labour market, consumer analysis and predictions. Call centres will be replaced by empathic and empirically responsive AI applications that will replace all officials. It is estimated that only in China in the next decade AI and automation will lead to the replacement of 100 million workers, after setting e record high in terms of venture capital investments in 2016 (KPMG, 2017) and after issuing guidelines on developing AI in China *with the goal of becoming a global innovation center in this field by 2030* (The State Council of the People's Republic of China, 2017).

Last year, Unilever made an international campaign to respond to any queries related to hair care and haircuts - All Things Hair - by mail or phone. At the end of the campaign, the one answering any type of question turned out to be a smart computer. At the University of Georgia, in the United States, Professor Watson was voted the best professor of 2016. The professor responded by mail and telephone to any question from the students. In fact, Watson was an artificial intelligence program for IBM (Esoteric Detective, 2016).

Start-ups in the agrifood industry have developed "food intelligence," ranging from "Star Trek" replicators, to the use of nantes or the creation of "digital beverages". Smart food appeared and will develop using AI. Another application of science is the use of artificial intelligence and the "big data" - massive gathering of information - in order to be able to make the best vegetarian dishes. A team of Chilean researchers has developed a company called The Not Company, which uses algorithms to create herbal foods entirely based on herbs. Their virtual assistant, called Giuseppe, "learns" about the molecular composition of various dishes, including taste and texture, and then uses this data to recommend herbal recipes that taste similar to those made from animal products. Another company that makes such products is Impossible Foods, which produces "meat" and "cheese" without being based on animal products, but only on plants. The company is based in Redwood City, Silicon Valley. According to the United Nations, livestock farms account for 14.5% of planetary greenhouse gases and require huge amounts of water and animal feed, according to The Economist. For example, producing one kilogram of beef requires 10 kilograms of fodder, and for the same quantity of pork 5 kilograms of food is needed. Finally, for a pound of chicken, 2.5 kilograms of food is needed (The Economist, 2015).

5. Transports of the Future

Future transport will be linked to automatic piloting through artificial intelligence. Unmanned cars will begin to conquer the roads around the world over the next decade. The biggest changes will occur in freight transport - where trucks without a driver will make "caravans" on the motorways of the world. The first truck carrying 50 thousands beers down a Colorado highway was a reality in 2016 (Wired, 2016). The disappearance of professional drivers will lead only in the United States - where professional driving is the most widespread occupation - to unemployment of over 3 million people. Operational safety and efficiency will increase significantly in delivering fresh food and not only with driverless trucks.

Romania - in the absence of the interurban infrastructure, it could have the future chance of investing in the organization of intelligent cities - in which the transport would be organized as for the third millennium. At the same time we can revitalize a modern naval transport on the Danube and the sea for the agrifood industry.

6. Virtual Reality

Virtual Reality (VR) has been in some form or another around us for decades, but only now technology has reached a point where it is actually cost efficient. VR is not just for online gamers / gamblers, but is trying to become an entrepreneurial tool that will help people work more efficiently and train employees and customers remotely from around the world, agriculture included here-in. Technology significantly eliminates physical barriers in exploring the world and food by humans.

For example, using Google Earth VR, it allows users to fly over a city, stand on the mountains and even go into space. Food art and design are re-imagined and education becomes more interactive. So is the augmented reality of your farmer job or of virtual agriculture or of virtual crops and monitoring or of the rural areas, even for weather forecasts or checks on livestock or veterinary diagnostic (Affinity VR, 2016; Intersog, 2016). McDonald's is offering VR tours of its farms to explain how the food is grown and prepared (Farmers Guardian, 2016). The future applications for VR in agriculture have great development potential and can even revolutionize the agrifood industry in a smart farming.

In agriculture, it will be possible to know how many wheat grains are harvested at the exact time desired throughout Central Asia, or it will be possible to know how many corn grains are preserved at that time in all the localities, for example, in Vaslui County.

7. Blockchain System

Any message sent over the Internet bears the signature of the transmitter. In addition, a copy remains on the Internet. The Blockchain system abolishes this way of sending information: through encryption procedures, the message is sent anonymously, originally, without a copy, to a receiver.

It's exactly what was needed for financial transactions – such as the Bitcoin - where you do not have to show your identity for any purchase you pay. The system is ideal for making the notary firms or public finance agencies disappear - contracts are passed through Blockchain directly without the need for intermediaries. Some big economic powers aim to partially transform the national currency into Bitcoin for the next few years (the value of a Bitcoin last year was $500 and this year got to more than $4000). However, the system needs to be improved, the Internet can encrypt a limited number of Bitcoin, and the emergence of supercomputers may lead to decryption of any code, thus the advantage of payments in Bitcoin or in any other cryptocoin disappears.

8. Internet of Things

All objects and people on the Globe tend to connect through the World Wide Web. Therefore, this connection of billions and billions of information, data, objects and people can be used for business, macroeconomic goals, for controlling pollution, terrorism, etc., to make life easier for every inhabitant of the Planet by streamlining all personal actions in a general context.

From 2015, the European Union has begun building a Single European Digital Market. All major economic powers have extensive research programs to improve interactions between the components of the Internet of Things.

In small, at the level of every citizen, things will be radically transformed through the Internet of All Things - a written note left on the refrigerator with the hour and the lunch menu will be read by smart home appliances. The refrigerator will thaw exactly the ingredients required for the meal, after choosing the ideal recipes from the Internet, the cooker will preheat at the required time and the family members will be personally notified by lunch time.

In general, an application called Macroscope (the descendant of the Telescope or Microscope) - which will be the computer connected to everything in the world - will be able to give the synthesized data about anything - for example, about how many coffees are being drank at that moment worldwide or just in the applicant's neighbourhood (IBM, 2017).

In agriculture, it will be possible to know how many wheat grains are harvested at the exact time desired throughout Central Asia, or it will be possible to know how many corn grains are preserved at that time in all the localities in Vaslui County.

9. Resources of the Future

Some “classical” natural resources - oil, coal - are already dethroned by countries with resources needed for the new technologies. For example, all of the future batteries are using lithium - so countries with lithium-rich subsoil - Argentina, Portugal, and Zimbabwe - might become rich countries in the future.

To gain resources for tomorrow’s technologies, China has “conquered” some Asian, African and South Asian countries, exporting mining know-how, concluding agreements with countries rich in special minerals - yttrium, ytterbium, terbium, etc. essential for high-tech products. China controls 90% of the world’s rare earth resources (Energy Skeptic, 2013). Large countries with rarely important technological resources - beryllium, palladium, iridium, osmium, tantalum, cobalt, etc. are: Canada, USA, Mexico, Congo, Brazil, Russia, Australia. Romania has the chance of rare ore deposits in various mixtures with gold, silver or uranium - germanium, titanium, cobalt, gallium etc. In the near future, rare mineral resources are very limited and it is intended to extract the ore from space asteroids using unmanned spatial equipment.

In the near future there will be a reversal of the priorities related to natural resources. Oil is today and will continue to be in a “crisis” of overproduction, given the alternative energy sources for its replacement. The oil will quickly return to the role of “building material” - for a car, oil is used not only as fuel, but also for making wheels, dashboard, paint, etc. Along with alternative energies, there are now projects that produce fuel - for example - genetic engineering. Scientists who have deciphered the human genome and made the first synthetic cells have now put Exxon Mobil into a huge project where microbes and algae produce fuel. Such fuel-producing farms are a potential income-earning opportunity for farmers in Romania as well (ExxonMobil, 2017). The well-known $5 billion project of Elon Musk - Solar City - the largest solar panel manufacturer in the world is another example (WIVBTV, 2014). Agriculture which is a resource economic sector, during the next time will be influenced by the food products innovated under the influence of the digital revolution. This evolutions will change the conventional production structure in accordance with the solutions (“resources”) given by the digital environment.”

10. Labour Force of the Future

People working, but also business managers and responsible states will be subjected in the years to follow to a huge centrifugal force exerted by replacing the human factor by technology. The computer - which has made physical work unnecessary in many areas - has brought many women into the companies who have been able to do the same work as men. Today, robotization and artificial intelligence make it unnecessary for any person outside a few operators / supervisors of the production process.

There is still some pressure from trade unions and protectionist states that delay the replacement of the human labour force, but in the face of the fierce competition of “unscrupulous” collapse of productivity and bankruptcy of entire economic branches, states, trade unions and companies will sell and buy technology. This will replace workers and civil servants.

In this context, an emerging “unworkable class” - a term introduced by anthropologist Yuval Harari in 2011 (translated in English in 2014) - will have no jobs and will be supported by the state aware of their inability to find an occupation. Financial aid in this direction has already been introduced in Finland and Switzerland, and there is a strong debate about the minimum guaranteed income.

The unworkable class must be given the necessary time to spend time. Thus, there can be a real increase in tourism, the "leisure" industry - just like some retired people in the society.

People with more free time may be more involved in community activities - be promoters of civic initiatives. There is still a chance for these people. The disappearance of factory work (perhaps even the disappearance of factories) can structure the framework of small intra-community industries that produce objects / provide ultrapersonal services to other community members, other than those produced by the automated and impersonal industry. There have been initiatives of this kind in the world - because some people have already begun to become "useless", being replaced by technology.

Another factor that will influence the human workforce in the future, with important effects in the economy and future business plans is the demographic factor. Nations that will have a significant rate of demographic growth are on the rise in the world economy. So, more citizens mean more unworkable citizens, but also more active citizens. At present, the labour force from agriculture is known to be the most numerous and with a relatively lower level of qualification than in other economic sectors. In the future, on is probable that under the impact of the digital revolution these disparities regarding the labour force will be diminished.

CHALLENGE: THE COST OF PATENTING ROMANIAN INVENTIONS IS MORE THAN DOUBLE THAN IN GERMANY AND TRIPLE COMPARED TO POLAND

The complex dynamic of the future economy trough the technology disruptions described above in this chapter calls for a strong perspective based on research, innovation, development and trademarks to stay competitive in a highly competitive world. An important step from the action plan in this direction is to significantly reduce costs, but also the time required for the issuance of Romanian patents, facilitating the transfer into the real economy and the new high-tech Romanian start-ups emergence.

Paradoxically, the cost of patenting Romanian inventions proves to be more than twice as high as in Germany and three times higher than in Poland (Paul, 2017 a). The fees charged by the *Romanian State Office for Inventions and Trademarks* (OSIM) for patenting inventions are higher than those of neighbours in Slovakia, the Czech Republic or Serbia; even higher than in developed countries like Canada, more expensive than in China or in the Russian Federation (*Figure 1.*). Meanwhile, the procedures are twice as long as in South Korea. The time for processing protection claims is longer in Romania (4 years on average) than at the European Union, or in the US, Japan (3 years) or South Korea offices (2 years)[2].

OSIM's work deserves to be prioritized by decision-makers and reconsidered for several other reasons (Paul, 2017 a):

1. The Romanian inventions are world-wide recognised as excellent, but still the relevant research talents are insufficiently capitalized economically. Year by year, the Romanian inventors have been awarded with first class special distinctions - in 2017 as well with 35 gold medals and 8 silver medals - at the International Exhibition of Inventions in Geneva.
2. Technological readiness ranks Romania 51st world-wide in the Global Competitiveness Report 2017 – 2018, innovation ranks 96th and business sophistication is lagging Romania behind on the 116th out of 137 countries analysed (Schwab, 2017).

Figure 1.
Source: Paul, 2017 a

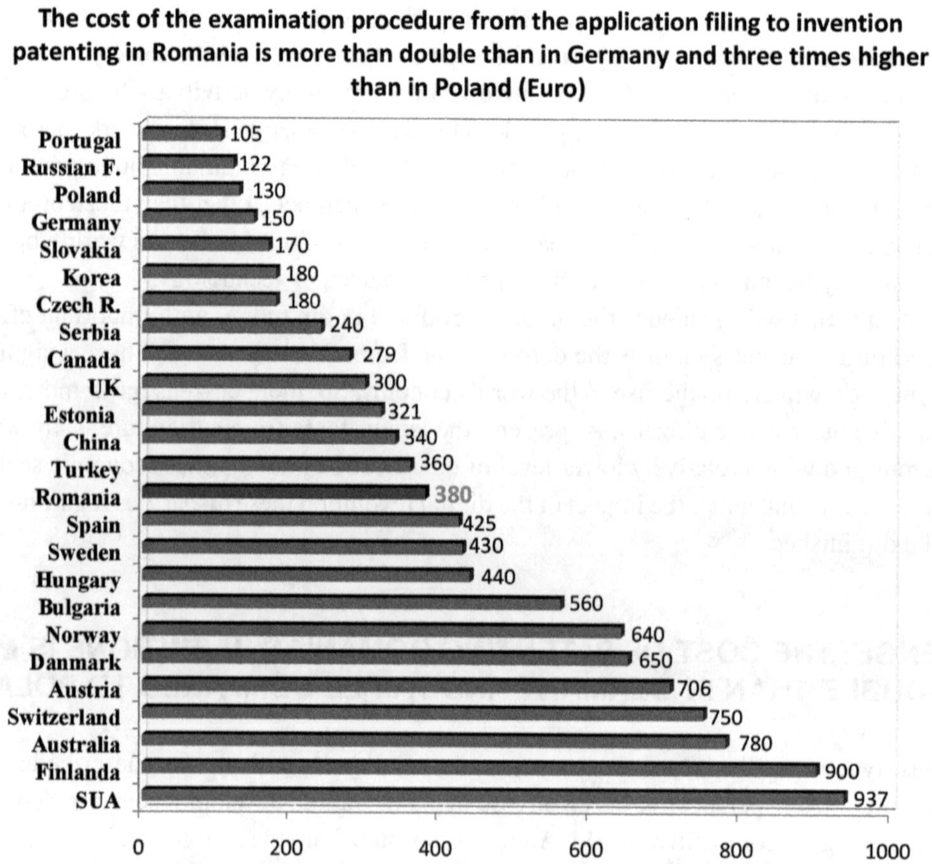

Table 1. Average processing period for protection claims, from registration to decision (in months)

	Patents	2009	2012	2013	2014	2015	2016
1	The Romanian State Office for Inventions and Trademarks	72.0	74.1	67.8	63.0	52.4	48.12
2	European Patent Office	41.4	40.6	43.9	45.3	46,9	41.7
3	Japanese Patent Office	31.6	31.8	31.8	32.4	33.9	35.3
4	US Patent Office	29.1	31.1	31.9	32.2	34.6	35.3
5	Korean Patent Office	28.8	25.5	19.7	15.0	17.4	22.2

Source: (Paul, 2017 a)

3. The incomes of those involved in Romanian research-development and innovation are low relative to the average earnings. A reduction in the cost of patenting their inventions is desirable. The inventor's personality is anti-bureaucratic. Procedures should be significantly shortened to world best practices in the field.
4. OSIM should not be considered as a collector of taxes for the national budget. In 2016, OSIM had a budget surplus, that is, it collected from taxes about 2 million lei more than it spent. Half of

OSIM's revenue comes from registration fees, and the other half from patenting fees. OSIM is not a Fiscal Authority. Even the fiscal authority is being restructured through the Prevention Law and by eliminating some burdensome taxes and procedures. OSIM must be restructured from the point of view of lowering the bureaucracy burden of examining files and of reducing the fees charged, at the threshold of the Industrial Revolution 4.0 at least.

5. The fees for the international protection of Romanian inventions must be rethought. The current real impact of public financial support for international patenting of Romanian inventions is only about 10%.
6. OSIM should become important inventions' and trademarks' promoter and protector in our treasury of national and international technology and digitalisation priorities. Alongside with strong actions and active venture capital mechanisms to stimulate the Romanian inventions absorption in the real economy for competitive investments, for high-value-added and better payee jobs.

This challenge calls for significant political and administrative decision-makers support, as well as for private decision-makers and public opinion support. Otherwise, an uneducated and unfriendly political and business environment for research and inventions may only give birth to an irrelevant context for the 3rd millennium, an uncompetitive and consuming-based economy on debt. We can reverse this trend by assuming five responsibilities of key public and private actors (Paul, a):

- An ambitious government program to reduce the costs and administrative of burden public procedures at OSIM.
- The Government's awareness and relevant actions for competitive remuneration of inventors and significantly higher RDI public budgeting support from the ridiculous current levels of 0.4% of GDP, one of the lowest in EU.
- By amending the law on inventions and trademark for correlation with European directives in the Romanian Parliament.
- A "quick starter" private association for capitalizing on Romanian inventions in the real economy.
- The active involvement of the media in the free promotion of Romanian inventions and innovations, trough more frequent bending towards science and technology in public debates to build the public technological and digital literacy, so as not to generate a wide social class of unworkable people in the future economy.

OPPORTUNITIES: ROMANIAN INSTITUTIONAL INNOVATION DEDICATED TO FORESIGHT, TECHNOLOGY SCOUTING AND COMPETITIVENESS

At the beginning of 2017, a think-tank called INACO - The Romanian Initiative for Competitiveness has been set up, as non-governmental research organization to be a generator of competitiveness for public institutions, local communities and business initiatives[3]. INACO is a non-profit community of more than 30 Romanian high-level professionals - university professors, researchers, managers, consultants, analysts, experts in economics, sociology, social shapers, global futurist, experts in branding, PR and communication etc. – dedicated to competitiveness improvements through analyses, investigations, researches, trainings, public debates and solutions proposals for the Romanian public institutions, for the businesses decision-makers and for youth competitiveness initiatives.

As a civic institutional innovation, INACO's main objective is to promote the competitiveness trough foresight, innovation and technology scouting as fundamentals for future economic development, following the failure of a legislative initiative filled in the Romanian Parliament by the author in 2015 – as an MP during those times - to build a public-funded national authority dedicated to this goals called the National Competitiveness Council / CONACO (L644, 2015), following the best practices world-wide.

INACO provides updated information about the Romanian competitiveness dynamics and recommendations to increase the quality of the economic and social public policies in Romania, to increase the competitiveness locally, regionally and nationally, develops specific programs of economic investigations, surveys, foresights and provides non-profit consulting/training/mentoring for a more competitive local, regional and national business environment.

INACO has four branches: Competitiveness Generator, InvestGate, ICOTI - Competitiveness Initiative for Youth and INACO INFO trying to contribute as much to preparing citizens, not only decision-makers, for the future economy and future jobs (INACO, 2017).

An integrated Romanian software for farmers called INOVAGRIA, launched by SIVECO's eAgriculture Department in 2016, has been promoted by INACO through InvestGate (Paul, 2017 b), in order to help farms' administration, to better control resources management, optimise the national and European subsidies, manage the lease contracts, updated with meteo forecasts in 2017 as mobile phone application (INOVAGRIA Meteo, 2017). Another example of investigation led by INACO was dedicated to sturgeon poachers (Paul, 2017 c) or to the access to resources and high-quality food for Romanians (Buzescu, 2017), as well the research dedicated to the costs of inventions patenting described in the previous sub-chapter (Paul, 2017 b). Last, but not least, INACO called for a national cohesion to create the Romanian Working Group for the Future Economy, addressed to the Government and all political parties and experts (Paul, 2017, d) and runs trainings with public and private decision-makers related to the impact of mega-technology trends and future economy.

SOLUTIONS AND RECOMMENDATIONS

In the Romanian specialized literature appears from time to time such an issue. During the last time (28 years!) the majority of papers were written taking into account the "perception" about phenomenon or process and we recommend the use of mathematical models within agricultural economy for answering fundamental challenges.

Romania is facing two-fold multiple challenges, as many other countries in the region.

On one hand, the technological developments, scientific research, industrial revolution (the fourth, historically speaking) challenges that should not catch our eyes closed or looking elsewhere. We must take action to catch the train of these accelerated changes. Everyone has to adapt - the state / entrepreneur / worker / unemployed. Regardless of age or gender, the ability to adapt to change must be maximal and permanent. Especially young people, at the beginning of the road in life, must be helped to start towards the future in a direction that has its own future. It is no longer the time of "tinkering and waiters" schools - as it was just a few years ago stated publicly from the Presidential level. McDonald's wants to remove kitchen workers from the restaurants in the United States, and even those who sell and replace them with robots.

On the other hand, the European Union "with more speeds" is a relevant productivity challenge - as long as the French farmer is given a grant per hectare more than three times that of the Romanian farmer (when, in fact, the proportion should have been reversed to reduce the gaps) – that calls for an economic Romanian development strategy compatible with the major economic powers - Germany, France etc.

What to do?

Creative Romania

What does Romania have to do - what does each of us have to do – to respond to both pressures that Romania is currently facing?

A key concept is CREATIVITY. Global trends will abolish, as we have seen, any industrial or agricultural work that can be replaced by the cheaper and cost-effective automation. Business people who deal with such businesses need to adapt technologically without delay, make the necessary investments, or they will soon be eliminated from the market. The only jobs and, therefore, the only businesses that will not disappear will be the creative ones. And "creative" does not mean art or advertising only, but creative in business. Being creative in the workplace or in the business is to provide original and saleable services and goods. It means adding value to what a car can do, even with artificial intelligence. A value-added to what a competitor could do on the same market or in the same job. Of course, there are areas where you cannot be creative - remember the disappearance of truck drivers. In reality, but also operating in a parabolic key, truck and taxi drivers will disappear, but racing and race drivers will multiply. So those creative ones, with added value added that cannot be replaced by automated ones, will resist.

Romania, in its entirety, could go to the creative industries that will resist the technology breakthroughs. Let's talk about the best example - Romanian IT, which started with the creativity and the power of adapting some entrepreneurs who wired the whole country with optical cable before most of the Western countries. The IT sector, now benefiting from high-speed Internet, is already behind the IT of Western countries, which are investing heavily in the fibre optic network. At this moment, the Romanian IT has no advantage, but creativity and bioinformatics where we can enter the league of those who matter globally. The maintenance of non-creative and poorly paid IT jobs and businesses is not a solution in the future economy. IT, to resist, has to make great strides in research, in interface areas with other domains - as we have shown, with bioengineering, with artificial intelligence. It is never too late as a Romanian creative idea like Facebook's, for example, doubles Romania's GDP on the short-run.

Digitalised Traditional Romania

In addition to creativity - specific to the Romanians, who had so many original inventors with vindictive ideas – there is a niche left free from both the pressure of technology globalization and the European perfection, called tradition. And it's not only about folk tradition or customs, but about a tradition, about something that - again historically - we have reached the level of excellence.

One of the fields is AGRICULTURE. The Romanian agriculture has begun to function well again and to provide substantial profits for its investors and entrepreneurs and not necessarily because of the state interventions with many bottlenecks in the field. If we combine the need to adapt to the changes that will come with tradition and creativity, we should start in agriculture from the best world-wide examples - we mentioned the use of autonomous tractors - robots in some areas in Romania - and no longer improvise in quenching small challenges fires, but to start a high-performing farming. A specific type of high value-added farming that would perfectly blend with the tradition would be organic.

And with organic agriculture that respects nature and man, it perfectly matches another important economic axis of traditional Romania, that is TOURISM. Transylvania was in 2016 the top destination for tourists of the world's largest tourist guide - Lonely Planet. Bucharest has become a place of leisure - at the European and world-wide level, similar to Berlin or London. Romania as a whole, is summer, is full of bicycle tourists who enjoy the preservation of a landscape and local traditions, practicing ecological tourism. So - organic farming and eco tourism could be the "spearhead" of economic development strategy in a project called Digitalised Creative and Traditional Romania.

How should it work?

On a personal level, we all have to choose an economic path for the future being always interested in the extremely dynamic permanent evolutions. And be prepared to adapt to change, to entering new territories, to accumulate knowledge from unknown domains - as a student, worker, entrepreneur or member of the Government.

On the national level, if Romania were to choose the dual path - creativity and tradition trough digitalisation - the directions of action would be simple and can give a chance to the project:

- **For Creativity:** Much improved education. Digital and technological literacy is compulsory as a societal attitude, alongside with state and private scholarships at the best universities in the world for upper and middle-level public and private decision-makers.
- **For Traditional Romania:** Infrastructure. Namely, land improvements for agriculture, roads for tourism, and accelerated digitalisation and automation according to the local priorities derived from the regional strategic reindustrialisation maps.

Only that, would say some. Of course not. But this would be a relevant start for the future economy. To become and stay competitive, a nation, a region or a business or even and individual have to understand the perspective of tomorrow's world, to understand yourself as a real and potential economic power, but then to bring more innovation and creativity relative to the others, through collective and collaborative actions. Without leaving behind those who cannot do it as well as you can and turn the latter challenge into a technological opportunity for social innovation as well. How to deal with this challenge should be future debate and research directions to prepare for the future.

Suggestions for Future Research

This chapter contributed significantly for building the public, private and individual awareness related to the need of a systematic on-going institutional - national, sectors and company - approach, for an effective and efficient technology scouting, which can be improved in future researches focusing more on the changes implications of these mega-technology trend identified and described here-in pragmatically in labour markets patterns and future education needs. Is the current education process prepared for the future economy and future Agrifood industry?

ACKNOWLEDGMENT

This research was supported by the Romanian think-tank INACO – The Initiative for Competitiveness and by the Faculty of International Business and Economics – The Academy of Economic Studies, Bucharest.

REFERENCES

L644/2015, Propunere legislativă privind organizarea şi funcţionarea Consiliului Naţional pentru Competitivitate. (n.d.). Retrieved from http://www.cdep.ro/pls/proiecte/upl_pck.proiect?idp=19474&cam=1

Aeronews. (2015). *The General Electric prints a 3D mini jet engine*. Retrieved from http://www.aeronewstv.com/en/industry/engine-manufacturers/2814-general-electric-prints-a-3d-mini-jet-engine.html

Affinity, V. R. (2016). *Is agriculture the next beneficiary of VR?* Retrieved from http://www.affinityvr.com/vr-aid-agriculture-in-future/

Agroinfo. (2014). *Robotul Banat - In Top 10 la Field Robot Event 2014*. Retrieved from https://www.agroinfo.ro/video/evenimente/robotul-banat-in-top-10-la-field-robot-event-2014

AiirSource Military. (2013). *AlphaDog, U.S. Marines Robot Pack Animal - Legged Squad Support System*. Retrieved from https://www.youtube.com/watch?v=cr-wBpYpSfE

AmsterdamSmartCity. (2013). *Amsterdam Smart City*. Retrieved from https://www.youtube.com/watch?v=FinLi65Xtik

Associated Press. (2016). *Mind-Controlled Drones Race to the Future*. Retrieved from https://www.youtube.com/watch?v=U08ma8KNOnY

Buzescu, A. (2017). *România Competitivă prin accesul la resurse şi hrană cu nutrienţii potriviţi. Ce face Franţa şi ce ar trebui să facem noi?* Retrieved from INACO: http://inaco.ro/rom%C8%83nia-competitiva-prin-accesul-la-resurse-si-hrana-cu-nutrientii-potriviti-ce-face-franta-si-ce-ar-trebui-sa-facem-noi/

CNH Industrial. (2016). *The CNH Industrial Autonomous Tractor Concept*. Retrieved from https://www.youtube.com/watch?v=ALmqer120qM

Committee on the Future Economy. (2017). *Report of the Committee on the Future Economy Pioneers of the next generation*. Committee on the Future Economy. Retrieved from https://sustainabledevelopment.un.org/content/documents/16265Committee_on_the_Future_Economy_Report.pdf

Descoperă.ro. (2016). *Oamenii de ştiinţă au imprimat 3D şi implantat maxilare, bucăţi de muşchi şi ''cartilaje în formă de ureche''. ''Este un avans important''*. Retrieved from http://www.descopera.ro/stiinta/15076770-oamenii-de-stiinta-au-imprimat-3d-si-implantat-maxilare-bucati-de-muschi-si-cartilaje-in-forma-de-ureche-este-un-avans-important-fotovideo

DGI. (2017). *DJI MG-1S - Agricultural Wonder Drone*. Retrieved from https://www.youtube.com/watch?v=P2YPG8PO9JU

EC DG for Agriculture and Rural Developments - FP 7. (n.d.). *Projects On Smart Agri-Food Systems*. Directorate H. General aspects of rural development and research H.5. Research and innovation. Retrieved from file:///C:/Users/Test/Downloads/Smartagri-foodprojectspdf.pdf

Energyskeptic. (2013). *China controls over 90% of rare earth mineral production*. Retrieved from http://energyskeptic.com/2013/china-prd-most-rare-earth-minerals/

Engadget. (2017). *Dubai's flying drone taxi service is lifting off*. Retrieved from https://www.youtube.com/watch?v=qArZGwtRINg

Esoteric Detective. (2016). *Students shocked! They were being taught university by advanced AI (IBM, Jill Watson)*. Retrieved from https://www.youtube.com/watch?v=vk9xuB2deIk

European Commission. (2017). *Europe on the Move. An agenda for a socially fair transition towards clean, competitive and connected mobility for all*. COM(2017) 283 final, Brussels, 31.5.2017. Retrieved from https://ec.europa.eu/transport/sites/transport/files/swd20170177-europe-on-the-move.pdf

ExxonMobil. (2017). *ExxonMobil and SGI renew their game-changing algae partnership*. Retrieved from https://www.youtube.com/watch?v=e40RaXZ4tUk

Farmbot. (2016). *Top Amazing Harvest Robot*. Retrieved from https://www.youtube.com/watch?v=mNB42SzUIxU

Farmers Guardian. (2016). *Young farmers using virtual reality in food education plan*. Retrieved from https://www.fginsight.com/vip/vip/young-farmers-using-virtual-reality-in-food-education-plan-13460

Global Commission on the Future of Work. (2017). *Launch of High-Level Global Commission on the Future of Work*. Speech by Prime Minister Stefan Löfven in Geneva 21 August 2017. Retrieved from http://www.government.se/speeches/2017/08/launch-of-high-level-global-commission-on-the-future-of-work/

Harari, Y. N. (2014). Sapiens: A Brief History of Humankind. Harvill Secker.

IBM. (2017). *Macroscopes will help us understand Earth's complexity in infinite detail*. Retrieved from https://www.youtube.com/watch?v=qKMugpYD6tA

INACO. (2017). Retrieved from the Initiative for Competitiveness: www.inaco.ro

INOVAGRIA. (2017). Retrieved from INOVAGRIA Meteo: https://www.youtube.com/watch?v=caCkq9Z5o3w

Intersog. (2016). *VR Applications in Agriculture*. Retrieved from https://intersog.com/blog/augmented-reality/vr-applications-in-agriculture/

Kiwi in Munich. (2016). *Real-life Robocop: Dubai to get android police force by 2017*. Retrieved from https://www.youtube.com/watch?v=MWYtNFb77H0

KPMG. (2017). *China VC investments set a record high in 2016, artificial intelligence a new focus, finds KPMG analysis*. Retrieved from https://home.kpmg.com/cn/en/home/news-media/press-releases/2017/01/vc-investments-record-high-artificial-intelligence.html

Ma-Dupont, V. (2016). *Sweden: a "Ministry of the Future" to think about tomorrow's public policy.* Retrieved from Responsive Public Management No. 82 – April 2016, Gestion Publique: https://www.economie.gouv.fr/files/files/directions_services/igpde-editions-publications/revuesGestionPublique/IGPDE_Reactive_Suede_avril_2016_En.pdf

MIDEA Robot Factory. (2016). Retrieved from https://www.youtube.com/watch?v=2dJewkJZ4PU

Motoman robots packing IKEA. (2012). Retrieved from https://www.youtube.com/watch?v=Duub_dO_TwU

NECIT Solutions. (2012). *NeoFace Access Control.* Retrieved from https://www.youtube.com/watch?v=Hsc9LDeRQlA

New China, T. V. (2014). *3D printers print ten houses in 24 hours.* Retrieved from https://www.youtube.com/watch?v=SObzNdyRTBs

New Scientist. (2016). *Farmers: These ARE the droids you're looking for.* Retrieved from https://www.youtube.com/watch?v=hjd5DaxkLhQ

PackagingDigest. (2017). *How smart packaging sensors safeguard foods and drugs.* Retrieved from http://www.packagingdigest.com/smart-packaging/how-smart-packaging-sensors-safeguard-foods-and-drugs-2017-04-13

Pambuccian, V. (2017). *De la Homo Faber la Robo Sapiens. A history of the Future written by cars.* Pwp Presentation, INACO Conference organized at the Academy of Economic Studies. Retrieved from http://inaco.ro/de-la-homo-faber-la-robo-sapiens-cum-vor-arata-afacerile/

Paul, A. (2017a). *Costurile brevetării invențiilor românești sunt de peste două ori mai mari decât în Germania și triple față de Polonia.* Retrieved from INACO: https://inaco.ro/costurile-brevetarii-inventiilor-romanesti-sunt-de-peste-doua-ori-mai-mari-decat-in-germania-si-triple-fata-de-polonia/

Paul, A. (2017b). *O aplicație IT 100% românească te poate ajuta să faci agricultură performantă.* Retrieved from INACO: https://inaco.ro/o-aplicatie-it-100-romaneasca-te-poate-ajuta-sa-faci-agricultura-performanta/

Paul, A. (2017c). *Statul apără braconierii de sturioni, nu sturionii! Mafia de milioane de euro.* Retrieved from INACO: https://inaco.ro/statul-apara-braconierii-de-sturioni-nu-sturionii-mafia-de-milioane-de-euro/

Paul, A. (2017d). *INACO propune crearea Grupului De Lucru Pentru Economia Viitorului (GLEV).* Retrieved from ClubEconomic: http://clubeconomic.ro/opinii/inaco-propune-crearea-grupului-de-lucru-pentru-economia-viitorului-glev-4088/

Plant Based News. (2016). *Meet The Meat that is Tipped to Save the World*. Retrieved from https://www.youtube.com/watch?v=VjEBDFuSoRw

PwC. (2017). *The Long View How will the global economic order change by 2050?* Retrieved from http://www.pwc.com/gx/en/world-2050/assets/pwc-the-world-in-2050-full-report-feb-2017.pdf

Ranasinghe, N. (2017). *Virtual Lemonade: Let's Teleport Your Lemonade*. Retrieved from https://www.youtube.com/watch?v=w8gTsGr2JpA&feature=youtu.be

Reuters. (2015). *Real-life Star Trek 'replicator' prepares meal in 30 seconds*. Retrieved from http://www.reuters.com/article/us-israel-meals-on-demand-tracked-idUSKBN0NQ1PG20150505

Rohrbeck, R. (2007). *Technology Scouting - a Case Study on the Deutsche Telekom Laboratories*. Available from: https://www.researchgate.net/publication/24114558_Technology_Scouting_-_a_Case_Study_on_the_Deutsche_Telekom_Laboratories

ABC Rural. (2016). *Meat products from 3D printer could be the new food for aged care homes and restaurants*. Retrieved from http://www.abc.net.au/news/rural/2016-12-02/meat-ink-3d-printing/8086060

Schwab, K. (2017). *The Global Competitiveness Report 2017–2018*. Retrieved from World Economic Forum: http://www3.weforum.org/docs/GCR2017-2018/05FullReport/TheGlobalCompetitivenessReport2017%E2%80%932018.pdf

Sheeran, K., Ackerman, F., Enelow, N., Goodstein, E., Hahnel, R., Power, T. M., & Schor, J. (2014). *The Future Economy. An Annotated Analytical Framework for Case Studies*. Economics for Equity & Environment. Retrieved from http://futureecon.com/wp-content/uploads/future-economy-framework_new.pdfhttp://futureecon.com/

Tech Medium. (2017). *First Real Flying Car in Dubai Ehang 184*. Retrieved from https://www.youtube.com/watch?v=UAbquGXnXmQ

TechCrunch. (2017). *Cafe X Opens In San Francisco, Bringing Robots To The Coffee*. Retrieved from https://techcrunch.com/2017/01/30/cafe-x-opens-in-san-francisco-bringing-robots-to-the-coffee-shop/

The Economist. (2015). *Silicon Valley gets a taste for food*. Retrieved from https://www.economist.com/news/technology-quarterly/21645497-tech-startups-are-moving-food-business-make-sustainable-versions-meat

The Economist. (2017). *The growth of industrial robots*. Retrieved Sep. 15, 2017, from https://www.economist.com/blogs/graphicdetail/2017/03/daily-chart-19

The State Council of the People's Republic of China. (2017). *China issues guideline on artificial intelligence development*. Retrieved from http://english.gov.cn/policies/latest_releases/2017/07/20/content_281475742458322.htm

VentureRadar. (2015). *5 Food Technology Startups to Watch*. Retrieved from http://blog.ventureradar.com/2015/11/17/5-food-technology-startups-to-watch/

Vocea.biz. (2017). *A apărut "mâncarea inteligentă". Cum arată și ce conține*. Retrieved from https://vocea.biz/economic/2017/may/20/main-a-aparut-mancarea-inteligenta-cum-arata-si-ce-contine-video/

Wall Street Journal. (2014). *Glowing Plants to Light Your Home? Here They Come!* Retrieved from https://www.youtube.com/watch?v=AM1q0ifx91A

Wired. (2016). *Uber's Self-Driving Truck Makes Its First Delivery*. Retrieved from https://www.youtube.com/watch?v=sIlCR4eG8_o

WIVBTV. (2014). *SolarCity investing $5B in Buffalo, creating 3,000 jobs*. Retrieved from https://www.youtube.com/watch?v=bj2cSTHGFFk&t=43s

World Economic Forum. (2017), *How can we maximize the benefits of science and technology for society?* Center for the Fourth Industrial Revolution, San Francisco: https://www.weforum.org/center-for-the-fourth-industrial-revolution/home

World Economic Forum. (2017). *New Vision for Agriculture*. Retrieved from https://www.weforum.org/projects/new-vision-for-agriculture

World Science Festival. (2016). *Miguel Nicolelis Helps Paralyzed Patients Learn to Move Again*. Retrieved from https://www.youtube.com/watch?v=MHT6AGdzEaU

KEY TERMS AND DEFINITIONS

3D Printing: Reproduction of physical objects or human organs trough three-dimensional digital thin successive layers using digital models. It is also called additive manufacturing.

Artificial Intelligence: Simulation of intelligent human behavior by computer and machines.

Biotechnology: Genetic engineering of biological processes for industrial, health, and other purposes.

Blockchain System: The information system linked and secured using cryptography.

Internet of Things: The interconnection through the internet of smart devices enabling them to send and receive data.

Inventions Patenting: The official process to recognize and protect an invention that gives an exclusive right to make, use or sell an invention for a specified period of time and in a specified region.

Robot: A programmable by a computer machine capable of carrying out actions automatically.

Technology Scouting: Looking for the technology trends, current inventions and innovations applied in the real economy, understand their practical usage and foresee their potential, for strategic purposes.

Virtual Reality: An artificial three-dimensional computer-generated image or environment or reality, each of them interacted with the person using special electronic equipment with sensors.

ENDNOTES

1 "This year, China is aiming for breakthroughs in high- performance processors and other key technologies to build the world's first prototype exascale supercomputer, the Tianhe-3, said Meng Xiangfei, the director of application at the National Super Computer Tianjin Center. The prototype is expected to be completed in early 2018. "Exascale" means it will be capable of making a quintillion (1 followed by 18 zeros) calculations per second. That is at least 10 times faster than the world's current speed champ, the Sunway TaihuLight, China's first supercomputer to use domestically designed processors. That computer has a peak speed of 125 quadrillion (1 followed by 15 zeros) calculations per second".
China Daily (2017, feb. 2), *China to jump supercomputer barrier*, by Zhang Zhihao, http://www.chinadaily.com.cn/china/2017-02/20/content_28259294.htm:

2 Details about OSIM fees and deadlines can be found here: http://www.osim.ro/legislatie/Taxe-proprietate-industriala.pdf

3 The author of this chapter, Andreea Paul, is one of the three founding members and the president of INACO. The coordinator of this book, Gabriel Popescu, is a volunteer member of INACO responsible for the agriculture competitiveness strategy.

Compilation of References

(2002). Recommendation of the Council on Improving the Environmental Performance of Public Procurement. *Improving the Environmental Performance of Public Procurement: Report on Implementation of the Council Recommendation of the Council on Improving the Environmental Performance of Public Procurement, C*, 3.

A.N.P.M. (2010). Raport anual – Starea factorilor de mediu în România. In *Cap. 5. Protecţia Naturii şi Biodiversitatea.* Retrieved from http://www.anpm.ro/anpm_resources/migrated_content/uploads/48584_5.Cap.5.Protectia%20Naturii%20si%20Biodiversitatea.2010.pdf

A.N.P.M. (2010). Raport anual – Starea factorilor de mediu în România. In *Cap.4. Utilizarea terenurilor.* Retrieved from http://www.anpm.ro/anpm_resources/migrated_content/uploads/48599_4.Cap.4.Utilizarea%20Terenurilor.pdf

ABC Rural. (2016). *Meat products from 3D printer could be the new food for aged care homes and restaurants.* Retrieved from http://www.abc.net.au/news/rural/2016-12-02/meat-ink-3d-printing/8086060

Abugamea, G. H. (2008). A Dynamic Analysis for Agricultural Production Determinants in Paletsine: 1980-2003. In *Proceedings of International Conference on Applied Economics, ICOAE* (pp. 3-10). Academic Press.

Aceleanu, M. I. (2016). Sustainability and Competitiveness of Romanian Farms through Organic Agriculture. *Sustainability Journal, 8*(245), 4–6.

Advisory Consulting. (2017). *Potentialul dezvoltarii sectorului agricol din Romania.* Retrieved August 28, 2017, from https://www.juridice.ro/wp-content/uploads/2017/03/Raport_PwC-agricultura.pdf

Aeronews. (2015). *The General Electric prints a 3D mini jet engine.* Retrieved from http://www.aeronewstv.com/en/industry/engine-manufacturers/2814-general-electric-prints-a-3d-mini-jet-engine.html

Affinity, V. R. (2016). *Is agriculture the next beneficiary of VR?* Retrieved from http://www.affinityvr.com/vr-aid-agriculture-in-future/

Agroinfo. (2014). *Robotul Banat - In Top 10 la Field Robot Event 2014.* Retrieved from https://www.agroinfo.ro/video/evenimente/robotul-banat-in-top-10-la-field-robot-event-2014

AiirSource Military. (2013). *AlphaDog, U.S. Marines Robot Pack Animal - Legged Squad Support System.* Retrieved from https://www.youtube.com/watch?v=cr-wBpYpSfE

Albu, L., Caraiani, P., & Iordan, M. (2012). *Perspectivele pieţei muncii din România în contextul Strategiei Europa 2020.* Report within the project Îmbunătăţirea capacităţii instituţionale, de evaluare şi formulare de politici macroeconomice în domeniul convergenţei economice cu Uniunea Europeană a Comisiei Naţionale de Prognoză, cod SMIS 27153. Retrieved from www.cnp.ro/inovatie/docs/seminar-studii-25-06-2012/Rezumat%20studiu%20Piata %20muncii.pdf

Alliance International of Cooperative Property. (1998). *Statistics and information on European cooperation*. Retrieved from http:/www.ica.coop/europe

Altbach, P. G., & Salmi, J. (Eds.). (2011). *The road to academic excellence: The making of world-class research universities*. World Bank Publications. doi:10.1596/978-0-8213-8805-1

Altman, M. (2010). History and Theories of Cooperatives. In H. K. Anheier, S. Toepler, & R. List (Eds.), *International Encyclopedia of Civil Society* (pp. 563–570). New York: Springer Science –Business Media.

AmsterdamSmartCity. (2013). *Amsterdam Smart City*. Retrieved from https://www.youtube.com/watch?v=FinLi65Xtik

Anderl, M. (2009). Emisiile în aer produse de agricultură în Austria şi România. *Pro Environment*, *2*, 84–92.

Andrei, T., & Bourbonnais, R. (2008). *Econometrics*. Bucharest: Economica Publishing House.

Andrei, T., Stancu, S., Iacob, A. I., & Tusa, E. (2008). *Introduction to econometrics using Eviews*. Bucharest: Economica Publishing House.

Antonescu, C. (1979). *Albinele şi noi*. Bucureşti, România: Redacţia publicaţiilor apicole.

Antonie, I. (2016). Honey resources of Avrig City (Sibiu county) and economic relevance. *Scientific Papers. Series Management, Economic, Engineering in Agriculture and Rural Development*, *16*(4), 37–41.

Ashenfelter, O., & Storchmann, K. (2016). Climate change and wine: A review of the economic implications. *Journal of Wine Economics*, *11*(1), 105–138. doi:10.1017/jwe.2016.5

Associated Press. (2016). *Mind-Controlled Drones Race to the Future*. Retrieved from https://www.youtube.com/watch?v=U08ma8KNOnY

Asteriou, D., & Hall, S. G. (2007). *Applied econometrics: A modern approach using eviews and microfit* (revised edition). Palgrave Macmillan.

Avetisian, G. A. (1978). *Apicultura*. Bucureşti, România: Editura Apimondia.

Bădescu, I., Buruiană, C., Şişeştean, Gh., & Abraham, D. (Eds.). (2011). *The peasants and the New Europe*. Bucharest: Ed. MicaValahie.

Bannon, S., Ford, K., & Meltzer, L. (2011). Understanding Millennials in the workplace. *The CPA Journal*, *81*(11), 61.

Bansil, P. (2002). *Economic Problems of Indian Agriculture* (6th ed.). New Delhi: Techno Economic Research Institute Publication.

Bara, S., Zaharia, A., Teodor, C., & Preda, E. (2017). *Possible ways to stimulate the human resource within romanian agricultural and forestry research*. Paper presented at "The International Conference Competitiveness of Agro-Food and Environmental Economy", CAFEE 2017, Bucharest, Romania.

Barnosky, A. D., Koch, P. L., Feranec, R. S., Wing, S. L., & Shabel, A. B. (2004). Assessing the Causes of Late Pleistocene Extinctions on the Continents. *Science*, *306*(5693), 70–75. doi:10.1126cience.1101476 PMID:15459379

Bendre, G. (2017). *Romania has significant potential in agriculture, but low productivity, says PwC Romania study*. Retrieved from http://www.business-review.eu/news/romania-has-significant-potential-in-agriculture-but-low-productivity-says-pwc-romania-study-132113

Bennet, G. (2010). *Food identity preservation and traceability*. Boca Raton, FL: CRC Press.

Bercovitz, J., & Feldman, M. (2006). Entrepreneurial universities and technology transfer: A conceptual framework for understanding knowledge-based economic development. *The Journal of Technology Transfer*, *31*(1), 175–188. doi:10.100710961-005-5029-z

Berkes, F. (2008). *Sacred Ecology* (2nd ed.). New York, NY: Routledge.

Best, H. (2008). Organic agriculture and the conventionalization hypothesis: A case study from West Germany. *Agriculture and Human Values*, *25*(1), 95–106. doi:10.100710460-007-9073-1

Bijman, J., & van Dijk, G. (2009). *Corporate Governance in agricultural cooperatives: a Dutch perspective*, Wageningen University.

Boakye, M., & Akpor, O. (2012). Community Participation in Water Resources Management in South Africa. *International Journal of Environmental Sciences and Development*, 511–516. doi:10.7763/IJESD.2012.V3.277

Bogdan, A. T., Pop, V., Diaconescu, D. N., Chelmu, S. S., & Străteanu, A. (2009). *Despre eco-economie şi economie verde în România*. Referat ştiinţific, Academia Română, I.N.C.E. Retrieved from http://www.enciclopedia-economica.ro/upload/files/AT_2_DIMECOECONOMICE_2-5-1-prezentareprofBogdan.pdf

Boisot, M. (2002). The creation and sharing of knowledge. *The strategic management of intellectual capital and organizational knowledge*, 65-77.

Bradbear, N. (2009). *Bees and their role in forest livelihoods*. Rome: Food and Agriculture Organization of the United Nations.

Bran, M. (2005). *Ecotehnica exploatatiei vegetale*. Editura Printech.

Bran, M. (2012). *Cercetări privind dimensiunea eco-economică a biodiversităţii vegetale din Regiunea de Dezvoltare Sud Muntenia, în condiţiile schimbărilor climatice şi de ecotehnică, în concordanţă cu dezvoltarea durabilă (Lucrare de cercetare ştiinţifică pentru finalizarea Programului Postdoctoral, Contract POSDRU/89/1.5/S/63258)*. INCE.

Brook, B., & Bowman, D. M. J. S. (2004). The uncertain blitzkrieg of Pleistocene megafauna. *Journal of Biogeography*, *31*(4), 517–523. doi:10.1046/j.1365-2699.2003.01028.x

Brown, L. (2001). Eco-economie. Crearea unei economii pentru planeta noastră. Editura tehnică.

Brugmann, J., & Prahalad, C. (2007). Cocreating business's new social compact. *Harvard Business Review*, *85*(2), 80–90. PMID:17345682

Buckwell, A., Matthews, A., Baldock, D., & Mathijs, E. (2017). *CAP - Thinking Out of the Box: Further modernisation of the CAP – why, what and how?* Brussels: RISE Foundation.

Bura, M. (2010). Pe piaţa europeană se poate comercializa doar miere de calitate, *Revista Ferma*, 5. Retrieved from https://www.revista-ferma.ro/articole/apicultura/pe-piata-europeana-se-poate-comercializa-doar-miere-de-calitate

Bura, M., & Patruică, S. (2005). *Tehnologie apicolă*. Timişoara, România: Editura Solness.

Burja, C. (2012). Determinants of the agricultural productivity growth among Romanian regions. *Annales Universitatis Apulensis Series Oeconomica*, *1*(14).

Buzescu, A. (2017). *România Competitivă prin accesul la resurse şi hrană cu nutrienţii potriviţi. Ce face Franţa şi ce ar trebui să facem noi?* Retrieved from INACO: http://inaco.ro/rom%C8%83nia-competitiva-prin-accesul-la-resurse-si-hrana-cu-nutrientii-potriviti-ce-face-franta-si-ce-ar-trebui-sa-facem-noi/

Byatnal, A. (2011). 38 more 'ideal' villages to come up in Maharashtra by June. *The Hindu*. Retrieved from http://www.thehindu.com/news/national/other-states/38-more-ideal-villages-to-come-up-in-maharashtra-by-june/article2006895.ece

Camera de Comert si Industrie a Romaniei. (2014). *Statistica – Agricultura*. Retrieved August 21, 2017, from https://ccir.ro/wp-content/uploads/2014/11/statistica-agricultura.pdf

CAP Context indicators. (2017). *CAP Context indicators*. Retrieved June 23, 2017, from https://ec.europa.eu/agriculture/sites/agriculture/files/cap-indicators/context/2016/indicator-table_en.pdf

Carney, D. (2003). *Sustainable livelihoods approaches: progress and possibilities for change*. London: Department for International Development.

Carson, R. (1962). Silent Spring. *The New Yorker Magazine*.

Carvalho, F. P. (2006). Agriculture, pesticides, food security and food safety. *Environmental Science & Policy*, *9*(7), 685–692. doi:10.1016/j.envsci.2006.08.002

CEDEFOP. (2017). *Employment trends*. Retrieved August 17, 2017, from http://www.cedefop.europa.eu/en/publications-and-resources/data-visualisations/employment-trends

Cereal Science and Technology. (2016). *The economic importance of cereal grains*. Retrieved from http://cereal-scientech.blogspot.ro/2011/12/economic-importance-of-cereal-grains.html

Chambers, R. (1989). Editorial introduction: Vulnerability, coping and policy. *IDS Bulletin*, *20*(2), 1–7. doi:10.1111/j.1759-5436.1989.mp20002001.x

Chambers, R., & Conway, G. (1992). *Sustainable rural livelihoods*. Brighton, UK: Institute of Development Studies.

Chirilă, A., & Patruică, S. (2004). *Tehnologii apicole moderne. Stupăritul pastoral*. Bucureşti, România: Agenţia Naţională de Consultanţă Agricolă.

Cîrnu, L., & Roman, Gh. (1986). *Din viaţa albinelor*. Bucureşti, România: Editura Ceres.

CNH Industrial. (2016). *The CNH Industrial Autonomous Tractor Concept*. Retrieved from https://www.youtube.com/watch?v=ALmqer120qM

Cobb, D., Feber, R., Hopkins, A., Stockdale, L., O'Riordan, T., Clements, B., ... Macdonald, D. (1999). Integrating theenvironmental and economic consequences of converting to organic agri-culture: Evidence from a case study. *Land Use Policy*, *16*(4), 207–221. doi:10.1016/S0264-8377(99)00023-X

Cociu, A. (2016). Long-term effects of tillage systems on winter wheat, maize and soybean grain yield and yield stability under rainfed conditions in the eastern Romanian Danube plain. Romanian Agricultural Research, 33, 97-109.

Cociu, Al., & Alionte, E. (2017). The effect of different tillage systems on grain yield and its quality of winter wheat, maize and soybean under different weather conditions. *Romanian Agricultural Research*, (34): 59–67. Retrieved from http://www.incda-fundulea.ro/rar/nr34/rar34.8.pdf

Cofacerating. (2008). *Coop de France*. Report Bétail et Viande.

COGECA- European agri-cooperatives. (2014). *Development of agricultural cooperatives in the EU*, Brussels, PUB (14) 112:3.

Cohen, B., & Winn, M. I. (2007). Market imperfections, opportunity and sustainable entrepreneurship. *Journal of Business Venturing*, *22*(1), 29–49. doi:10.1016/j.jbusvent.2004.12.001

Colding, J. (1998). Analysis of hunting options by the use of general food taboos. *Ecological Modelling*, *110*(1), 5–17. doi:10.1016/S0304-3800(98)00038-6

Colding, J., & Folke, C. (1997). The relations among threatened species, their protection, and taboos. *Conservation Ecology*, *1*(1), 6. doi:10.5751/ES-00018-010106

Committee on the Future Economy. (2017). *Report of the Committee on the Future Economy Pioneers of the next generation*. Committee on the Future Economy. Retrieved from https://sustainabledevelopment.un.org/content/documents/16265Committee_on_the_Future_Economy_Report.pdf

Communication from the Commission to the Council and the European Parliament Stimulating Technologies for Sustainable Development. (2004). Bruxelles: An Environmental Technologies Action Plan for the European Union.

Coop de France. (2009). *Repère Economique*. Author.

Coop de France. (n.d.). Retrieved from www.coopdefrance.coop

Cooperation. (n.d.). *Responsible finish ownership*. Retrieved from www.pellervo.fi

Copa Cogeca. (n.d.). Retrieved from www.copa-cogeca.be

Copenhagen Economics. (2007). *Study on Reduced VAT applied to Goods and Services in the Member States of the European Union*. Academic Press.

Cornescu, V., Marinescu, P., Curteanu, D., & Toma, S. (2004). *Management de la teorie la practică*. Bucureşti, Romania: Editura Universităţii.

Crinnion, J. W. (2010). Organic food contain higher level of certain nutrients, lower level of pesticides and may provide health benefits for the consumer. In *Environmental Medicine* (pp. 4-12). Academic Press. Retrieved on 20 August 2017 from http://www.altmedrev.com/publications/15/1/4.pdf

Cristache, S. E. (2003). *Statistical Methods of Calculating and Analyzing Economic Efficiency in Commerce* (Doctoral Thesis). ASE Bucharest Publishing House.

Crucefix, D. (2012). 10 years back and 10 years forward. In Let the good market flow- Global Organic Market Access in 2012 and beyond (pp 10-13). UNCTAD, FAO & IFOAM.

Crutzen, P. J., & Stoermer, E. F. (2000). The 'Anthropocene'. *Global Change Newsletter*, *41*, 17–18.

Daia, P. (2017). ZE Agribusiness: How Can Romanian Farmers Get Out of the Raw Exports Race? Presentation at the *Financial Journal Conference*.

Daly, H. E. (1990). *Toward Some Operational Principles of Sustainable Development* (Vol. 2). Ecological Economics.

Da-Rocha, J. M., & Restuccia, D. (2006). The Role of Agriculture in Aggregate Business Cycle. *Review of Economic Dynamics*, *9*(3), 455–482. doi:10.1016/j.red.2005.12.002

Dean, T. J., & McMullen, J. S. (2007). Toward a theory of sustainable entrepreneurship: Reducing environmental degradation through entrepreneurial action. *Journal of Business Venturing*, *22*(1), 50–76. doi:10.1016/j.jbusvent.2005.09.003

Dearden, L., Fitzsimons, E., & Wyness, G. (2014). Money for nothing: Estimating the impact of student aid on participation in higher education. *Economics of Education Review*, *43*, 66–78. doi:10.1016/j.econedurev.2014.09.005

Dennett, D. C. (2017). *From Bacteria to Bach and Back: The Evolution of Minds*. New York, NY: W. W. Norton & Company.

Descoperă.ro. (2016). *Oamenii de ştiinţă au imprimat 3D şi implantat maxilare, bucăţi de muşchi şi ''cartilaje în formă de ureche''. ''Este un avans important''*. Retrieved from http://www.descopera.ro/stiinta/15076770-oamenii-de-stiinta-au-imprimat-3d-si-implantat-maxilare-bucati-de-muschi-si-cartilaje-in-forma-de-ureche-este-un-avans-important-fotovideo

DGI. (2017). *DJI MG-1S - Agricultural Wonder Drone*. Retrieved from https://www.youtube.com/watch?v=P2YPG8PO9JU

Di Gregorio, D., & Shane, S. (2003). Why do some universities generate more start-ups than others? *Research Policy, 32*(2), 209–227. doi:10.1016/S0048-7333(02)00097-5

Diamond, J. M. (2005). *Collapse: How Societies Choose to Fail or Succeed.* New York, NY: Wiking Penguin.

Diao, X., Nwafor, M., Alpuerto, V., Akramov, K., Rhoe, V., & Salau, S. (2012). Nigeria. In X. Diao, J. Thurlow, S. Benin, & S. Fan (Eds.), *Strategies and Priorities for African Agriculture: Economywide Perspectives from Country Studies* (p. 231). Washington, DC: International Food Policy Research Institute.

Dincu, I., & Bran, M. (2002). *AgroEcoFitotehnie* (Vol. I). Bucureşti, Romania: Editura Meteor Press.

Dîrja, M. (2000). *Combaterea eroziunii solului.* Cluj-Napoca, Romania: Editura Risoprint.

Dobre, I. (2011). Methodologies for analysis and evaluation of risk in agricultural exploitations. Internal Auditing & Risk Management, 4(24), 89-97.

Dorward, A., Anderson, S., Clark, S., Keane, B., & Moguel, J. (2001). *Asset functions and livelihood strategies: a framework for pro-poor analysis, policy and practice.* Academic Press.

Douglas, K. (2010). *Romania and the Common Agricultural Policy.* Denver University. Retrieved August 23, 2017, from http://sar.org.ro/wp-content/uploads/2013/01/Romania-and-the-Common-Agricultural-Policy.pdf

Drydakis, N. (2016). The effect of university attended on graduates' labor market prospects: A field study of Great Britain. *Economics of Education Review, 52*, 192–208. doi:10.1016/j.econedurev.2016.03.001

Duch meat association. (n.d.). Retrieved from www.cov.nl

Dungumaro, E., & Madulu, N. (2003). Public participation in integrated water resources management: the case of Tanzania. *Physics and Chemistry of the Earth, Parts A/B/C,* 28(20-27), 1009-1014.

Durkheim, E. (1982). *The Rules of Sociological Method.* The Free Press.

Durkheim, E. (1995). *The Elementary Forms of Religious Life*. New York, NY: The Free Press.

Dyson, T. (1996). *Population and Food: Global Trends and Future Prospects*. London, UK: Rout-ledge.

E.P.A. (n.d.). *Greenhouse Gas Emissions.* Retrieved from https://www.epa.gov/ghgemissions/sources-greenhouse-gas-emissions#agriculture

EC DG for Agriculture and Rural Developments - FP 7. (n.d.). *Projects On Smart Agri-Food Systems.* Directorate H. General aspects of rural development and research H.5. Research and innovation. Retrieved from file:///C:/Users/Test/Downloads/Smartagri-foodprojectspdf.pdf

Eco-procurment, the path to a greener market place, European Project – Environmental Relief Potential of Urban Action on Avoidance and Detoxification of Waste Streams through Green Public Procurement Buying Green Handbook. (2007). *Collection of statistical information on Green Public Procurement in the EU.* Academic Press.

Edmond, C. (2017, July 18). *Why Japan's streets are spotless.* Retrieved September 4, 2017, from https://www.weforum.org/agenda/2017/07/why-japanese-dont-litter/?utm_content=buffera55a5&utm_medium=social&utm_source=facebook.com&utm_campaign=buffer

Ekbom, A. (1998). Some determinants to agricultural productivity: An application to the Kenyan highlands. In *World Conference of Environmental Economics* (pp. 25-27). Academic Press.

EKOenergy. (2013). Retrieved from www.ekoenergy.org/criteria/

El-Hage Sclalabba, N. (2013). Organic Agriculture's Contribution to Sustainability. *Crop Management, 12*. doi:10.1094/CM-2013-0429-09-PS

Elkington, J. (1997). *Cannibals with Forks. The Triple Bottom Line of 21st Century Business*. Oxford, UK: Capstone Publishing Limited.

Ellingson, T. (2001). *The Myth of the Noble Savage*. Berkeley, CA: University of California Press. doi:10.1525/california/9780520222687.001.0001

Elrod, T., Russell, G. J., Shocker, A. D., Andrews, R. L., Bacon, L., Bayus, B. L., ... Mazanec, J. A. (2002). Inferring market structure from customer response to competing and complementary products. *Marketing Letters, 13*(3), 221–232. doi:10.1023/A:1020222821774

Ene, A. E. (2015). *EU Funds for beekeepers*. Retrieved December 28, 2017, http://www.informativ.info/economic/237-fonduri-nerambursabile-pentru-apicultori

Energyskeptic. (2013). *China controls over 90% of rare earth mineral production*. Retrieved from http://energyskeptic.com/2013/china-prd-most-rare-earth-minerals/

Engadget. (2017). *Dubai's flying drone taxi service is lifting off*. Retrieved from https://www.youtube.com/watch?v=qArZGwtRINg

Esoteric Detective. (2016). *Students shocked! They were being taught university by advanced AI (IBM, Jill Watson)*. Retrieved from https://www.youtube.com/watch?v=vk9xuB2deIk

ESPON Database Portal. (2016). Retrieved from http://database.espon.eu/db2/home;jsessionid=b7c0e44c4a7260687a2df61e4e09

EU Commision. (2010). *Europa Agenda*. Retrieved November 14, 2015, from https://ec.europa.eu/research/innovation-union/pdf/innovation-union-communication_en.pdf

EU Monitor. (2016). *Education statistics*. Retrieved March 23, 2017, from https://ec.europa.eu/education/sites/education/files/monitor2016_en.pdf

Euliss, N. H. Jr, Smith, L. M., Liu, S., Feng, M., Mushet, D. M., Auch, R. F., & Loveland, T. R. (2010). The need for simultaneous evaluation of ecosystem services and land use change. *Environmental Science & Technology, 44*(20), 7761–7763. doi:10.1021/es102761c PMID:20809588

Eurobarometer conducted by European Commission. (2016). *Europeans, agriculture and the Common Agricultural Policy*. Retrieved August 25, 2017 from https://ec.europa.eu/agriculture/survey_ro

European Commission (2014a). *Agriculture and Rural Development, The CAP, Overview, The EU explained: Agriculture*. Retrieved from http://ec.europa.eu/agriculture/index_en.htm

European Commission (2014b). *Agriculture and Rural Development, Romania, Common Agricultural Policy*. Retrieved from http://ec.europa.eu/agriculture/cap-in-your-country/index_ro.htm

European Commission (2016a). *International Cooperation and Development, Food and agriculture, Food and nutrition security*. Retrieved from https://ec.europa.eu/europeaid/sectors/food-and-agriculture/food-and-nutrition-security_en

European Commission (2016c). *Sustainable Development indicators, Climate change, Greenhouse gas emissions by sector.* Retrieved from http://ec.europa.eu/eurostat/data/database

European Commission (EC). (2017a). *Recognised-research-entities.* EUROSTAT. Retrieved from http://ec.europa.eu/eurostat/documents/203647/ 771732/Recognised-research-entities.pdf

European Commission (EC). (2018). *Implementing the Sustainable Development Goals through the next Multi-Annual Financial Framework of the European Union.* Advisory report to the European Commission by the Multi-Stakeholder Platform on the Implementation of the Sustainable Development Goals in the EU, March 2018. Retrieved from https://ec.europa.eu/info/sites/info/files/adopted-position-paper-on-the-mff_en.pdf

European Commission. (2003). *Report from the Commission to the European Parliament, the Council, the European Economic and Social Committee and the Committee of the Regions on the implementation of Regulation (EC) No 1782/2003, Publication no. 1435/2003 of the Council of 22 July 2003 on the Statute of the European Cooperative Society.* ECS.

European Commission. (2010). *Europe 2020 - a European Strategy for Smart Sustainable and Inclusive Growth.* Retrieved August 23, 2017, from http://ec.europa.eu/eu2020/pdf

European Commission. (2010). Europe 2020 - A European strategy for smart, sustainable and inclusive growth. Brussels: EC.

European Commission. (2011). *Analysis associated with the Roadmap to a Resource Efficient Europe.* Part I, Commission staff working paper, Brussels, 20.9.2011, SEC(2011) 1067 final, Accompanying the document Communication from the Commission to the European Parliament, the Council, the European Economic and Social Committee and the Committee of Regions, Roadmap to a Resource Efficient Europe, COM(2011) 571 final, SEC(2011) 1068 final.

European Commission. (2013). *A European Strategy for Smart, Green and Inclusive Growth* (Communication from the European Commission, 2020). Retrieved August 23, 2017 http://eurlex.europa.eu/LexUriServ/LexUriServ.do?uri=

European Commission. (2016). *Let us understand the Policies of the European Union: Agriculture* (the Directorate General for Citizen Communication and Information). Retrieved September 10, 2017, from https://europa.eu/european-union/topics_en

European Commission. (2016b). *Climate Action-Policies-Greenhouse gas emission.* Retrieved from http://ec.europa.eu/clima/policies/g-gas/index_en.htm

European Commission. (2016d). Overview of CAP Reform 2014-2020. Agricultural Policy Perspectives Briefs, no.5, December 2013, pp 1-10. Retrieved from http://ec.europa.eu/agriculture/policy-perspectives/policy-briefs/index_en.htm

European Commission. (2017). *Agriculture. A Partnership between Europe and Farmers (CAP).* Retrieved August 23, 2017, from http://europa.eu/pol/pdf/flipbook/en/agriculture_en.pdf

European Commission. (2017). *Europe on the Move. An agenda for a socially fair transition towards clean, competitive and connected mobility for all.* COM(2017) 283 final, Brussels, 31.5.2017. Retrieved from https://ec.europa.eu/transport/sites/transport/files/swd20170177-europe-on-the-move.pdf

European Commission. (2017b). *Access to microdata. Overview.* EUROSTAT. Retrieved from http://ec.europa.eu/eurostat/web/ microdata/overview

European Commission. (n.d.). Retrieved from http://ec.europa.eu/agriculture/quality/door/list.html

European Parliament Briefing. (2017). *Agricultural education and lifelong training in the EU.* Retrieved December 29, 2017, from http://www.europarl.europa.eu/RegData/etudes/BRIE/2017/608788/EPRS_BRI(2017)608788_EN.pdf

European Union (EU). Council Regulation (EC) No. 834/2007 of 28 June 2007 on organic production and labeling of organic products and repealing Regulation (EEC) No 2092/91. (2007). Retrieved December 28, 2017, from: http://eur-lex.europa.eu/oj/direct-access.html

European Union. (2010). *Horizon 2020.* Retrieved September 5, 2017, from http://ec.europa.eu/eurostat/web/europe-2020-indicators/europe-2020-strategy/main-tables

Europeană, C. (2010), *EUROPA 2020 - O strategie europeană pentru o creștere inteligentă, ecologică și favorabilă incluziunii.* Retrieved from http://www.fonduri-structurale.ro/Document_Files//fs2014-2020/00010115/ohsc1_Strategia%20Europa%202020.pdf

Eurostat Database. (n.d.). Retrieved from ec.europa.eu

Eurostat. (2016). Retrieved from http://appsso.eurostat.ec.europa.eu/nui/submitViewTableAction.do

Eurostat. (2017). *Statistics_on_rural_areas in_the_EU.* Retrieved from http://ec.europa.eu/eurostat/statistics explained/index.php/Statistics_on_rural_areas in_the_EU

EViews. (2010). QMS Quantitative Micro Software (Version 7.0) [software]. Available from http://www.eviews.com/EViews7/EViews7Student/evstud7.html

ExxonMobil. (2017). *ExxonMobil and SGI renew their game-changing algae partnership.* Retrieved from https://www.youtube.com/watch?v=e40RaXZ4tUk

Fareed, F. (2015). To develop Saansad Adarsh Gram Yojana villages, Centre cites three models. *The Indian Express.* Retrieved from http://indianexpress.com/article/cities/lucknow/to-develop-saansad-adarsh-gram-yojana-villages-centre-cites-three-models/

Farm Accountancy Data Network Database. (n.d.). Retrieved from http://ec.europa.eu/agriculture/rica/database/database_en.cfm

Farmbot. (2016). *Top Amazing Harvest Robot.* Retrieved from https://www.youtube.com/watch?v=mNB42SzUIxU

Farmers Guardian. (2016). *Young farmers using virtual reality in food education plan.* Retrieved from https://www.fginsight.com/vip/vip/young-farmers-using-virtual-reality-in-food-education-plan-13460

Federatia Nationala a Grupurilor de Actiune Locala (FNGAL). (2017). *Module cursuri formare.* Retrieved from http://www.fngal.ro/images/module/CURSURI_FNGAL.pdf

Fetea, I. (2015). România, un rai al albinelor. *România rurală, 19*(3), 10-12.

Fibl &IFOAM- Organics International. (2017). *The world of organic agriculture 2017.* Retrieved August 25, 2017, from www.organic-world.net

Fischler, F. (2012). Keynote speech: The role of organic farming in our societies. In Let the good market flow- Global Organic Market Access in 2012 and beyond. (pp. 9-10). UNCTAD, FAO & IFOAM.

Fleming, E., Mounter, S., Grant, B., Griffith, G., & Villano, R. (2014). The New World challenge: Performance trends in wine production in major wine-exporting countries in the 2000s and their implications for the Australian wine industry. *Wine Economics and Policy, 3*(2), 115–126. doi:10.1016/j.wep.2014.12.002

Florea, N. (2003). *Degradarea, protecţia şi ameliorarea solurilor şi a terenurilor.* Bucureşti, Romania: Editura Universitară.

Food and Agriculture Organization (FAO). (n.d.). Retrieved from http://www.fao.org/faostat/en

Food and Agriculture Organization of the United Nations (FAO). (2017). *2017 The State of Food and Agriculture- Leveraging Food Systems for Inclusive Rural Transformation*. Rome: Food and Agriculture Organization of the United Nations.

Fromhold-Eisebith, M. (1992). Regionalwirtschaftliche Effekte des Wissens-und Technologietransfers der Rheinisch-Westfälischen Technischen Hochschule Aachen. *Geographische Zeitschrift*, 230–244.

Fromhold-Eisebith, M., & Werker, C. (2013). Universities' functions in knowledge transfer: A geographical perspective. *The Annals of Regional Science*, *51*(3), 621–643. doi:10.100700168-013-0559-z

Gafsi, M., Le Tron, S., & Mouchet, C. (2010). Organic farming is it a sustainable agriculture. Development issues for sustainable organic farming in Midi-Pyrénées Région. In' Innovation and Sustainable Development in Agriculture and Food, June, 28.

Gavrilescu, D., & Giurcă, D. (Eds.). (2000). Economie agroalimentară. editura Expert şi Bioterra, 111-113.

George, T. (Ed.). (1999). India 1000 to 2000 (1st ed.). Chennai: Express Publications (Madurai) Ltd.

Gibb, A. A. (1996). Entrepreneurship and small business management: Can we afford to neglect them in the twenty-first century business school? *British Journal of Management*, *7*(4), 309–322. doi:10.1111/j.1467-8551.1996.tb00121.x

Giescke & Nhi. (2010). *A General Framework for Measuring VAT Compliance Rates*. General Paper No. G-206 August 2010 Centre of Policy Studies and the Impact Project.

Gill, S. (2008). When Rural Development Succeeds? Experience from the Two Punjabs and West Bengal. In S. Bhaumik (Ed.), *Reforming Indian Agriculture: Towards Employment Generation and Poverty Reduction: Essays in Honour of G.K. Chadha* (pp. 293–294). New Delhi: SAGE Publications India Pvt Ltd.

Giogia, M. (in press). Adevărate nestemate în coroana regală a timpului. *Ziarul Timpul, 34*. Retrieved August 10, 2017, from http://www.fiitea.org/foundation/files/martag/34Rmg.pdf

Giovannucci, D., Scherr, S. J., Nierenberg, D., Hebebrand, C., Shapiro, J., Milder, J.,Wheeler, K. (2012). *Food and agriculture: The future of sustainability. The sustainable development in the 21st century (SD21)*. Report for Rio, 20.

Girardin, P., Bockstaller, C., Werf, H.V.D. (1999). Indicators: tools to evaluate theenvironmental impacts of farming systems. *Journal Sustain. Agric.*, *13*(4), 5–21. 0310.1300/J064v13n04

Global Commission on the Future of Work. (2017). *Launch of High-Level Global Commission on the Future of Work*. Speech by Prime Minister Stefan Löfven in Geneva 21 August 2017. Retrieved from http://www.government.se/speeches/2017/08/launch-of-high-level-global-commission-on-the-future-of-work/

Gobin, A., Jones, R., Kirkby, M., Campling, P., Govers, G., Kosmas, C., & Gentile, A. R. (2004). Indicators for pan-European assessment and monitoring of soil erosion by water. *Environmental Science & Policy*, *7*(1), 25–38. doi:10.1016/j.envsci.2003.09.004

Goldsmith, E. (1974). *Religion - the control mechanism of a stable society, a talk given before the Farmington Trust, 1974*. Retrieved September 4, 2017, from http://www.edwardgoldsmith.org/695/religion-the-control-mechanism-of-a-stable-society/

Gopalakrishnan, T. (2007). *Vegetable crops. Pitam Pura*. New Delhi: New India Pub. Agency.

Government of Romania. (2014). *Government Decision no.929/2014 approving the National Strategy for Research, Development and Innovation 2014-2020*. Official Monitor, Part I no. 785 of 28.10.2014. Retrieved from http://www.cdi2020.ro/wpcontent/uploads/2014/10/ HOT%C4%82R%C3%82RE-nr-929-din-21-octombrie-2014-privind-aprobarea-Strategiei-na%C5%A3ionale-de-cercetare-dezvoltare-%C5%9Fi-inovare-2014-2020.pdf, accessed on 9 October 2017

Gradstein, F. M., & Ogg, J. G. (2012). The Chronostratigraphic Scale. In F. M. Gradstein, J. G. Ogg, M. D. Schmitz, & G. M. Ogg (Eds.), *The Geologic Time Scale 2012* (pp. 31–42). Oxford, UK: Elsevier BV. doi:10.1016/B978-0-444-59425-9.00002-0

Gradstein, F. M., Ogg, J. G., Schmitz, M. D., & Ogg, G. M. (Eds.). (2012). *The Geologic Time Scale 2012*. Oxford, UK: Elsevier BV.

Gu, G., & Zhang, Ch. (2002). Analysis on the Structure of Honey Production and Trade in the World. *Apiacta, 2*, 2002. Retrieved from http://www.fiitea.org/cgi-bin/index.cgi

Guitierrez, M. S. (2005). A cooperative Synthesis of 20th Century Agricultural Cooperative Movements in Europe. *Journal of Rural Cooperation.*

Gunderson, L. H. (2000). Ecological resilience—in theory and application. *Annual Review of Ecology and Systematics, 31*(1), 425–439. doi:10.1146/annurev.ecolsys.31.1.425

Guvernul României & Ministerul Mediului şi Dezvoltării Durabile. (2008). Strategia Naţională pentru Dezvoltare Durabilă a României. *Orizonturi 2013-2020-2030*, 19-22. Retrieved from http://www.insse.ro/cms/files/IDDT2012/index_IDDT.htm

Halberg, N. (2012). Assessment of the environmental sustainability of organic farm-ing: Definitions, indicators and the major challenges. *Canadian Journal of Plant Science, 92*(6), 981–996. RetrievedFebruary102017. doi:10.4141/cjps2012-035

Halberg, N., Verscuur, G., & Googlass, G. (2005). Farm level environmentalindicators; are they useful? An overview of green accounting sys-tems for European farms. *Agriculture, Ecosystems & Environment, 105*(1–2), 195–212. doi:10.1016/j.agee.2004.04.003

Hall, J. K., Daneke, G. A., & Lenox, M. J. (2010). Sustainable development and entrepreneurship: Past contributions and future directions. *Journal of Business Venturing, 25*(5), 439–448. doi:10.1016/j.jbusvent.2010.01.002

Hall, J., & Vredenburg, H. (2003). The challenges of innovating for sustainable development. *Sloan Management Review, 45*(1), 61–69.

Harari, Y. N. (2014). Sapiens: A Brief History of Humankind. Harvill Secker.

Harari, Y. N. (2015). *Sapiens: A Brief History of Humankind.* New York, NY: Harper & Collins.

Harari, Y. N. (2016). *Homo Deus: A Brief History of Tomorrow.* New York, NY: Harper & Collins.

Harold, V. D. M. (2007). *Exploring agricultural taxation in Europe.* Wageningen University.

Harris, M. (1974). *Cows, Pigs, Wars, and Witches: The Riddles of Culture.* New York, NY: Random House.

Harris, M. (1978). *Cannibals and Kings. The Origins of Culture.* Fontana, CA: Collins.

Hatekar, N., Kulkarni, S., & Mehta, P. (2015). Culture, Community and Institutions Voluntary Provision of Public Goods in Maharashtra. *Economic and Political Weekly, L*(46-47).

Hazell, P., & Wood, S. (2008). Drivers of change in global agriculture. *Philosophical Transactions of the Royal Society of London. Series B, Biological Sciences, 363*(1491), 495–515. doi:10.1098/rstb.2007.2166 PMID:17656343

Henrich, J. (2015). *The Secret of Our Success: How Culture Is Driving Human Evolution, Domesticating Our Species, and Making Us Smarter.* Princeton, NJ: Princeton University Press.

Hitchcock, D., & Willard, M. (2009). *The Business Guide to Sustainability. Practical Strategies and Tools for Organizations.* Earthscan.

Hiware Bazar Panchayat Records. (2016). *Hiware Bazar. Implementing Water Resources Management: Information brief.* United Nations. Retrieved 1 October 2017, from http://www.un.org/waterforlifedecade/waterandsustainabledevelopment2015/images/implementing_water_eng.pdf

Hristea, C. L. (1943). *Stupăritul*. Bucureşti, România: Cartea Românească.

Hristov, H., & Kuhar, A. (2015). Subjective knowledge as a determinant of young adult consumers wine behaviour. *British Food Journal, 117*(12), 2930–2946. doi:10.1108/BFJ-04-2015-0163

Hunn, E. (1993). What is traditional ecological knowledge? In N. M. Williams & G. Baines (Eds.), *Traditional Ecological Knowledge: Wisdom for Sustainable Development* (pp. 13–15). Canberra: Centre for Resource and Environmental Studies, Australian National University.

Ianoş, I., & Heller, W. (2006). *Spaţiu, economie şi sisteme de aşezări*. Bucureşti, Romania: Editura Tehnică.

Ianoş, I., & Humeau, J. B. (2000). *Teoria sistemelor de aşezări umane*. Bucureşti, Romania: Editura Tehnică.

IBM. (2017). *Macroscopes will help us understand Earth's complexity in infinite detail.* Retrieved from https://www.youtube.com/watch?v=qKMugpYD6tA

ICOS Cooperative Statistics. (2008). *114th Annual Report and accounts.* Irish Cooperative Organisation Society Ltd.

Igual, J., Francisco, J., & Meliá Martí, E. (2008). Social Economy and the Cooperative Movement in Europe: Contributions to a New Vision of Agriculture and Rural Development in the Europe of The 27 CIRIEC-España. *C.I.R.I.E.C. España*, (62), 147–172.

INACO. (2017). Retrieved from the Initiative for Competitiveness: www.inaco.ro

Indira, R. (2004). Taxing agriculture in a developing country: a possible approach. Academic Press.

INOVAGRIA. (2017). Retrieved from INOVAGRIA Meteo: https://www.youtube.com/watch?v=caCkq9Z5o3w

Institutional Characteristics of the German Agro-Food-ChainmW. (n.d.). Grosskopfm University of Hohenheim, Stuttgart.

International Federation of Beekeepers' Associations (Apimondia). Apimondia. (2016). Retrieved December 28, 2017, from: https://www.apimondia.com/en

Intersog. (2016). *VR Applications in Agriculture*. Retrieved from https://intersog.com/blog/augmented-reality/vr-applications-in-agriculture/

Ion, T. C. (in press). 36,000 investment projects in agriculture worth 2.16 billion euros. *Cuget liber newspaper.* Retrieved September 2017, from https: www.cugetliber.ro/stiri-economie, on-line.

Jaba, E. (2003). *Statistics* (3rd ed.). Bucharest: Economic Publishing House.

Jitea, I-M. (2010). *Personal Agricultural Income Taxation. A Comparative Analysis between Several New Member States*. Academic Press.

Jordan, N., & Warner, K. D. (2010). Enhancing the multifunctionality of US Agriculture. *Bioscience, 60*(1), 60–66. doi:10.1525/bio.2010.60.1.10

Kandari, L. S., Bisht, V. K., Bhardwaj, M., & Thakur, A. K. (2014). Conservation and management of sacred groves, myths and beliefs of tribal communities: A case study from north-India. *Environmental Systems Research, 3*(1), 16. doi:10.118640068-014-0016-8

Kayne, J. (1999). *State entrepreneurship policies and programs*. Academic Press.

Kellner, K., & Bosch, O. J. H. (2003). The value of local and indigenous knowledge for the development of information systems for conservation management. In Conserving Biodiversity in Arid Regions (pp. 105-118). Springer US. doi:10.1007/978-1-4615-0375-0_8

Kenkel, J. L. (1995). *Introductory Statistics for Management and Economics* (4th ed.). South-Western Publishing Group.

Kiwi in Munich. (2016). *Real-life Robocop: Dubai to get android police force by 2017.* Retrieved from https://www.youtube.com/watch?v=MWYtNFb77H0

Kobets, M. I. (2004). Organic farming in the context of sustainable development. Kyiv: Agrarian Policy for Human Development, 22.

Kotler, Ph., Jain, D., & Maesincee, S. (2009). *Marketing Moves: A New Approach to Profits, Growth, and Renewal.* Bucharest: Meteor Press Publishing House.

KPMG. (2017). *China VC investments set a record high in 2016, artificial intelligence a new focus, finds KPMG analysis.* Retrieved from https://home.kpmg.com/cn/en/home/news-media/press-releases/2017/01/vc-investments-record-high-artificial-intelligence.html

Kumar, A. (2012). P. Sainath on "Inequality, livelihood and agrarian crisis" in India. *New Socialist Alternative*. Retrieved 11 September 2017, from http://www.socialism.in/index.php/p-sainath-on-inequality-livelihood-and-agrarian-crisis-in-india/

Kummamuru, B. (2016). *WBA Global Bioenergy Statistics*. Retrieved from http://worldbioenergy.org/uploads/WBA%20Global%20Bioenergy%20Statistics%202016.pdf

Kunke, R. (2012). One village. 60 millionaires. The miracle of Hiware Bazar. *Development Dialogue*. Retrieved 17 October 2017, from http://agrariancrisis.in/2013/08/17/one-village-60-millionaires-the-miracle-of-hiware-bazar/

L644/2015, Propunere legislativă privind organizarea şi funcţionarea Consiliului Naţional pentru Competitivitate. (n.d.). Retrieved from http://www.cdep.ro/pls/proiecte/upl_pck.proiect?idp=19474&cam=1

Laland, K. N., & Brown, G. R. (2002). *Sense and Nonsense: Evolutionary Perspectives on Human Behaviour*. Oxford, UK: Oxford University Press.

Laven, A., & Boomsma, M. (2012). *Incentives for sustainable cocoa production in Ghana*. Amsterdam: Royal Tropical Institute.

Law 1/2011. (n.d.). Retrieved from http://www.fen.ro/legea_educatiei_final.pdf

Lenton, P. (2015). Determining student satisfaction: An economic analysis of the National Student Survey. *Economics of Education Review*, *47*, 118–127. doi:10.1016/j.econedurev.2015.05.001

Leten, B., Landoni, P., & Van Looy, B. (2014). Science or graduates: How do firms benefit from the proximity of universities? *Research Policy*, *43*(8), 1398–1412. doi:10.1016/j.respol.2014.03.005

Levy, F., & Murnane, R. J. (1992). US earnings levels and earnings inequality: A review of recent trends and proposed explanations. *Journal of Economic Literature*, *30*(3), 1333–1381.

Lewis, S. L., & Maslin, M. A. (2015). Defining the Anthropocene. *Nature*, *519*(7542), 171–180. doi:10.1038/nature14258 PMID:25762280

Lipton, M. (1994). *Growing points in poverty research: labour issues*. International Institute for Labour Studies.

Lombardi, P., Dal Bianco, A., Freda, R., Caracciolo, F., & Cembalo, L. (2016). Development and trade competitiveness of the European wine sector: A gravity analysis of intra-EU flows. *Wine Economics and Policy*, *5*(1), 50–59. doi:10.1016/j.wep.2015.12.002

Lopez, R. (1994). The environment as a factor of production: The effects of economic growth and trade liberalization. *Journal of Environmental Economics and Management*, *27*(2), 163–184. doi:10.1006/jeem.1994.1032

Low, B. S. (1996). Behavioral Ecology of Conservation in Traditional Societies. *Human Nature - An Interdisciplinary Biosocial Perspective, 7*(4), 353-379.

M.A.D.R. (2016). *Raport de monitorizare a reziduurilor de pesticide*. Retrieved from http://www.madr.ro/reziduuri-de-pesticide-in-plante-si-produse-vegetale/raport-de-monitorizare-a-reziduurilor-de-pesticide.html

Ma-Dupont, V. (2016). *Sweden: a "Ministry of the Future" to think about tomorrow's public policy*. Retrieved from Responsive Public Management No. 82 – April 2016, Gestion Publique: https://www.economie.gouv.fr/files/files/directions_services/igpde-editions-publications/revuesGestionPublique/IGPDE_Reactive_Suede_avril_2016_En.pdf

Magurran, A. E. (1998). *Ecological Diversity and Its Measurement*. Princeton, NJ: Princeton University Press.

Malhotra, K. N., & Peterson, M. (2006). *Basic Marketing Research*. Englewood Cliffs, NJ: Prentice Hall.

Mamatzakis, E. C. (2003). Public infrastructure and productivity growth in Greek agriculture. *Agricultural Economics*, *29*(2), 169–180. doi:10.1111/j.1574-0862.2003.tb00155.x

Maquet, A., Siderer, Y., & Anklam, E. (2005). Need to research to support consumer confidence in the growing organic food market. *Trends in Food Science & Technology*, 332–343.

Marcuello, C., & Lopez, A. (2003). *Agricultural Cooperatives: a wealth, employment, and social involvement generator*. Academic Press.

Marin, A. (2014). *Education and specialized training - Ways to increase performance in agriculture*. Paper presented at the 5th edition of the international symposium Agrarian Economy and Rural Development – Realities and Perspectives for Romania, Bucharest, Romania. Retrieved from http://mpra.ub.uni-muenchen.de/61725/

Marin, A. (2016). *Short chain - A solution for current capitalization of vegetable production*. Paper presented at the 7th edition of the international symposium Agrarian Economy and Rural Development – Realities and Perspectives for Romania, Bucharest, Romania.

Marin, A. (2017b). *Considerations regarding the association/cooperation of small and medium fruit producers*. Paper presented at the 8th edition of the international symposium Agrarian Economy and Rural Development – Realities and Perspectives for Romania, Bucharest, Romania.

Marin, A. (2017a). *Analysis of the association forms of small and medium producers in the fruit and vegetables sector*. Bucharest: ASE Publishing House.

Marin, A., Chetroiu, R., & Iurchevici, L. (2016). *Study regarding the situation of the romanian vegetables sector and the analysis of the branch of vegetables production valuation*. Bucharest: Romanian Academy Publishing House.

Marin, A., Chetroiu, R., & Iurchevici, L. (2017). *Researches on the growth of the production of fruit*. Bucharest: Romanian Academy Publishing House.

Marin, A., Chetroiu, R., Iurchevici, L., Ion, R. A., Dobre, I., Bran, M., & Costaiche, M. (2016). *Study on the situation of vegetables sector national and community*. Bucharest: ASE Publishing House.

Marin, A., Chetroiu, R., Iurchevici, L., & Tudor, V. (2017). *Study on the situation of the fruit sector at national and community level.* Bucharest: ASE Publishing House.

Marin, A., Ion, R. A., Chetroiu, R., & Iurchevici, L. (2017). *Designing and experimenting with the development models of short chains to capitalize on the production of vegetables.* Bucharest: ASE Publishing House.

Marin, A., Ion, R. A., Dobre, I., Chetroiu, R., Iurchevici, L., Pop, R. E., & Costaiche, M. (2016). *Design and experimentation of the development models of the short cutting values of the vegetables production.* Bucharest: ASE Publishing House.

Maslin, M. (2017). *The Cradle of Humanity: How the changing landscape of Africa made us so smart.* Oxford, UK: Oxford University Press.

Mateescu, C. (2015). Cercetarea, baza dezvoltării apiculturii româneşti. *România rurală, 19*(3), 13-23.

Menzani, T., & Zamgni, V. (2009). Cooperative networks in the Italian economy. *Enterprise and Society*, 4.

Meyer-Rochow, V. B. (2009). Food taboos: Their origins and purposes. *Journal of Ethnobiology and Ethnomedicine, 5*(1), 18. doi:10.1186/1746-4269-5-18 PMID:19563636

MIDEA Robot Factory. (2016). Retrieved from https://www.youtube.com/watch?v=2dJewkJZ4PU

Miller, A. S., & Kanazawa, S. (2000). *Order by Accident. The Origins and Consequences of Conformity in Contemporary Japan.* Boulder, CO: Westview Press.

Ministerul Agriculturi si Dezvoltarii Rurale. (2016). *Programul de Guvernare 2016-2020. Sprijinirea fermierilor romani.* Retrieved 20 August, 2017, from http://www.madr.ro/docs/agricultura/Programul_de_Guvernare_pentru_agricultura.pdf

Ministerul Agriculturii si Dezvoltarii Rurale- Direcţia Generală de Dezvoltare Rurală. (2015, May 26). *Programul Naţional de Dezvoltare Rurală 2014-2020 (PNDR).*

Ministerul Agriculturii si Dezvoltarii Rurale. (2015). *Date generale despre agricultura Romaniei.* Retrieved August 21, 2017, from http://www.madr.ro/docs/agricultura/agricultura-romaniei-2015.pdf

Ministerul Dezvoltării Regionale şi Administraţiei Publice. (2016). *Dezvoltare regionala, Programe europene 2007-2013, Programul Operational Regional 2007-2013, Regiuni de dezvoltare.* Retrieved from http://www.mdrap.ro/dezvoltare-regionala/-2257/programul-operational-regional-2007-2013/-2975

Ministry of Agricultural, Food and Forestry Policies (Italy). (n.d.). Retrieved from www.politicheagricole.it

Ministry of Agriculture and Rural Development (MARD). (2015). *National Program for Rural Development 2014 – 2020.* General Direction for Rural Development (MA -NRDP), Program adopted by EC on 26.05.2015

Ministry of Agriculture and Rural Development (MARD). (2017). *Situatia proiectelor depuse in data de 14 septembrie 2017, PNDR 2014-2020.* Retrieved from http://www.pndr.ro/situatia-proiectelor-depuse-2014-2020/item/279-situatia-proiectelor-depuse-in-data-de-14-septembrie-2017-pndr-2014-2020.html on 20 sept.2017

Ministry of Agriculture and Rural Development (MARD). (2017). *The stage of implementation of NRDP 2014-2020 to date.* General direction for rural development (MA-NRDP), Report adopted in 14.09.2017

Ministry of Agriculture and Rural Development of the Republic of Poland. (n.d.). Retrieved from www.minrol.gov.pl

Ministry of Agriculture and Rural Development Policies (Romania). (n.d.). Retrieved from www.madr.ro

Ministry of Agriculture and Rural Development. (2015). *Sector development strategy medium and long-term agri-food horizon 2020-2030.* Retrieved September 10, 2017, from http://www.madr.ro/docs/agricultura/strategiaagroalimentara-2020-2030.pdf

Ministry of Agriculture and Rural Development. (2017). *Organic farming*. Retrieved 20 August, 2017, from www.madr.ro/ro/agricultura-ecologica.html

Ministry of Agriculture and Rural Development. (n.d.). Retrieved from http://www.madr.ro

Ministry of Agriculture and Rural Development. (n.d.). *The National Rural Development Program (NRDP)*. Retrieved from www.pndr.ro

Ministry of Agriculture. (2012). *National Rural Development Plan 2014-2020*. Retrieved September 10, 2017, from http://www.madr.ro/pndr-2014-2020/instrument-financiar-de-creditare.html

Ministry of Public Finance. (2014). *The evaluation of macroeconomics effects of the VAT reduction for flour and bread*. Author.

Ministry of Research and Innovation (MRI). (2017). *Institute Naţionale de Cercetare Dezvoltare (INCD) în coordonare*. Retrieved from http://www.research.gov.ro/ro/articol/4514/sistemul-de-cercetare-institute-nationale-de-cercetare-dezvoltare-in-coordonare#Prezentare

Morgan, R. P. C. (2005). Soil Erosion and Conservation (3rd ed.). Blackwell Publ.

Morgan, K., & Murdoch, J. (2000). Organic vs. conventional agriculture: Knowl-edge, power and innovation in the food chain. *Geoforum Elsevier*, *31*(2), 159–173. doi:10.1016/S0016-7185(99)00029-9

Morgenstern, M. (in press). Artificial intelligence: The impact on jobs. Automation and anxiety. *The economist*.

Morris, M., & Western, B. (1999). Inequality in Earnings at the Close of the Twentieth Century. *Annual Review of Sociology*, *25*(1), 623–657. doi:10.1146/annurev.soc.25.1.623

Motoman robots packing IKEA. (2012). Retrieved from https://www.youtube.com/watch?v=Duub_dO_TwU

Müller, D., Leitão, P. J., & Sikor, T. (2013). Comparing the determinants of cropland abandonment in Albania and Romania using boosted regression trees. *Agricultural Systems*, *117*, 66–77. doi:10.1016/j.agsy.2012.12.010

Mundle, S. (2017). Agrarian crisis: the challenge of a small farmer economy. *Livemint*. Retrieved from http://www.livemint.com/Opinion/Y3Fp6CcumJhRIEwl2WeUMM/Agrarian-crisis-the-challenge-of-a-small-farmer-economy.html

Muntean, L. S., Roman, G. V., Borcean, I., & Axinte, M. (2003). Fitotehnie. Editura Ion Ionescu de la Brad.

Musila, W., Todt, H., Uster, D., & Dalitz, H. (2005). Is Geodiversity Correlated to Biodiversity? A Case Study of the Relationship Between Spatial Heterogeneity of Soil Resources and Tree Diversity in a Western Kenyan Rainforest. In B. A. Huber, B. J. Sinclair, & K. Lampe (Eds.), *H. African Biodiversity*. New York: Springer-Verlag. doi:10.1007/0-387-24320-8_40

N.I.S. (2012-2016). Retrieved from http://www.insse.ro/cms/

Naldi, L., Nilsson, P., Westlund, H., & Wixe, S. (2015). What is smart rural development? *Journal of Rural Studies*, *40*, 90–101. doi:10.1016/j.jrurstud.2015.06.006

Nasution, H. N., Mavondo, F. T., Matanda, M. J., & Ndubisi, N. O. (2011). Entrepreneurship: Its relationship with market orientation and learning orientation and as antecedents to innovation and customer value. *Industrial Marketing Management*, *40*(3), 336–345. doi:10.1016/j.indmarman.2010.08.002

National Agency for Fiscal Administration. (n.d.). Retrieved from www.anaf.ro

National Bank of Romania. (n.d.). Retrieved August 23, 2017, from http://www.bnr.ro/Cursul-de-schimb-3544.aspx

National Institute of Statistics (NIS). (2017a). *Statistical Yearbook of Romania*. NIS.

National Institute of Statistics (NIS). (n.d.). *Data sets*. Retrieved from www.insse.ro

National Institute of Statistics. (1990). *Romanian Statistical Yearbook*. Bucharest, Romania: Author.

National Institute of Statistics. (2015). *Tempo/line*. Retrieved August 23, 2017 from http://statistici.insse.ro

National Institute of Statistics. (2017). *Tempo/line*. Retrieved August 23, 2017 from http://statistici.insse.ro

National Institute of Statistics. (2017b). *Coordinates of living standard in Romania. Population income and consumption, in 2016*. Bucharest: Author.

National Institute of Statistics. (2017c). *Tempo online*. Author.

National Rural Development Programme. (2014). *National Rural Development Programme*. Retrieved December 10, 2014, from http://madr.ro/docs/dezvoltare-rurala/programare-2014-2020/PNDR_2014_-_2020_01.07.2014.pdf

National Statistics Institute. (2016), Tempo online, Educație. *Studenți și cursanți înscriși în învățământul superior, pe grupe de specializări*. Retrieved July 12, 2017, from http://statistici.insse.ro/shop/index.jsp?page=tempo3&lang=ro&ind=SCL103L

NECIT Solutions. (2012). *NeoFace Access Control*. Retrieved from https://www.youtube.com/watch?v=Hsc9LDeRQlA

Neely, C., Sutherland, K., & Johnson, J. (2004). *Do sustainable livelihood approaches have a positive impact on the rural poor? A look at twelve case studies*. Academic Press.

Negulescu, M., Vaicum, L., & Pătru, C. (1995). *Protecţia mediului înconjurător*. Bucureşti: Editura Tehnică.

New China, T. V. (2014). *3D printers print ten houses in 24 hours*. Retrieved from https://www.youtube.com/watch?v=SObzNdyRTBs

New Scientist. (2016). *Farmers: These ARE the droids you're looking for*. Retrieved from https://www.youtube.com/watch?v=hjd5DaxkLhQ

No, R. 106/2008 on a Community energy-efficiency labelling programme for office equipment. (n.d.). Retrieved from RELIEF Project Information webpage: www.sustainable-procurement.org/about-us/past-projects/relief/

Nowak, A., Kijek, T., & Domańska, K. (2015). Technical efficiency and its determinants in the European Union agriculture. [Zemědělská Ekonomika]. *Agricultural Economics*, *61*(6), 275–283.

OECD. (2012). Environmental Outlook to 2050: The Consequences of Inaction. Paris: OECD.

OECD. (2017). *Skills beyond school*. Retrieved July 12, 2017, from http://www.oecd.org/education/skills-beyond-school/

Olesen, J. E. (2006). *Climate change as a driver for European agriculture. SCAR-Foresight in the field of agricultural research in Europe*. Expert paper.

Olesen, J. E., & Bindi, M. (2004). Agricultural impacts and adaptations to climate change in Europe. *Farm Policy Journal*, *1*(3), 36–46.

Open Working Group. (2014). *Open Working Group proposal for sustainable development goals*. New York: Author.

Order 724/1082/360/2013 –traditional food attestation

Ordinance no. 37/2005 on the recognition and functioning of producer groups for the marketing of

Organisation for Economic Co-Operation and Development (OECD). (2006, October). Reinventing Rural Policy. *Policy Brief.*

PackagingDigest. (2017). *How smart packaging sensors safeguard foods and drugs*. Retrieved from http://www.packagingdigest.com/smart-packaging/how-smart-packaging-sensors-safeguard-foods-and-drugs-2017-04-13

Palicica, M., Gavrilă, C., & Ion, L. (2007). *Pedagogy for agricultural education*. Timisoara, Romania: Mirton Publishing House.

Pambuccian, V. (2017). *De la Homo Faber la Robo Sapiens. A history of the Future written by cars*. Pwp Presentation, INACO Conference organized at the Academy of Economic Studies. Retrieved from http://inaco.ro/de-la-homo-faber-la-robo-sapiens-cum-vor-arata-afacerile/

Patra, T., & Poshakwale, S. (2006). Economic variables and stock market returns: Evidence from the Athens stock exchange. *Applied Financial Economics*, *16*(13), 993–1005. doi:10.1080/09603100500426523

Paturot, D., Mellbye, K., & Brys, B. (2013). *Average Personal Income Tax Rate and Wedge Progression in OECD Countries*. Academic Press.

Paul, A. (2017a). *Costurile brevetării invențiilor românești sunt de peste două ori mai mari decât în Germania și triple față de Polonia*. Retrieved from INACO: https://inaco.ro/costurile-brevetarii-inventiilor-romanesti-sunt-de-peste-doua-ori-mai-mari-decat-in-germania-si-triple-fata-de-polonia/

Paul, A. (2017b). *O aplicație IT 100% românească te poate ajuta să faci agricultură performantă*. Retrieved from INACO: https://inaco.ro/o-aplicatie-it-100-romaneasca-te-poate-ajuta-sa-faci-agricultura-performanta/

Paul, A. (2017c). *Statul apără braconierii de sturioni, nu sturionii! Mafia de milioane de euro*. Retrieved from INACO: https://inaco.ro/statul-apara-braconierii-de-sturioni-nu-sturionii-mafia-de-milioane-de-euro/

Paul, A. (2017d). *INACO propune crearea Grupului De Lucru Pentru Economia Viitorului (GLEV)*. Retrieved from ClubEconomic: http://clubeconomic.ro/opinii/inaco-propune-crearea-grupului-de-lucru-pentru-economia-viitorului-glev-4088/

Petrescu, C. (Ed.). (2013). *Cooperatives from Romania. Actors of socio-economic development*. Bucharest: Polirom Publishing.

Petrişor, A. I. (2008). Ecologie urbană, dezvoltare spaţială durabilă şi legislaţie. O abordare ecologică a relaţiilor dintre om, spaţiul construit şi mediu. Editura Fundaţiei România de mâine.

Philip, Mc., & George, R. (2007). *The Blackwell Companion to Globalization*. New York: Wiley-Blackwell.

Piasecki, A., & Jurasz, J. (2015). Development of water and sewage infrastructure on rural areas in Poland. *Development*, *15*(3).

Plant Based News. (2016). *Meet The Meat that is Tipped to Save the World*. Retrieved from https://www.youtube.com/watch?v=VjEBDFuSoRw

Pomarici, E., Lerro, M., Chrysochou, P., Vecchio, R., & Krontalis, A. K. (2017). *One size does (obviously not) fit all: using product attributes for wine market segmentation*. Wine Economics and Policy.

Popescu, C. R. (2017). The Role of Total Quality Management in Developing the Concept of Social Responsibility to Protect Public Interest in Associations of Liberal Professions. *Amfiteatru Economic, 19*(Special No. 11), 1091-1106. Retrieved from http://www.amfiteatrueconomic.ro/temp/Article_2685.pdf

Popescu, C. R., Popescu, G. N., & Popescu, V. A. (2017). Assessment of the State of Implementation of Excellence Model Common Assessment Framework (CAF) 2013 by the National Institutes of Research – Development – Innovation in Romania. *Amfiteatru Economic, 19*(44), 41-60. Retrieved from http://www.amfiteatrueconomic.ro/temp/Articol_2593.pdf

Popescu, C. R., Popescu, V. A., & Popescu, G. N. (2014). The entrepreneur's role in the performance growth of the financial audit activity in Romania. *Amfiteatru Economic, 17*(38), 232-251. Retrieved from http://www.amfiteatrueconomic.ro/temp/Article_2382.pdf

Popescu, G. (2014). The common agricultural policy-in the traps of land property. *Calitatea, 15*(8), 8-15.

Popescu, G. (2007). *Cooperation in agriculture, from the land market to the transfer of knowledge*. Iasi: Terra Nostra.

Popescu, G. (2013). *Old issues, new relationships in agriculture*. Bucharest, Romania: Romanian Academy Publishing.

Popescu, G. (2013). *Old problems, new relation in agriculture*. Bucharest: Romanian Academy Publishing.

Popescu, G. (2014). *Cooperația în agricultură de la argumentul istoric la transferul de cunoaștere*. București.

Popescu, G. (2014). *Cooperation in agriculture from the historic argument to the knowledge transfer*. Bucuresti: Editura Academiei Romane.

Popescu, G. (2014). *Cooperation in agriculture, from the historical argument to transfer of knowledge*. Bucharest: Romanian Academy Publishing.

Popescu, G., & Bara, S. (2017). *The Knowledge Transfer Economy (Unpublished Lecture Notes)*. Bucharest: Bucharest University of Economic Studies.

Popescu, G., & Zaharia, A. (2015). Analysis of technical progress, energy consumption and value added in European Union's agriculture. *2nd International Multidisciplinary Scientific Conference on Social Sciences and Arts*, 3, 49 - 56.

Potocnik, J. (2017). Preface. In CAP - Thinking Out of the Box: Further modernisation of the CAP – why,what and how? (pp. 4-5). RISE Foundation.

Potter, R., Conway, D., Evans, R., & Lloyd-Evans, S. (2012). *Key concepts in development geography*. London: SAGE. doi:10.4135/9781473914834

Powell, W. W., & Snellman, K. (2004). The knowledge economy. *Annual Review of Sociology, 30*(1), 199–220. doi:10.1146/annurev.soc.29.010202.100037

PriceWaterHouse Coopers. (2009). *Significant si Ecofys*. Author.

PricewaterhouseCoopers (PwC). (2017). *Potential of agricultural sector development in Romania*. PricewaterhouseCoopers Management Consultants. Retrieved September 10, 2017, from https://www.juridice.ro/wp-content/uploads/2017/03/Raport_PwC-agricultura.pdf

PricewaterhouseCoopers. (n.d.). *Potential of agricultural sector development in Romania*. PricewaterhouseCoopers Management Consultants SRL. Retrieved from https://www.juridice.ro/wp-content/uploads/2017/03/Raport_PwC-agricultura.pdf

PVE product boards for livestock; meat and eggs. (n.d.). Retrieved from www.pve.nl

PwC. (2017). *Potentialul dezvoltarii sectorului agricol din Romania*. Retrieved 20 august 2017, from https://www.juridice.ro/wp-content/uploads/2017/03/Raport_PwC-agricultura.pdf

PwC. (2017). *The Long View How will the global economic order change by 2050?* Retrieved from http://www.pwc.com/gx/en/world-2050/assets/pwc-the-world-in-2050-full-report-feb-2017.pdf

Quinton, J. N., Govers, G. V., Oost, K., & Bardgett, R.D. (2010). The impact of agricultural soil erosion on biogeochemical cycling. Nature Geoscience, 3, 311 – 314.

Quiroz, D., & van Andel, T. (2015). Evidence of a link between taboos and sacrifices and resource scarcity of ritual plants. *Journal of Ethnobiology and Ethnomedicine*, *11*(1), 5. doi:10.1186/1746-4269-11-5 PMID:25573058

R193 - Promotion of Cooperatives Recommendation. (2002). *Recommendation concerning Promotion of Cooperatives*. Geneva, 90th ILC session (No. 193).

Ramakrishnan, P. S., Saxena, K. G., & Chandrashekara, U. M. (Eds.). (1998). *Conserving the Sacred for Biodiversity Management*. New Delhi: Oxford and IBH Publishing.

Rametra, A. (2016). *India Together: Rays of hope for the 'local' in Meghalaya*. Retrieved 13 October 2017, from http://www.indiatogether.org/rays-of-hope-for-the-local-in-meghalaya-agriculture

Ranasinghe, N. (2017). *Virtual Lemonade: Let's Teleport Your Lemonade*. Retrieved from https://www.youtube.com/watch?v=w8gTsGr2JpA&feature=youtu.be

Rankin, M., & Nogales, E. G. (2016). *Public-private partnerships for agribusiness development, A review of international experiences*. Rome: Food and Agriculture Organization of the United Nations.

Raport, C. O. M. (2016). *776 al Comisiei către Parlamentul European şi Consiliu cu privire la punerea în aplicare a măsurilor privind sectorul apicol prevăzute de Regulamentul (UE) nr. 1308/2013 al Parlamentului European şi al Consiliului de instituire a unei organizări comune a pieţelor produselor agricole*. Retrieved from http://ec.europa.eu/transparency/regdoc/rep/1/2016/RO/COM-2016-776-F1-RO-MAIN-PART-1.PDF

Rappaport, R. A. (1984). *Pigs for the Ancestors. Ritual in the Ecology of a New Guinea People*. New Haven, CT: Yale University Press.

Redford, K. (1991). The Ecologically Noble Savage. *Cultural Survival Quarterly*, *15*(1), 46–48.

Regulamentul (CE) nr. 1221/97 al Consiliului din 25 iunie 1997 de stabilire a normelor generale de punere în aplicare a măsurilor de îmbunătăţire a producerii şi comercializării mierii, JO L 173, 1.7.1997, pp. 1.

Rendleman, C. M., Hoemmen, G. A., Altman, I., Taylor, B., Moon, W., & Smith, S. (2016). Wine Industry Competitiveness: A survey of the Shawnee Hills American Viticultural Area. *Wine Economics and Policy*, *5*(1), 4–13. doi:10.1016/j.wep.2016.03.002

Rennings, K., & Wiggering, H. (1997). *Steps towards indicators of sustainable development - linking economic and ecological concepts* (Vol. 20). Ecological Economics.

Reuters. (2015). *Real-life Star Trek 'replicator' prepares meal in 30 seconds*. Retrieved from http://www.reuters.com/article/us-israel-meals-on-demand-tracked-idUSKBN0NQ1PG20150505

Rigby, D., Woodhouse, P., Young, T., & Burton, M. (2001). Constructing a farm levelindicator of sustainable agricultural practice. *Ecologic. Econ. Elsevier*, *39*(3), 463–478. doi:10.1016/S0921-8009(01)00245-2

Rigg, J. (2016). *Challenging Southeast Asian Development: The Shadows of Success*. New York: Routledge.

Rohrbeck, R. (2007). *Technology Scouting - a Case Study on the Deutsche Telekom Laboratories*. Available from: https://www.researchgate.net/publication/24114558_Technology_Scouting_-_a_Case_Study_on_the_Deutsche_Telekom_Laboratories

Rolland-May, C. (2000). *Evaluation des territoires. Concepts, modèle, méthodes*. Editura Hermes Science Publications.

Rolston, A., Jennings, E., & Linnane, S. (2017). Water matters: An assessment of opinion on water management and community engagement in the Republic of Ireland and the United Kingdom. *PLoS One*, *12*(4), e0174957. doi:10.1371/journal.pone.0174957 PMID:28369136

Romanian Land Fund Law no 18/1991, published in Official Gazette from 20.02.1991

Romanian Land Fund Law no 18/1991, republished, with subsequent amendments and completions. (n.d.). Retrieved from http://www.dreptonline.ro/legislatie/legea_fondului_funciar.php

Romanian Law no 1/2005 on the organization and operation of the cooperation, published in Official Gazette of 03.03.2005.

Romanian Law no. 1/2000 for restoring ownership of agricultural land and forestry

Romanian Law no. 31/1990 on commercial society, publish in Official Gazette no.1066/17.11.2004

Romanian Law no. 338/2005 for approving OG 37/2005 Report on the recognition and operation of producer groups for the marketing of agricultural and forestry products, Official Gazette no.1098/12.05.2005.

Romanian Law no. 36/1991 concerning agricultural societies and other forms of Association in agriculture, published in the Official Gazette of 06/05.1991.

Romanian Law no. 45/2009 regarding the organization and functioning of the Academy of Agricultural and Forestry Sciences Gheorghe Ionescu-Sisesti and of the research and development system in the fields of agriculture, forestry and food industry. (2009). Published in Official Montor of Romania.

Romanian Law regarding the Organization and encouraging agriculture, published in Official Gazette no. 67 of 22.03. 1937. (n.d.). Retrieved from http://www.cdep.ro/pls/legis/legis_pck.htp_act_text?idt=27536

Romanian National Agency for Environmental Protection (2016). *Areas-Soil*. Retrieved from http://www.anpm.ro/sol-subsol

Romanian National Institute of Statistics. (2016a). Retrieved from http://statistici.insse.ro/shop/?page=tempo3&lang=ro&ind=PMI114B

Romanian National Institute of Statistics. (2016b). Retrieved from http://statistici.insse.ro/shop/

României, P. (2004). Legea nr. 58 din 13 iulie 1994 pentru ratificarea Convenţiei privind diversitatea biologică, semnată la Rio de Janeiro la 5 iunie 1992. Monitorul Oficial, 199(1).

Rundgren, G. (2012). Future: what will happen, what should happen. In Let the good market flow- Global Organic Market Access in 2012 and beyond (pp. 9-10). UNCTAD, FAO & IFOAM.

Sachs, J. D. (2015). *Age of Sustainable Development*. New York: Columbia University Press. doi:10.7312ach17314

Salmi, J. (2009). *The challenge of establishing world-class universities*. World Bank Publications. doi:10.1596/978-0-8213-7865-6

Sanderson, S. K. (2001). *The Evolution of Human Sociality. A Darwinian Conflict Perspective*. Lanham, MD: Rowman & Littlefield Publishers, Inc.

Santos, H. C., Varnum, M. E. W., & Grossmann, I. (2017). Global Increases in Individualism. *Psychological Science*, *28*(9), 1228–1239. doi:10.1177/0956797617700622 PMID:28703638

Sârbu, C. N. (2006). *Locuirea în România: o abordare-cadru*. Bucureşti, Romania: Editura Universitară Ion Mincu.

Sarojam. (2016). *The Village Of Millionaires - Hiware Bazar - Spiritual Discussion*. Retrieved 3 February 2018, from https://www.speakingtree.in/discussion/the-village-of-millionaires-hiware-bazar

Satish, A. B., Bhaveshananda, S., & Sengupta, B. (2012). Organic agriculture: Way towards sustainable development. *International Journal of Environmental Sciences*, *3*(1), 209.

Schäufele, I., & Hamm, U. (2018). Organic wine purchase behaviour in Germany: Exploring the attitude-behaviour-gap with data from a household panel. *Food Quality and Preference*, *63*, 1–11. doi:10.1016/j.foodqual.2017.07.010

Schwab, K. (2017). *The Global Competitiveness Report 2017–2018*. Retrieved from World Economic Forum: http://www3.weforum.org/docs/GCR2017-2018/05FullReport/TheGlobalCompetitivenessReport2017%E2%80%932018.pdf

Schweitzer, S., & Alexander, L. (2015). *Access to Asia*. Hoboken, NJ: John Wiley & Sons, Inc.

Scoones, I. (1998). *Sustainable rural livelihoods: a framework for analysis*. Academic Press.

Sen, A. (1975). *Employment, Technology and Development*. Oxford, UK: Clarendon Press.

Seufert, V. (2012). Organic agriculture as an opportunity for sustainable agricultural development. *Research to Practice Policy Briefs. Policy Brief*, 13.

Sheeran, K., Ackerman, F., Enelow, N., Goodstein, E., Hahnel, R., Power, T. M., & Schor, J. (2014). *The Future Economy. An Annotated Analytical Framework for Case Studies*. Economics for Equity & Environment. Retrieved from http://futureecon.com/wp-content/uploads/future-economy-framework_new.pdfhttp://futureecon.com/

Simoons, J. F. (1994). *Eat Not This Flesh. Food Avoidances from Prehistory to the Present* (2nd ed.). Madison, WI: The University of Wisconsin Press.

Skills Panorama. (2017). *Skills Panorama*. Retrieved July 18, 2017, from http://skillspanorama.cedefop.europa.eu/en/

Špička, J. (2014). The regional efficiency of mixed crop and livestock type of farming and its determinants. *AGRIS On-Line Papers in Economics and Informatics*, *6*(1), 99–109.

Spring, J. (2015). *Economization of Education: Human Capital, Global Corporations, Skills-Based Schooling*. New York: Routledge.

Stanciu, M. (2013). Study concerning the capitalization on local level of traditional products obtained from sheep. Academic Press.

Syvitski, J. (2012). Anthropocene: An epoch of our making. *Global Change*, *8*(78), 12–15.

Szczepaniak, I., & Tereszczuk, M. (2016). *Regional and Traditional Products as a Way to Improve agricultural and forestry products*. Competitiveness of Polish Food Producers in Foreign Markets, Thematic publication number 38 – Traditional foods, Ministry of Agriculture and Rural Development. Retrieved from http://madr.ro/docs/dezvoltare-rurala/rndr/buletine-tematice/PT38.pdf

Tech Medium. (2017). *First Real Flying Car in Dubai Ehang 184*. Retrieved from https://www.youtube.com/watch?v=UAbquGXnXmQ

TechCrunch. (2017). *Cafe X Opens In San Francisco, Bringing Robots To The Coffee*. Retrieved from https://techcrunch.com/2017/01/30/cafe-x-opens-in-san-francisco-bringing-robots-to-the-coffee-shop/

Teryomenko, H. (2008). *Farm Size and Determinants of Agricultural Productivity in Ukraine* (Doctoral dissertation). National University.

The Council of State. (1977). *Status of agricultural production cooperative of 28 September 1977*. Official Gazette no. 105 of 30 September 1977. Retrieved from http://www.monitoruljuridic.ro/act/statutul-din-28-septembrie-1977-cooperativei-agricole-de-productie-emitent-consiliul-de-stat-publicat-n-monitorul-oficial-20811.html

The Economist. (2015). *Silicon Valley gets a taste for food.* Retrieved from https://www.economist.com/news/technology-quarterly/21645497-tech-startups-are-moving-food-business-make-sustainable-versions-meat

The Economist. (2017). *The growth of industrial robots.* Retrieved Sep. 15, 2017, from https://www.economist.com/blogs/graphicdetail/2017/03/daily-chart-19

The German Cooperatives in Europe DGRV – DeutscherGenossenschafts- und Raiffeisenverbande V. (n.d.). *German Cooperative and Raiffeisen Confederation.* Retrieved from www.dgrv.de

The Romanian Ministry of Agriculture and Rural Development. (2016). *National Rural Development Programme 2014 – 2020, The Romanian village has future!* Retrieved from http://www.madr.ro/pndr-2014-2020/implementare-pndr-2014-2020/documente-aprobate.html

The State Council of the People's Republic of China. (2017). *China issues guideline on artificial intelligence development.* Retrieved from http://english.gov.cn/policies/latest_releases/2017/07/20/content_281475742458322.htm

The World Co-operative Monitor. (2016). *The World Co-operative Monitor Database: Exploring the Co-Operative Economy.* Retrieved from http://old.ica.coop/sites/default/files/WCM_2016.pdf?_ga=2.176310509.978773152.1506020642-1176687090.1506020642

Tocco, B., Davidova, S., & Bailey, A. (2014). labor adjustments in agriculture: Evidence from Romania. *Studies in Agricultural Economics (Budapest), 116*(2), 67–73. doi:10.7896/j.1406

Toy, T. J., Foster, G. R., & Renard, K. G. (2002). *Soil Erosion: Processes, Prediction, Measurement, and Control.* John Wiley and Sons.

Treaty, L. (2007, December 17). Treaty of Lisbon amending the Treaty on European Union and the Treaty establishing the European Community. *Official Journal of the European Union.*

Ţuţuianu, O. (2011). *Evaluarea şi raportarea performanţei de mediu. Indicatori de mediu ediţia a III-a.* Bucureşti, Romania: Editura AGIR.

Udmale, P., Ichikawa, Y., Manandhar, S., Ishidaira, H., Kiem, A., Shaowei, N., & Panda, S. (2015). How did the 2012 drought affect rural livelihoods in vulnerable areas? Empirical evidence from India. *International Journal of Disaster Risk Reduction, 13*, 454–469. doi:10.1016/j.ijdrr.2015.08.002

United Nations (UN). (2015). *Transforming our world: the 2030 Agenda for Sustainable Development.* Resolution adopted by the General Assembly on 25 September 2015. Author.

United Nations Environment Programme (UNEP). (2016). *Food Systems and Natural Resources. A Report of the Working Group on Food Systems of the International Resource Panel* . Author.

United Nations. (2017). *Milenium Development Goals.* Retrieved July 25, 2017, from http://www.un.org/millennimumgoals/2015_MDG_Report/pdf/MDG%202015%20rev%20(July%201).pdf<

V.I.B. (n.d.). *Effect of genetically modified crops on the environment.* Retrieved from http://www.vib.be/en/news/Documents/vib_fact_genetisch%20gewijzigde%20gewassen_ENG_2016_LR.pdf

Van der Werf, H.M., Petit, J.(2002). Evaluation of the environmental impact of agriculture at the farm level: a comparison and analysis of 12 indicator-basedmethods. *Agric. Ecosyst. Environ., 93*(1), 131–145. (01)00354-1.10.1016/S0167-8809

Vasile, E., & Bran, Ş. D. (2017). Lignocellulose bio resources and renewable energy. Internal Auditing & Risk Management, 46(S1), 23-30.

Vătămanu, V. (2012). *Acumularea productiei vegetale si factorii care o conditioneaza*. Retrieved from http://www.agrimedia.ro/articole/acumularea-productiei-vegetale-si-factorii-care-o-conditioneaza

Velikova, N., & Dodd, T. (2016). Sustainability of the wine market though emerging consumer segments: The case of US Hispanic consumers. *Agriculture and Agricultural Science Procedia*, *8*, 81–87. doi:10.1016/j.aaspro.2016.02.011

VentureRadar. (2015). *5 Food Technology Startups to Watch*. Retrieved from http://blog.ventureradar.com/2015/11/17/5-food-technology-startups-to-watch/

Vezzoli, C., & Manzini, E. (2008). *Design for Environmental Sustainability*. London: Springer Science & Business Media.

Vijayasarathy, K., & Ashok, K. R. (2015). Climate Adaptation in Agriculture through Technological Option: Determinants and Impact on Efficiency of Production. *Agricultural Economics Research Review*, *28*(1), 103. doi:10.5958/0974-0279.2015.00008.7

Vitousek, P. M., Mooney, H. A., Lubchenco, J., & Melillo, J. M. (1997). Human domination of Earth's ecosystems. *Science. New Series*, *277*(5325), 494–499.

Vlachos, V. A. (2017). *A macroeconomic estimation of wine production in Greece*. Wine Economics and Policy.

Vocea.biz. (2017). *A apărut "mâncarea inteligentă". Cum arată și ce conține*. Retrieved from https://vocea.biz/economic/2017/may/20/main-a-aparut-mancarea-inteligenta-cum-arata-si-ce-contine-video/

Voicu, R., & Dobre, I. (2003). *Organizarea şi strategia dezvoltării unităţilor agricole*. Bucureşti: Editura ASE.

Von Braun, J. (2010). *Globalization of Food and Agriculture and the Poor*. Oxford University Press.

Vreja, L. O., & Talpau, M. (2012). Intangible assets as strategic resources of the knowledge based Organization. *Metalurgia International*, *17*(1), 175–180.

Wall Street Journal. (2014). *Glowing Plants to Light Your Home? Here They Come!* Retrieved from https://www.youtube.com/watch?v=AM1q0ifx91A

Walsh, B. (2012). WholeFood Blues: Why Organic Agriculture May Not Be So Sustainable. *TIME.com*. Retrieved 26 January 2018, from http://science.time.com/2012/04/26/whole-food-blues-why-organic-agriculture-may-not-be-so-sustainable/

Weber, M. (2015). Science as a vocation. In P. Lassman & I. Velody (Eds.), *Max Weber's 'Science as a Vocation'* (pp. 3–32). New York, NY: Routledge.

Wheelan, R. (1999). *Wild in Woods: The Myth of the Noble Eco-Savage*. London: IEA Environment Unit.

Wiebensohn, J., & Jackenkroll, M. (2013). Evaluation and modelling of a standard based spatial data infrastructure for precision farming. In *Proceedings of the EFITA-WCCA-CIGR Conference* (*Vol. 2427*, p. C0107). Academic Press.

Williams, C. (2002). Nutritional quality of organic food: Shades of grey or shades of green? *The Proceedings of the Nutrition Society*, *61*(1), 19–24. doi:10.1079/PNS2001126 PMID:12002790

Wired. (2016). *Uber's Self-Driving Truck Makes Its First Delivery*. Retrieved from https://www.youtube.com/watch?v=sIlCR4eG8_o

WIVBTV. (2014). *SolarCity investing $5B in Buffalo, creating 3,000 jobs*. Retrieved from https://www.youtube.com/watch?v=bj2cSTHGFFk&t=43s

World Commission on Environment and Development (WCED). (1987). *Our Common Future*. Oxford, UK: Oxford University Press.

World Economic Forum. (2017), *How can we maximize the benefits of science and technology for society?* Center for the Fourth Industrial Revolution, San Francisco: https://www.weforum.org/center-for-the-fourth-industrial-revolution/home

World Economic Forum. (2017). *Future of Jobs.* Retrieved August 20, 2017, from http://www3.weforum.org/docs/WEF_Future_of_Jobs.pdf

World Economic Forum. (2017). *Global Competitiveness Report 2017-2018, in the framework of the System Initiative on Shaping the Future of Economic Progress.* Retrieved from http://www3.weforum.org/docs/GCR2017-2018/05FullReport/TheGlobalCompetitivenessReport2017%E2%80%932018.pdf

World Economic Forum. (2017). *New Vision for Agriculture.* Retrieved from https://www.weforum.org/projects/new-vision-for-agriculture

World Science Festival. (2016). *Miguel Nicolelis Helps Paralyzed Patients Learn to Move Again.* Retrieved from https://www.youtube.com/watch?v=MHT6AGdzEaU

WWAP (United Nations World Water Assessment Programme). (2015). *The United Nations World Water Development Report 2015: Water for a Sustainable World.* Paris: UNESCO.

Xiangzheng, D., Zhihui, L., Jikun, H., Qingling, S., & Yanfei, L. (2013). A Revisit to the Impacts of Land Use Changes on the Human Wellbeing via Altering the Ecosystem Provisioning Services. *Hindawi Publishing Corporation Advances in Meteorology, 2013*, 907367. doi:10.1155/2013/907367

Zacharis, A. L., Bygrave, W. D., & Shepherd, D. A. (2000). *Global Entrepreneurship Monitor: National Entrepreneurship Assessment.* United States, 2000 Executive Report. Kauffman Center for Entrepreneurial Leadership at the Ewing Marion Kauffman Foundation, 3.

Zaharia, A., Antonescu, A. (2014). Agriculture, greenhouse gas emissions and climate change. *14th International Multidisciplinary Scientific GeoConference SGEM*, 3.

Zahiu, L. (1999). *Management agricol.* Editura Economică, Bucureşti.

Zalasiewicz, J., Crutzen, P. J., & Steffen, W. (2012). The Anthropocene. In F. M. Gradstein, J. G. Ogg, M. D. Schmitz & G. M. Ogg (Eds.), The Geologic Time Scale 2012 (pp. 1033-1040). Oxford, UK: Elsevier BV.

Zimmermann, D. (2015). *Europe's pesticide addiction.* Greenpeace report. Retrieved February 10, 2017, from https://www.greenpeace.org/archive-international/en/press/releases/2015/Europes-pesticide-addiction-threatens-our-environment---Greenpeace-report/

Zoomers, A. (2013). Rural livelihoods. In V. Desai & R. Potter (Eds.), *The Companion to Development Studies* (2nd ed.). New York: Routledge.

About the Contributors

Sergiu Bălan is Associate Professor at the Bucharest University of Economic Studies.

Mariana Bran is Professor at Bucharest University of Economic Studies. She is engineer licensed in the agrarian profile, specialized in agriculture and she earned PhD title in the field of Agronomy. In this sense, the research area is agro-economic and environmental issues, and the subject matters taught are technological. The subject matters are Comparative agricultural technologies, Management of the processing and quality of agro-food products, Technology of crops, Agrarian and food technologies. She published various articles and books in the agricultural field. Within this area, she has taken part in numerous research projects both as a researcher or project manager.

Daniela Stefania Bran is PhDs in Economics at Bucharest University of Economic Studies. She promotes about bioeconomy, agricultural vegetal waste as a resource for risk mitigation in global warming condition, crop protection and management of chemicals in farms. All of these are parts of various articles.

Henriette Călin was born in 1992 in Romania. She received the bachelor degree in Agrofood and Environmental Economics from the Bucharest University of Economic Studies in 2014, and the master degree in Ecologic Economics. At the same faculty she is in progress with the Ph.D. studies started in 2016, with the title The role of public-private partnership in the modernization of agriculture.

Diana Caprita, PhD Student in Economics, Graduated Management of International Projects- Master Program and Licensed in Commerce. Working as a parliamentary counselor in the Senate of Romania, over the past two and a half years I have researched the issue of circular economy, one of the priorities of economic development in Romania. In this context, during the research period, I was able to participate in numerous conferences organized at national and international level, focusing mainly on ways to implement the principles of the circular economy, on the take-up of good practice models among the European states, and also on creating an efficient legislative framework in order to contribute to the development of the recycling industry in Romania.

Boboc Dan is a PhD professor at The Bucharest Academy of Economic Studies and from 2012 is Dean of Faculty of Agrofood and Enviromental Economics. He has graduated the Faculty of Horticulture from USAMV Bucharest and has two PhD diplomas in: Economics and in Agronomics. As well, he is consultant at SC METRO responding to quality management for purchasing food products. Mr BOBOC DAN research has focused on management of food products, agricultural products processing, etc., research that have resulted in a series of publications and research projects, giving him functions in several of professional organizations such as The Association of Authorised Tasters from Romania – President. It mentioned that Mr BOBOC Dan was appointed as delegate expert from Romania for the Health and Safety Commission as well as for Justice and Commerce in OIV, International Organisation of Wine. He is also hight expert from Romania in FAO, specialized in food security. Mr. Prof. BOBOC DAN, PhD is the author of 11 specialty books, of which 3 books as the first author which treat various problems in reference to food security, food safety, models for organizing the food markets, matching the profile of economic organizations agrifood with EU requirements, etc.. Mr. Prof. BOBOC DAN has participated in 11 research projects, from which 4 international (two as director, "The design of the information system for monitoring traceability chain wine products in Romania," Grant Agreement 142050/08.10.2008 and „The assessment of risk caused by the enviromental conditions in romanian vineyards", and two as member - Solutions and Interventions for the technological transfer and the innovation of the agro-food sector in South East regions - TECH.FOOD in South East Europe Programme, Priority Axis 1: Facilitation of innovation and entrepreneurship, AOI 1.1: Develop technology and innovation networks in specific fields,. nr.1969/08.12.2008. and Designing an information system for monitoring traceability in pork production, Grant Agreement 141.921/13.08.2008.), and the rest representing national projects.

Anca Dinu was born in Campina, Romania, in 1992. She received the bachelor degree in Agrifood and environmental economics from the Bucharest University of Economic Studies in 2014, and the master degree in Economics and administration of agrifood business. At the same faculty she is in progress with the Ph.D. studies started in 2016, with the title "The role of taxation in the process of intensifying exchanges on the agrifood market". In 2017 she joined the Institute of agricultural economics of the Romanian Academy, as a research assistant. Her current research interests include taxation in agriculture, the power of taxes for the national development, the power of crisis, the measurement of taxes.

Iuliana Dobre is Associate Professor at Bucharest University of Economic Studies, Faculty of Agro-food and Environmental Economics. Scientific contributions concerned the theory in farm management and practices to improve on them. Mentioned work is completed by articles and books in the farm management field, such as management of production structure, optimal resources allocation in farms and economic analysis of farms. Participation in the doctoral committee, research projects evaluation and reviewer in various scientific papers in the farm management field are also parts of the professional activity.

Ladaru Georgiana-Raluca is vice dean of the The faculty of Agrifood and Environmental Economics (EAM) withing the Bucharest University of Economic Studies, Romania.She has graduated the Faculty of Agro-Food and Environmental Economics in 2008 and she is PhD in economics since December 2011. The title of the PhD thesis was "The wine chain competitiveness in Romania". She is teaching activities for the following disciplines: The research of agricultural markets, Agromarketing. She is Vice Dean of The Faculty of Agrifood and Environmental Economics. She participated in many research projects, among which: 1. The design of the information system for monitoring traceability chain wine products in Romania," Grant Agreement 142050/08.10.2008. 2. Solutions and Interventions for the technological transfer and the innovation of the agro-food sector in South East regions - TECH.FOOD in South East Europe Programme, Priority Axis 1: Facilitation of innovation and entrepreneurship, AOI 1.1: Develop technology and innovation networks in specific fields,. nr.1969/08.12.2008. - International project. 3.

Modelling the response of farms to the integration of the economic environment through sustainable management of soil resources, CEEX Program, Agreement number: 56/2006, with the Institute of Agricultural Economy of Romanian Academy, 4. Solutions for increasing the chain competitiveness, Research and Development Sectorial no. 6.1.3., Ministry of Agriculture, Forests and Rural Development, Agreement number: 367.1 /08.12.2006, with the Research Institute for Agricultural Economy and Rural Development, 100.000 lei 5. Establish a system for developing business plans, production strategies, feasibility studies and other useful analysis in decision-making crop production, Sectorial Program, 6.2.2., Ministry of Agriculture, Forests and Rural Development. She published four books, as author or co-author, with the following title: 1. Strategic directions for improving competitiveness in the wine chain in Romania 2. Performance of wine chain 3. The performance of agriculture and rural development 4. Food products market in Romania - traditional and organic.

Luminita Ghita is Head of Sustainable Development Department at Ministry of Environment. Scientific contributions concerned the theory in environmental economics, sustainable development, green procurement.

Ancuta Marin has been an economist since 1996, PhD in agronomy and a master in rural development. Author of thirteen books in the field of economy and rural development.

Aishwarya Narayan is rooted in Chennai, Tamil Nadu. She is currently in her final year of undergraduate studies at Symbiosis School of Economics, Pune. She was the President of the Debating society and is currently an active member of the student government in her capacity as a member of the placement committee. She has been an ardent leader in organizing various events in the college. Although versatile in various fields, her interests lean towards development economics; she hopes to in the future end the vicious cycle of poverty traps existent in Indian slums. To help her achieve her vision, she intends to pursue Masters in Economics following her graduation.

Sushma Nayak is a Visiting Faculty at Symbiosis School of Economics (SSE), a department of Symbiosis International (Deemed University). In addition to M.Phil. (Economics), she also holds a Post Graduate Diploma in Supply Chain Management. She has 8.5 years of teaching experience and has taught 14 different courses relating to Economics, Banking, Finance and Political Economy. Presently, she teaches Indian Banking & Financial System and Economic Journalism at SSE. She is a key member of the Center for Academic Writing at SSE, which assists students in writing essays, editorials, columns, commentaries, papers and reports of varied types. Furthermore, she guides students toward making of Annual Art Magazine and Yearbook. She also works towards the making of Oeconomica Scriptis — an in-house publication of SSE — which is a compilation of outstanding academic writings by students. She has been honored at the hands of Dr. Bibek Debroy, Economist and Member of National Institution for Transforming India (NITI Aayog) — the think tank of Indian Government, for her contribution to the Center for Academic Writing at SSE. Presently, she is one of the Editors for a book titled Maintaining Financial Stability in Times of Risk and Uncertainty, anticipated to be released in 2019 by IGI Global. She is also serving as a Guest Editor to Special Issues in 3 journals: International Journal of Comparative Management (IJCM), Inderscience Publishers Ltd; International Journal of Business Analytics (IJBAN), IGI Global; Journal of Electronic Commerce in Organizations (JECO), IGI Global. She is further serving as an interim reviewer for three IGI Global journals viz. International Journal of Civic Engagement and

Social Change (IJCESC), International Journal of Information Systems and Supply Chain Management (IJISSCM) and International Journal of Distance Education Technologies (IJDET). Her subjects of interest include Population Studies, Agricultural Economics, Economics of Social Sector, Banking & Finance, Managerial Economics, Public Finance, Theory of Political Economy and Supply Chain Management.

Shashikant Nehul has completed his Masters in Environmental Sciences and is presently pursuing his Ph.D in Air Pollution and Climate change. Mr. Shashikant Nehul is a research scholar at Department of Environmental Sciences, Savitribai Phule Pune University. He is presently associated with National Air Monitoring Program funded by Central Pollution Control Board, Government of India. He was also awarded with Erasmus mundus scholarship funded by European Union for his doctoral studies at Uppsala University, Sweden. He has a teaching experience of 7 years and is serving as a Visiting Faculty for Symbiosis School of Economics (SSE) and Symbiosis Institute of Health Sciences (SIHS), a constituent of Symbiosis International University (SIU).

Simona Roxana Pătărlăgeanu is a PhD in economics since 2001. In 1995 she was employed as an Assistant Lecturer at Bucharest University of Economic Studies, and currently works as Associate Professor. She writes and presents papers on economic, agro-food and environmental issues.

Andreea Maria Paul (Vass) is Associate professor, PhD, at the Department of International Business and Economics, Bucharest University of Economic Studies Romania and ex-senior researcher at National Scientific Research Institute for Economy (2001-2008), ex-economic adviser for the Romanian Presidency (2005-2007), for a MP of the European Parliament (2007-2008), for the Romanian Government, of two Prime-Ministers (2009-2012) and an ex-Member of the Romanian Parliament (2012-2016), having more than 16 years of experience in socio-economic research and public policy making. Currently she is the founder and president of the Romanian think-tank INACO - Initiative for Competitiveness (2017). She holds a PhD diploma in International Business and Economics from 2008 and relevant publishing activity. She has a rich experience in academic teaching and economic research, teaching courses as international economics, European economic governance and global economic governance. The main areas of interest include among others: competitiveness analysis, technology scouting, gender economics, international macroeconomics, analysis and assessments of public policies etc.

Cristina Raluca Popescu is Associate Professor PhD. PhD. at the Faculty of Business and Administration, University of Bucharest and PhD. Supervisor at the Economic I Doctoral School at The Bucharest University of Economic Studies. The field of expertise includes microeconomics, macroeconomics, European business environment, contemporary business environment, business process management, commercial management, sales force management, human resources management.

Gheorghe N. Popescu is Professor PhD. at the Faculty of "Accounting and Management Information Systems", the Department of "Management Information Systems", at The Bucharest University of Economic Studies, Bucharest, Romania and PhD. Supervisor at the "Accounting" Doctoral School at The Bucharest University of Economic Studies, Bucharest, Romania. He holds a PhD. in Economics ("Management Information Systems") and a vast working experience: 14 years in military informatics; 27 years in academic and research fields; 17 years as external auditor of financial / statutory audit, agreed procedures and related activities; 15 years as internal auditor; 7 years as internal public auditor, with experience in research and investment in aviation and marine. He has a wide range of activity both in the academic and research fields, due to the activities held at The Bucharest University of Economic Studies, Bucharest, Romania, Commercial Academy "Satu-Mare", Satu-Mare, Romania, Ministry of National Defense, Body of Expert and Licensed Accountants of Romania – CECCAR, Chamber of Financial Auditors of Romania – CAF, Body of Fiscal Consultants of Romanian – CCF, National Institute of Aerospace Research "Elie Carafoli" – INCAS, National Institute of Research and Development for Environmental Protection – INCDPM and National Institute of Research "Cantacuziono". He was the Head of the Department Services for Members of the Chamber of Financial Auditors of Romania (CAFR) and coordinator of the "Working Group" set up in this department, during 2003-2014. Since 1995, he is also CEO and partner of an accounting, auditing and consulting private firm in Romania. He attended numerous training and documentation session in Romania and abroad (France – University of Toulouse, Scotland, Italy). He currently provides scientifically and editorial consultancy to several Publishing Houses in Romanian (Lumen Publishing House and Gestiunea Publishing House). He is a member of the Scientific Advisory Board at the "International Journal of Economic Behavior – IJEB", a member of the Review Board at the "International Journal of Health Economics – IJHE" and a member of the Editorial Board at the international journal "Manager", the Faculty of Business and Administration, the University of Bucharest, Romania and a member of the Scientific Advisory Board of the International Business Information Management Association (IBIMA) Publishing, USA, for the IBIMA annual editions of its International Conferences – IBIMA Proceedings are indexed by Web of Sciences since 2006 and by SCOPUS since 2005. He was a member the editorial board of the "Financial Audit" international journal during 2003-2008. He is a member of the following professional associations: The Body of Expert and Licensed Accountants of Romania (CECCAR) (since 1994); Chamber of Financial Auditors of Romania (CAFR) (since 2000) and The Body of Fiscal Consultants of Romanian (CCF) (since 2007). He is a member of the research body Research Center "Analysis and Regional Policy", The Bucharest University of Economic Studies, Bucharest, Romania. In terms of research, he is a single author or coauthor of over 30 books/study materials, 10 Web of Sciences scientific articles, more than 150 articles, studies and conference papers published in Romania and abroad and a director or member of 16 research projects. In terms of the audit and accounting experience, during the time period 2003 till present he conducted over 150 analyzes and financial audits – the client portfolio includes entities activating in the industry, agriculture, commerce, research fields, state owned / majority state-owned entities or non-profit entities. His fields of interest are Economics, Accounting, Audit, Management Information Systems, Analysis and Risk Control, Risk Management, Controlling and Assisting Decision.

Elena Preda is Lecturer in Ecology at The Department of Agrifood and Environmental Economics. She has scientific expertise in assessment of ecosystems functions and services, biodiversity indicators and water quality assessment. Her research activities concern: i) contributions to identification and assessment of ecosystem services provided by different ecosystems through a successful stakeholders involvement; ii) contributions to development of new scientific and technical basis for assessment of climatic changes impacts on groundwater dependent ecosystems (e.g. development of indicators sets for vulnerability assessment of groundwater dependent ecosystems); ii) studying the litter decomposition process for understanding the impact of different drivers within climate changes context; iii) assessment the role of different communities in ecosystems functioning.

Mirela Stoian is a PhD professor at the Department of Agro-Food and Environmental Economics, Bucharest University of Economic Studies. During his university career, he also carried out important research, which resulted in the publication of university courses, specialized books and a large number of studies and articles: articles indexed Web of Science; articles with original scientific contributions, published in indexed journals in at least two international databases; articles/studies published in the volumes of the Web of Science Proceedings or international indexed conferences held in Romania or abroad. It also participated in a number of 11 research contracts, nine of which were national contracts, obtained through competition, and two were contracts with the business environment and coordinated, as a director of research contracts, of which two national and one with the business environment.

Ioana Todirica was born in Bucharest, Romania, in 1992. She received the bachelor degree in Agro-food and environmental economics from the Bucharest University of Economic Studies in 2014, and the master degree in Rural and Regional Projects Development Management. At the same faculty she is in progress with the Ph.D. studies started in 2016, with the title Traditional foods market considering the industrialization of agro-food production processes.

Carmen Lenuta Trica is Associate Professor at Bucharest University of Economic Studies, Faculty of Agro-food and Environmental Economics. Scientific contributions concerned the theory in environmental economics, environmental management and assessment of natural capital. These add practices to improve on them. Mentioned works are completed by articles and books in the environmental economics field. Participation in the research projects and reviewer in various scientific papers in the environmental field are also parts of the professional activity.

Lucia Ovidia Vreja is an Associate Professor at the Bucharest University of Economic Studies, Romania.

Alina Zaharia is a PhD in economics since 2017. Currently, she works as an Assistant Professor at the Department of Agri-food and Environmental Economics from the Bucharest University of Economic Studies. Her interests are in environmental economics, renewable energy policies, energy equity, climate change, sustainable use of resources, green buildings, applied econometrics and economic modelling.

Index

E

F

G

H

I

K

L

M

N

Purchase Print, E-Book, or Print + E-Book

IGI Global books are available in three unique pricing formats:
Print Only, E-Book Only, or Print + E-Book. Shipping fees apply.

www.igi-global.com

Recommended Reference Books

ISBN: 978-1-5225-3079-4
© 2018; 660 pp.
List Price: $245

ISBN: 978-1-5225-3064-6
© 2018; 230 pp.
List Price: $155

ISBN: 978-1-5225-3176-0
© 2018; 395 pp.
List Price: $235

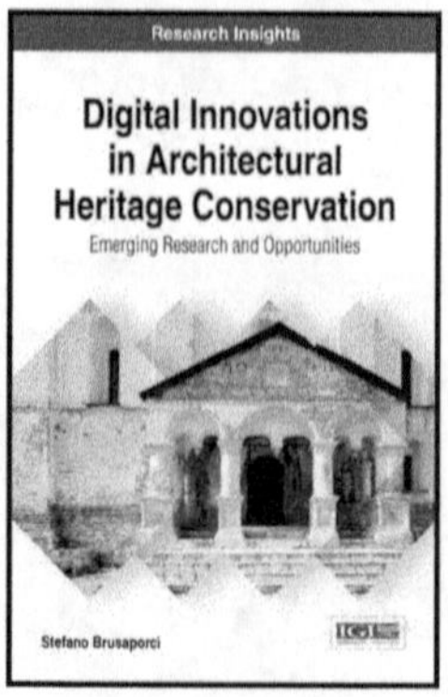

ISBN: 978-1-5225-2434-2
© 2017; 152 pp.
List Price: $115

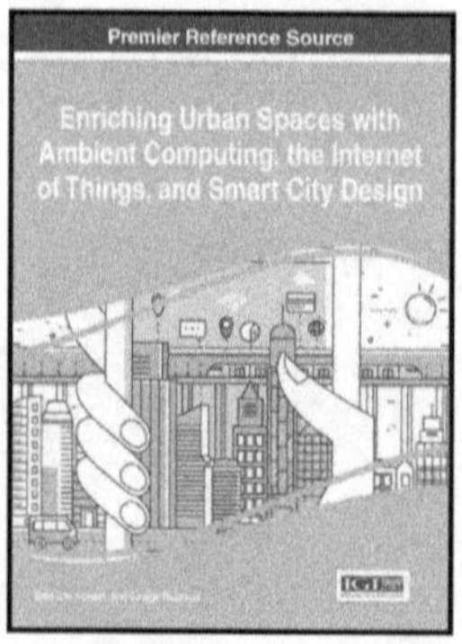

ISBN: 978-1-5225-0827-4
© 2017; 323 pp.
List Price: $210

ISBN: 978-1-5225-1677-4
© 2017; 417 pp.
List Price: $350

Do you want to stay current on the latest research trends, product announcements, news and special offers?
Join IGI Global's mailing list today and start enjoying exclusive perks sent only to IGI Global members.
Add your name to the list at **www.igi-global.com/newsletters.**

Publisher of Peer-Reviewed, Timely, and Innovative Academic Research

IGI Global
DISSEMINATOR OF KNOWLEDGE

www.igi-global.com Sign up at www.igi-global.com/newsletters facebook.com/igiglobal twitter.com/igiglobal linkedin.com/igiglobal

Ensure Quality Research is Introduced to the Academic Community

Become an IGI Global Reviewer for Authored Book Projects

The overall success of an authored book project is dependent on quality and timely reviews.

In this competitive age of scholarly publishing, constructive and timely feedback significantly expedites the turnaround time of manuscripts from submission to acceptance, allowing the publication and discovery of forward-thinking research at a much more expeditious rate. Several IGI Global authored book projects are currently seeking highly qualified experts in the field to fill vacancies on their respective editorial review boards:

Applications may be sent to:
development@igi-global.com

Applicants must have a doctorate (or an equivalent degree) as well as publishing and reviewing experience. Reviewers are asked to write reviews in a timely, collegial, and constructive manner. All reviewers will begin their role on an ad-hoc basis for a period of one year, and upon successful completion of this term can be considered for full editorial review board status, with the potential for a subsequent promotion to Associate Editor.

If you have a colleague that may be interested in this opportunity, we encourage you to share this information with them.

IGI Global
DISSEMINATOR OF KNOWLEDGE
www.igi-global.com

Celebrating 30 Years of Scholarly Knowledge Creation & Dissemination

CELEBRATING 30 years EST. 1988 Academic Publishing Excellence

InfoSci®-Books

A Collection of 4,000+ Reference Books Containing Over 87,000 Full-Text Chapters Focusing on Emerging Research

This database is a collection of over 4,000+ IGI Global single and multi-volume reference books, handbooks of research, and encyclopedias, encompassing groundbreaking research from prominent experts worldwide. These books are highly cited and currently recognized in prestigious indices such as: Web of Science™ and Scopus®.

Librarian Features:

- No Set-Up or Maintenance Fees
- Guarantee of No More Than A 5% Annual Price Increase
- COUNTER 4 Usage Reports
- Complimentary Archival Access
- Free MARC Records

Researcher Features:

- Unlimited Simultaneous Users
- No Embargo of Content
- Full Book Download
- Full-Text Search Engine
- No DRM

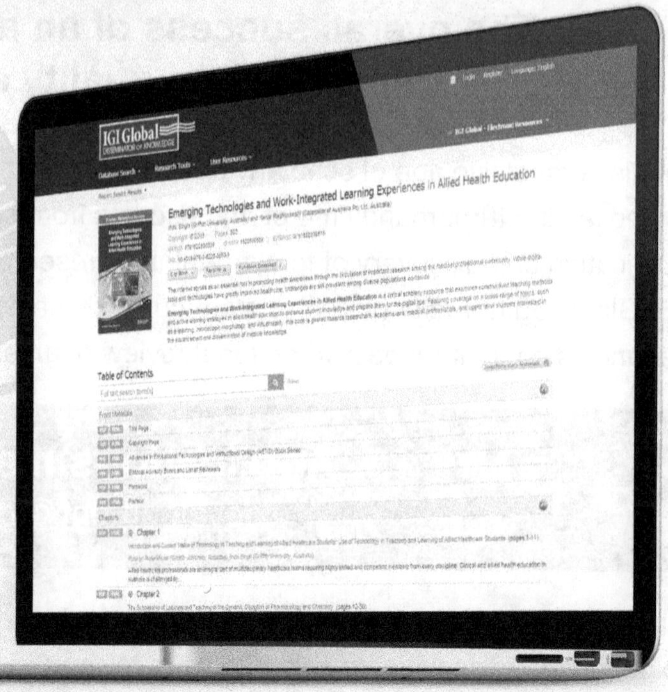

To Find Out More or To Purchase This Database:

www.igi-global.com/infosci-books

eresources@igi-global.com • Toll Free: 1-866-342-6657 ext. 100 • Phone: 717-533-8845 x100

IGI Global
DISSEMINATOR OF KNOWLEDGE
www.igi-global.com

IGI Global Proudly Partners with

eContent Pro International

Expert Copy Editing

eContent Pro International copy editors, with over 70 years of combined experience, will provide complete and comprehensive care for your document by resolving all issues with spelling, punctuation, grammar, terminology, jargon, semantics, syntax, consistency, flow, and more. In addition, they will format your document to the style you specify (APA, Chicago, etc.). All edits will be performed using Microsoft Word's Track Changes feature, which allows for fast and simple review and management of edits.

Additional Services

eContent Pro International also offers fast and affordable proofreading to enhance the readability of your document, professional translation in over 100 languages, and market localization services to help businesses and organizations localize their content and grow into new markets around the globe.

IGI Global Authors Save 25% on eContent Pro International's Services!

Scan the QR Code to Receive Your 25% Discount

The 25% discount is applied directly to your eContent Pro International shopping cart when placing an order through IGI Global's referral link. Use the QR code to access this referral link. eContent Pro International has the right to end or modify any promotion at any time.

Email: customerservice@econtentpro.com

econtentpro.com

INTERNATIONAL
EST. 1987

Information Resources Management Association

Advancing the Concepts & Practices of Information Resources Management in Modern Organizations

Become an IRMA Member

Members of the **Information Resources Management Association (IRMA)** understand the importance of community within their field of study. The Information Resources Management Association is an ideal venue through which professionals, students, and academicians can convene and share the latest industry innovations and scholarly research that is changing the field of information science and technology. Become a member today and enjoy the benefits of membership as well as the opportunity to collaborate and network with fellow experts in the field.

IRMA Membership Benefits:

- **One FREE Journal Subscription**
- **30% Off Additional Journal Subscriptions**
- **20% Off Book Purchases**
- Updates on the latest events and research on Information Resources Management through the IRMA-L listserv.
- Updates on new open access and downloadable content added to Research IRM.
- A copy of the Information Technology Management Newsletter twice a year.
- A certificate of membership.

IRMA Membership $195

Scan code or visit **irma-international.org** and begin by selecting your free journal subscription.

Membership is good for one full year.

www.irma-international.org

IGI Global
DISSEMINATOR OF KNOWLEDGE
www.igi-global.com

Available to Order Now

Order through www.igi-global.com with **Free Standard Shipping**.

The Premier Reference for Information Science & Information Technology

100% Original Content
Contains 705 new, peer-reviewed articles with color figures covering over 80 categories in 11 subject areas

Diverse Contributions
More than 1,100 experts from 74 unique countries contributed their specialized knowledge

Easy Navigation
Includes two tables of content and a comprehensive index in each volume for the user's convenience

Highly-Cited
Embraces a complete list of references and additional reading sections to allow for further research

Included in:
InfoSci-Books

Encyclopedia of Information Science and Technology Fourth Edition

A Comprehensive 10-Volume Set

Mehdi Khosrow-Pour, D.B.A. (Information Resources Management Association, USA)
ISBN: 978-1-5225-2255-3; © 2018; Pg: 8,104; Release Date: July 2017

For a limited time, receive the complimentary e-books for the First, Second, and Third editions with the purchase of the *Encyclopedia of Information Science and Technology, Fourth Edition* e-book.**

The **Encyclopedia of Information Science and Technology, Fourth Edition** is a 10-volume set which includes 705 original and previously unpublished research articles covering a full range of perspectives, applications, and techniques contributed by thousands of experts and researchers from around the globe. This authoritative encyclopedia is an all-encompassing, well-established reference source that is ideally designed to disseminate the most forward-thinking and diverse research findings. With critical perspectives on the impact of information science management and new technologies in modern settings, including but not limited to computer science, education, healthcare, government, engineering, business, and natural and physical sciences, it is a pivotal and relevant source of knowledge that will benefit every professional within the field of information science and technology and is an invaluable addition to every academic and corporate library.

Scan for Online Bookstore

New Low Pricing!

Hardcover: **$2,848****
~~List Price: $5,695~~

E-Book: **$2,848****
~~List Price: $5,695~~

Hardcover + E-Book: **$3,448****
~~List Price: $6,895~~

Both E-Book Prices Include:
- *Encyclopedia of Information Science and Technology, First Edition E-Book*
- *Encyclopedia of Information Science and Technology, Second Edition E-Book*
- *Encyclopedia of Information Science and Technology, Third Edition E-Book*

*Purchase the Encyclopedia of Information Science and Technology, Fourth Edition e-book and receive the first, second, and third e-book editions for free. Offer is only valid with purchase of the fourth edition's e-book.

**50% discount cannot be combined with any other offer except the free first, second, and third editions e-book offer. Discount is valid on the Encyclopedia of Information Science and Technology, Fourth Edition when purchased directly through the IGI Global Online Bookstore and may not be used by distributors or book sellers.

Recommend this Title to Your Institution's Library: www.igi-global.com/books

www.ingramcontent.com/pod-product-compliance
Lightning Source LLC
LaVergne TN
LVHW060403300926
849162LV00006B/1330